T0092990

THE EMPIRE OF CLIMATE

The Empire of Climate

A HISTORY OF AN IDEA

DAVID N. LIVINGSTONE

PRINCETON UNIVERSITY PRESS
PRINCETON & OXFORD

Published by Princeton University Press
41 William Street, Princeton, New Jersey 08540
99 Banbury Road, Oxford OX2 6JX

press.princeton.edu

Library of Congress Control Number: 2023947075

ISBN 9780691236704
ISBN (e-book) 9780691236711

British Library Cataloging-in-Publication Data is available

Editorial: Eric Crahan, Whitney Rauenhorst
Jacket: Katie Osborne
Production: Danielle Amatucci
Publicity: Alyssa Sanford (US); Kate Farquhar-Thomson (UK)

Jacket Credit: Thomas Cole, *The Course of Empire: Destruction,* Oil on canvas, 1836, 39½ × 63½ in. Collection of The New-York Historical Society, 1858.4. Photograph by Explore Thomas Cole / Wikimedia Commons

This book has been composed in Arno

Printed in the United States of America

10 9 8 7 6 5 4 3 2 1

for

Esther, Daniel & Marcus

the coming generation

CONTENTS

PREFACE

IN HER inaugural lecture as Humanitas Visiting Professor at the University of Oxford in 2013, Lorraine Daston began by reflecting on the history of ideas as an intellectual preoccupation. "Until very recently, and perhaps still," she began, "in the corner of the academic map that I come from, to call someone a historian of ideas is tantamount to calling them a tax evader or a cat murderer!" In her view, historical scholarship in the previous couple of decades had been characterized by "work with the historical microscope, a probing of the local details, the filigree texture of the everyday, the contextual." These revelations, she remarked, had brought enormous benefits. But, she continued, "sometimes it's pleasant to combine the work of the microscope with the telescope. That is what the history of ideas does. The telescope perhaps will not deliver such a treasure trove of personal and quirky stories, but it can reveal patterns that would be invisible at the pixilated level of the microscope."[1]

To change the metaphor from telescopes and microscopes to territories and maps, but with the same ambitions, my aim in what follows is to provide an outline chart of the realm I refer to as "the empire of climate." Since ancient times, the idea that the climate exerts a determining influence on minds and bodies, health and well-being, customs and character, war and wealth has attracted a long line of committed followers. Recent concerns over climate change has stimulated a renewed fascination with the role of climate in shaping human affairs and a penchant for forecasting the apocalyptic consequences of global warming. Taking the measure of this impulse over the longue durée is my quarry here. So the cartographic figure I have in mind is more like a continental map than a town plan, more like a sketch-map than an architect's site drawing. It is intended as an introductory guide to a vast terrain and, for that reason alone, no doubt suffers from the weaknesses of every mapmaking venture: silences, selectiveness, subjectivity.

Like all mapmakers and explorers, I am deeply beholden to many friends and colleagues who have provided sustenance along the way, identified key landmarks on the horizon, filled in blanks in the landscape, and encouraged me to just keep going. A Major Research Fellowship funded by the Leverhulme Trust facilitated the early stages of this program of research and provided freedom from other duties to begin the task of surveying the terrain. John Hedley Brooke, Nicolaas Rupke, Michael Ruse, Tristan Sturm, and John Stenhouse all read parts of the text in one form or another and took the trouble to send along helpful comments and wise advice. Over the years, I have been enormously enriched by conversations with John Agnew, John Brewer, Georgina Endfield, Richard English, Diarmid Finnegan, Frank Gourley, Andrew Holmes, Mike Hulme, Nuala Johnson, Stephen Kelly, Mark Noll, Caroline Sumpter, Stephen Williams, and Charlie Withers, all of whom have contributed, in more ways than I can ever hope to tell, to the completion of this, and other, projects. I have also benefited from the stimulus of two paleoecologists, Keith Bennett and Graeme Swindles—colleagues at different times who showed a great interest in the history of thinking about climatic influence. I am grateful too to audiences who engaged with parts of my argument, notably at the universities of Bergen, British Columbia, Dublin, Harvard, Nottingham, St. Andrews, and Tallinn. I owe a particular debt of gratitude to George Bain, who demonstrated his enduring interest in this project by reading every chapter as I completed it and offering a wide range of astute comments and challenges from which I have greatly profited. Justin Livingstone likewise scrutinized chapter after chapter with the eye of a textual critic and provided me with numerous insights, imaginative suggestions, and good reasons for why I should say some things in a different and hopefully more compelling way. It has also been a great pleasure to work with Christie Henry, Eric Crahan, and Whitney Rauenhorst at Princeton University Press, and with Angela Piliouras at Westchester Publishing Services. Most of all I remain indebted to Frances, who has once again read every word of this work and managed to do so with good humor, honesty, and an enviable capacity to make criticism sound like commendation.

Despite the best endeavors of these friends and colleagues, of course, the flaws that critics will be sure to identify show how bad a listener I can be . . . even while nodding in congenial agreement to the wisest pieces of advice!

THE EMPIRE OF CLIMATE

Introduction

1

A Matter of Degree

A REMARK by the eighteenth-century *philosophe*, Charles-Louis de Secondat, Baron de Montesquieu, has provided me with a name for the conceptual territory I want to explore in this inquiry. In his famous treatise, *The Spirit of the Laws*, which first appeared in French in 1748, Montesquieu declared that the "empire of the climate is the first, the most powerful of all empires."[1] It was a controversial claim. In the early 1750s, the Sorbonne, a theological college of the University of Paris that, on royal authority, had to approve all religious publications, identified this proposition as one of the passages that called for censure. One problem was that Montesquieu had attributed to climate too great an influence on human affairs when accounting for the manners and mores of nations. Another was his observation that religious belief systems could not be satisfactorily transferred from one climatic regime to another, and therefore that climate had set the bounds of both Christianity and Islam.[2] Clearly, the climatic empire of which Montesquieu spoke exercised its power over a far-flung realm—much to the annoyance of the religious establishment.

Since Montesquieu's day, the empire of climate has continued to extend its influence over more and more regions of human life and culture. And in the wake of prevailing concerns over the consequences of rapid climate change, it looks set to maintain its imperial rule into the indefinite future. Across the worlds of technical research, academic scholarship, popular science, and climate journalism, as conventionally understood, a common resort to the elemental power of climate's agency is plainly on view. And it is for this reason that, in what follows, I routinely pass across the fuzzy borders between these cultural domains, thereby transgressing distinctions between specialist science and pop-science, learned erudition and lay sentiment. Before outlining the itinerary we will follow in our travels through this empire, it will therefore be useful to demonstrate something of the explanatory hold climate retains among professional historians, science journalists, and popular writers alike, as well as reflect on a couple of conceptual issues that weave their way through the narrative that follows.

History by Degree

When *National Geographic* premiered its television show "Six Degrees Could Change the World," the columnist Joanne Ostrow contrasted it with Al Gore's *An Inconvenient Truth*. Under the strapline "*Six Degrees* charts climatic apocalypse in HD television," she reported that while Gore's Oscar-winning documentary film felt like a mere lecture, *National Geographic*'s offering was more "like a cool disaster movie, complete with fantastic animation and scary predictions." At just 2 degrees Celsius warmer, she observed, pine beetles will kill off the forests and "tiny Pacific islands will sink under the ocean"; at 1 degree more, category 6 hurricanes will become "regular occurrences" and thousands of species will disappear; at 4 degrees, "hordes of climate refugees" will flee famine, with the number reaching into the tens of millions at 5 degrees. When 6 degrees are reached, the world resembles "dinosaur time"—the Cretaceous Period.[3] Accompanying the original *National Geographic* broadcast was an interactive website inviting visitors to explore the earth's future at different degrees of global warming.[4] Enter the site, and a map of the world with what could be called a "disaster dial" along the margin appeared on the screen. In line with the program, viewers were confronted with the same dire warnings. On clicking the dial for 2 degrees warmer, urban Bolivians were seen moving into rural areas in search of water; at 3 degrees, floods appeared in New York City; at 4 degrees, deserts arrived in southern Europe; and point 5 on the dial disclosed worldwide political upheaval, economic disaster, and armed conflict.

This way of foretelling humanity's destiny has become increasingly commonplace. In point of fact, *National Geographic*'s project was inspired by a 2007 book by the journalist Mark Lynas, *Six Degrees: Our Future on a Hotter Planet*. Drawing heavily on the contributions of climate modelers and repeatedly underscoring the massive significance of critical tipping points, Lynas presented a history of the future degree by degree. He developed what could be described as a calculus of catastrophe as the temperature ratchets up notch by notch in a globally warming world. The book's eschatological ethos, prefixed by an epigraph from Dante's *Inferno* on entering the First Circle of Hell, certainly caught the attention of newspaper reviewers: the *Daily Mail* found it—all too ironically—"chilling," the *Sunday Times* thought it "terrifying," and the

Financial Times dubbed it an "apocalyptic primer."[5] Why? The worldwide catalogue of impending calamity offered no hiding place.

In these storylines, there is a strong sense of historical predestination when certain climate changes come about. Humanity's future is dictated literally by degree. Prophecy, of course, is a precariously uncertain business, not least on account of the wide range of variables involved. Indeed, none other than Sir John Houghton, former chair of the Intergovernmental Panel on Climate Change, issued a word of caution when giving evidence to the United Kingdom's House of Lords Select Committee on Economic Affairs on January 18, 2005: "when you put models together which are climate models added to impact models added to economic models, then you have to be very wary indeed of the sort of answers you are getting and how realistic they are."[6]

It is not just climatic futurists, however, who see in climate the fate of human civilization. In recent decades, historians of different stripes have sought to find in climatic circumstances the driving force of humanity's story. Blockbuster books with deep historical and wide geographical reach have done much to reposition human history more firmly in the context of its diverse natural and climatic environments. *Guns, Germs and Steel* (1997) by the biogeographer and historian Jared Diamond is a case in point. A virtuoso performance taking his readers to well-nigh every corner of the globe, Diamond inserted into his narrative subjects as diverse as the history of plant and animal domestication, the genesis of food production, the evolution of writing, and the impact of pathogens on the fate of human cultures, ever stressing the fundamental role played by climatic geography in choreographing humanity's global drama. Throughout, he intended external forces to triumph fully over any resort to innate cultural or ethnic essentialism.[7] Mention too might be made of *The Wealth and Poverty of Nations* (1998) by the economic historian David Landes. Taking nothing less than universal history as his remit, Landes looked to climate and environment as primary agents of historical change as he plotted a global geography of history's "winners" and "losers."[8] Even the titles on the shelf of books produced by the prolific archaeologist-anthropologist Brian Fagan disclose the place he accorded to climate's influence in the unfolding of human affairs. A sampling: *The Long Summer: How Climate Changed Civilization, El Niño and the Fate of Civilizations,* and *The Little Ice Age: How Climate Made History*. And while he conceded that "environmental determinism may be intellectually bankrupt," he nonetheless declared that "the great warming propelled humanity" across the Behring Strait and into an uninhabited continent. Flips in the North Atlantic

Oscillation, he told his readers, and the advent of El-Niños "caused civilizations to collapse."[9]

At least in part, what has fostered the resurgence of interest in the role of the weather in sculpting the shape of human history has been the increasing availability of paleoenvironmental data on climate variations over the centuries. The distinguished French historian Emmanuel LeRoy Ladurie, celebrated for his work on the history of peasantry, environmental history, and much else besides, had already recognized well over half a century ago the significance of merging climate records and archival inquiry. When his *Times of Feast, Times of Famine*, the English translation of his work on history and climate, made its appearance in 1971, it revealed just how serious was his engagement with the research of the climate scientists of the time. He was careful, however, to resist facile attributions of climatic causes for historical events. For he found the earlier "climatic interpretation of human history," of the kind championed by environmental determinists such as Ellsworth Huntington, to be "highly dangerous."[10] And so he frequently paused to identify the limitations of climate's explanatory power. Much more recently, the same caution has been voiced by the environmental historian Dagomar Degroot, who warns that "we come no closer to understanding its [climate's] impact when we view it as a straightforward cause of human events."[11]

Over the past few decades, research on dendrochronology, volcanic ash traces, ice sheet sediments, pollen analysis, lacustrine deposits, and the like has mushroomed, generating vast quantities of proxy data about past climatic environments and the changes they have undergone over extended periods of time.[12] Records such as these have encouraged historians to pose new questions about the dynamics of historical explanation and, together with access to a wider range of archival materials, have prompted many to read history more exclusively through the prism of climatology. Of course, large climate-related data sets have huge significance for the historian's craft. Rich information of this kind can illuminate some of the silences and uncertainties in the archive and address old issues in new ways. Big data's promises seem endless.[13] And yet, their very availability can suggest that, as Jo Guldi and David Armitage put it, "we are locked into our history, our path dependent on larger structures that arrived before we did," thereby opening the door to what they call "reductionist fictions about our past and future merely masquerading as data-supported theories."[14] For every promise large data sets offer, it seems, there is some peril lurking in the neighborhood. Besides the reductionism that Guldi and Armitage identify, what might be dubbed "data-overdetermination" is another snare.

The very presence of rich data on climate over millennia has the potential to seduce many into giving them a controlling explanatory role that favors causation over correlation, assertion over argument, declaration over documentation. There is nothing necessary about this pitfall, of course, but it does carry a cautionary warning about the lure of epistemic zealotry.

Climate Determinisms and Histories of the Future

These proposals, both historical and futuristic, may be taken as emblematic of a renewed resort to climatic readings of the human story. And they bear witness to two sets of issues that require some attention before providing an outline sketch-map of the way in which I intend to traverse the empire of climate. First, explanatory accounts of both past and future scenarios are frequently intertwined and mutually reinforcing. Historians typically seek to mobilize the results of their inquiries into the influence of climate on earlier societies for contemporary and future purposes. For their part, climate futurists routinely contextualize their prophetic forecasts by reference to historical reconstructions of life in times of climate stress. In one way or another, chronicles of the past and histories of the future merge in claims about the impact of climate and climate change on human society. Second, advocates of a climatic reading of retrospective and prospective history have aroused anxiety among some over what they see as a rejuvenation of a discredited environmental determinism. Despite routine disavowals by proponents of climate-driven history, the charge of determinism, or sometimes reductionism, continues to be made.[15] Apart from anything else, this alerts us to a fluidity in the use of the term "determinism" and to suspect that determinisms come in a variety of different shades.[16]

The first of these propensities—the futurizing of history—has become increasingly commonplace. Indeed, Lynda Walsh has gone so far as to suggest that scientists, including climate scientists, are frequently called upon to perform the role previously allocated to the prophet and to manufacture rhetorical certainty out of empirical complexity.[17] A couple of the authors mentioned above illustrate, in somewhat different ways, the climatic melding of past, present, and future. As he reached toward the conclusion of *Floods, Famines and Emperors*, for example, Fagan paused to observe: "History tells us that El-Niños have sometimes provided the knockout punch that topples states and great rulers. How infinitely greater these kinds of stresses are in our overpopulated and polluted world! If the scientists are right . . . then the fate of entire nations could lie in the unrelenting punches" of climatic events.[18] As for humankind's fragility in the face of a tyrannical climate, he remarked that we moderns "have not erased our vulnerability but merely traded it upscale."[19] Such comments catch the eye of reviewers and blurb writers who hungrily fasten on contemporary lessons to be drawn from historical inquiries. The dust jacket of *The Great Warming*, for example, used as advertising copy the *Publishers Weekly* remark that "looking backward, Fagan presents a well-documented warning to those who choose to look forward."[20]

With an eye to current preoccupations, another prominent historian, Wolfgang Behringer, began his *Cultural History of Climate* by calling attention to how historical investigations can inform contemporary practice. The experience of the Little Ice Age in Europe during the sixteenth and seventeenth centuries, he insisted, "may be regarded as a trial run for *global warming*." "We shall learn from it," he continued, "that even minor changes in the climate may result in huge social, political and religious convulsions." One of the prevailing motifs that wound its way through this narrative was the thought that climatic challenges inspire human creativity. To put it another way, Behringer's was not a chronicle of despair but rather a hermeneutics of hope. "The world did not collapse" during the Little Ice Age, he noted. "Instead, the crisis provoked a flexible cultural response, and even a lasting improvement in living conditions."[21] History, in this vein, is once again staged as a sequence of cautionary parables intended to encourage us to imagine a more creative public conversation by taking the possibilities as seriously as the perils of climate change.

For both advocates and detractors alike, the second concern—the specter of determinism—haunts the landscape of climatic theories of history and society. Diamond, for instance, claimed to be fully aware of the nasty aroma that has long clung to geographical determinism in various guises—an ugly racism, cultural supremacism, and the like. But these, he reckoned, were merely unfortunate contingent accretions to climatic causation; they were not intrinsic to it. Its sense of fatalism, too, has troubled many, but Diamond insisted that necessitarian defeatism was merely a confusion "of an explanation of causes with a justification or acceptance of results."[22] Not everyone has found his self-defense compelling. One critic insisted that Diamond had callously relegated the fate of large numbers of people to their natural environments and at the same time given the public a seemingly "scientific" justification for Eurocentric prejudice—whatever his claims to the contrary.[23] Another charged him with what might be dubbed causal impressionism. "In order to demonstrate the veracity of a cause-and-effect explanation," Andrew Sluyter insisted, "every link in the chain—from putative first cause right up to final effect—requires serious attention"—something that was just impossible when the trails of influence are strung out over five centuries. Even more troubling to Sluyter were the "harmful policies" *Guns, Germs and Steel* had the uncanny potential to unleash, and he thus warned his readers in no uncertain terms that Diamond's "junk science" demanded "vigorous intellectual damage control."[24] Others were convinced that Diamond had factored out of historical explanation the role of "human consciousness, desire, political power" and suchlike "in the distribution of wealth and power." In the end, they concluded, his cardinal error was conceptual—elevating "limiting factors to the level of causation."[25]

Closely connected has been the temptation to succumb to the enticements of what Mike Hulme has labeled epistemological slippage. By this, he means the inclination to transfer "predictive authority from one domain of knowledge to another without appropriate theoretical or analytical justification." Because scientists make authoritative claims about historical and prospective climates, there is a tendency to imagine the past and future as "climate-shaped" and to gloss over social, cultural, and political atmospherics. And something similar underlies Hulme's suspicion that a good deal of climatic futurology is erected on what he calls climate reductionism. That is a process in which "climate is first extracted from the matrix of interdependencies that shape human life within the physical world" and then "elevated to the role of dominant predictor variable." What has so enormously facilitated the "eschatological rhetoric" characteristic of this turn to "climate-driven destiny" is the "hegemony" exerted by predictive natural sciences like climate modeling. In Hulme's view, the development of computer-based simulation models by a powerful epistemic community of climate scientists "has allowed a form of climate reductionism to dominate contemporary analysis and thinking about the future."[26]

If Hulme's analysis is in the right neighborhood, then it will be wise to pay close attention to Mary Midgley's remark that if the aim of reductionism is to simplify explanation, a "mere exercise in logical hygiene" as she has it, then we need to be sure that we are not operating with a false economy. For alongside this appeal to epistemic frugality is the ironic, if common, inclination to engage in a gross expansion of the territory over which the explanation is claimed to have jurisdiction. These are the "lush speculative outgrowths" that Midgley regards as "designed to stimulate the imagination" to travel "in unexpected directions rather than to discipline it."[27] In many ways, this chimes with the observation of Philip Smith and Nicolas Howe that "more and more aspects of planetary ecology and social experience are tethered in one way or another" in the rhetoric of climate change.[28]

The explanatory powers attributed to climate, of course, come in various shapes and sizes. Sometimes it is the direct influence of climate on human bodies or historical events that advocates have in mind. Sometimes climate's influence is believed to be indirect, modulated by or operating through a range of other agents. On some occasions, the causal chain is short and immediate; on others, it is long and circuitous. Sometimes climatic explanations are monocausal; sometimes they operate in conjunction with other mechanisms. Plainly, advocates of climatic causation have adopted different stances—some softer, some harder—on just how climate influences history and society. And indeed there has been long-standing fluidity about what the designation "determinism" actually names. Frequently, those most often dubbed environmental or climatic determinists eschew the label for one reason or another. At the same time, there are those, like Robert Kaplan, who warmly embrace determinism with all its fatalistic overtones. Having found in Europe's temperate zone "the perfect degree" of environmental challenge to stimulate its denizens "to rise to greater civilizational heights," he told his readers of their need to engage with geographical determinists who "make liberal humanists profoundly uneasy." These figures, he went on, "were hardly philosophers: rather, they were geographers, historians and strategists who assumed the map determined nearly everything, leaving relatively little room for human agency."[29] All of this cautions against seeking terminological exactitude in dealing with a suite of concepts that are related more by family resemblance than analytical specificity. In my view, it is a mistake to impose conceptual clarity on ideas that display historical opacity. Rather than operating with a stipulative definition or seeking philosophical precision, we would be better advised to work empirically and endeavor to map the different shapes and forms "climatic determinism" has assumed. That at least is the perspective that will guide us on the journey on which we are about to embark.

Explorations in the Empire of Climate

Exploring the empire of climate is my quest in this book. It is an expansive domain extending both backward and forward in time, as well as in the scope of items that are believed to come under its sway. There is no one correct way to navigate this conceptual space any more than there is one way to arrange books on a shelf or paint a landscape. My approach has been to begin from a series of contemporary concerns about the influence of climate and climate change on human life and to set these preoccupations in a much wider historical context. Accordingly, I have chosen four arenas where the impact of weather and climate continues to be a source of distress or intrigue: health, mind, wealth, and war. I make no claim to have comprehensively mapped this terrain. Specialists in various fields will no doubt discern silences and absences, born as much of ignorance as of necessity, in my cartography. My aim, however, is to be suggestive rather than exhaustive, indicative rather than all-inclusive. But I hope nonetheless that the four major baselines that structure how I have sought to navigate *The Empire of Climate* will throw light on contemporary anxieties, encourage others to continue exploring this realm, and perhaps even better equip our own society for the climatic challenges that we all face. I am quite persuaded by Hulme's telling observation that the recent "phenomenon of climate-change is not a decisive break from the past, neither it is a unique outcome of modernity." Rather, climate change should be seen, as he puts it, "as the latest stage in the cultural evolution of the idea of climate, an idea which enables humans to live with their weather through a widening and changing range of cultural and material artefacts, practices, rituals and symbols."[30]

Part I begins with the anticipated effects of contemporary and future climate change on human health. Cardiovascular, neurological, respiratory, and many other disorders are anticipated to sharply rise with climate change, thereby putting unbearable pressure on public health systems. Allusions to the writings of the ancient Greek physician Hippocrates are not uncommon among these assessments, and accordingly, we initially inspect the Hippocratic *On Airs, Waters and Places* and its ongoing legacy.[31] The modern revitalization of this tradition owes much to the seventeenth-century contributions of Thomas Sydenham, whose Baconian outlook reinforced the role he allocated to climate in a range of seasonal disorders. Other medical authors during the period of the Scientific Revolution bear further witness to the vitality of the Hippocratic inheritance at the time. The development of a burgeoning literature on medical geography during the eighteenth and nineteenth centuries continued this tradition and had the politically potent effect of dividing the world into sickly and salubrious spaces according to climatic regime. The resulting medical cartography encouraged those with particular conditions, and who could afford it, to move temporarily or permanently to sites where the climate was deemed beneficial to their health.[32] Such concerns also deeply affected those Europeans who, for one reason or another, spent long periods of time in the tropical world because they were on military duty, were part of colonial officialdom, engaged in business enterprises, or answered a missionary calling. They and their families, believing themselves to be in danger from the malign influence of a tropical sun, sought refuge away from the "infested" lowlands in hill stations where efforts were made to replicate European sites of domesticity and cultural landscapes.

So significant was this latter enterprise to the imperial powers that I devote chapter 3 to the invention and reinvention of the tropical world as a medical *and* a moral domain. Concerns about the direct influence of the sun's rays on European constitutions stimulated a brisk trade in handbooks of tropical medicine designed to offer prophylactic advice to imperial travelers on how to remain well in the low latitudes. Often the recommendations turned sermonic, with advice being offered on moral hygiene in its widest sense. Whether or not Europeans could acclimatize to these conditions deeply divided medical and political opinion for on its resolution the very nature of empire crucially depended. Along the way, local people and places often, but not always, found themselves represented as diseased and dangerous and dissolute. To meet these challenges, a whole range of strategies—textile, sartorial, architectural, medicinal, behavioral, symbolic—were put in place to protect Europeans from climatic evils. All of these came under the influence of the empire's potent climate.[33]

In this connection, Amitav Ghosh is doubtless entirely correct to remind us, in *The Great Derangement*, that to look at the current "climate crisis through the prism of empire is to recognize . . . that the continent of Asia is conceptually critical to every aspect of global warming: its causes, its philosophical and historical implications, and the possibility of a global response to it."[34] But it is equally vital to recall that the peoples of Asia, Africa, and other parts of "the empire" were often on the receiving end of climate's conceptual condemnation long before the recent threats of climate change were registered. For the denunciation of tropical climates went hand in hand with the denigration of tropical peoples—medically, morally, and mentally. At the same time, indigenous peoples were sometimes thought to be superior to white colonizers in coping with tropical conditions, and this served to provide a naturalistic apologia for labor exploitation and, in some cases, to build a political system on the philosophy of sun and slavery. By the same token, the trope of tropical vulnerability among Europeans provided critics, some African and Asian, with the opportunity to present medical and other scientific evidence aimed to discourage aspiring colonial settlers and to cast serious doubts on long-term imperial success. Either way, traces of tropicality persisted well into the twentieth century, particularly among those with a passion to racialize climatology.

The final section of part I further follows this residual legacy by dwelling on the development of the science of biometeorology and the mobilization of its neo-Hippocratic health philosophy for eugenic ends by some of its pioneers. Here medical practitioners, enamored of finding causal connections between weather and well-being, and frequently engaging in tropical disparagement, found a comfortable home. The expression of these sentiments took many forms. Some sought to Darwinize the entire enterprise by translating their obsession with meteorology and medicine into a lexicon of fitness for environment, struggle for survival, and natural selection. Huge amounts of medical-meteorological data, reaching into thousands of pages, were accumulated by enthusiasts like William F. Petersen, who, during the 1930s and 1940s, sought to link the passing of cyclonic weather systems with seasonal patterns in the occurrence of vascular disease, gallbladder disorders, tuberculosis, coronary thrombosis, thyroid disease, and many more. Some found in this bioclimatic synthesis grounds for tightening American regulations on immigration to ensure that the eugenic health of the population was not compromised by the influx of foreigners from so-called inferior climatic regimes. Those zones were home to the lowest rungs of human potential and were reported to be deficient in vital energy. Others dwelt on the climatic circumstances that pertained during early human evolution and marshaled their findings to insist that there were climatic optima—temperature in particular—for reproduction and birth, as indeed for physical well-being, intellectual development, and mental stability. Convictions of this sort found their way into eugenic catechisms designed to instruct prospective parents on doing the best thing for their offspring. The idea of an optimal season for conception, the causes of differential birth rates within a population, and calls to further advance eugenic selection fed into these recommendations. There were implications too for a host of dysgenic anxieties about criminality, idiocy, and insanity that were thought to correlate with birth month. More recently, in the context of unprecedented anthropogenic climate change, popular writings keeping this picture before the minds of the general public demonstrate the lingering resonances of these earlier biometeorological preoccupations.

Part II takes as its center of gravity the influence of weather and climate on the human mind. In the first of the two chapters comprising this section, attention falls on the early evolution of the human species and its cognitive powers. The current enthusiasm for identifying climate as a critical causal factor in the emergence of the human brain can be traced back to at least the early twentieth century, when Darwinian natural scientists identified changes in climate as the fundamental driver of evolutionary development. To communicate the patterns of vertebrate geography that ensued, a number of students of historical zoogeography resorted to the visual rhetoric of cartography to convey the evolutionary story they espoused. The focus here is initially on a set of writers who turned to a north polar projection of the globe in order to visualize the chronicle of human development. This projection, by relegating hot climates to the fringes of the map and by giving optical dominance to northern climes, facilitated a narrative that gave pride of place to what they described as "Caucasian" and "Mongolian" races while relegating "Negroes" and "Australians" to the edge of global space and to the margins of human significance. Climate-inspired cartography in this register, frequently expressed in the anthropometric language of head form and cephalic index, fed an evolutionary imagination structured by a racialized hermeneutic of human history.

While expressing dissatisfaction with this racial rendering of human evolution, recent students of paleoanthropology have nonetheless frequently resorted to climate change as the major evolutionary forcing-agent in the emergence of large-brained hominids. Different explanations for causally connecting climate change and human evolution have been put forward. Some have linked their bioclimatic model of head-form evolution with crucial behavioral changes in the anthropoid move from tree-dwelling to grassland habitation. Others have dwelt on the development of flexible cognitive capacities that successful hominids evolved to meet the challenges of rapid climate changes. Those who developed a "brain for all seasons," as it has been styled, enjoyed selective advantage in the Darwinian struggle for life. Yet others turned to the connections between head form and thermoregulation in search of an explanation for human brain development and have not been slow to link their tale of early human evolution with latter-day challenges arising from global warming. While contemporary paleoanthropologists routinely distance themselves from what are widely regarded as the excesses of climatic determinism, there are those who continue to wonder if these more recent accounts of climate, crania, and cognition have successfully jettisoned the racial resonances of their early twentieth-century predecessors. At the same time, others, noting that climate causation chimes with present anxieties over global warming, have pondered whether the research agenda of paleoanthropology has been steered as much by contemporary concerns about rampant climate change as by the environmental record of humanity's deep past.

Chapter 6 follows another route into the realms of mind and meteorology by charting something of the different ways the human psyche has been reported to be responsive to the dictates of atmospheric conditions. The impact of climate change on individual and community mental health has become a subject of much contemporary concern, whether on account of extreme weather, oppressively high temperatures, or the experience of flooding, drought, and hurricanes. Early roots of the impulse to find in the weather the source of psychological dispositions and disorders, however, are to be found in the ethnographic components of the Hippocratic *On Airs, Waters and Places*. The modern revitalization of this tradition owed much to the physiological theory of mind advanced by Montesquieu in the eighteenth century, which allowed him to associate national character with the earth's climatic zones. Later writers continued to find causal relations between temperature and temperament, and they expanded the meteorological forces that shaped local psychologies to include the effects of wind on regional psyche. As these and other neo-Hippocratics saw it, climate reached into the deepest recesses of the human soul. Moods, morals, and mindsets were all believed to be subject to the imperatives of the weather.

One particular dimension of mental life that attracted growing attention during the eighteenth century and well beyond was the association between meteorology and melancholy. Whether portrayed as nervous exhaustion, neurasthenia, depression, or just low spirits, numerous writers fastened on the climate as a major determinant of this mental health syndrome. This condition, frequently discussed in the context of the effect of hot weather on European constitutions, was typically taken as a colonial ailment par excellence. But the underlying premise was often expanded into a general climate theory of psychic constitution. The correlation between mind, metabolism, and month of the year prompted a number of writers to identify seasonal patterns in instances of suicide, prevalence of crime, periodic mood swings, and the performance of schoolchildren. One of the more conspicuous conditions to emerge from this tradition of inquiry was seasonal affective disorder, for which a number of different, if related, modes of climatic influence have been identified. Closely associated is a rapidly expanding body of research on the causal influence of high temperatures on aggression, violence, and anger among individuals and communities inhabiting particular climatic regimes. These findings, however, have been contested as other researchers relate higher temperatures to more positive mood swings. Either way, thermic theories of temperament encouraged some to construct global climatic geographies of crime, cognition, and mental constitution. The possibilities afforded by this branch of psychological research for future-casting in the context of rapid climate change have proven to be extensive.

Some of the themes that wend their way through these first two parts of *The Empire of Climate* continue to manifest themselves in the chapters on wealth that constitute part III of my account. The first of these, chapter 7, introduces the theme of "weather, wealth, and zonal economics" by drawing attention to a spate of recent online publications pondering the influence of climate on national economies. These journalistic reports, while sometimes contested, glean support for their claims in the writings of economists, one a Nobel laureate, who find distance from the equator—a surrogate for climate zone—to be a major determinant of the wealth of nations. The direct reference in some of this work to the earlier zonal history of civilization by the fourteenth-century Islamic scholar Ibn Khaldūn and later Montesquieu prompts a reexamination of their thinking on zonal wealth and the physiology of economic performance. In their writings, and in those of a number of other political thinkers, travelers, and philosophers, economics was all-of-a-piece with a climatic philosophy of civilization. Such naturalistic theories of economy and culture were often constructed on the physiology of the human body and its responses to climatic forces. Increasingly, industrial progress, intellectual accomplishment, aesthetic sensibility, legislative structures, and technological achievements were brought within climate's sphere of influence.

The impact of Montesquieu's zonal geo-philosophy was both deep and lasting. It was picked up by Immanuel Kant in his reflections on the different human races, by David Hume in his essay on national character, and by Johann Gottfried Herder, who, though rather more ambiguous, nonetheless insisted that local geography irresistibly impressed itself on human cultures. In the writings of European geographers like Alexander von Humboldt and Friedrich Ratzel, this zonal portrayal of civilization and the wealth of nations continued to flourish. Moreover, Realpolitik of the most practical kind flowed from the zonal *mentalité*, not least in debates about the naturalization of slavery and the problems of colonial labor and wealth generation in tropical empires.[35]

My second cut at the bonding of weather and wealth, chapter 8, continues the theme of what I call the "sun, soil, and slavery" nexus. The diffusion across the Atlantic of the zonal geopolitical model of the world's wealth, notably by the Swiss geologist and geographer Arnold Guyot, further reinforced in the United States a long-standing resort to a climatic apologia for the slave labor that powered the plantation agriculture of the Old South. Even while voicing support for abolition, Guyot nonetheless typecast the temperate zone as fitting its inhabitants to legislate, the hot zone to labor—one to govern, the other to grind. There had, of course, already been moves to justify American slavery on climatic grounds. Thomas Jefferson, for example, had earlier connected climate, African physiology, and slave labor. At much greater length, and with yet more extreme language, nineteenth-century racial supremacists like John Van Evrie opposed any thought of emancipation, juxtaposing the mind of the white races to the muscle of the black. Climate had apparently dictated the racial distribution of brain and brawn. During the final decades of the nineteenth century and on into the twentieth, prominent American intellectuals enamored of racial science found scientific support for interpreting the wealth of the South's plantation economy as the consequence of its temperature and topography. Here the moral ecology of the weather-and-wealth school of economic geography appears in its darkest light.

Weather and wealth, of course, could be causally conjoined in other ways too. The exceptionally popular mid-nineteenth-century English historian Henry Thomas Buckle is a case in point. Buckle's understanding of wealth and civilization was erected foursquare on his philosophy of food. A positivist ethos and more than a smattering of determinism pervaded his *History of Civilization in England*. The Comtean philosophy of social physics supported Buckle's conviction that history obeyed fixed laws, and chief among the forces of nature that governed human actions was the influence of climate on food production and supply. The different dietary habits of populations in hot and cold climatic zones prompted him to find in the chemistry of food ingestion the roots of national temperaments, the foundation of social institutions, and the chief cornerstone of economic life.

The physiological economics that writers like these frequently espoused had implications for the management of imperial empires across the world. Frequently, the argument was made that because the climate had rendered impossible any responsible self-governance of hot regions, the temperate world must take up the task of overseeing the tropics' rich natural resources. Whether on account of inherent inability, climate-induced lethargy, or because of what was seen as long-standing inclinations toward despotism, European powers debated how best to advance their colonial aspirations in the face of these obstacles and the vulnerability of white bodies in tropical heat. From political judgments of this ilk emerged the idea of tropical trusteeship, by which the resources of the tropical regions could be administered—and exploited—from afar. This mode of governance was naturalized in talk of the difficulties of white acclimatization, the medical tyranny of the climate, and the imperatives of evolutionary progress. Ideas of this stripe, far from gathering dust on the shelves of academia or in the book rooms of gentlemen's dining clubs, were adopted by a whole suite of British imperial officialdom who fastened on them to support their advocacy of colonial stewardship.

Under the title "Climate, Capital, Civilization," the final chapter of part III extends the narrative of weather and wealth into the twentieth century and beyond. Here we find a perpetuation of the zonal theory of economic performance; the elaboration of hypotheses about the links between temperature, energy, and worker efficiency; and the continuing interest in tracing the rise and fall of civilizations to the vicissitudes of climate. Besides these, supporters of a climatic explanation of national wealth generation have embarked on projects to demonstrate the response of the market to changing weather conditions, to identify the relationship between economic cycles and cyclical patterns of rainfall, and to rehabilitate physiological economics by placing it on the more secure empirical footing of the human body's thermoregulatory system. The prognosticated effects of these processes, as climate change increasingly grips the human species, continue to catch the eye of popular science journalists and contributors to a wide range of social media platforms.

We pause first at what I call the atmospherics of employment. Here we inspect efforts to measure worker efficiency and intellectual prowess at different daily and seasonal degrees of temperature. This endeavor furnished not only industrial data but also, in aggregate, a global map of human efficiency, worker productivity, and climatic energy that reinforced long-standing stereotypes. It had the advantage too of fitting snugly with the taken-for-granted assumption that attributed wealth and civilization to the influence of the climate. At the same time, the cognate idea of climatic comfort zones was mobilized, not least in Australia, in efforts to inform government economic strategy and, more controversially, immigration policy. It was a short step to dividing the world into a climatically determined hierarchy of spaces appropriate for white colonial settlement. An impressive visual rhetoric of econographs, homoclimes, and climographs accompanied the aspiration to wrest citizenship from the sphere of the humanities and transfer it into the realm of the natural sciences. Advocates of these moves could also call upon efforts to naturalize economics by examining the influence of meteorology on the stock market. During the early decades of the twentieth century, efforts to correlate rainfall cycles with financial crises, trade patterns, and the general price index were advanced to make sense of market fluctuations. Some of these endeavors caught the eye of ecologists interested in biological cycles, investment managers, and writers of climatology textbooks. More recently, and with a seemingly broad readership, numerous books have appeared resurrecting, if not transforming, the old idea that the fate of civilization is to be found in the climate. Here too, past, present, and future merge in projects to use archaeological inquiries about ancient civilizations as morality tales for the Anthropocene future. All of these, in one way or another, can be seen as a continuation of Montesquieu's vision of climate as the greatest of all empires. So, too, can the return to what has been called physioeconomics, an approach to macroeconomics that attributes growth and stagnation to the effects of temperature on the body's success or failure to maintain homeostasis in the different climate zones.

Two chapters on climate and conflict constitute part IV—the final section—of my analysis. Chapter 10 begins by drawing attention to a spate of popular publications tracing the outbreak of hostilities, both present and prophesied, to the effects of global warming. While analysts of these climate wars tend to dwell on climate *change* as a dominant catalyst for internecine and international war, the impulse to find in atmospheric conditions the cause of warfare has a much deeper history. The humoral theories of temperament adopted by the ancient Hippocratics meant that some peoples were believed to be more prone to warlike behavior than others who were portrayed as possessing a gentler disposition. Proposals of this sort continued to attract supporters and flourished in the wake of Montesquieu, who remarked that while cold climates induced bravery, the inhabitants of warmer climes were fearful. The geography of human temperament that such portrayals delivered was explained by reference to climate and the workings of the cardiovascular system.

One conspicuous application of this neo-Hippocratic philosophy to civil conflict appeared in John William Draper's mid-nineteenth-century *History of the American Civil War*. Draper attributed the commercial and cultural differences between the northern and southern states to the agency of climate acting on human physiology. To him, cold climates promoted abolition; hot climates resisted it. Later writers, too, regarding slavery as a product of the differing climatic and physiographic geographies of these zones, found in climate and landscape the prime movers of the Civil War. In doing so, they easily managed to bypass any talk of moral accountability or recrimination on either side. Yet more recently, paleoenvironmental research on the Little Ice Age during the sixteenth to nineteenth centuries has encouraged many writers to find in the harsh climatic conditions of the time the genesis of social unrest, the incitement to insurrection, the outbreak of hostilities, and the advent of civil strife. That case has been prosecuted with different degrees of determinative intensity and has certainly attracted its fair share of critics. But, even if in different ways, that idea has gripped the imagination of many who harbor deep concerns about the intensification of bloodshed that climate change is expected to trigger in the years to come.

Finally, chapter 11 pursues this line of thinking into the late twentieth and twenty-first centuries by examining something of how climate change has become a matter of national security concern. A proliferating literature has appeared over the past couple of decades arguing for a strong causal relationship between global climate change and civil conflict. Some of these are synthetic overviews making generalized correlations between weather and war, some have focused on the effects of El Niño, others present data covering several centuries, and still others focus on conflict in particular venues—China, Darfur, Kenya, South Africa. In pursuit of support for the weather-and-warfare thesis, advocates have frequently resorted to historical episodes for illustration and corroboration, though sometimes with rather contradictory results. The inferences drawn from paleoenvironmental data from thirteenth-century Mongolia during the time of Genghis Khan (Chinggis Khaan) are a case in point. Numerous news outlets reported in 2014 that a temporary period of warm, wet conditions, ascertained from tree-ring records, had spurred the rise of the Mongol empire. By contrast, early twentieth-century explanations for the success of Genghis Khan laid emphasis on what at the time were believed to be increasingly arid conditions in Central Asia. Somewhere along the lines, the need to attend to counterfactuals would seem to have been overlooked.

Nevertheless, the corpus of work on climate and conflict has resulted in calls for climate change to be located at the heart of national security agendas, and a range of Pentagon-orientated reports on the subject have appeared under the auspices of military personnel. Needless to say, critics have not altogether welcomed these developments. Many have challenged the wisdom of espousing the neo-Malthusian struggle for resources that frequently underlies the climate-and-conflict thesis, not least on account of its fatalist sense of naturalistic inevitability.[36] Alex de Waal, human rights activist and research professor at Tufts University, for example, expresses concern at the simplistic harnessing of climate change as the cause of the Darfur crisis. Climate change may indeed be a factor in the outbreak of conflict, he concedes, but adds that "social institutions can handle these conflicts and settle them in a non-violent manner—it is mismanagement and militarization that cause war and massacre."[37]

I have taken health, mind, wealth, and war as the cardinal points of the map I have constructed of the empire of climate. Others will survey this territory differently using different coordinates and foregrounding different landscape features. Besides, there is no doubt many subregions have escaped my cartographic endeavors. All of this is simply because there are multiple routes across any piece of terrain and different horizonscapes to catch the eye. It will be obvious too from what follows that the streams of thought and pathways of influence that I identify crisscrossed in complex and sometimes unexpected ways, converging and diverging across medical, mental, monetary, and military landscapes. Some figures will appear in several different sections of the book on different, but related, themes. This is because medical climatology, for example, intersected in key ways with judgments about weather and wealth. Ideas about the influence of climate on the human psyche were swept into theories about market performance, stock prices, and economic cycles. The impact of climate on the human body was mobilized to justify racial hierarchy, slave labor, and imperial supervision. Those interested in the environmental origins of war sometimes relied on a latitudinal geography of temperament and emotion, sometimes on the different labor systems the climate supposedly fostered, and sometimes on a Malthusian struggle for diminishing resources. Proposals causally linking climatic conditions with the evolution of the human brain can be found constructing racial hierarchies, judging colonial success or failure, and dwelling on the climatic determination of the human psyche. These intersecting streams of thought demonstrate something of the manifold tributaries and rivulets that zigzag their way across the empire of climate. But by focusing on health, mind, wealth, and war, my aim has been to excavate a history of the present. I do not mean this in the presentist sense of Whiggish history; my intention is rather to underscore the ongoing fusion of past, present, and future horizons so as to expose the multifarious, contradictory, and contested ways in which the story of climate and humanity has been, and continues to be, told.

PART ONE

Health

2

Heirs of Hippocrates

"CURB GLOBAL warming or face a health catastrophe." So readers of the *Times* newspaper were informed on Tuesday, June 23, 2015. "Climate change," the article began, "is creating a 'medical emergency.'" Severe drought, it predicted, could result in a billion-and-a-half more medical incidents every year. With extreme heat and excessive rainfall, the number could soar to between two and three billion. Reporting the results of research carried out by an interdisciplinary team of European and Chinese collaborators for the 2015 *Lancet* Commission on Health and Climate Change, the newspaper quoted its authors' conclusion: "climate change could have 'potentially catastrophic effects' for human health."[1]

As headlines like these indicate, matters of climate are increasingly colonizing the conversation about human health. The Atlanta-based Centers for Disease Control and Prevention—the national public health institute of the United States—for example, announced in 2010, "Weather and climate have affected human health for millennia. Now, climate change is altering weather and climate patterns that previously have been relatively stable. . . . These changes have the potential to affect human health in several direct and indirect ways."[2] Elevated temperatures, extreme weather events, and ecosystem modifications are set to bring about increased incidences of respiratory allergies, cardiovascular disease, neurological disorders, cancer, and vector-borne illnesses. Less directly, the federal agency points out, climate change will exacerbate chronic ailments through interruptions in the provision of health care, population dislocation, and infrastructure damage.

Numerous other organizations periodically issue comparable warnings to visitors to their websites on how climate is a clear and present health danger. The United Kingdom's national weather service, the Met Office, for instance, reports that many areas will witness "a dramatic deterioration in their environment which will impact in a negative way on people's health."[3] Unsafe drinking water, insufficient food, and infections transmitted by other organisms are just some of the ruinous consequences of a changing climate. Felicity Liggins, a senior climate change consultant with the Met Office, further dramatizes the issue. The 2003 heat wave, she recalls, resulted in 35,000 deaths in Western Europe through heat stroke and related heart attacks. "Modelling here at the Met Office Hadley Centre," she continued, "has shown that by 2040 such a summer could be an average one and by 2060 the temperature we experienced back in 2003 could represent a cool summer."[4] Besides this grim forecast, she points to the soaring incidences of asthma and rhinitis in urban heat islands; to the escalation of cholera, typhoid, and gastroenteritis after severe flooding; and to the ways in which the geographical distribution of infectious diseases like malaria, Lyme disease, and dengue could significantly change.

The diagnoses of numerous other bodies could easily be elaborated. "From the tropics to the arctic," the World Health Organization announced in a 2005 online factsheet, "both climate and weather have powerful impacts, both direct and indirect, on human life."[5] Hypothermia, respiratory conditions, and heart disease were identified as just a few of the acute adverse health effects of short-term fluctuations in weather. The U.S. Environmental Protection Agency likewise elaborates on the impact of heat waves, extreme weather events, and changes in air quality arising from increases in ozone and fine particulate matter. Aggravated asthma, increased allergenic pollen, lung disease, intestinal illnesses, and food-, water-, and animal-borne infections are all highlighted.[6] The Natural Resources Defense Council adds another dimension: the health-related costs of extreme weather incidents, which are anticipated to increase "by billions more dollars." Small wonder, the report notes, that "some people have likened the effects of climate change to 'weather on steroids.'"[7]

In 2010, several of these and other similar environmental health agencies joined forces to produce a U.S. government report identifying crucial research questions urgently in need of resolution on the human health effects of climate change.[8] Focused rather single-mindedly on the implications for the United States, the document itemized around a dozen research needs, each prefaced by a summary state-of-the-art assessment of the ways in which particular climatic conditions directly and indirectly endanger health. In large part, the motivation was to secure increased funding, for the interagency working group noted that "the complicated relationships between climate change, the environment, and human health have not represented high priorities for scientific research in the United States."[9] According to this assessment, the disease burden that climate will impose on humanity is heavy indeed. Changes in air quality and ground-level ozone are expected to bring about a sharp rise in respiratory diseases and cardiovascular problems. Higher ambient temperatures, stratospheric ozone depletion, and flood-related runoff from land with toxic pollutants will increase the incidence of various forms of cancer. Extreme heat will act as a stressor for preexisting cardiovascular ailments as well as intensifying rates of foodborne gastroenteritis and salmonellosis infections. The list could go on: birth defects, spontaneous abortions, developmental irregularities, neurological disorders, waterborne and zoonotic diseases—all these and many more—are predicted to escalate on account of the changing climate. Urgent action, the report insists, is needed. And all the more so because of the additional challenge climate will pose to the nation's health care infrastructure. "Natural systems adapt to environmental changes or they fail," it concludes in Darwinian-speak. "Climate change threatens many of the health and built systems that protect and preserve our nation's health." It goes on: "The infrastructure we have put in place to protect health and to provide well being in the United States is extremely diverse and includes hospitals, clinics, public health agencies, trained personnel. . . . Threats to these systems from climate change range from damage to natural and built physical infrastructure to damage to intangible organizational structures (human and capital) that are required to maintain resilience to environmental threats."[10]

Underlying these alerts is a large body of specialist research invested in medicalizing climate and climate change. So extensive has this industry become that it has spawned a flourishing literature over the past couple of decades coordinating for different readerships the findings of countless medical investigations.[11] One of the most extensive of these was the 2003 World Health Organization's lengthy report on *Climate Change and Human Health*. Synthesizing a remarkably expansive body of medical-meteorological research, it presented its readers with synoptic overviews of such themes as the health effects of climate extremes, climate change and infectious diseases, the medical consequences of stratospheric ozone depletion, and ultraviolet radiation.[12] What characterized many of these reviews was the resort to particular case studies designed to have local empirical bite. The correlation between the 1997–1998 El Niño and the major outbreak of falciparum malaria in Kenya, for example, was highlighted, as was the increased incidence of dengue in Vietnam during El Niño years. Or again, the association of cholera in Bangladesh with the monsoon season, the peaking of cyclospora infections in Peru during the summer months, the seasonality of Saint Louis encephalitis in South Florida, and the role of warm winters followed by hot, dry summers in the spread of West Nile virus in the Americas were all used to illustrate the climate sensitivity of infectious diseases. Given such a chronicle, it is not surprising that the report concluded its lengthy review by underscoring its fundamental message: "Climate is an important determinant for human health. Both weather and climatic variables can be seen as human exposures that directly or indirectly impact on human health."[13]

In what follows, I aim to chart something of the deeper history of the various ways in which climate has been mobilized in the service of medicine to explain a range of disorders and as a tool for discriminating between healthful and harmful locations. A whole swathe of medical geographies that mapped moral and mental, as well as medical, conditions across the globe found a willing readership, not least among those who could afford to travel to venues that promised health and healing. The wider intellectual contexts within which these concerns were cultivated will attract our attention, as will the cultural interests that were frequently propagated among practitioners of the art of medical climatology.

Hippocratic Gestures

A couple of things are conspicuous about the recent surge of interest in environmental medicine and the likely health consequences of climate change. First, among the works I have already identified, Hippocrates' name routinely surfaces as the anchor point for the entire enterprise of tracing the medical impacts of climate change. In the published version of the Harben Lecture for the Royal Institute of Public Health in 2005, for example, Sir Andy Haines and his team began by announcing, "It has been known for thousands of years, at least since the time of Hippocrates that climate has wide-ranging impacts on health."[14] Howard Frumkin and his associates followed suit in their 2008 review of climate change and public health. "Weather and climate have been known to affect human health since the time of Hippocrates," they proclaimed.[15] These observations echoed the introduction to the 2003 World Health Organization report *Climate Change and Human Health* by Anthony McMichael of the Australian National University, who staged the contemporary climate-health crisis as "an old story writ large." After sketching in something of the dimensions of the current climate-change challenge, he returned to the "ancient struggle" between climate and health. Hippocrates was the first port of call. The recognition of how human well-being could be affected by climate change, he conceded, was "a recent development." But the fundamental realization that human health and disease were closely linked to climate was clearly identified by "the Greek physician Hippocrates," who charted the connection between epidemics and seasonal variation and instructed his readers to attend to ecological conditions when establishing settlements.[16] These self-conscious references to the Hippocratic tradition might seem merely ornamental, but as we will see, it has significantly informed the thinking of prominent advocates of medical climatology who produced works directly reengaging with the detail of the Hippocratic corpus.

Second, these commentaries on the effects of climate on human health collectively bear witness to the salience of what might be called "the revenge of the particular." Charles Rosenberg has drawn attention to the move in health care away from the generic toward the specific, a departure signaling a return to the understanding of "the classical body" as "situated, not abstracted and generalised." The Hippocratic tradition, he reminds us, is grounded in the vision of a clinician "who is an obligate climatologist, geographer, political scientist and ethnographer as well as healer"—a conception of the medical arts that persisted up to the modern era. By the mid-twentieth century, however, what he calls "the epidemiology of place" had become dramatically decentered in Western medicine in favor of laboratory-driven diagnostics. But in recent years, he discerns an increasing dissatisfaction with such "abstracted bodies" and a recovery of "bodies situated in specific places."[17] Rosenberg dates this paradigm shift to the late 1990s. So too does the World Health Organization. Prior to the mid-1990s, it declares, the "epidemiologists' limited conventional approach to environment health" meant that "there was very little awareness of the risks posed to the health of human populations by global climate change."[18] It was the publication of the Second Assessment Report of the Intergovernmental Panel on Climate Change in 1996 that brought the subject to the fore and its Third Report in 2001 that highlighted the need to turn the spotlight on health impacts at more regional scales.

Concerns about the health challenges of climate change in particular localities are not restricted to the world of medical specialists of course. In various ways, wider audiences have become attuned to this revival of medical localism. Consider what might be labeled postcode prognostication, an online means of checking health risks by address. The website of the U.S. Natural Resources Defense Council, for example, provides a facility for keying in your postal code and receiving a report itemizing the health threats arising from climate change for particular neighborhoods. Information is provided for the occurrence of ozone and ragweed, average number of days per year of extreme heat, the incidence of positive reporting of dengue fever and mosquito vector species, average number of days of low watershed flow, and the like. When viewed in composite, the site provides a cartography of climate health risk from area code to continental scale. The resulting pathological geography reinscribes long-held perceptions about what is often called tropicality, as we shall later see; the humid subtropical southern states are conspicuously at greater risk from infectious diseases.

At the same time, particularity calls attention to the ways in which place-specific features may moderate how the health effects of climate change should be understood. As Pim Martens put it in the *American Scientist*, "Does an extra foot of water correspond to a certain number of deaths? Surely this depends on whether it happens in a populated region and whether the community has the economic and technological means to respond to the crisis. What may be manageable for one region may be overwhelming for another."[19] Plainly, geography authorizes relativity.

The Hippocratic sense of place, particularity, and pathology also comes strongly through in what Al Gore described as "a landmark book," aimed at a nonspecialist audience, entitled *Changing Planet, Changing Health: How the Climate Crisis Threatens Our Health and What We Can Do about It*. Cowritten with the journalist Dan Ferber, the book centered on the life and thought of the Harvard physician Paul Epstein, who passed away in 2011, the year the volume appeared. The take-home message was crystal clear: "climate change is hazardous to our health." Because "climate change harms health now" and because "it could devastate public health by mid-century," he warned, "we must transform the way we power society and organize our economy to preserve a liveable planet."[20] To give feet to this *cri de coeur*, Epstein adopted a quasi-autobiographical stance and wove his analysis around a sequence of episodic ethnographic-like portraits of his experience of climate-related disease outbreaks in various locations around the world. The changing geographies of climate change and cholera, malaria, dengue fever, asthma, and amnesia were thus charted through Epstein's narrative of his personal travels, significant encounters with individual people in particular places, and stories of humble folk struck down with preventable illnesses. His experience of confronting cholera in Mozambique, for example, led him to suspect a link between El Niño and cholera epidemics and thus to the research of the environmental microbiologist Rita Colwell, later director of the National Science Foundation. His encounter with cerebral malaria in the foothills of Mount Kenya, in an area markedly different from the lower-lying tropical zones and hitherto free from malaria, prompted him to ponder the potential connections between global warming and the spread of the disease. It was this intuition that led him to Andrew Githeko at the Kenya Medical Research Institute, who had been working on how to use local weather maps to predict epidemics and to take prophylactic actions. In Honduras, he reported how six and a half thousand people lost their lives in the wake of Hurricane Mitch in 1998 as hundreds of neighborhoods were denied access to public water and as respiratory and water-borne infections soared when three-fourths of the republic's hospitals suffered major water problems. The story of a four-year-old in Harlem gasping for breath in the throes of an asthma attack illustrated the kind of thing that routinely happens on what's called ozone alert days. The confusion and short-term memory loss of a number of Canadians on Prince Edward Island in 1987 on account of shellfish poisoning caused by the blooming of a particular species of algae provided him with the opportunity to elaborate on the breakdown of marine ecosystems precipitated in part by gradually warming seas.

Epstein's idiographic narrative, however, did not rest content with particularistic ethnography. Rather, he used individual stories like these to build a cumulative case culminating in a comprehensive diagnosis. The earth is in an ailing condition; it is in a febrile state; it has "fallen ill." The geography of droughts, downpours, and diseases; charts of storms and sickness; maps of infectious ailments and El-Ninõ effects: all these were assembled in the cause of a grander verdict. "In less than three decades," he concluded, "climate change" moved from merely being "a worry for a handful of climate scientists to a clear and present danger."[21] Nor did Epstein ignore the cultural politics of climate knowledge at the turn of the new millennium. For interspersed with these verdicts was a variety of episodes rehearsing his efforts to get health issues on the climate change agenda, his struggles against climate deniers and recalcitrant policymakers, and his involvements with the Intergovernmental Panel on Climate Change. While Epstein worked hard to locate the health consequences of climate change in the wider setting of global political economy, not everyone was persuaded. One reviewer insisted that, for all the consciousness raising about the dangers of climate change, exposure to disease, "now and in the future, has everything to do with patterns of persistent social inequality that must be properly understood and tackled if public health is to be truly public."[22]

For an even wider public readership, the journalist Pat Thomas, later editor of *The Ecologist* magazine, provided an overview in 2004 of how the weather and climate affect health. Rooting her concerns in the writings of Hippocrates and such latter-day successors as Ellsworth Huntington, William F. Petersen, and S.W. Tromp, she deplored the way in which the weather's influence on human health had been sidelined in medical science in favor of more predictable causes of disease such as germs and genes. By contrast, she presented the "human barometer" as "extraordinarily sensitive . . . to electrical forces, heat, water and wind." "Who we are, how we behave and the state of our health," she went on, cannot be separated from these atmospheric influences. Her book was therefore dedicated to providing a conspectus of the impact of seasonal variation, solar and lunar cycles, wind, storms, and temperature on the human body. The impression readers were left with was that the range of weather-induced ailments was virtually limitless. Birth defects, cardiovascular-related illnesses, eating disorders like bulimia, and the onset of respiratory problems were reported as following seasonal patterns. Changes in climatic conditions induced rheumatoid arthritis, fibromyalgia, and other pain syndromes. Stormy weather could bring on migraine and an increased incidence of heart attacks; high temperatures damaged skin, eyes, and the immune system, while at the same time, the avoidance of natural light resulted in vitamin deficiency. In light of these revelations, it is not surprising that Thomas echoed the opinion of an editorial in the *Lancet* that failure to limit greenhouse gases to curb climate change constituted an "act of biopolitical terrorism."[23]

The collective effect of these testimonies—whether in the form of academic research, works of popular medicine, institutional statements, or official reports—is plain: matters of climate are now firmly on today's health agenda. But for all the urgency injected into these commentaries and for all the novel challenges that are said to arise from current anxieties over climate change, this is no new departure, as many of these writers are well aware. Tracing connections between weather and well-being has long been a human preoccupation, and my purpose in sketching something of that trajectory here is to locate these current anxieties in a much wider cultural and historical context.

On Airs, Waters and Places and Its Legacy

Right to the present day, pronouncements on the intimate connections between climate and health frequently root their concerns in the ancient Hippocratic corpus of the fifth century BC. As Frederick Sargent put it in 1982, "Most present day writers on weather and health could easily, and without prejudice, be set within the framework established by Hippocrates approximately 2400 years ago."[24] Sometimes allusion to the Hippocratics is simply gestural, but often enough, it is with a stronger sense of historical continuity. While Hippocrates himself is only now known through legend, the Hippocratic *On Airs, Waters and Places*, likely written to assist Greek medical practitioners in anticipating the diseases they could expect to encounter in a new location, is typically regarded as the first systematic attempt to catalogue the medical effects of climate.[25] At the same time, the later sections of the treatise elaborated more widely on the power of climate to mold the human body differently in different places and to regulate human deeds and dispositions. In Europe, for example, climatic variation meant greater diversity of bodily forms compared with elsewhere: "the figures of Europeans differ more than those of Asiatics," it declared.[26] In Europe's mountainous regions, inhabitants were tall, enterprising, and pugnacious, while those dwelling in poorly ventilated lowlands were fleshy, dark complexioned, and lacking in courage. For all that, the thrust of the treatise was more particularly to demonstrate the influences of the seasons, prevailing winds, and water quality on health. In so doing, Hippocrates and his followers did much to move medical investigation toward natural explanations of ill health and thus away from the idea that disease was to be seen as punishment for wrongdoing or could be diagnosed through the arts of augury. Elevating themselves above "dabblers in divination," Roy Porter observes, "the Hippocratics posited a *natural* theory of disease aetiology."[27]

The Hippocratic catalogue of weather-related ailments was extensive. In cities exposed to hot winds from the South, for instance, women were considered sickly, prone to excessive menstruation, and to suffer frequent miscarriages, while the men were subject to dysentery, diarrhea, and pleurisies. Men dwelling in cities with cold winds from the North, by contrast, were afflicted with lung disorders, ophthalmic problems, nosebleeds, and severe—though rare—attacks of epilepsy. In those same locations, women experienced difficult births and problems with breastfeeding.[28] Important too in this Hippocratic register was the pattern of seasonal weather. As the Hippocratic *Aphorisms* make clear, many conditions revealed a distinctive calendrical distribution: dysentery, asthma, and sciatica occurred more frequently in autumn; respiratory illnesses and chest pains tended to be winter ailments; hemorrhages, arthritic flare-up, and skin irritations dominated springtime; and fevers, diarrhea, and mouth ulcers were characteristic of summer.[29] Besides, because the body was attuned to annual rhythms, when the weather deviated from its seasonal norm, more sickness was likely to occur, with violent changes around the time of the equinox or solstice being particularly unfavorable.[30] Seasonal regularity, by contrast, meant that diseases came readily to the point of crisis.

What animated this tradition of environmental medicine, of course, was the fundamental role that the Hippocratics accorded to the fluids known as humors, which were held in balance in the healthy body.[31] Two in particular were connected with illness—yellow bile and phlegm. Alongside blood and black bile, these four humors were correlated with the four elements—air, fire, earth, water. In consequence, they each were associated with a different season of the year when one tended to predominate over the others. In winter, the cold and wet weather meant that phlegm increased, and with it, bronchial conditions became more common. During spring, blood was in the ascendency, and this induced fevers, nosebleeds and the like. In the hotter and drier summer months, yellow bile had dominance, while during autumn, black bile came to the fore. This holistic schema, which held tightly together the human organism and its environment, enabled various medical conditions to be anticipated as inferences from prevailing climatic conditions. Prognosis was its prime concern, and the Hippocratics practiced it in naturalistic ways that departed from earlier forms of prognostication.

And yet while climate occupied a prominent place in the Hippocratic project, its practitioners did not succumb to any crude climatic reductionism. Theirs was a qualified determinism that allowed for the forces of social institutions to play their part, thereby leaving some space at least for human culture to resist the imperatives of environment. For all that, as we shall see, Hippocrates' successors found in *On Airs, Waters and Places,* as well as other Hippocratic texts, resources in which to ground a rather more forceful determinist medical meteorology.

No doubt many instances of the appropriation of either the anthropological or medical components of the Hippocratic corpus could be traced through ancient and medieval writings. Aristotle (384–322 BC), for example, pondered on why people living in extremes of cold or heat were brutish in character and why cowardice and courage had distinct climate-related distributions.[32] Posidonius (ca. 135–ca. 51 BC), the polymathic Greek Stoic philosopher, likewise believed that human culture was significantly shaped by climate and that the mixing of airs affected the activity of the body, which in turn conditioned the operations of the mind.[33] And perhaps most conspicuously, the writings of the Roman student of architecture, Marcus Vitruvius (fl. first century BC), strongly resonated with the Hippocratic writers, likely derived from Posidonius, in depicting the health dangers of warm winds and urging that temperate conditions best facilitated a wholesome balancing of the humors. Architecturally, this meant that street alignment in cities, for example, should be such as to ensure protection from harmful airs.[34] At the same time, he recommended that marshy terrain should be avoided on account of the "unwholesome effluvia" that arose from marsh creatures. Similarly, he registered the differential influences of heat and cold, commenting that "those who change a cold for a hot climate, rarely escape sickness, but are soon carried off; whereas, on the other hand, those who pass from a hot to a cold climate, far from being injured by the change, are thereby generally strengthened."[35] During the Middle Ages, Hippocratic texts were translated into Latin and used by medical practitioners in Salerno in southern Italy during the eleventh century, subsequently spreading much further afield. Under the influence of Constantinus Africanus (ca. 1020–87), numerous Greek texts that had been translated into Arabic began to be Latinized, including Galen's commentaries on the Hippocratic *Aphorisms*. The Islamic physician Muhammad ibn Zakariya al-Razi (865–925) in particular had used the Hippocratic writings in the construction of his own scheme of medical treatment and identified how climate and season influenced human health.[36] Through his extensive writings, Rhazes, as al-Razi was known in the West, exerted a powerful influence on the teaching of medicine in the centuries that followed. As a consequence, many medieval medical practitioners introduced "a pot-pourri of Hippocratic treatments," as Roy Porter put it, into their therapeutic portfolios.[37]

The modern revitalization of Hippocratic medicine, however, is of more recent provenance.[38] And it owed a good deal to the work of the seventeenth-century "English Hippocrates," Thomas Sydenham (1624–1689), who "revived the Hippocratic idea that the seeds of disease lie both within and outside the body—that disease, especially when large numbers of people are affected, finds its origins in man's habitat."[39] Sydenham, a veteran of the English Civil War serving with the parliamentary militia for Dorset, was engaged in medicine in London from around 1655 and focused his attention in particular on smallpox and other fevers. His custom was to carefully observe patient symptoms and monitor a range of therapies, all the while affecting disdain for medical textbooks. In so doing, Sydenham prioritized experience over conventional scholarly authorities prizing, in emphatically Baconian style, observation over speculation. This precept, common among early advocates of the new natural philosophy, was doubtless attractive to John Locke and Robert Boyle, both of whom served apprenticeships with him.[40] At the same time, his zeal for therapeutic empiricism was grounded in a moderate providential naturalism that gave pride of place to a divine architect who consistently worked through natural law.[41]

For all his declared scorn for medical writers, the one authority to whom Sydenham had perennial recourse was indeed Hippocrates, "the Romulus of medicine," who had "laid the solid and immoveable foundation for the whole superstructure of medicine." As he further put it in the preface to the third edition of his *Medical Observations Concerning the History and Cure of Acute Diseases* in 1666, "How much the ancients, and pre-eminently amongst these Hippocrates, performed is known to all. It is to these, and to the compilers from their writings, that we owe the greater part of our skill in therapeutics."[42] Certainly he did depart from canonical Hippocratism in some ways, not least by seeking particular remedies for particular maladies. The use of cinchona (known as Peruvian or Jesuit's bark) for dealing with ague (fever) is a case in point. In so doing, he rejected the practice of using standard remedies to eliminate pernicious humors. Instead, Sydenham worked on the principle that, as Porter describes it, "diseases were specific entities," thereby moving diagnosis toward newer conceptualizations.[43] Nevertheless, Sydenham never abandoned Hippocratic modes of thinking, relying in particular on humoralism in his treatment of chronic ailments and in identifying environmental conditions as the cause of epidemic diseases in London during the 1660s and 1670s. In accounting for the ague epidemic, an intermittent fever, which occurred between March and July, for example, he looked to the influence of the sun's heat on the humors that had amassed in the blood over the winter months.

Throughout his diagnoses, Sydenham remained attentive to the ways in which certain diseases "through some mysterious instinct of Nature, follow the seasons as truly as plants and birds of passage." In consequence, he was sure that the intermittent fevers of spring were very different from their autumnal counterparts. Each fever had its own "genius," its own "constitution," and needed treating appropriately for its particular season of the year. Failing to provide therapies in accordance with the natural rhythms of specific ailments was thus highly dangerous. As he explained,

> I have seen . . . spring tertians so maltreated by undue bloodletting, by undue purging, and by unsuitable regimen, that they have spun out their existence until the period of the autumnal ones; and then as the time of the year is diametrically opposed to the genius of the disease in question, it annihilates it at once. The patient, nevertheless, has in the interim been so worn out by the repetition of the fits, and the protraction of his ailment, that he seems to be in the last extremity.[44]

Fevers, then, assumed the characteristics of their season of occurrence. In recording his findings for the years from 1661 to 1675, supported no doubt by the meteorological observations amassed by associates like Robert Hooke, Christopher Wren, John Locke, and Robert Boyle, he thus reported that he had identified "five sorts of constitutions, or five peculiar dispositions of the atmosphere, and as many peculiar species of epidemics." In consequence, "pleurisies, quinsies, and other like ailments, generally prevail when long and severe cold is rapidly succeeded by sudden heat."[45] And more generally, the different seasons directly influenced the condition of the blood, with the accumulation of phlegmatic humors in winter and a certain excitation in summer.

At the same time, Sydenham was sure that such conditions were the consequence of how "morbific" or "peccant" matter, emanating as effluvia from an atmosphere "stuffed full of particles which are hostile to the economy of the human body," interacted with the individual patient, the season, and the specific "constitution" of the epidemic.[46] Maladies, as he put it in the very first paragraph of his *Medical Observations*, "arise partly from the particles of the atmosphere, partly from the different fermentations and putrefactions of the humours." Some operated by insinuating themselves into the "juices" of the body, disturbing them, and mingling with the blood. In due course, the whole bodily frame was contaminated with "the contagion of disease." Other types were retained within the body for "longer than they ought to be, its powers having proved incompetent, first to their digestion, afterwards to their excretion."[47] Such concerns with "the continuous flux of particles" reveal that Sydenham's determinations were a synthesis of Hippocratic environmentalism and the new natural philosophy.[48] As his biographer Kenneth Dewhurst put it, "He adopted Boyle's corpuscular theory of epidemics, and grafted it on to the Hippocratic concept of an epidemic constitution. He believed that fevers changed their characteristics according to the particular constitution of the year, and according to the prevailing epidemic."[49]

A comparable retooling of Hippocratic modes of thought to conform to the imperatives of current natural philosophy was forthcoming from the pen of the Scottish physician, satirist, and political pamphleteer, John Arbuthnot (1667–1735). By the time his *Essay Concerning the Effects of Air on Human Bodies* first appeared in 1733, toward the end of his life, Arbuthnot was a well-known public figure not least on account of his best-selling John Bull pamphlets, which presented the portly matter-of-fact Bull as a national personification of England. Arbuthnot was a close friend of Jonathan Swift, George Frideric Handel, and Alexander Pope, as well as an associate of such prominent natural philosophers as John Flamsteed, Isaac Newton, and Edmund Halley at the Royal Society, and was appointed physician to the royal household.[50]

While Sydenham's Hippocratism was reconstructed out of Boyle's corpuscular particles, Arbuthnot's was fixed to Newtonian moorings. The ascendancy of Newtonian philosophy during the early eighteenth century meant, as Janković observes, that the "association of air and the mechanical body was a dominant physiological assumption," and Arbuthnot, alongside George Cheyne, did much to translate the mathematicization of respiratory lung motion, the circulation of the blood, and "digestive pneumatics" into medical practice.[51] Thus, in the early pages of what Golinski calls "the most widely read theoretical discussion of the atmospheric causes of disease,"[52] Arbuthnot drew his audience's attention to a range of mathematical measurements of the features of the air—its temperature and moisture levels as well as its gravity, elasticity, and pressure—so as to keep medical readers abreast of the findings of natural philosophers, chemists, mathematicians, and agricultural teachers.

This survey completed, Arbuthnot turned his attention to the influences of the air on respiration, medical disorders of one sort or another, and pestilential fevers. Again, the intellectual paternity of Hippocrates was lauded. As "the first Founder of our Art, the great Hippocrates" had demonstrated, he wrote, how the air played a critical role in "the Oeconomy of Diseases."[53] Moist weather, for instance, relaxed the fibers to such a degree that the human body was subject to many maladies. To Arbuthnot, as to many others in the eighteenth century, the concept of "fiber" had a pivotal role to play in medical theory. Fibers—filaments or fine threads—were believed to be the body's fundamental building units and were essential to the circulation of fluids throughout an organism. To perform this physiological role, their elasticity had to be maintained, and when their oscillations were impaired, frequently by changed atmospheric conditions, vital movements in the body's ecosystem could not be carried out.[54] And so warm humid conditions, by relaxing the fibers, meant that some of their "Elasticity or Force for circulating the Fluids" was lost and that victims experienced a whole raft of aches and pains in those regions of the body where the circulation of the "juices" was less than perfect, as in the scars of old wounds and in "luxated or bruised Parts." Variable weather with frequent changes from hot to cold, as well as from wet to dry, produced "all the Distempers of the Catarhous Kind, Arthritick Diseases and . . . Leprosies." By contrast, "Cold Air, by its immediate Contact with the Surface of the Lungs, is capable of abating or stopping the Circulation of the Blood, and bringing them into an inflammatory State, and by producing Catarrhs and Coughs, is productive of all the Effects of such Defluxions upon the Lungs, Ulcerations, and all sorts of pulmonick Consumptions."[55] Within the kingdom of medicine, the weather ruled over a large territory indeed.

In Arbuthnot's mind, moreover, the empire of climate extended far beyond the strictly physiological. Given his political instincts and his penchant for sketching national archetypes, it is perhaps not surprising that in his account of the effects of air, he would turn his gaze toward the ways in which it had sculpted the distinctive traits of the various nations. Both reason and experience bore witness to the ways in which "the Air operates sensibly in forming the Constitutions of Mankind, the Specialities of Features, Complexion, Temper, and consequently the Manners of Mankind, which are found to vary much in different Countries and Climates."[56] Because the "Temper and Passions" reflected the fluctuations of the air, feelings of being "sprightly, dejected, hopeful, despairing," and suchlike were determined by the weather. And because there were days when the faculties of the mind, memory, and imaginations were more vigorous, it seemed to Arbuthnot highly likely "that the Genius of Nations depends upon that of their Air." Not surprisingly, this genius of nations, as he described it, displayed its own climatic geography:

> Arts and Sciences have hardly ever appeared in very great or very small Latitudes: The Inhabitants of some Countries succeed best in those Arts which require Industry and great Application of Mind: Others in such as require Imagination; from hence some Countries produce better Mathematicians, Philosophers, and Mechanicks; others better Painters, Statuaries, Architects, and Poets, which, besides the Rules of Art, demand Imagination. It seems to me, that Labour is more tolerable to the Inhabitants of colder Climates, and Liveliness of Imagination to those of hot.[57]

For Arbuthnot, as for Hippocrates, medical meteorology embraced mind and matter, clinical practice and cultural accomplishment. And the self-same principles that governed these perceptions translated into even wider global generalities encompassing disposition, temperament, and character. "Extremes of Heat and Cold in great Latitudes," he affirmed, by "relaxing and constringing the Fibers by turns," aroused "an Activity and Tolerance of Motion and Labour, in dry frosty Weather, more than in hot." In the tropics, by contrast, people were trapped in a constant "State of our hottest Weather." The implications were apparently clear. The oscillatory motion of the fibers in northern nations stimulated "greater Activity and Courage"; those relentlessly exposed to excessive heat, whose fibers were decidedly less subject to fluctuations, were "lazy and indolent" and therefore predisposed to "follow naturally a slavish Disposition, or an Aversion to contend with such as have got the Mastery of them."[58] Here Arbuthnot echoed the sentiments of the French scholar and diplomat Abbé Jean-Baptiste Du Bos.[59] For in his 1719 *Réflexions Critiques sur la Poësie et sur la Peinture,* du Bos contended that genius owed its genesis less to such *causes morales* as education and governance than to the character of the air and the prevailing climate. Because artistic brilliance only flourished in regions with suitable weather regimes, he concluded that the temporal rise and decline of the creative impulse must have been on account of secular climate change.[60]

The production of Hippocratic treatises of this sort at once stimulated, and was stimulated by, the increasing availability of works recording, sometimes over extended periods of time, weather conditions and associated medical disorders. It was a dreary business. Clifton Wintringham (1689–1748), for example, confessed to the "continual tediousness" that accompanied his own efforts in this direction.[61] Wintringham, a medical practitioner in York, was the author of a number of works on pestilential fevers and endemic illnesses, but his most significant publication was the journal he brought out in 1727 under the title *Commentarius Nosologicus,* which recorded nearly twenty years' observations on weather, disease, and epidemics in his own city.[62] In similar fashion, John Huxham (1692–1768), a provincial doctor in Plymouth well known for his *Essay on Fevers* (1750), had begun to keep weather observations from 1724 in response to the initiative put in place by James Jurin (1684–1750)—English physician, ardent Newtonian, and a secretary to the Royal Society—to find volunteers willing to keep daily weather records.[63] Soon Huxham embarked on identifying correlations between "Constitutions of the Atmosphere" and "epidemic Distempers" and published—initially in Latin—his findings for the years 1728 to 1748. In doing so, Huxham was convinced that he was following the "noble Example" of "our most ancient and best Master in Physic, *Hippocrates*."[64]

Later, the physician and naturalist, John Rutty (1687–1775), did much the same in Dublin, drawing on his personal diaries to produce a forty-year conspectus of weather and disease. Rutty, a student of Hermann Boerhaave (1668–1738) at the University of Leiden and a dedicated Quaker who also kept a spiritual journal, had already published a work on the medical dimensions of mineral waters, an analysis of the results of his experiments on milk, and a two-volume natural history of County Dublin.[65] Now, in 1770, he brought out his *Chronological History of the Weather and Seasons* in part to refute those "who are crying down any attempts to trace a connection between the state of the air and weather, and the prevailing epidemics."[66] The volume, in predictable style, reviewed Dublin's seasonal weather for each year together with recorded illnesses and information from the bills of mortality. But he also paused to comment on conditions "in Asia and other hotter countries" where "the humors are more disposed to putrefaction, and the fibres more crisp and contracted, ardent, pestilential, and contagious fevers are much more frequent, and even endemial," providing supporting information, drawn from diverse sources, on the East Indies, Egypt, Minorca, Barbados, and so on.[67]

"Considering that they often made for extremely dull reading," Jan Golinski observes, "the volume of such publications in the eighteenth century is remarkable."[68] Indeed. Nonetheless, they bear witness to the reinvigoration of Hippocratic medicine in the period. In part, this revival was stimulated by the translations of Hippocratic works that became available in the eighteenth and on into the nineteenth centuries.[69] Francis Clifton, a devotee of Boerhaave, for example, produced an English translation of *On Airs, Waters and Places* in 1734 designed for practitioners. Victor Amedée Magnan did the same for French audiences with his 1787 translation. In 1787 and 1792, the medical historian Kurt Sprengel brought out his two-volume *Apologie des Hippokrates und seiner Grundsätze,* which included German translations of several works in the Hippocratic corpus. And then in 1800, the Greek physician Diamantios Coray produced a new French translation of *On Airs, Waters and Places,* as did Émile Littré in 1840 as part of this ten-volume edition of the complete oeuvre of Hippocrates.

Place, Pathology, and Weather

The availability of such works during the eighteenth and nineteenth centuries fostered the further flourishing of the Hippocratic system, sometimes in association with other schemes of natural philosophy and often in the form of medical topographies.[70] The cartographic impetus that this neo-Hippocratic revival cultivated delivered national and regional catalogues of localities branded as healthy or harmful, baleful or beneficial. Place and pathology were inseparably conjoined. In the United States, for example, Scottish physician Lionel Chalmers (ca. 1715–1777), who trained in Edinburgh before settling in South Carolina, recorded meteorological observations and medical conditions for a decade between 1750 and 1760 and subsequently published, in 1776, his *Account of the Weather and Diseases of South-Carolina,* followed a year later by his *Essay on Fevers.* Having charted the climatic and topographic conditions in the state, Chalmers highlighted the ways in which human health was affected by "putrid effluvia," "vapours" arising from "the sun's rays penetrating the miry soil," and other weather-related circumstances.[71] Subsequently, he embarked on elucidating for his readers the effects of seasonal weather, extreme temperatures, changeable conditions, and variations in moisture on a range of disorders—fevers, consumptions, and infections of various kinds. Crucial to his thinking was the influence he believed the weather exerted on the body's fibers and fluids. Cold, moist air, for instance, tended "to augment the elasticity of the solids, and condense the fluids" and exert a "pernicious" influence on the human body.[72] In keeping with his neo-Hippocratic cast of mind, Chalmers devoted lengthy parts of his treatise to specifying various therapeutic treatments and medicinal preparations for the numerous ailments he described.

In comparable vein, William Currie (1754–1828), a Philadelphia physician who originally studied for the Episcopal ministry and worked for a time as an army surgeon during the Revolutionary period, produced a work on medical topography in 1792 in which he surveyed the soil, climate, and epidemics of every state of the Union. Later, in the wake of the bouts of yellow fever that afflicted Philadelphia in the 1790s, he actively participated in public debates on the subject and published his own records of the epidemic in 1798.[73] Even if, as his obituarist put it, he wrote in a "dry and clumsy style," Currie displayed much industry in his collation of data gathered through extensive correspondence with local physicians and the scouring of countless medical and geographical works. His aim was to vindicate his conviction that "climate and situation have a remarkable influence upon the constitution and health, as well as upon the complexion and figure of mankind."[74] Taken in the round, Currie's *Historical Account of the Climates and Diseases of the United States* was a rather unsystematic compendium of disparate materials with extensive extracts excerpted from a wide range of sources: medical writers and natural philosophers, letters from influential physicians, meteorological records, local bills of mortality, and the like. Among the information he accumulated were sundry lists of diseases, records of age of death in different places, monthly temperature and rainfall patterns, and catalogues of individual cases of disease and treatments such as the use of laudanum, bleeding, saline purges, tartar emetics, alkaline salts, and Seneca root. In some ways, the volume was an exercise in the logging of particulars in classical Baconian mode. As he himself declared, "In the investigation of the nature and causes of diseases, and in determining the effects of remedies, I have been uniformly governed by actual experience, never by hypothesis or plausible conjectures." Where "reasoning or theory" did enter his diagnoses, these were, he insisted in true Baconian fashion, "founded on cautious and judicious induction."[75]

For all these reasons, the work maintained an intensely local health focus as he surveyed climatic and medical conditions in each of the states, their cities and subregions. And it was precisely this emphasis that enabled him to subject various vicinities to medical judgment. In the low marshy regions of New Jersey, for instance, Currie's local informant issued the finding, drawn from personal observation, that the inhabitants of these districts were generally sickly. The women, he reported, were "frequently Chloriotic [*sic*], are generally pale, squalid, and have carious teeth."[76] But in the high, dry, and mountainous parts of the state, they remained healthy and enjoyed much greater longevity. For Delaware, readers learned that the hills of Brandywine and Christiana were among the most wholesome districts comparable to the best in the whole country. In particular, the borough of Wilmington was superior in health and beauty to any town of Currie's acquaintance between New England and the southern border of Virginia. Kent, however, was condemned as the sickliest of the three counties of Delaware, despite enjoying the most fertile soil. Dover, the county town and state capital, was dismissed as "truly unhealthful," with the air suffering "exceedingly from stagnation." As emblematic of the whole, Currie's Hippocratic treatment of North Carolina gives a sense of the local verdicts he handed down on one location after another:

> The Country and Climate to the Westward is in general healthy, the low Grounds along the Rivers alone being otherwise, particularly those of the Roanoak, occasioned by exhalations arising from the damp Soil, Stagnated Waters, and by the putrescent particles with which the Atmosphere is replete, whose free circulation and purification is prevented in such situations, by lofty thick Woods, impervious to the Sun's rays, under which a dismal Gloom for ever dwells.[77]

If Currie's medical-meteorological diagnostics were, to some degree at least, an extended exercise in pathologizing particular places, that did not prevent him from issuing a bullish finale to the work as a whole. Its tone was reminiscent of Jefferson's buoyant confidence in the superiority of the New World's products compared with the denigration of the French naturalist Buffon.[78] While the new United States could not boast of a climate superior to other nations on the same latitude, Currie was nonetheless sure that few countries exceeded it in fertility of soil, and none equaled its political advantages. Indeed, the residents of the middle and particularly the northern states lived in a healthier environment and survived on average to a greater age than the peoples of Europe. American diseases were more simple and uniform than elsewhere, and the country as a whole was exempt from many of the most destructive infestations that plagued other parts of the globe. Small wonder that in his valedictory proclamation, Currie announced in suitably patriotic tones, "North-America is the only portion of this spacious globe where man can live securely, and enjoy all the privileges to which he has a native right."[79]

However dreary this Enlightenment enterprise might seem to modern readers, the medical geography venture, replete with detailed charts of daily temperature, wind direction, rainfall, humidity, and the like, was nothing short of "a kind of early 'big science.'"[80] Garnering data from a myriad locations, it sought to impose conceptual order on the chaotic particularities of widely dispersed places and to bring the vast terrain of the new continent within the uniting embrace of medical science. And so, in the years that followed, medical geography as an enterprise flourished. As Ronald Numbers put it, "During the first two-thirds of the nineteenth century, medical geography, broadly conceived, reigned as the queen of the medical sciences."[81]

For the United States, no one did more to bring the geography of the continent under medical surveillance than New Jersey physician, social reformer, and natural philosopher Daniel Drake (1785–1852), who later became president of the Medical College of Ohio, Cincinnati, of which he was one of the founders. A prolific author, Drake also made contributions to geology, botany, and meteorology and was elected to membership of the American Philosophical Society. In 1850, he put together his monumental *Systematic Treatise* cataloguing "the principal diseases of the Interior Valley of North America" and their manifestation "in the Caucasian, African, Indian, and Esquimaux Varieties of its Population." He had been planning the work for nearly thirty years and from time to time published materials relevant to it, though he was apparently frequently diverted by other duties.[82] His first scientific publication, for example, was on the medical topography of his hometown, Mays Lick in Kentucky,[83] and shortly after it appeared, he published, in 1810, *Notices Concerning Cincinnati*, a work that reviewed the city's local geography—topography, geology, and, in much greater detail, climate and weather—as a prelude to charting its prevailing diseases. Perhaps predictably, given the environmental conditions of the town, miasmata, from which a variety of fevers were believed to emanate, headed his list of endemic causes of disease. Because noxious exhalations from marshes could be offset by arboreal absorption, Drake urged that the forest should be preserved in "the most sacred manner." Variations in temperature were also considered a stimulus to illness. Here he explained that neither heat nor cold, by themselves, produced many ailments "at this place," but he did contend that "their sudden alteration is a most fruitful source of disease." Following Benjamin Rush (1746–1813), he reported that diurnal variations, which were greater in spring and autumn than in summer, tended "to excite intermitting and remitting fever."[84] But that was inconsequential compared with the effects of temperature variability in inducing rheumatism, tooth decay, pneumonia, pleurisy, consumption, and croup. In charting the distribution of such conditions, it was only to be expected that the cartography of disease in the region closely followed local environmental geography.

During the 1840s, Drake began to attend more single-mindedly to the business of producing a much fuller treatise and embarked on the task of gathering medical-meteorological information based on his travels in Alabama, Missouri, Illinois, and the Upper Mississippi River region, as well as from West Virginia, Pennsylvania, and western New York State. Throughout the work, the influence of the French *philosophe* and orientalist Comte de Volney, particularly his *View of the Soil and Climate of the United States of America*, which came out in English translation in 1804, clearly revealed itself even if Drake departed from some of Volney's diagnoses. Drake's *Systematic Treatise* thus reflected the Hippocratic caste of French hygenism, particularly in the work of the celebrated Montpellier school. That tradition, mediated through the writings of René La Roche in the *North American Medical and Surgical Journal* during the 1820s, reinforced the need for physicians to adapt their practices to local circumstances and inspired the compilation of regional medical topographies.

Drake devoted the first and longest part of his account to the topographic, climatic, and hydrological conditions of the interior of the continent and to the social circumstances that he was certain were shaped by such environmental particulars. He organized his observations around the hierarchy of watersheds in the southern Mississippi, St. Lawrence, Hudson, and Arctic basins. A crucial element in this portrait was the role of climate as a cause of various ailments. As he explained, "As no fact in etiology is more universally admitted, than the influence of climate in the production of diseases, it follows that he who would understand the origin and modification of the diseases of a country, must study its meteorology." In Drake's account, the effects of the climate were "predisposing and exciting," and they thus acted in both "direct and indirect" ways.[85]

The second book of the *Treatise* dealt exclusively with febrile diseases, "autumnal fever" being his particular focus of concern. Here, while reaffirming the significance of atmospheric conditions, humidity and temperature in particular, on the pattern of the disease, he was careful to discriminate between climate as a necessary but not sufficient condition for the onset of fevers of various kinds—autumnal, intermittent, congestive, miasmatic, marsh, ague, and so on—in the Interior Valley. He thus assigned a prominent but not overdominating role to atmospheric conditions in the outbreak of these afflictions. Having charted, and presented in tabular form, the number of attacks of autumnal fever in different quarters of each year in different locations, he concluded that "solar heat plays an indispensable part, in every hypothesis which has been proposed to explain the origin of autumnal fever." Still, while high temperature was a prerequisite for the onset of fever, Drake was sure that heat alone was not the sole explanation since its effects were moderated in different geographical contexts. At the same time, he remained convinced that seasonal variability, while sometimes challenging to the constitution, provided a "stimulus" that was "preferable, in its influences on the constitution, to long-continued and intense heat or cold."[86] Such valorizing of climatic variation perfectly suited Drake's more general cultural creed.

Drake's medical geography was domiciled in a wider moral and political universe. His long-standing commitment to the Temperance movement, for example, manifested itself not least in his periodic comments on the various deleterious effects of alcohol consumption.[87] He thus devoted a section of his *Treatise* to the range of alcoholic beverages available and to the patterns of imbibing in the continental interior. His findings were alarming, and he insisted that if alcoholic drinks had indeed been a necessity for the human species, "they would have been made productions of nature; but they are not the productions of nature; and, therefore, they are not necessary." The Creator, ever beneficent toward His creatures, had scattered around the human race all the physical stimulants the body required for its full perfection, and it was disturbing to witness humanity's turning instead to alcohol, "this factitious and baneful stimulant." Regional character also fell within the compass of Drake's medical climatology with its foundations in Hippocratic cosmology. The physiological modification that the four races inhabiting North America—"Caucasian," "African," "North-American Indian," and "Mongolian" or "Esquimaux"—were undergoing evinced the compound influence of intermarriage, climate shifts, diet, and "change of political, moral, and social condition."[88] The Anglo-Saxon race above all showed that the temperate zone of northern Europe had supremely fitted it for the varied climatic conditions of North America's continental interior. Other racial types were less suitable, as he demonstrated in several ways. He supported the idea that Indian tribes should be removed from Ohio, for example, and likewise lamented the increasing number of free blacks entering the state.[89] It was plain, therefore, that in this venture, Drake was engaged in the project of documenting the prevailing conditions and desirable constituents of the new society being formed in the American West, one that reinforced the prejudices of patriotic Western Republicans. This was a vision entirely in keeping with his admiration for Thomas Jefferson and his support for the French Revolution and reflected, as Dorn puts it, the "heavy draughts of progressive republicanism" Drake had long imbibed.[90]

In other national settings, comparable works attributing diseases, to one degree or another, to the vicissitudes of climate could readily be enumerated. Alfred Haviland (1825–1903), a medical officer for several English counties and a teacher at St. Thomas' Hospital in London, for instance, produced his conspectus *Climate, Weather and Disease* in 1855.[91] His review of classical medical literature led him to expand on the value of the study of climatology for medical practitioners and to expound the iatro-meteorological method of the Hippocratics. Haviland gave pride of place to "the assemblage of meteoric phenomena" that explained "the most remarkable diseases that prevailed" during the seasons of three different years.[92] These declarations served as prefatory to his elucidation of a host of ailments whose roots lay in the vagaries of the weather. It was an extensive undertaking causally connecting seasonality, wind, temperature, precipitation, and humidity in various ways with fever, consumption, mumps, ophthalmia, dysentery, paralysis, smallpox, and so on. And a lengthy commentary on the climate of Greece, presented as the foundation stone of the Hippocratic "pestilential constitution," confirmed in Haviland's mind both the direct and indirect ways in which climatic geography governed health in particular and the human constitution more generally.[93]

As part of his more general project, Haviland gathered disease and mortality statistics for a number of years—for England and Wales, for the English Lake District, for Ceylon from army records, and so on—in hopes of uncovering general laws governing the geographical patterns of disease.[94] In his account of the spatial distribution of heart disease and dropsy (edema), he ventured the generalization that over areas where the sea air had uninterrupted access, the death rate from heart disease and dropsy was low, whereas in places "where the tidal wave has no access," there the highest levels of mortality were to be found.[95] And later, when he brought out a second edition of his *Geographical Distribution of Disease in Great Britain* (1892), he took the view that diseases that in the past had been regarded as having a merely coincidental association with the environment were now being shown to have a much tighter causal connection with climate and soil.[96] Just as the English ethnologist and philologist Robert Gordon Latham (1812–1888) was convinced that there was "something" in the climate that connected racial types to climatic regimes, so Haviland was certain that "in that '*something*' we expect to find the link connecting the earth's airs, waters and soils with our bodies."[97]

Concluding his analysis of the geographical distribution of heart disease, cancer, and pulmonary tuberculosis, he observed that these three major causes of mortality had been influenced both by the general weather characteristics of a country as well as by the specifics of "local climates." In his scrutiny of the patterns of death rates from cancer, for instance, he correlated mortality rates with clay and limestone regions, stressing "the necessity of studying local climates, not only in relation to man's immediate requirement as regards temperature, winds, weight of atmosphere, etc., but with due regard to the requirements of lowly vegetable organisms" (see figure 2.1).[98] Here he had in mind the bacterial parasites that Sir James Paget (1814–1899), the English pathologist, believed might well be causally connected with cancerous diseases. In this way, Haviland found it possible to combine Hippocratic environmentalism with the new parasitology by dwelling on the climatic conditions under which microphytic organisms could flourish. What is no less notable is that Haviland located his medical geography in the context of Britain's racial history during the Stone, Bronze, and Iron Ages using the contrast between brachycephalic and dolichocephalic features as defining types.

It was, in part, the availability of regional medical geographies of this kind that fostered the writing of more general works on medical climatology, which sought to synthesize the ways in which climate governed human health. Edward Smith (1819–1874), an English physician who worked on the physiology of nutrition, is a case in point.[99] A public health worker, advocate of prison reform, pioneer in experimental physiology, and fellow of the Royal Society, Smith focused on seasonal cycles as part of his more general scrutiny of the impact of cyclical changes—daily, weekly, life course, and generational—on the human system. And from his experiments on periodicities in pulse rate, urea excretion, respiration, and the like, he developed practical proposals regarding the care of patients, appropriate times for drug administration, and the organization of night work.[100] While studying for his initial medical degree, he revealed his penchant for natural theology with an 1839 prize-winning

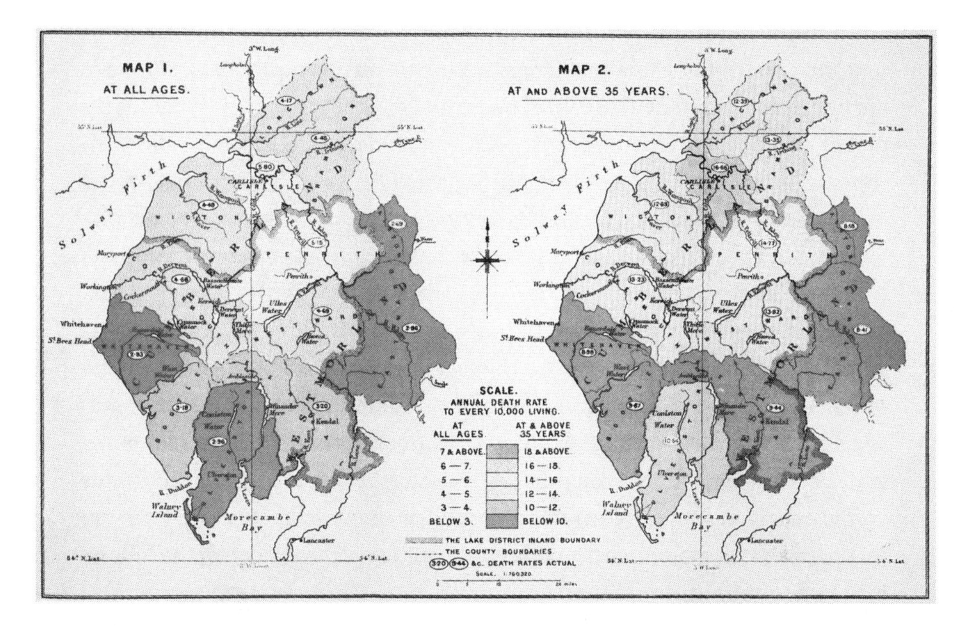

FIGURE 2.1. Haviland's Maps of the Geographical Distribution of Cancer (Females) in the English Lake District
Source: Alfred Haviland, *The Geographical Distribution of Disease in Great Britain* (London: Swan Sonnenschein, 1892)

essay on how the aortic system manifested the wisdom, power, and goodness of God.[101] And later, while practicing medicine in Birmingham, he undertook the task of traveling in northeastern Texas as an advance agent for a firm concerned with selling overseas lands to British emigrants.

Smith rooted his thinking about human health and seasonality in the "cyclical changes in the animal economy from season to season"—a principle that would later come to greater prominence, as we will see, among those preoccupied with birth season and its eugenic possibilities. For Smith was convinced that the month of the year in which people were born had a significant bearing on what he called their "viability." Those born in the winter months, apparently, had the greatest viability; those born in the summer had least. In Smith's opinion, there was "no subject which influences the conduct of mankind so much as the instinctive knowledge and common observation connected with season." For that very reason, he embarked on the task of canvassing the wisdom of the ancients—Hippocrates in particular—who had devoted much energy to elucidating seasonal patterns of health. Like them, Smith was sure that diseases that were regular and mild occurred during periods of the year marked by uniform conditions, while "in inconstant and unseasonable times the diseases are uncertain and difficult of cure." Autumn brought the most pernicious illnesses, while spring was the healthiest season and most free from fatalities. In support of this lawlike generalization, Smith set out to gather empirical detail on the role of seasonal cycles in stimulating a range of diseases using data he gathered between 1858 and 1861. Because various physiological operations—pulse rate, muscular force, and what he called "vital action"—displayed seasonal patterns, the human body was more or less susceptible to specific illness at particular times of the year. "Brain diseases prevail in the cold season," he declared, while "eruptive diseases" manifest themselves at the period of seasonal change when people are exposed to "the confluence of two sets of causes which are antagonistic to each other."[102] Such discoveries had implications for medical therapies too, and Smith elaborated on his belief that in the same way that particular seasons engendered certain ailments, the change of the following season was frequently curative.

Identifying seasonal cycles as the foundation on which to erect a comprehensive medical system could lead in rather conflicting directions. On the one hand, it connected medicine to the world of natural history through identifying commonalities between human and animal flourishing. In that guise, medicine incarnated itself as a kind of physiological ecology. On the other, the allure of natural rhythms and climatic cycles could prompt advocates, like Smith himself, to turn toward the occult arts of the stars. And he could assert that the "beneficial action of the rotation of the seasons is the true foundation of expectant methods of medical treatment." At the same time, he felt that it also provided grounds for homoeopathy and for a sympathetic reading of early natural astrology, which, he explained, whatever its errors, was based on "knowledge of the effect of season" and of the influence of astronomical phenomena.[103]

As the enterprise continued to flourish, numerous works were added to the canon of medical geography throughout the nineteenth century. Arthur Bordier (1841–1910), who specialized in the races and diseases of the Malay Archipelago, brought out *La Géographie Médicale* in 1884. Émile-Léon Poincaré (1828–1892), who had a specialist interest in diabetes and diseases of the nervous system, made pioneering contributions to industrial hygiene, and was, incidentally, father of the famous mathematician and philosopher Henri Poincaré, published his *Prophylaxie et Géographie Médicale* the same year. And the *Geographical Pathology* of Andrew Davison (1836–1918), the Edinburgh expert on tropical diseases, appeared in 1892.[104] But the pinnacle of the entire medical geography venture is often considered August Hirsch's (1817–1894) *Handbuch der Historisch-Geographischen Pathologie,* first published between 1859 and 1864 and later translated into English, from the second edition, by Charles Creighton during the 1880s.[105]

Hirsch had spent several years as a medical practitioner in Elbing (Elbląg) and Danzig prior to his appointment as a medical professor at the University of Berlin in 1863. Here he pursued his research on epidemics and served on a number of medical commissions appointed by the German government.[106] But it is the *Handbuch* for which he remains best known. And while this and comparable works were evidently of wider scope than medical climatology, discussion of climatic explanations for various medical conditions frequently came to the fore.

Hirsch's *Handbuch* was consciously conceived as a Hippocratic venture. While he positioned his own endeavors in that tradition, however, he was equally insistent that Hippocratic medicine had only more recently been placed on a properly scientific footing. It was only in the sixteenth century, with the opening up of the globe to European eyes, that the firsthand empirical interrogation of the natural world began to displace dogmatic speculations. As a consequence, he reckoned, "naturalists and physicians again endeavoured to find matter for scientific investigation in the changing aspects of organic life at various parts of the globe."[107] At the same time, he placed his own endeavors in the lineage of medical geography that figured prominently on its horizon such practitioners as Leonard Ludwig Finke (1747–1837), Alfred Mühry (1811–1888), Jean Christian Marc François Boudin (1806–1867), and Henri-Clermond Lombard (1803–1895).[108] And while all these authors departed in various ways from the particular enthusiasms of each other, their collective works constituted a citationary reserve to which medical writers concerned with connecting health to the particularities of place had regular recourse.

Hirsch's resort to weather and climate as explanatory factors in the genesis and distribution of disease was prosecuted with rather greater circumspection than often was the case. In many instances, he specifically rejected climate as a cause or condition of illnesses of one sort or another. In the case of influenza, for example, he insisted that "climatic and telluric influences . . . do not help us much to explain the cause and origin of the malady," that is, he wryly quipped, "unless indeed we are content with conjectures that have no foundation and hypotheses for which there is no proof."[109] The incidence of scarlet fever, too, was entirely independent of climate or season or weather both with respect to its geographical distribution and the intensity of an attack. It was the same for cholera, syphilis, yaws, cancer, epilepsy, and many more. Nevertheless, the range of disorders over which climate held imperial sway was extensive. The origin of dengue fever, he reported, depended on "atmospheric influences proper to the climate, particularly the temperature." The frequency and extent of measles outbreaks were under the influence of the weather and the seasons, even though their distribution was independent of climate. In explaining the eruption and circulation of malaria, he considered that "climatic and telluric conditions hold the first place."[110] The list could readily be extended: yellow fever, dysentery, hyperemia of the liver, kidney disease, respiratory disorders, croup—all of these bore the marks of climate's governance of the human body.

The mushrooming of the medical climatology enterprise during the eighteenth and nineteenth centuries reinforced the localism that had long been embedded in Hippocratic medicine by its extensive accumulation of regional medico-meteorological inventories. As often as not, these works were suffused with matters spiraling well beyond medical diagnoses narrowly conceived. For their authors were as interested in moral and cultural affairs as in illness and disease while they issued judgments on places ranging from the nourishing to the noxious. At the same time, by taking the particularities of place seriously, they embedded medical practice in a wider ecological framework that sought to treat patients in holistic ways. Moreover, it would be mistaken to consider this genre wholly backward looking in its orientation. The incorporation of the new parasitology by some practitioners of medical geography discloses a certain dynamism in the prosecution of the project. That there were also therapeutic possibilities in the medical mental maps that these geographies generated was obvious and opened up remedial possibilities for those who could afford to benefit from them.

Climate and Health Tourism

Taken in the round, the increasing availability of such medical geographies fostered the idea that the globe could be divided, at various scales, into a mosaic of healthy and unhealthy spaces largely circumscribed by climatic conditions. In different parts of the world, different medico-meteorological regimes held sway. Accordingly, some zones were categorized as sickly, others as salubrious; some pathological, others restorative; some hazardous, others wholesome. Given this global medical map, the perceived health benefits arising from short- or long-term residence in more propitious climes soon began to gain traction and to spawn a range of migration therapies. In Germany, for example, Adolf Mühry (1810–1888), a medical practitioner and Hanoverian court physician, produced several works of medical geography. Among his multifaceted inquiries were his investigations into the role of climatic factors in explaining the medical conditions of different peoples, which led him to urge patients inhabiting one climatically controlled disease regime to move to a more healthy climatic zone. In large measure, this was on account of his conviction that the distribution of diseases followed a lawlike pattern and that certain health conditions were bounded by climate.[111] What gave added authority to such medical cartographics was the immense popularity of the geographical writings of the Prussian polymath Alexander von Humboldt. Mühry, for example, dedicated his work on the foundations of nosogeography to Humboldt and his later volume on medical climatology to "one of Britain's most Humboldtian of scientists, Sir John Herschel."[112] He borrowed from Humboldt the idea of the isotherm in his delineation of the classification scheme he used for his nosogeographical mapwork, which plotted the global distribution of disease in relation to climatic conditions. Crucial too in Muhry's conception of medical cartography was the plant geography of Humboldt. For just like plants, temperature-sensitive infectious diseases were distributed in distinctive zones. The northern limit of malaria, for example, was identified as the 40°F isotherm, while the southward range of typhoid did not extend beyond the 74°F isotherm.

Altitude, no less than latitude, was critically important in Humboldt's representations of how different plant species occupied distinct climatic zones at increasing elevations. And this was picked up by Humboldtian medical geographers. In his *Medizinische Geographie* of 1853, for example, Caspar Friedrich Fuchs specified entero-mesenteric, catarrhal, and dysenteric regions according to latitude and altitude.[113] Like the distribution of vegetation, disease tended to follow the contours of altitudinal gradient. Mühry too was attentive to altitude and commented on the absence of certain diseases at particular elevations, noting that phthisis (pulmonary tuberculosis) did not present itself in elevated regions. In his opinion, the formation of tubercles in the lungs was made more difficult by the low atmospheric pressure and the reduction in the density of oxygen.[114] Such portrayals chimed with those pointing to the therapeutic benefits to be gained from visiting mountainous, often Alpine, venues. In 1859, for example, Hermann Brehmer (1826–1889) opened his climate institutes for tuberculosis treatment, an event that has sometimes been seen as the inauguration of what has come to be known as "climate tourism," in the Alps and other mountainous zones. He had earlier made a visit to the Himalayas and on his return sought to reproduce the high-altitude climatic conditions that he believed had cured him of the illness. To him, such elevated locations were immune spaces distant from the breeding grounds of disease in the lowland cities. His thinking about sanitoria—places of healing—was swept into German national planning in the wake of the country's rapid industrialization during the late nineteenth and early twentieth centuries and the challenges of tuberculosis.[115]

Earlier elevated resorts for the treatment of various ailments had already been in existence, of course, notably in India. Robert Baikie (1799–1889), for instance, insisted in 1834 that early stage tuberculosis could be successfully treated in the Nilgiri Hills, where he himself was the medical officer based in Ootacamund.[116] And in Europe itself, during the 1840s, Louis-André Gosse (1791–1873) and Johann Jakob Guggenbühl (1816–1863) recommended higher elevations for the treatment of "cretinism"—congenital hypothyroidism.[117] At the same time, in South America, Johann Jakob von Tschudi's (1818–1889) elucidation of the geography of disease, alongside that of Archibald Smith (1820–1870), identified the absence of pulmonary tuberculosis in high altitudes.[118] But whatever the precise origins, the idea of resorting to mountainous climates to treat illnesses of one sort or another certainly caught on. During the second half of the nineteenth century, Henri-Clermond Lombard (1803–1895) told the readers of his essay on medicine and mountain climates that high altitudes brought phthistic immunity, a view that was promulgated in Latin America by Denis Jourdanet (1815–1892), who pioneered research on altitude sickness and hypoxia.[119]

This drift in medical opinion soon created an audience for guidebooks to locations with suitable climatic conditions for patients with various ailments.[120] In 1862, for example, a treatise on medicine and meteorology carried the subtitle "A topographical and meteorological description of the localities resorted to in winter and summer by invalids of various classes." *Medical Climatology* was the work of a medical practitioner and fellow of the Royal Society of Edinburgh, Robert Edmund Scoresby-Jackson (1835–1867), who had written a thesis on "Climate, Health and Disease" for his MD degree in 1857. Scoresby-Jackson was also well known as the biographer of his uncle, the Arctic explorer and clergyman-scientist William Scoresby. Besides this and his other duties, he acted as chair of the medical department of the Scottish Meteorological Society.[121]

Appropriately enough for a work dedicated to the Scottish geographer and mapmaker Alexander Keith Johnston—"as a slight acknowledgment of the advantages derived from his stupendous works by the members of the medical profession"—Scoresby-Jackson devoted the first substantive chapter of *Medical Climatology* to elucidating the nature and causes of local climates, dwelling on the role of distance from the equator, elevation, relative position of land and sea, oceanic currents, state of cultivation, and so on, in determining weather conditions. What is conspicuous, moreover, is that he domesticated this introduction to climatology for medical practitioners in the context of natural theology, prefacing his remarks with the confessional declaration that "the Great First Cause has distributed His creatures according to the counsel of His will; He has supplied them severally with constitutions, food, raiment, and even medicines suitable to the regions in which they are placed; and the order of His original design is yet maintained by 'fire and hail, snow and vapours; stormy wind fulfilling His word.'" The global distribution of living things followed a divine ordering of the natural world. Lions, for example, were not made to occupy icebergs, nor were reindeer to graze in "the region of palms." Correspondingly, "Esquimaux would perish where negroes revel in the exuberance of nature's gifts. . . . Nor can the inhabitant of temperate regions, with all his superior advantages, ultimately abrogate the laws of climate." Theologically and geographically, there were fixed boundaries that people transgressed at their medical peril. Clearly, what bounded Scoresby-Jackson's thinking throughout his medical climatological work was that different human races were, by divine appointment, suited to different places and that permanent migration into a different climatic regime was usually deleterious. All of this meant that it was only "a *temporary* and not a *permanent* residence in a foreign land" that enabled climate to act as a curative agent, to fulfill its role as "the handmaid of medicine."[122]

With this in mind, *Medical Climatology* was designed to assist doctors in answering the question posed by patients aware of climate's potentially restorative powers: "Where must I go?"[123] The book was thus, in the main, a kind of medical gazetteer. But before getting to an inventory of the atmospheric conditions of locations to which Britons, at least of a certain class, typically resorted, Scoresby-Jackson identified the framework governing his medical climatology project. The scaffolding to which he fixed his therapeutic meteorology was what he called "Hygenics." In so labeling his treatise, he intended to position it in the long Hippocratic tradition that recognized the value of a change of climate as a "remedial agent" and thus of hygienics as preventive medicine. The task was to match ailment to atmosphere, well-being to weather, medical complaint to climatic condition. Early medical-meteorological diagnosis was critical. In the case of pulmonary consumption, for instance, if a change of climate was not recommended by a medical consultant until every other remedy had been tried and failed, it would be an act of unpardonable cruelty to send the patient away from home when the condition had become so far advanced. All this meant that medical climatology should operate with a set of adjectives akin to such common meteorological descriptors as "torrid, temperate, frigid, hot, cold, dry, humid, and so forth"; "exciting, irritant, bracing, sedative, relaxing, and such like," he believed, captured the "peculiar influence" that different climates exercised on "the constitutions of persons inhabiting them, whether permanently or for a season."[124]

Now it simply remained for Scoresby-Jackson to devote the remainder of his lengthy text to sketching those diseases for which changing a patient's climate was recommended as a therapeutic agent and to surveying a wide range of locations suitable for travelers within the imperial arc. An alphabetical listing of complaints from ague and anemia to rachitis (rickets) and scrofula served as a preface to some four hundred pages of regional portraiture—interlaced here and there with commentary on mineral spas—beginning with Algeria and ending with the West Indies. In each case, he provided his readers with basic geographical information and meteorological data, as well as commentary for voyagers on modes of transportation and sights to be visited, before identifying those ailments that would receive benefit from a period of residence elsewhere. His commentary on Algiers is illustrative:

> As a resort from the inclement seasons of northern Europe for persons threatened with pulmonary consumption, Algiers, in my opinion, is deservedly in good reputation. The climate is far from being of a relaxing character; on the contrary, it combines with its usual mildness and equability a decidedly bracing and tonic influence. Consumptive patients, in whom there is a well-marked deposit of crude tubercle, may pass one *or more* winters in Algiers with advantage, under circumstances which afford Nature the most ample leisure for repairing the disorganised structure.[125]

And that became the general pattern. Those traveling to Australia from Europe found there rejuvenating "springs of life." The air was "remarkably elastic," he added, quoting the prolific Anglo-Irish author and civil servant Montgomery Martin (ca. 1801–1868), who served as the colonial treasurer of Hong Kong. According to Martin, Scoresby-Jackson reported, older European visitors arriving in the Australian colonies found "much of the hilarity of youth restored to them."[126] Closer to home, when surveying potential winter residences in France, he paused to confirm that the Mistral was exceptionally hostile to patients with respiratory diseases, especially those with tubercular conditions, and went on to disabuse readers of the mistaken idea that Montpellier enjoyed a salubrious climate. The climate of the southeast of France, notably Languedoc and Provence, was dry, hot, and unsettled and was therefore "one of the worst in the world for consumptive invalids." Pau, by contrast, was to be recommended for patients needing "a sedative and somewhat relaxing climate" which afforded a much-needed "soothing and beneficial" atmosphere.[127]

Scoresby-Jackson extended this self-same medico-meteorological imagination to Germany, Belgium, Switzerland, Spain, Ireland, Italy, and so on, providing for medical colleagues and patients alike a mental map of the globe comprising an intricate mosaic of healthy and harmful locations, a portfolio of places that the health-conscious traveler should seek out or shun. This therapeutic cartography was intended to serve the needs of doctors and their patients. But built into the very fabric of the enterprise were also judgments about people and places, weather and welfare, that traveled far beyond medical terrain and into the realm of imperial realpolitik.

In the midst of our current concerns over the health consequences of climate change, there has been a conspicuous recovery of the Hippocratic vision as the foundation stone of responsible medical planning. Whether or not references to Hippocrates in the pronouncements of major health organizations and prominent medical scientists are more decorative than constitutive, there can be little doubt that the ancient Greek physician at least serves as a symbolic figurehead for the need to take with utmost seriousness the challenges that global warming, environmental change, and extreme weather events present to the human species. This revival, however, is only the latest in a series of reappropriations of the Hippocratic corpus for different purposes in different situations. The revitalization of Hippocrates' therapeutics in the mid-seventeenth century, for example, took place in the context of Baconian natural philosophy with its emphasis on the triumph of empiricism and experience over authority and a priori reasoning. By "Baconianizing" Hippocrates, it was possible to wed place-based medical environmentalism to the inductive injunctions of the new natural philosophy. In the next generation, Hippocratic medicine could be attached to the Newtonian philosophy with its mechanical metaphors and mathematical methods. In the nineteenth century, the further retrieval of the Hippocratic system was called upon to resource the production of geographical inventories that fed into a range of therapeutic measures tied to place and space. In this tradition of work, the scrutiny of diseases was, as Michael Osborne tellingly characterizes it, "a medical analogue of cartography."[128]

This cartographic impulse was rarely restricted to the mapping of bodily ailments and disorders in different climate regimes, however. Medical geographies were frequently interspersed with moral judgments, evaluations of character, depictions of temperament, and assessments of intellectual aptitude. From its earliest days, Hippocratic medicine incorporated an anthropological dimension that highlighted the ways in which different climates produced cultures that displayed cowardice or courage, or engendered behavior that was judged to be brutish or polished. This inclination continued to manifest itself. In the early eighteenth century, for instance, artistic flair and aesthetic sentiment were attributed to the condition of the atmosphere, while in the climatic governance of the body's fibers, some found the key to national dispositions toward indolence or industry. To preoccupations of this kind, we will later return. Meanwhile, practitioners of medical topography were often fast to pass judgment on the morbidity of particular districts. And at the same time, medical geography could be used as a platform to trumpet the health virtues of entire continents. If pathology pulled in one direction, patriotism could pull in another. For these writers of medical climatology, disease ecology and moral cartography were much closer than distant cousins.

The proclivity to split the globe into a mosaic of spaces labeled as either wholesome or harmful, restorative or destructive, benign or malevolent manifested itself at very different scales. Some ranged widely over whole continental landmasses, while others operated at a more detailed scale to distinguish between low marshy districts where residents were sickly and upland regions where a bracing climate kept inhabitants in rude health. In different locations, the geographical distribution of diseases was charted to reveal spaces of vulnerability at fine scales. Concurrently, others kept an eye on the health significance of entire latitudinal zones and were mindful of global patterns of illness at the national scale. Those sensibilities in turn fostered the production of therapeutic guides for patients in search of healthful locations and stimulated the construction of sanitoria in what were considered to be favorable venues. Enterprises such as these delivered a mental map of global health and sickness that taught readers, to use John Willinsky's phrase, how "to divide the world."[129]

Fred Sargent once remarked that whatever "their methodological significance, the medical topographies and geographies of the nineteenth century were conceptually disappointing."[130] He may well have been right about that. But conceptual disappointment should not be equated with cultural insignificance. For the implications of breaking the world into salubrious and sickly spaces—places of vigor and vulnerability—had far wider political resonances. These medical climatologies, erected as they were foursquare on Hippocratic foundations, had the effect of pathologizing whole regions of the globe by typecasting them as pestilential. And nowhere did this more conspicuously surface than in the way in which they served to sustain the cultural politics of one dominant geographical imaginary lodged deep in the European colonial psyche: the tropical world.

3

Tropical Terrain

MORAL AND MEDICAL

IN THEIR construction of worldwide medical climatologies, many writers devoted their energies to elucidating conditions in one of the globe's terrestrial zones in particular—the tropics. At least in part, what animated what might be termed "tropical anxiety" was the fact that these regions were frequently the site of European colonial adventures and thus raised questions about the health of the white race in the tropics.[1] In a profound sense, this was actually a matter of unease about climate *change*: as colonial settlers, administrators, military personnel, and the like moved across the globe, they experienced changing climates and thus the health consequences believed to arise from tropical exposure. What is conspicuous about much writing on health in tropical regimes during the nineteenth century is the ease with which moral and medical concerns were merged. This fusion facilitated the tendency to gather together many different places and peoples under the general rubric of the "tropical" and fed the propensity to typecast this constructed world as a site of sickness and sin, a space of medical jeopardy and moral peril. In what follows, the impact such a mind-set exerted on medical discourse, imperial aspirations, and material practices will be a primary point of reference.

Tropical Tyranny

Many of those who have just been at the forefront of our considerations paused in their accounts of medical geography to issue judgments on the human body in hot climates. In 1861, for example, Edward Smith referred to "the ready susceptibility of those living in hot climates to the influences of morbid agencies."[2] The following year, Scoresby-Jackson, noting that moderately elevated locations were always more healthy than those at sea level, insisted that in "hot climates there are usually residences erected for summer use, and for invalids at all seasons, in lofty situations." The Sanitarium of Darjeeling, in the Presidency of Bengal, he pointed out, was "an example of a cool and salubrious retreat for British soldiers."[3] Rather more extensively, Hirsch directed his readers' attention to a range of chronic diseases afflicting tropical zones. Anemia, he explained, was a characteristically "morbid phenomenon of tropical regions" and was mostly, though not exclusively, to be found among white residents. Calling on the authority of John Sullivan, author of *The Endemic Diseases of Tropical Climate*,[4] Hirsch reported that there could scarcely be any doubt "that the peculiar influence of tropical climate which is called 'relaxing' and which cannot be otherwise defined in physiological terms, is, for white people not habituated, a very material factor in the development of this disorder of nutrition." Moreover, its effects were not restricted to the immediate victim. The "anaemic habit of body in white residents of the tropics communicates itself to their offspring," and therein lay "the chief obstacle to the acclimatization of the white race in the tropics." Here the politics of disease plainly surfaced. This is not to say that Hirsch attributed tropical illness solely to the malign influence of climate. He was at pains to insist that many ailments—leprosy, scurvy, beriberi, and scrofula, for example—had distributions that were incompatible with the claim that climate was the cause of the condition. But others did fall under the control of climate, notably malaria, yellow fever, ascariasis, and yaws, which he described as "an exquisitely tropical" affliction.[5]

Such characterizations of the tropical world, of course, were part and parcel of European imperial projects in various parts of the world and often sprang from anxieties about the exposure of colonial settlers and servicemen to a changed climate. It was European fear of hot climates during their experience of colonial enterprises from the seventeenth to the nineteenth centuries that transformed the study of sickness and space. As Valenčius explains, "The very organization of medical geography was shaped by its colonial context: medical geography was created as a field primarily by military physicians and those connected with militarily-enforced settlement. Priorities reflected the demands of troop deployment and military expansion."[6]

These preoccupations long predated the surveys of Smith, Scoresby-Jackson, and Hirsch, of course. Indeed, the influence of tropical regimes on human health—frequently on the welfare of Europeans in hot climates—had been the specific subject of medical inquiry for several generations. The Scottish physician James Lind (1716–1794), for example, published his *Essay on Diseases*

Incidental to Europeans in Hot Climates in 1768. A champion of naval hygiene who had served as a surgeon's mate on voyages off the coast of West Africa and in the West Indies prior to practicing medicine in Edinburgh, Lind did pioneering work on the cure for scurvy.[7] The well-being of navy personnel was his abiding passion, and in 1757, a few years after his *Treatise on Scurvy* made its appearance, he brought out *An Essay on the Most Effectual Means of Preserving the Health of Seamen in the Royal Navy*, in which he pressed the case for better naval hygiene and more humane treatment of crews.

Lind's treatise on the health of Europeans in hot climates, dedicated to Philip Stephens, the secretary of the Admiralty, was a manual for tropical settlers and soldiers alike. It was a popular work going through some six editions by 1811 and appearing in German, Dutch, and French translations. Here he surveyed a wide range of literature on tropical ailments and provided a port-by-port review of health conditions in each locality. Neo-Hippocratic themes crystallized again and again as he identified, like other miasmatists before him, zones of hot swampy environments that exuded pestilential vapors and noxious effluvia. Ever concerned for the everyday health of seamen, the book, as one biographer observes, "can be seen chiefly as a set of environmental, public health recommendations."[8]

What stimulated Lind's inquiries were the health effects of climate change. When settlers and regiments found themselves stationed in the globe's remotest locations, they necessarily exchanged "their native for a distant climate" and thus required the utmost care and attention to ensure their continued good health in a manner "somewhat analogous to that of plants, removed into a foreign soil." It was a task of enormous significance, for Lind was convinced that the locations that Europeans frequented well beyond their own shores were extremely unhealthy and that those climates often proved fatal to colonizers. The fifteenth- and sixteenth-century Portuguese experience in southern Africa provided indubitable evidence. Along the coast of Guinea, he reported, sickness was far more threatening to the settlers' well-being than shipwreck and decidedly more menacing than wars with native peoples. Securing territory was as much a medical as a military achievement. To Lind, the long-term legacy of that experience was racial degeneration and cultural degradation. In many of the places the Portuguese formerly settled, he lamented, "We can hardly trace any vestige of their posterity, but such as are of the Mulatto breed." What remained even of their language was corrupted. It was much the same in Jamaica, where the numbers of English settlers sacrificed to the climate were "hardly credible."[9]

Given these circumstances, Lind set himself the task of providing medical advice to European travelers, and he devoted a good deal of his text to surveying the conditions that prevailed, and the therapies to be practiced, in different locations. Taken in the round, Lind assumed the mantle of a medical magistrate ruling on the health condition of the regions of the world—Europe, the Americas, Africa, the East Indies—and advising Europeans on how to cope with hot climates. The state of health of those onboard company ships that annually visited the banks of Newfoundland was proof positive that even intense degrees of cold, when appropriate safeguards were put in place, generated few illnesses. In South Carolina, by contrast, diseases were "obstinate, acute and violent" and comparable to the effects of West Indian climates, which induced fevers that had proven so fatal to newly arrived Europeans. And while the wholesomeness of parts of North Africa could not be doubted, he judged the English settlements on the Senegal and Gambia rivers to be in a "remarkably unhealthy" condition.[10] On the west coast of India, he found the climate of Bengal (Mumbai) to be lethal for European residents, while at Surat in Gujarat and Tellicherry (Thalassery) in Kerala, they generally enjoyed good health.

Seeking out salubrious spots for European colonists in hot climates was his chief concern, and the safe spaces he identified bore the stamp of his neo-Hippocratic sympathies.[11] During what he called the "sickly seasons," Europeans were advised to reside at some distance from "unwholesome marshes and foul shores." They were encouraged too to retreat from the "sultry heats" of the lowlands to temperate hill sides. By taking up residence there, as on the island of Dominica, they could escape the attacks of malaria and other fevers and as a result could "enjoy as good a state of health and constitution as if they were in France." It was precisely the same in "the hot, southern, and less healthy parts of West Florida," where a number of elevated situations that experienced dry weather and were exposed to the winds afforded "a safe and certain retreat" from the illnesses that were common during the summer months. Mountaintops were not advised, however. In Jamaica, for instance, withdrawing to the "barren, cold and bleak summits of the Blue mountains" only invited trouble as "the sudden transition from the scorching heats in the vallies or woods, to so intense a degree of cold, must be injurious to the constitution."[12]

Meanwhile, the English physician William Hillary (1697–1763), already known for his *Practical Essay on the Smallpox* (1740) and his writings on the medicinal values of spa water, had turned his attention to ailments of what he called the "torrid zone" in the West Indies. Hillary was of Quaker background and had studied with Hermann Boerhaave in Leiden prior to practicing medicine in Ripon, Bath, and Somerset. In 1752, he moved to Barbados, where he remained for half a decade, during which he directed his energies toward understanding the medical effects of "changes of the air" on Barbados and the West Indian islands more generally.[13] His *Observations on the Changes of the Air, and the Concomitant Epidemical Diseases in the Island of Barbadoes* and his *Treatise on Such Diseases as are the Most Frequent in, or are Peculiar to the West-India Islands, or the Torrid Zone* were later republished in the United States with additional notes by Benjamin Rush in 1811. In these treatises, Hillary catalogued weather conditions over a series of months alongside "the variations of the concomitant epidemical diseases" in order to determine whether the ailments were "influenced, caused, or changed, by those variations of the weather."[14] Of critical importance was his belief that on their first visit to the West Indian islands, Europeans and North Americans, and more especially Britons, were usually seized with a fever not long after their arrival on account of the greatly increased heat of the climate. And so he saw it as his task to delineate the trajectory of the diseases they encountered and to provide guidance on tried-and-trusted treatments for conditions such as yellow fever, dysentery, tetany, elephantiasis, and Guinea worm. Throughout these assessments, Hillary drew on the authority of that "wise father and prince of physicians," the "great Hippocrates."[15]

Taken together, the writings of Lind and Hillary reveal that these medical practitioners had come to believe that they were facing a distinctively Caribbean disease environment different from anything that they had known in Europe. Earlier writers like Thomas Trapham (d. 1692), who practiced in Jamaica a generation earlier, were of the opinion that medical conditions there were better than in England, while Hans Sloane (1660–1753) believed that in Jamaica, he never came across an illness that he had not previously witnessed in Europe.[16] But now, just a few decades later, Hillary was certain that he was observing diseases—such as yellow fever—virtually unknown in Europe, as was Lind who, while he considered Barbados to be relatively benign, noted the destructive effects of yellow fever on the other sugar islands. As Kiple and Ornelas put it, these medical practitioners "accustomed to combating European illnesses" now "faced an unknown enemy in African diseases" transported to the Caribbean courtesy of the slave trade.[17] The contrast between temperate and tropical was stark. Like Lind, John Huxham, whose medical meteorological records have already been noted, attributed the presence of widespread fevers to the moist atmosphere arising from swampy lowlands or from the "Continuance of cold, rainy, thick Weather," while Charles Bisset (1717–1791), like Hillary, drew attention to changes in the condition of the air, particularly nocturnal land breezes carrying harmful vapors, in his scrutiny of the common "causes of health, and endemic diseases, in the West-Indies."[18]

Taken in the round, the conventional wisdom that treatises of this sort embodied delivered an image of the tropical world as a space of vulnerability requiring constant vigilance. In the main, their focus was on how, if at all, white European bodies could cope with the rigors of a hot climate for colonial settlement or military occupancy. Seeking out safe spaces in the torrid zone that more closely resembled climatic conditions in European homelands, frequently at higher altitudes and away from coastal lowlands, was a major strategy that medical practitioners recommended expatriates to follow. The fear that such depictions engendered certainly prompted some to ponder whether the risks were worth it. And nowhere, perhaps, were these more clearly visible than in the debates that raged over the possibility of white acclimatization to tropical nature.

Acclimatization: Aspiration and Anxiety

Dark though these latter characterizations of tropical ailments undoubtedly were, many medical practitioners retained a confidence in the capacity of European migrants to adapt to the prevailing climatic and disease conditions. Lind's *Essay on Diseases Incidental to Europeans in Hot Climates*, for example, was animated, as Mark Harrison puts it, by "an underlying optimism about European acclimatization in the tropics and the possibility of colonization."[19] To Lind, European constitutions could soon become seasoned to East and West Indian climates provided their physiology was not impaired by "the repeated attacks of sickness upon their first arrival." Europeans, when thus habituated, were no more vulnerable to disease than their fellow countrymen and women who stayed at home. Convinced as he was that it was a sudden climatic change that brought the greatest danger of sickness, he was nonetheless certain "that a seasoned constitution in any part of the world is chiefly to be acquired by remaining there for some length of time."[20] In the West Indies, twelve months were sufficient for newcomers to be perfectly seasoned to the new conditions. Concurrently, Bisset was assuring his readers that in healthful places not exposed to noxious effluvia, disease was not particularly taxing on properly acclimated Europeans. Indeed, he went so far as to suggest that "healthy persons have a more constant and uniform flow of good spirits in the West-Indies, than in Great-Britain; especially after being seasoned to the hot climates."[21]

The comfortable, if not complacent, optimism that attended these casual remarks, however, was soon to be replaced by a darker sense of the tyranny that climate exerted on humanity's migratory impulses.[22] In the early years of the nineteenth century, for example, James Johnson (1775–1845), an Irish naval surgeon, prolific author, and later medical practitioner in London, brought out his *Influence of Tropical Climates on European Constitutions*—a work in which he perpetuated the miasmatic theory that swamps and marshes universally gave rise to intermittent and remittent fevers, "varying in degree and danger, according to the heat, rains, and other circumstances of the season." The lethal influence of an atmosphere "impregnated with marsh effluvia, on the human frame is in some places astonishing," Johnson warned his readers.[23] First appearing in 1813, the volume enjoyed considerable longevity. The sixth 1841 edition was published with the assistance of James Ranald Martin (1796–1874), who had occupied the post of presidency surgeon at the Presidency General Hospital in Calcutta until 1840 and in years to come would receive a knighthood for his services. Johnson himself could draw on a wealth of personal experience, having occupied duty stations in China and India, as well as Nova Scotia and the Mediterranean, by 1814 and for forty years thereafter editing the *Medico-Chirurgical Review*.

Right from start, readers were told that the "tender and innocent sheep, when transported from the inclemency of the north, to pant under a vertical sun on the equator will, in a few generations, exchange its warm fleece of *wool* for a much more convenient coat, of *hair*." But it was a very different story for the human species. For Johnson was certain that "the tender frame of man" was unable to endure those effects of exposure to the vicissitudes of climate, which could speedily bring about "a change in the structure, or, at least, the exterior, of unprotected animals."[24] To that degree, humankind was less efficient than the animal kingdom in offering protection from the somatic influences of climatic extremes. Whether they succumbed as early victims of a tyrannical climate or whether it was their progeny who would gradually, but inevitably, degenerate through exposure to an alien sun or the evils of interracial breeding, Johnson believed that Europeans were ill-advised to attempt permanent settlement in tropical India. The idea that time could season humans against disease had proven to be a fallacy. It was the very weakness of the human constitution that rendered people much less capable of coping with new climatic regimes than animals and that made the study of climate and the practice of hygiene all the more crucial, even for the briefest of stays in the tropical world. What was also notable about Johnson's thinking was the way he linked the effects of climate with what he called "sympathetic connections." The idea was that influences on one part of the body could provoke "sympathetic" responses in other parts of the anatomy through a variety of means. For instance, Johnson and Martin both perceived linkages between the skin and the gastrointestinal tract, the skin and the liver, and the liver and the brain. An immediate implication of such associations, as we will subsequently see, was that the impact of atmospheric conditions on the surface of the body had repercussions on other parts of the human system, including the psyche, frequently through excessive perspiration and other accumulated effects of tropical heat.

Other voices soon raised similar concerns. Arthur Saunders Thomson (1816–1860), for example, an assistant surgeon stationed with the 17th Foot at Poona (now Pune) in Maharashtra, India, assessed the theory of acclimatization in an article for the *Madras Quarterly Medical Journal* in 1840. Thomson had access to the *Statistical Reports of the Health of Troops in the Colonies* and used the data contained therein to ascertain whether mortality increased or decreased with the length of residence in the tropical world. The news was not good. In Madras alone, the annual death rate among military personnel was 3 percent higher than in Britain, while in Bengal, the risk of death increased with every year of residence among civil servants and soldiers alike. His conclusion was clear: "no length of seasoning will diminish the deleterious influence of a tropical climate on the European constitution."[25] In fact, the longer Europeans remained in a tropical climate, the more powerful its pernicious influence proved to be. Three years later, and now with the 14th Light Dragoons, he addressed the Medical and Physiological Society of Bombay on the following question: "Could the natives of a temperate climate colonize and increase in a tropical country and vice versa?" His answer, drawing on demographic statistics from across the tropical world, was pessimistic. The price of European settlement was high, in fact too high. Colonial ventures in hot climates brought sickness for the settlers themselves and offspring that were physically and mentally degenerate. Supporters of acclimatization, he reckoned, were obviously motivated by partisan politics, not medical research.[26]

Further reinforcement for an uncompromising stance on acclimatization came from the pen of the infamous Robert Knox (1791–1862), army surgeon, philosophical anatomist, and ethnologist, who attained notoriety for his association with the body-snatchers Burke and Hare. Hissing xenophobic vitriol, the Edinburgh-based anatomist and ethnologist opposed acclimatization in every shape and form as he expanded on the sinister long-term consequences, both medical and moral, of tropical colonization.[27] Knox was already infamous on account of his contacts with the "resurrectionists"—graverobbers—from whom he procured subjects for his anatomical dissections. His racial proclamations, notably in *The Races of Men*, which first appeared in 1850, only added to his reputation. Here, rejecting all thought of human adaptation to climate, he declared acclimatization a myth. "Can a race of men permanently change their locality"? he asked. "Can a Saxon become an American? Or an African? Can an Asiatic become a European? Can any race live and thrive in all climates?" Knox declared it simply impossible to take such questions seriously. Just posing the question provided the answer. The European inhabitants of Jamaica, of Cuba, of Hispaniola, of the Windward and Leeward islands had made no acclimatizing progress whatsoever. "Cease importing fresh European blood," Knox warned, "and watch the results." Infertility, sickly children, and cultural degeneration followed as surely as night follows day. As he put it, "The European . . . cannot colonize a tropical country; he cannot identify himself with it, hold it he may, with the sword, as we hold India . . . but inhabitants of it, in the strict sense of the term, they cannot become."[28]

Pessimism about white acclimatization persisted.[29] Robert Slater Mair (1826–1920), a military surgeon and deputy coroner in Madras, for example, warned the readers of his *Medical Guide for Anglo-Indians* (1874) that however "wonderful" was the European race's powers of adaptation, long-term mental and physical deterioration, digestive disorders, and feeblemindedness in offspring were common among Anglo-Indians. Moreover, rather than facilitating acclimatization, prolonged residence only made these afflictions worse.[30] Daniel Henry Cullimore (ca. 1847–1892), a London doctor and Indian Army surgeon who had acted as consulting physician to the king of Burma, likewise took up the subject in a piece for *The Medical Press* in 1888. Summarizing the effects of sunstroke on blood degeneration, lung, liver and kidney disorders, and reviewing the fate of inhabitants of the temperate latitudes traveling in the tropical world where they found themselves enervated "by the influence of a debilitating climate and a generous soil," he insisted that long-lasting acclimatization could not be achieved.[31] Later, in 1890, he brought out *The Book of Climates*. It consisted of a catalogue of various medical-meteorological regimes around the world and included a good deal of material recycled from a variety of earlier publications in medical outlets. This survey afforded him the opportunity to confirm his conviction that he had now furnished the strongest possible evidence both of the debilitating effects of tropical atmospheres on the indigenous peoples and of "the impossibility of their permanent occupation and colonization by the races of temperate Europe." In large part, the reason lay in the inexorable effects high temperatures exerted on the bodies of settlers: the "vital capacity and respiratory movements become lowered, the pulse loses its tension, the plumpness, rotundity, and firmness of the body diminish, and the excretions, as the urea and carbonic acid, decrease."[32]

Like many others, Cullimore extolled the medical virtues of hill stations. In the upland climates of India, he reported that military mortality from all causes was dramatically decreased and that the number of deaths was comparable to the figures for England. Victims of leprosy, he added, "should be set in hill stations, and not, as at present, located in such damp situations as Port Blair, Malabar coast, or the island of Molokai." Besides, these higher-altitude locations also provided recuperating venues for those suffering from overwork or simply from heat exhaustion on the plains. Mountain climates were particularly appropriate for "delicate women and children" and brought much improvement to victims of "nervous depression and irritability," as well as to those suffering from "leucorrhoea, uterine weakness," and "intractable skin diseases."[33] And so when he came to his weather survey of India, he paused to describe such hill sanatoria as Simla, Kussowlie, Dugshai, Mussourie, and Darjeeling, drawing freely on the work of the Scottish physician Thomas Graham Balfour (1813–1891), who collated sanitary statistics of European military in India. Hill stations—essentially British enclaves—had become an important feature of imperial landscapes, evoking what has been described as "a racial and spatial category that symbolized superiority and difference" by presenting themselves as safe spaces distanced from the ceaseless medical strife of infested tropical lowlands.[34]

By now, concerns about tropical acclimatization were shifting from India toward Africa, and Cullimore made sure that in the second, 1891, edition of this volume, he introduced a new chapter on "The Climate of Africa as it Affects Europeans," reporting the views of Surgeon-General Thomas Heazle Parke (1857–1893), an Irish medical practitioner and naturalist who served in many campaigns and accompanied Henry Morton Stanley on his 1887 expedition.[35] Parke focused his attention on regional weather conditions and disease, pausing to contrast zones of threat with localities where the conditions were restorative and invigorating. Alexandria, for example, was "no place for the consumptive, asthmatic, or rheumatic to linger," on account of the steamy and suffocating conditions there. In Zanzibar, both intermittent and remittent fevers were "frequently fatal." Cairo, by contrast, particularly in its outer reaches, enjoyed an atmosphere that was "pure, fresh, dry, warm and invigorating" air. Cullimore's gaze was not narrowly restricted to the needs of medical inventory, however;

he also kept an eye on imperial politics, insisting that white settlers could only maintain good health "by the utilization of the natives to do the work in the field" and by returning "to Europe every two or three years."[36]

Much the same skepticism about the possibility of white acclimatization to tropical Africa came through that same year in the writings of Sir William Moore (1864–1944), surgeon-general to the government of Bombay and well known for his views on malaria and opium, as well as for his opposition to the new germ theory.[37] It was in India where Moore first addressed the acclimatization question. But now, in 1891, he posed the question "Is the Colonisation of Tropical Africa by Europeans Possible?" to members of the Epidemiological Society. Racial degeneration, infant mortality, hybrid sterility, and the malign physiological effects of heat provided the answer. Africa, lying in the "torrid zone" with, in certain regions, "perhaps the most unhealthy climate for Europeans in the world," represented "the acme of a climate inimical to the European constitution."[38] Besides, because manual labor was disastrous for Europeans in the tropics, and because the "indolent African" resisted agricultural work, colonization was impossible, and the sooner the Epidemiological Society made itself heard on the subject, the better. In Africa, tropical disease, agricultural adversity, and native temperament all conspired against the European imperialist.

By the end of the nineteenth century, skepticism about the possibility of white acclimatization to the tropics could draw on the supporting testimony of a substantial body of medical opinion. And imperial geographers were certainly aware of this corpus of commentary. Thus, Ernst Georg Ravenstein (1834–1913), an English-German cartographer and geographer who served in the Topographical Department of the British War Office from 1855 to 1875, just took the judgments of the antiacclimatization partisans as conventional wisdom when he came to draw his map of those lands of the globe that were "still available for European settlement."[39] Demographic decline of the British in India, the Dutch in Java, and the Portuguese in Brazil clinched the argument for him. Moreover, when the Africanist and British administrator Arthur Silva White (1859–1932) produced in the same year, 1891, a comparable map of the "Comparative value of African lands," he just built a climatic component into his assessment, for, he declared, it was only in the subtropical or temperate zones of Africa that successful European colonization was possible. Discriminating between what he called "Areas of highest resistance against the European Domination" and "Areas of highest relative value to the European Powers," White's map was a work of imperial rhetoric unveiling the value of colonial possessions. Those of highest value lay close to the coastline, while those of lowest worth were regions where climatic or political conditions militated against economic development. In constructing his map, White presented himself as a European diplomat, looking into Africa's imperial future and all the while nonchalantly declaring that "all humanitarian motives may be set aside as not being pertinent to the present inquiry."[40]

Moral Medicine

Over the *longue durée,* sermonic pronouncements by medical practitioners casting the tropical world as *morally* perilous as well as medically hazardous were frequently interwoven with climatic diagnoses.[41] So, when James Johnson appended a section on "Tropical Hygiene" to the 1821 edition of *The Influence of Tropical Climates on European Constitutions,* he concluded with a suite of observations on what he called the conduct of the passions:

> The monotony of life, and the apathy of mind, so conspicuous among Europeans in hot climates, together with the obstacles to matrimony, too often lead to vicious and immoral connexions with Native females, which speedily sap the foundation of principles imbibed in early youth, and involve a train of consequences, not seldom embarrassing, if not embittering every subsequent period of life! It is here that a taste for some of the more refined and elegant species of literature, will prove an invaluable acquisition for dispelling ennui, the moth of mind and body.[42]

A quarter of a century later, when Martin, another member of what has been called the therapeutic clergy, embarked on recasting and rewriting the entire work, these self-same sentiments remained. Now Martin added the comment, citing the French physiologist Georges Cabanis, that he hoped the potent benefits of education when conjoined with the example of their betters—the European governors—would enable "the natives of India" to "conquer the influences of climate, and of the depraving religious and political habits of ages." But in prefacing the extract above from Johnson's original text, he added as a cautionary note that there could be no doubt that the habits and character of the host population among whom colonists dwelt "must have a powerful influence" on the morals of European settlers. This put the European colonist in extreme danger. For the removal of religious restraint and the ready temptations to vice were real causes of trouble. In "respect to the effects of licentious indulgences between the tropics," he warned, "the reader may be assured that he will find, perhaps when too late, how much more dangerous and destructive they are than in Europe."[43]

In the meantime, Martin had brought out his own *Notes on the Medical Topography of Calcutta* in 1837 and an enlarged edition under the title *Official Report on the Medical Topography and Climate of Calcutta* two years later. Alongside reviewing the history and climate of Calcutta and assessing its medical geography district by district, Martin explained why charting the "morals of the natives of a country" was integral to the work of "a medical topographer." On this obvious principle, he explained, "is founded the axiom of medical topography 'that a slothful squalid-looking population invariably characterizes an unhealthy country.'"[44] For Martin, external appearance and internal state mirrored each other. Land, body, and spirit were all of a piece. So, in compiling his own moral register, Martin could judge that the Bengali were "utterly devoid of pride, national or individual." Whatever the benefits of European education and moral instruction, a penchant for falsehood and perjury could readily be detected among the Bengali. So too could their indifference to the sentiments of others, their "perfection in timidity," their disposition toward cruelty and ferocity, their litigious disposition, their "physical uncleanliness and obscene worship." All these were central to the moral verdict Martin meted out. "If the Bengallee be in easy circumstances," he mused, "the whole day is passed in eating, smoking, and chewing the pawn [paan], reclining and sleeping." It was a sorry tale. For if Bengali peoples were to be classed among the Caucasians, "the standard of the human race," Martin conceded that "the effects of climate and locality must indeed be great and remarkable." Atmospheric conditions plainly trumped racial constitution. These findings, moreover, could easily be generalized. "When we reflect on the habits and customs of the natives, their long misgovernment, their religion and morals, their diet, clothing, &c, and above all, their climate," Martin announced, "we can be at no loss to perceive why they should be what they are."[45] In these circumstances, European troops in India needed to take special care. Constant vigilance was vital, temperance was compulsory, and stern discipline was necessary not simply to punish wrongdoing but to foster good conduct.

The urgency for engaging in personal moral management was stimulated by the long-standing conviction that tropical climates were reflected in the moral character of their human occupants. Thus, Robert Felkin (1853–1926), a medical missionary and explorer with a theosophical-style interest in occultism, typified the people living in "flat, hot Bengal" as "timid, servile, and superstitious" while those occupying the high tableland of Mysore he portrayed as "brave, courteous but passionate."[46] Representations of this stripe had long been commonplace. Thirty years earlier, for example, the Scottish physician and diplomat John Crawfurd (1783–1868) told his hearers at the Ethnological Society of London that Australia's inauspicious climates had produced only "the feeblest . . . hordes of black, ill-formed, unseemly, naked savages," while in Africa, "the races of man . . . correspond with the disadvantages of its physical geography."[47] A few months later, James Hunt (1833–1869), a speech therapist who founded the Anthropological Society of London, could tell the same audience that in the tropics, "there is a low state of morality, and . . . the inhabitants of these regions are essentially sensual"; by contrast, the "temperate regions," he insisted, were characterized by "increased activity of the brain."[48] Race and region, ethology and ethnology, the moral and the material, were thus tightly, very tightly, tied together. Again, Edinburgh naturalist and explorer Joseph Thomson (1858–1895), recounting his African travels, reported in 1886 that "as the traveller passes up the river [Niger] and finds a continually improving climate . . . he coincidentally observes a higher type of humanity—better-ordered communities, more comfort, with more industry. That these pleasanter conditions are due to the improved environment cannot be doubted. To the student with Darwinian instincts most instructive lessons might be derived from a study of the relations between man and nature in these regions."[49] Viewing tropical climate through the lens of moral judgment was a common occupation.

Even those more hopeful for successful acclimatization paused from time to time to issue moral prescriptions and even to incorporate what they deemed to be appropriate ethical standards into the very definition of effective climatic adaptation. Thomas J. Hutchinson (1820–1885), an Anglo-Irish explorer who had served as senior surgeon on the Niger expedition prior to taking up the post of consul in several imperial sites, including the Bight of Biafra, is a case in point. While supporting Johnson's ideas about climatic influence and "sympathy" between skin and liver, and attributing the unhealthiness of certain localities as much to the vicissitudes of temperature as to the presence of malaria, Hutchinson was much more benign in his judgments of native peoples, castigating those who magnified for "our Christian Public" the "animality of the negro race a little too much." Still, he was sure that cultivating appropriate moral practices was essential for maintaining the health and welfare of sailors in the tropics. Permitting them to participate in public acts of worship and encouraging them to take every opportunity to walk along shorelines, where the topography permitted, were conducive to producing "the most salubrious effects, both mentally and physically, in exploring voyages as well as on board stationary trading vessels." Similar sentiments could be stimulated by fostering, "at sight of lofty ground," recollections of mountainous terrain associated with childhood. Religious observance, coastal walking, and landscape memory were, to Hutchinson, powerful channels of well-being in colonial settings. At the same time, he turned his guns on the demon drink and its malign influence in hot climates. There were few more powerful or predisposing causes of African fever, he insisted, than the inebriety of Europeans. Indulging in alcoholic consumption "beneath an intertropical sun" was, for Europeans, "like trying to put out a fire by heaping coals upon it." Indeed, for Hutchinson, abstemiousness was a precondition for successful adjustment to the rigors of a tropical sun. "A good constitution may become acclimatized," he wrote, "by abstinence from ardent spirits" as well as by thoughtful attention to diet, clothing, and skin care. Healthy exercise and a cheerful confidence in the performance of these prophylactic practices to obviate the "enervating influence of climate" were crucial to tropical acclimatization. Indeed, "fear and anxiety," "intemperance," and "want of attention to cleanliness of person" were among the conditions that indicated a proclivity toward contracting African fever.[50]

By the 1880s, with matters of imperial governance increasingly foregrounded, Cullimore took the opportunity to supplement his own backing of native labor as a colonial strategy with direct moral prescription. Whatever efforts were taken to escape the rigors of a tropical sun, he was certain these would be of no avail without practicing strict sobriety and following a temperate way of life. Neither the "habitual tippler, nor heavy opium or tobacco smokers, with weakened internal viscera, nor those slaves to other vices, can hope to live long in tropical climates," he announced. But even with the most sedulous precautions, Cullimore was sure the countries of equatorial and tropical Africa remained "unsuited to the first degree of colonization." Any attempted settlement by laboring agricultural emigrants from northern Europe was bound to end in failure unless the colonists were prepared to intermarry with the native races. The overwhelming fatigue induced by hard work in the punishing heat, especially in the absence of a cold season, would prove "fatal alike to their comfort, their health, and fertility." Second-degree colonization was possible elsewhere, he reckoned, notably in Central Africa, where conditions were not too dissimilar to places like Mauritius, North Queensland, or Louisiana. Here, of course, "the dark races" worked the fields, and for that reason, the planters, merchants, and overseers could enjoy good health and bring up their children much as they did at home.[51]

The medical and moral pathologization of tropical climates was no short-lived fixation.[52] Even with the triumph of the new tropical medicine dating, in large measure, from the publication of Patrick Manson's *Tropical Diseases* in 1898, it did not evaporate.[53] Indeed, the first edition of Manson's text was adorned with a miniature icon of "La Belle Sauvage" on the frontispiece.[54] The germ theory of disease, which displaced the older miasmic ways of thinking, plainly did not banish the tropes of tropicality from medicine's domain. This is conspicuous in the persistence of medico-moral meteorology in the writings of the Italian English physician Luigi Westenra Sambon (1865–1931), who joined Manson at the London School of Tropical Medicine soon after its establishment in 1899 and emerged as a dedicated advocate of the germ theory.[55] It was the evidence that he accumulated, along with George Carmichael Low, in a highly malarious region of Italy—the Roman campagna—that provided crucial evidence in support of Manson's mosquito transmission theory of malaria. The matter of human acclimatization to the tropical world was a major concern to Sambon, and he addressed several audiences on the topic, publishing his own ideas in several different outlets.[56]

Throughout, Sambon vigorously supported the view that Europeans could rapidly and fully adjust in a short space of time to the tropical world. He insisted that tropical climate had been the victim of prejudice and unwarranted medical stereotyping and that the real difficulties standing in the way of colonization were not to do with climate but rather with parasites. Acclimatization was, in the main, "a mere question of hygiene." Nevertheless, it was clear to Sambon that medical hygiene encompassed moral hygiene as well. "Personal habits are of the utmost importance," he declared; "temperance and morality are powerful weapons in the struggle for life." The reason was simple: "Sexual immorality under the influence of a tropical climate, and in the presence of a native servile and morally undeveloped population, rises to a climax unknown amid the restraints of home life, and becomes one of the most potent causes of physical prostration."[57] For Sambon, the Darwinian struggle against tropical parasites called for the self-same moral circumspection that earlier advocates of environmental medicine had demanded.[58] In this way, the long-standing penchant for moralizing tropical regimes persisted even among those of a colonial disposition who were critical of climate determinism and staunch advocates for acclimatization. Besides, however vigorously he opposed those who attributed illness to the perils of a tropical sun and however strenuously he supported the new germ theory, he nevertheless invented several items of tropical clothing to protect wearers from actinic rays.[59] Here practice outran theory; convention triumphed over investigation.

These interventions reveal just how prevalent moral evaluations of the tropical world were in discussions about acclimatization in the period. Indeed it became commonplace to actually build a moralistic component into the very definition of acclimatization; that is, to insist that successful acclimatization had to incorporate the maintenance of assumed white moral excellence.

Tropicality's Tactics

Taken in the round this constellation of themes connecting climate, medicine, and morality in the tropical world constitutes what David Arnold christened "tropicality." This invented category, he argues, requires us to think of the tropics "as a conceptual and not just physical, space" because "calling a part of the globe 'the tropics' (or by some equivalent term, such as the 'torrid zone') was a Western way of defining something culturally and politically alien, as well as environmentally distinctive, from Europe and other parts of the temperate zone." In large measure, Arnold's depictions stress the "otherness" of the tropics as pertaining to health and disease, and not least to the ways in which medical practitioners supplemented the reports of scientific experts by investing their descriptions of the tropical world of nature with "a pathological potency that marked them out from milder, more temperate lands."[60] Seen in this light, tropicality was a medical condition and one that, as Arnold puts it elsewhere, "was an especially potent and prevalent form of othering."[61] But it was also a moral space inasmuch as inhabitants and visitors alike were subject to the rule of a tropical regimen that stamped its dubious morality—and indelible inferiority—on them.

The ripples emanating from tropicality spread far and wide and were channeled in different settings for different purposes. As we have already seen, versions of medical tropicality could underwrite the construction of hill stations in various locations in Africa, India, and the Caribbean.[62] This impulse, frequently wedded to architectural nostalgia for distant homelands, gave spatial expression to the sense that European colonial administrators, military personnel, and the like needed to retreat to healthful citadels away from the infested lowlands in a zone insulated from diseased places and decadent people. The perils of lowland debility, and what was often called tropical inertia, provided nebulous justification for British officialdom to construct such hill residences. But as Dane Kennedy has shown, escaping to the "magic mountains" was more often than not a cultural and political judgment about native life and the desire for Britons to experience the reinvigorating presence of their fellow countrymen in a setting that mimicked life back home.[63] Hill stations were located on a landscape sculpted every bit as much by moral precept as by physical force.[64]

But hill stations were not the only way in which what might be called architectural tropicality manifested itself. Successive housing designs that European colonists adopted in India, for instance, mirrored changing ideas about the influence of climate on disease and the racial economy of imperial life. The early appropriation of the indigenous bungalow and the adoption of local mechanisms for controlling interior heat such as verandas and punkhas (manually operated ceiling fans) gave way, in the early decades of the nineteenth century, to architectural forms designed to protect Europeans from exposure to miasmatic exhalations. Thus, in the early days, the "tropicalization" of architecture was a consequence of reciprocal exchanges between indigenous practices and colonial culture as imperial buildings acquired local features. Later, with health concerns biting ever more deeply, prefabricated housing, designed to insulate Europeans from what were seen as noxious vapors, was shipped directly from Britain to the colonies. This was especially the case in the design of tropical barracks for military personnel. But the prototype soon spread to local hospitals under the influence of the Barrack and Hospital Improvement Commission appointed in 1861. These constructions were pregnant with symbolic associations that served to cast the tropics as a pathological site and to inscribe in space a racialized distancing of local inhabitants from colonial settlers. Tropical architecture, as Chang and King put it, was the product of "a power-knowledge configuration inextricably linked to asymmetrical colonial power relations."[65]

The fusion of natural necessity and landscape iconography conspicuously manifested itself in the thinking of Thomas Roger Smith (1830–1903), an English architect who went on to hold the professorship of architecture at University College London. He had gone out to Bombay in 1864 to oversee the building of the European Hospital from designs he had submitted to the government of Bombay. While there, he drew up plans for exhibition buildings (which were contracted but had to be abandoned due to a downturn in the production of cotton) and for the General Post Office and the Residency at Gunersh Kind, which were both completed.[66] Smith championed the idea that European building styles should be spread across the Empire—India in particular—and he engaged in a lengthy public debate on the subject, rejecting the views of those advocating the adoption of "Indic" styles.[67] In April 1868, Smith addressed the Royal Institute of British Architects on the construction of domestic buildings for European settlers in the tropics. Here he opened his remarks in naturalistic vein. "There exists nothing of a physical nature which causes such an entire revolution in our feelings and habits, in ourselves and in our surroundings," he proclaimed, "as the addition or withdrawal of a few degrees of heat." Appropriate buildings were essential in a climate where the blistering ferocity of the sun was so powerful that "nothing but English pluck prevents the attempt to work being altogether given up" and where it was virtually impossible for Europeans to expose themselves safely to its rays.[68]

Yet for all that, the iconic force of architecture was never far from his mind. Just as European administrations exhibited proper standards of "justice, order, law, energy, and honour—and that in no hesitating or feeble way," he mused, so the buildings constructed by colonists "ought to hold up a high standard of European art." In Smith's opinion, imperial architecture should be "a rallying point" for colonial identity as well as a distinctive symbolic mark of expatriate presence, "always to be beheld with respect and even with admiration by the natives of the country." Indeed, in this regard, Europeans were falling far behind for, as he put it, compared with "the best of Mahommedan buildings, which mark, as I should like ours to do, the residence of a conquering race," European architecture lacked the artistry of Islam's "nobler monuments."[69] That was something seriously in need of correction. But architectural symbolism, of course, came at a cost. For Smith was well aware that his bungalow design with its compound—"an ample walled enclosure"—was only conceivable on the assumption that low-rate native labor was readily available. As he realized, the number of attendants required was indeed very great. Mercifully, in all tropical countries, native labor, as a general rule, was both cheap and plentiful. Given these conditions, Smith advised that it was worth going to a good deal of trouble to ensure that both the stables and the servants' dwellings were carefully positioned so "that they shall not come to the windward of the building they belong to—that is, if the prevalent direction of the wind is known." The reason? "Various bad odours are likely to arise there."[70]

Tropical architecture, whether manifest in hill stations, in public buildings, or in European housing, was a compound product of climatic forcing, imperial symbolics, and native toil. In combination, they contributed mightily to the production and reproduction of Europe's tropical imagination. And health-conscious mindscapes shaped by its power found other material expressions too. Before embarking on a journey to the tropics, Victorian travelers spent a good deal of time and money on equipping themselves for the medical jeopardy into which they believed they would soon be plunged. As Ryan Johnson puts it, "ordinary items such as soap, clothing, foodstuffs and bedding became transformed into potentially life-saving items that required the fastidious attention of any would-be traveller." Care was needed to preserve European health in such unruly spaces, and merchandise such as desiccated vegetables, medical chests with compressed tablets, cholera belts or flannel binders, and pith helmets were widely advertised. How providers marketed their goods reveals a great deal about how they thought about tropical health, tropical hygiene, and tropical humanity. The manufacturer of one medicine chest by Burroughs Welcome & Co, for instance, portrayed the tropics as diseased and deadly: "A danger far worse than that of broken limbs, of cuts and gun-shots, hangs over the traveller in remote places, particularly in the Tropics," their advertising brochure explained. The worst menace visitors had to face was disease—"desolating ailments" that were reportedly "particularly fatal to the so-called white man who originates in temperate climates."[71]

As for tropical apparel such as flannel belts, undergarments, and spinal pads, advocates of different fabrics—linen, wool, cotton, silk—engaged in heated debates over which material best provided protection in the "torrid zone," not least for army personnel. The production of tropical clothing was thus encased within a complex web of colonial networks through which tropical anxieties coursed. Some powerful textile manufacturers fastened on one particular cloth as the best way forward; others resisted that inclination, urging that different tropical regimes required different materials. Money, meteorology, and medicine were thus as tightly interwoven in discussions over appropriate attire under a tropical sun as were questions of moral probity and racial identity. For anxieties over the right clothing to don in the tropics were the material manifestation of a deep-seated fear that Europeans could not survive and reproduce successfully in a world seemingly suited to darker-skinned peoples.[72]

The moral economy of tropical dress also manifested itself in concerns about head covering, particularly for military personnel. Pith helmets, or sola topi, were sun hats made from the pith of the stems of sola plants and were particularly favored by Europeans in India (see figure 3.1). Not unlike practitioners of tropical architecture, the Royal Commission on the Sanitary State of the Army in India was likewise concerned with heat, air, and light but focused attention on "the intimate scale of the soldier's body, particularly the head." The reason was that "excessive heat on the brain" was believed to produce "moral depression" and so the development of a double-layered helmet with internal ventilated spaces was welcomed for moral as well as biological reasons.[73] Given what was at stake, it is no surprise that in his 1858 examination

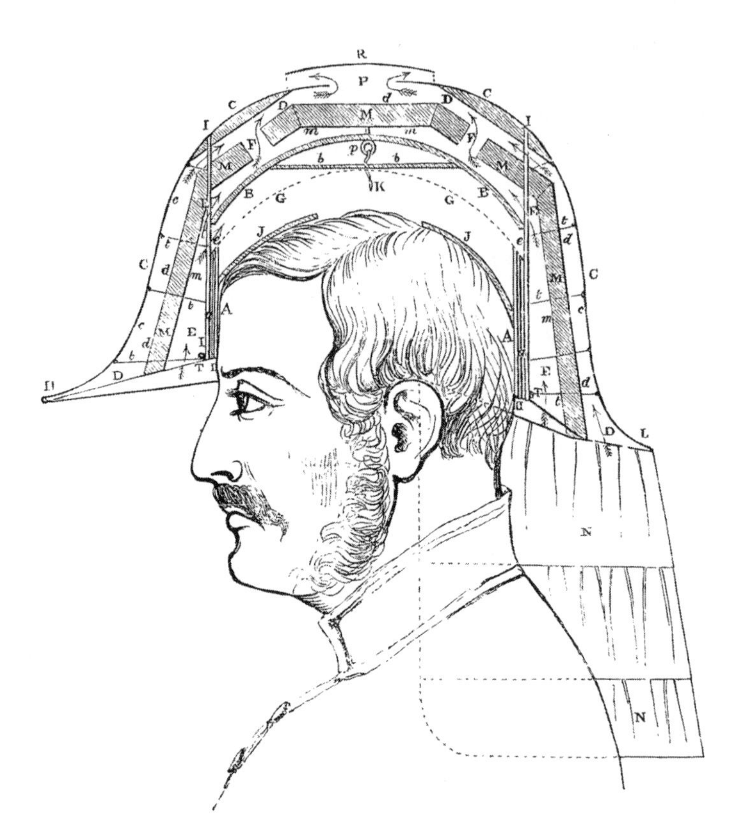

FIGURE 3.1. Proposed Helmet for Mounted Troops in India
Source: Julius Jeffreys, *The British Army in India: Its Preservation* (London: Longman, Brown, Green, Longmans, & Roberts, 1858)

of *The British Army in India: Its Preservation*, the surgeon Julius Jeffreys FRS (1800–1877), a pioneer of early air conditioning systems and inventor of the respirator, should devote more than thirty pages to explicating the relevance of the sciences of reflection, conduction, radiation, ventilation, exhalation, and evaporation in the construction of appropriate army headgear.[74] Indeed, Jeffreys' proposals for ways of maintaining military health were rooted in his concern to address the "grievous moral condition of the soldiery in India." The appalling drunkenness that prevailed among British regiments in India, not to mention the serious crimes and sexual indiscretions in which many were implicated, were reflective of the poor conditions under which they had to carry out their duties. Better housing, for example, would induce such improvements in the "moral habits" and health of the average British soldier that the existing blemish on Britain's national reputation could be diminished if not erased.[75]

Sartorial practices, plainly, had symbolic significance. To one degree or another, European dress codes in tropical colonies marked and maintained differences between colonized and colonizer, reinforcing the imperialists' sense of identity, authority, and dominion.[76] Yet at the same time, all these disclosed the perceived precariousness of the European body by constructing it as highly sensitive to climates beyond the temperate zones. In his introductory remarks to *The British Army in India*, Jeffreys commented on the "sickness of Europeans" and "its effect on the native mind." "The natives of India look upon us as white bears from the cold unhealthy North," he continued, "ferociously brave, but of sickly constitutions, disabling us from occupying their country without their aid."[77] So while clothing served to reinforce in European colonists a sense of superiority, at the same time, it advertised anxieties about their precariousness.[78]

European vulnerability in tropical climes soon became apparent in other venues too. Take, for example, the insurance industry. British insurance companies were, by the mid-nineteenth century, already well accustomed to incorporating the risk of hailstorms into agricultural insurance policies and increasingly provided cover for travelers to, and foreign residents in, the tropical world. Operating in an extra-scientific universe of climate statistics and actuarial data, insurance companies accumulated a good deal of expertise in handling climate risks, constructing, along the way, what has been called their own invisible meteorological data set.[79] What motivated this interest, of course, was the concern to determine climate's effects on illness and mortality and to provide suitable travel cover. One means of addressing the challenge was by latitude, namely, setting premiums for travelers according to the latitude to which they were moving. This strategy, however, was soon seen to be rather one-dimensional, and more sophisticated ways of disentangling the connections between climate and mortality were called for. In the mid-1830s, reports of fatalities among British troops in the colonies stimulated a concern with military vital statistics, and a Medical Board was appointed to investigate the health of soldiers in challenging climatic environments. These reports indicated that simple correlations between latitude, weather, and health were troublesome, and more pricing schedules were developed to meet concerns about the fate of policyholders traveling to the tropical world. Thus, Colonial Standard, a new company established in 1846, divided the world into a discrete suite of global regions and assigned each to one of four insurance classes. Europe, North America north of 38° but not west of the Mississippi, Australia, and New Zealand south of 30° found themselves allocated to class A. Class B was inhabited by America north of the 35° line of latitude (and north of 30° between November and June) but not west of the Mississippi, South Africa south of 20°, and Bermuda. Class C comprised India, Ceylon, Mauritius, and Chinese treaty ports, with the West Indies the sole occupant of class D. Other companies, in general terms, soon followed suit. In his volume on *Medical Examinations for Life Insurance*, for instance, Rush Medical College's Jonathan Adams Allen (1825–1890), former professor of physiology at the University of Michigan, began his analysis of "Residence in a Foreign Climate" with the observation, "Without exact reference to isothermal lines, natives of the zone extending from the thirtieth to the fiftieth parallels of latitude, may be considered as the best risks."[80]

The chief matter of climatic concern for the health insurance industry, not surprisingly, was movement between the temperate and tropical worlds. William Brinton (1823–1867), physician to the Royal Free Hospital, directly addressed this issue in his 1856 work *On the Medical Selection of Lives for Assurance.* When a native of a different climate came to reside in Britain, insurers considering taking on the risk were asked "whether his life is as good as that of one of our own countrymen." It was the same when an "Englishman is about to go abroad, and you have to determine whether his risks remain no greater than before." Brinton's answer was crystal clear: "The native of a warmer climate who seeks this country, runs a risk of pulmonic—and especially of tubercular—disease, which may be roughly regarded as hardly inferior to that of an hereditary tendency of ordinary intensity. This risk we may probably consider as less in the case of an immigrant from Southern Europe, than in one from a much hotter climate, such as Central Asia or Africa."[81] The isolines of insurance risk in the travel industry followed degrees of temperature and lines of latitude alike.

More than half a century later, the same outlook prevailed. When Thomas D. Lister (1869–1924), president of the Insurance Medical Society, directed the attention of readers of his *Medical Examination for Life Insurance* to the issue of "Foreign Lives and Residence," he recalled that insurance agents sometimes had occasion "to see a negro or an Indian life resident in England." From a risk perspective, such lives were judged not to be as good as the average English life. This was for the simple reason that these individuals were no longer living in the temperature zone "adapted for their race" and therefore that "there is what some actuaries consider to be an 'attenuated vitality' resulting from the previous generations of their ancestors having lived in a hot climate." Whatever the particular reason, Lister was sure that such cases were "not as good as those of the British born when they are living in the British Isles." In turn, when insurance examiners had to assess British residents in the tropics, they insisted that before any travel risk should be taken on, all "the usual precautions must be adopted, the special risks of the climate must be recognized, and particular attention must be bestowed on the age of the proposer, the time he has been in the country, the illnesses from which he has suffered, and the probability of their recurrence."[82]

Drawing on the collective wisdom of medical topographies enshrined in handbooks for life insurance examiners, such typologies of global regions succeeded not only in perpetrating prevalent ideas about tropicality but also in reducing medical-meteorological complexity to a simple metric: a cover price. In a literal sense, insurers put a premium on climate, and putting a price on it brought it within the sphere of the manageable. In this way, the complexity of climate could be tamed by indexing it to health risk, monetizing the result, and mapping a boundary between safe and risky latitudes.

The Rhetoric of Tropical Vulnerability

While the frailty of white bodies in the low latitudes found expression in the actuarial tabulations of health insurance risk, a sense of European infirmity in the tropical world could also provide certain writers with rich rhetorical resources for quite different political ends. Consider in this regard the writings of the Sierra Leone army officer and Edinburgh-trained physician of Ibo parentage, James Africanus Beale Horton (1835–1883), an African nationalist and fellow of the Royal Geographical Society. Jessica Howell's reading of Horton, one of a number of black doctors trained by the British War Office to replace white physicians in West Africa, is illuminating.[83] To her, Horton's record of his own vulnerabilities—he had been dispatched to a series of unhealthy posts in West Africa—was a critical means by which he sought to secure his own authority as a reliable witness and at the same time to subvert the idea that Africans were immune to tropical diseases. The climate of West Africa could be fatal for blacks, Horton insisted, as it certainly was for whites. And yet, he observed, Africans had been able to adapt and survive there, unlike European races who produced offspring suffering debilitating diseases caused by miasmas and climate and soon died out. In marked contrast to the likes of Richard Burton, Horton's diagnosis of the climatic perils of the imperial project was intended as a warning, not an encouragement, to would-be colonizers and thus as an expression of an emerging African nationalism.

Published in 1874, Horton's *Diseases of Tropical Climate and Their Treatment, with Hints for the Preservation of Health in the Tropics* bore the stamp of James Ranald Martin's influence. Not only was the volume dedicated to Martin, but his medical and moral advice was enshrined in Horton's text. Its appendix followed Martin's principles by elaborating rules for dress, diet, exercise, and the like in the tropics, particularly on alcohol consumption, bathing, and sexual morality. But what he most relied on to provide warrant for his observations was his own direct and all-too-embodied experience as an army doctor. Having spent a full fifteen years serving with troops "in some of the most deadly intertropical climates," he had found himself "continually in the vortex of Tropical Diseases." By contrast, he recollected how "Europeans fresh from Europe are very generally found to praise the climate in the tropics as the best in world." Time would tell a different tale. A combination of injurious climatic influences and their all-too-common temptation to "pooh-pooh any advice" offered to them by people whose long-standing occupancy and understanding of such environments should carry great weight inevitably meant they would soon "pay dearly for their folly."[84]

Horton had already made public a similar diagnosis in his 1867 *Physical and Medical Climate and Meteorology of the West Coast of Africa.* Here, in his prefatory statement to "The President and Members of the Executive and Legislative Council of the Central Government of Western Africa," Horton impressed on his readers that an insalubrious climate was the greatest danger would-be European colonists faced in pursuing their tropical dreams. The maladies endemic to tropical climates had the "mischievous effect" of curbing the progress of civilization in much of Africa. It was surely well known that on departing their home shores, the European was blighted with "a melancholy foreboding of a speedy termination of his existence." Indeed, even "his friends and relatives also reckon him, from the day of his embarkation, as amongst the dead." As if in confirmation of Galileo's declaration that the Book of Nature is written in the language of mathematics, Horton grounded these sentiments in the large amounts of meteorological data he had assembled for various West African locations. These in turn paved the way for depicting the effects of tropical heat on the human constitution and highlighting the toll exacted by a tropical climate on the mental powers of natives and European visitors alike. The consequences were devastating. No one, he reminisced, could ever forget the ravages that a pestilential and widespread outbreak of yellow fever brought to "the few European inhabitants of Bathurst in 1866." For reasons such as this, Horton sought to float the suggestion that the project of colonization should take racial constitution much more seriously. In support of this proposal, he called his readers' attention to the susceptibility of Europeans to disease in intertropical Africa and to the shocking waste of life due to the high rate of migrant mortality. In these circumstances, Horton urged that it would be highly advantageous to imperial governments if the colonial authorities were to "ascertain what races of men are best fitted to inhabit and develop the resources of the different colonies, and the limits within which they could properly thrive and increase."[85] While cast in the language of profit to the occupying powers, all of this was basically another way of saying that the African climate was the most powerful guardian of the continent's indigenous peoples.

Just what animated this analysis revealed itself with particular clarity in a volume Horton brought out the following year, 1868. It was a survey of West African countries and peoples, but, as much as anything else, was intended as "a Vindication of the African Race." Dedicated to Henry Venn of the Church Missionary Society in appreciation of his untiring zeal for the "moral, social, and Christian advancement of the African race," the work's preface set out Horton's motivations for producing an apologia for the African people. His aim was simply "to disprove those mischievous doctrines which have been promulgated to the detriment of the growing races of Western Africa." What he had in mind were the activities of the Anthropological Society of London, a scientific fellowship that was forever raking up "old malice" and encouraging its "agents abroad to search out the worst possible characteristics of the African, so to furnish material for venting their animus against them." The ambition of "these negrophobists," as he described them, was to prove that the African was "fitted only to remain a hewer of wood and drawer of water for the members of that select society." In what followed, Horton reprised the arguments of figures such as James Prichard, Sir William Temple, and Sir George Cornewall Lewis, who strenuously promoted the unity of the human race, against the likes of James Hunt, Karl Vogt, Charles Buxton, Lord Alfred Churchill, and that "most determined African hater" and vice president of the Anthropological Society, Captain Richard Burton. In contrast to the polygenists at the Anthropological Society, Horton claimed "the existence of the attribute of a common humanity in the African or negro race."[86] To him there were no radical discontinuities between black Africans and their "more civilised confrere." Whether in terms of moral integrity or intellectual endowment, nature had not discriminated between Africans and Europeans in dispensing these gifts. Where differences did arise, they were to be explained as a consequence of "the influence of external circumstances" rather than inherent constitution. And chief among those external circumstances were the actions of those who had sought to reduce African peoples "to the condition of wild beasts." Quoting "an old but true saying of a philosopher," Horton announced, "Treat men like beasts and you will make them such."[87] Society all too easily bestialized what nature had dignified.

Given these sentiments, the rhetorical possibilities afforded by the blending of climate, health, and tropicality for the future of African nationalism were of major proportions. European races could certainly survive for a brief period of time in any part of Africa, Horton believed. But in tropical Africa, he was just as certain that within a short span of time, they would inevitably die out. Of course, new immigrants would soon arrive to take their place, but he was no less certain that they too would very shortly "share the same fate." Aboriginal inhabitants, meanwhile, continued to increase and multiply, but the European race would ultimately be "annihilated." The path to that dismal destiny was strewn with disaster. Offspring "suffer seriously from birth to manhood from internal diseases, the result of miasmatic and climatic influences, and they must either be amalgamated with the white or black race, or they die out in about the second generation." The health effects of a tropical climate were not the chief subject Horton had in his sights. Rather, the implications of the fierce African sun for colonialism were on his radar. "In the tropical countries of Western Africa the idea of a permanent occupation by European settlers, if ever entertained, is impossible of realization," he proclaimed; "it is a mistake and a delusion." More in aspiration than achievement, Horton concluded, "The English Government is conscious of this; and the House of Commons Committee has now set on foot by resolution (and we hope it will soon be by actual practice) that great principle of establishing independent African nationalities as independent as the present Liberian Government."[88]

The idea that a perilous climate could serve as an ally against colonial powers was certainly commented on from time to time. Charles Darwin, for example, mused that when "civilised nations come into contact with barbarians the struggle is short, except where a deadly climate gives its aid to the native race."[89] And Ralph Abercromby (1842–1897), meteorologist and military man, quipped in his weather-hunting travelogue *Seas and Skies in Many Lands*, "The best protector of the African savage from European aggressors is the deadly climate of that dark continent."[90] In Horton's hands, however, European fragility in tropical regimes was harnessed much more directly in the cause of African nationalism. Climate served as a site for subverting the colonial ambitions of European powers altogether. His aim was to vindicate the African against the malign slurs of armchair anthropologists and greedy imperialists alike and to promote the virtues of African self-government.

Something similar is detectable in the autobiographical memoirs of Mary Seacole (1805–1881), a Jamaican nurse and hotelier who, as she put it herself, had "good Scotch blood" coursing through her veins.[91] From her mother, she had gleaned knowledge of Creole medicine and nursing practices and used these to good effect in combating cholera and yellow fever both in Jamaica and at Las Cruces in Panama. Later, during the mid-1850s, she traveled at her own expense to the Crimea, where she opened the British Hotel in Kadikoi on the outskirts of Balaklava and treated the sick and dying there.[92] While she was routinely compared with Florence Nightingale at the time, she soon faded from the public eye though she was later appropriated as an icon for various causes—civil rights, black nursing, and women's liberation.[93] Accordingly, in recent years, her popular autobiographical travel chronicle, which first appeared in 1857, has been the subject of considerable scholarly attention, with commentators pausing to underscore the ways in which she engaged in a range of self-authorizing practices to validate her medical expertise and to valorize the comparative resilience of the hybridized racial body in hostile climatic regimes.[94]

On the one hand, she extracted numerous excerpts from various sources commending her medical acumen and did not hesitate to advertise the superiority of her own methods over conventional European and American treatments. At Escribanos in New Granada, for instance, where "strangers to the climate suffered severely," she recalled that a surgeon sent out by the West Granada Gold-Mining Company was only too "glad to throw his physic to the dogs, and be cured in my way by mine." As for her service in the Crimea, she begged to "be excused for transcribing . . . from the columns of the Times" the approbation of the paper's war correspondent who bore witness to having "seen her go down, under fire, with her little store of creature comforts for our wounded men; and a more tender or skilful hand about a wound or broken limb could not be found among our best surgeons."[95] Indeed, she later received from the British government the Crimea Medal for her work among the sick and injured, an award that was supplemented by French, Russian, and Turkish decorations.[96]

At the same time, by referring to herself as "the yellow woman from Jamaica with the cholera medicine," Seacole distanced herself from white- and black-skinned people alike. She thus presented herself as the ideal "mixed-race subject" in possession of a remarkably robust constitution that permitted her to freely inhabit different disease zones and to attend to the health needs of colonizers by dispensing her own cures. By contrast, she casually lamented the early death of her husband Edwin Horatio Hamilton Seacole, a white British merchant, with the comment, "Poor man! he was very delicate; and before I undertook the charge of him, several doctors had expressed most unfavourable opinions of his health." Plainly, Mary Seacole found it possible to exploit the idea that her mixed racial heritage had gifted her with a more resilient constitution than purebred whites. Compared with the Cruces people in Panama, whom she typecast as "constitutionally cowardly," she happily quoted the testimonial of a medical officer from West Granada lauding her "peculiar fitness, in a constitutional point of view, for the duties of a medical attendant."[97] To this, her own valor in braving the climatically hostile lands through which she passed bore plain witness. No doubt it was that sense of constitutional self-confidence that encouraged her to pass off, as of no consequence, the only illness she had experienced—an attack in the Crimea that afflicted her just before Christmas 1855.

In describing the sickly white bodies under her care, Seacole resorted to the language of climate and constitution to confirm the unsuitability of the tropics for British colonizers. As Howell observes, Seacole's strategy was to demonstrate that Britons could only "survive temporarily in inimical climates" and that they were incapable of adjusting permanently to climatically challenging environments.[98] In this way, the fragility of the pureblood, when exposed to the cruelty of a tropical sun, stood in stark contrast to the hardiness of the hybrid. When Jamaica was struck by yellow fever in 1853, an outbreak that had "never made a more determined effort to exterminate the English in Jamaica than it did in that dreadful year," she underscored the inability of colonists to resist the terrifying power of the climate. It was dreadful to see the younger generation so mercilessly cut down in the bloom of life, she remarked, "not in battle with an enemy that threatened their country but in vain contest with a climate that refused to adopt them." Here, as indeed elsewhere, Seacole pointed out that the mother country paid dearly for "the possession of her colonies." Ironically, among the few gifts Nature had granted to imperial strangers among the Creoles were the virtues of care and compassion that the native peoples displayed in abundance: "all who are familiar with the West Indies will acknowledge that Nature has been favourable to strangers in a few respects, and that one of these had been in instilling into the hearts of the Creoles an affection for English people and an anxiety for their welfare, which shows itself warmest when they are sick and suffering."[99] What is clear is that for all the challenges the empire of climate posed to Europeans in the tropics, the indigenous peoples were well adjusted to the atmospheric conditions of the intertropical world. That was a blessing since both colonists and adventurers alike were massively reliant on the goodwill of the people among whom they traveled or settled.

Besides chronicling her nursing exploits on both sides of the Atlantic, Seacole kept a keen eye on the racial attitudes of the different nationalities she encountered. The Americans in particular were on the receiving end of her episodically biting commentary. Her experience of travel, she remarked, had not failed to teach her "that Americans (even from the Northern States) are always uncomfortable in the company of coloured people, and very often show this feeling in stronger ways than by sour looks and rude words." In Panama, she remarked on how the boatmen and muleteers were terribly bullied by the Americans "who would fain whop all creation abroad as they do their slaves at home." "Against the negroes, of whom there were many in the Isthmus," she went on, "the Yankees had a strong prejudice; but it was wonderful to see how freedom and equality elevate men, and the same negro who perhaps in Tennessee would have cowered like a beaten child or dog beneath an American's uplifted hand, would face him boldly here, and by equal courage and superior physical strength cow his old oppressor." More personally, when reflecting on her failure to secure backing from the managers of the Crimean Fund to underwrite her voluntary passage to the war zone, she wondered how likely it was that "American prejudices against colour had some root here? Did these ladies shrink from accepting my aid because my blood flowed beneath a somewhat duskier skin than theirs?"[100]

If the deadly challenge that a tropical climate posed to aspiring colonists could be marshaled in the cause of imperial subversion, the self-same rhetoric could be put to work for a very different geopolitical vision.[101] Take the case of Richard Burton (1821–1890), diplomat, explorer, scholar, and spy.[102] His own struggles with illness in the tropics have been scrutinized by a number of authors who have disclosed his proclivity for reading geographical locations through racial and moral lenses, as well as his strenuous efforts to reconcile the vulnerabilities of the white body with imperial ambitions.[103] Whatever the fragilities of European anatomy in the "dark continent," it never occurred to Burton that Africans were anything other than a separate, inferior species, and his portrayals of the enfeebled colonialist were casually interspersed with noxious portraits of local peoples as "hideous" and "bestial," intellectually slow and morally corrupt.[104] Perpetuating a fundamentally humoral understanding of hygiene, Burton's terrifying portrayals of West African climate were certainly *not* designed to discourage colonial investment but rather to demonstrate the strategies needed to ensure imperial success—racial segregation, careful medical cartography, residency at high altitude, and the like. Thus, while African nationalists mobilized climate in their political aspirations to retain sovereignty over their own soil, Burton faced its full terror with the confidence that, whatever the cost, Europe had the medical and strategic knowhow, as well as the intellectual and political will, to meet the challenge.

Burton's geopolitical predilections manifested themselves on the very first page—the dedication page—of his two volume *Wanderings in West Africa*, which appeared in 1863: "To The True Friends of Africa—not to the 'Philanthropist' or to Exeter Hall—These Pages are Inscribed." Within half a dozen pages, he was railing against the "perpetual imputation of 'improper indulgences' brought against Europeans by negro and negroid supporters of the ridiculous theory, 'Africa for the Africans.'" Burton hastily dismissed that allegation as the reason for the high rates of European mortality in Africa. No more persuasive, he claimed, was that "old story of deleterious climate" for, he insisted, the land was "as well fitted for northern constitutions as India is." Besides, the statistics had shown that the "sentimental squadron"—he likely had in mind the Royal Navy's West Africa Squadron tasked with suppressing the slave trade along the coastline—was "healthier than that of the Mediterranean." That apparently confirmed to Burton that the climate was not a "sufficient explanation" for colonial failures. And yet, for all the bluster, his accounts were strewn with records of his own frailty in the face of climate's imperial force and indeed the casualties of the white race more generally.[105] Mary Louise Pratt has not missed the irony here. For she reminds us that Burton's heroic narrative of his "discovery of Lake Tanganyika," emblematic as it was of what she calls "the monarch-of-all-I survey genre," was the work of an author "so ill that he had to be carried much of the way by African assistants."[106] Not surprisingly, he reported to his fiancée that he had faced more than twenty bouts of fever.[107]

Burton, of course, could rest in the knowledge that he was not alone.[108] European newcomers to Sierra Leone, he observed, "were attacked with climate-fever, which did not respect the doctors, and the settlers, after many quarrels and great insubordination, saw 800 of their little band carried to the grave." In Freetown, he noted that even the highest-ranking colonials succumbed with monotonous regularity to the deadly climate. Within the space of four years, three Spanish consuls had fallen victim "to a climate which has slain five captains-general, or governors, in five years." Small wonder that he himself was glad to escape from "the City of the Slain as from a slave-ship or from a plague hospital." It was even worse for imperial wives, who were overcome by tropical tyranny with even more alarming ferocity. There was no place where a wife was more needed, he lamented, than in the tropics; "but then comes the rub—how to keep the wife alive."[109]

Nonetheless, Burton remained invested in the idea that some degree of acclimatization was achievable. How else could the colonial project succeed? But he was no less sure that climate impressed itself on human populations in inexorable ways. In Madeira, for instance, he attributed mulatto skin tones, the "timidity" of thieves, and the fact that "the women rarely scold" to the "mildening effect of climate." The climate was well suited to patients with rheumatism, scrofula, pulmonary and bronchial problems, and a variety of infectious diseases. Yet expatriates did not fare well there. For even if they managed to survive the climate's initial physical onslaught, tropicality's psychological imprint ran deep. Madeira itself was illustrative. "Nostalgia is a disease as yet imperfectly recognised," he noted. "The only remedy . . . is constant occupation of mind if not of body, and this Madeira cannot afford. The *habitués* declare the climate hostile to work."[110] Elsewhere, the enervating nature of the weather, a life of tedium, a poor and frequently dreary diet, foul water, and the absence of cold weather, meant that "a man once 'down' remains so for a long time." As one colonial wife put it to Burton, "I felt amidst all the glory of tropic sunlight and everlasting verdure, a sort of ineffable dread connected with the climate."[111]

All of this echoed his earlier experience during his trek from Zanzibar to Tanganyika between 1857 and 1858. When he arrived at K'hutu, a place whose "hideousness" Burton did not hesitate to advertise, he announced, "The sensation experienced at once explains the apathy and indolence, the physical debility, and the mental prostration, that are the gifts of climates which moist heat and damp cold render equally unsalubrious and uncomfortable." In similar vein, when speaking of the rainy monsoon at Kawele, he declared, "The mind, enfeebled perhaps by an enervating climate, is fatigued and wearied by the monotony of the charms which haunt it; cloyed with costly fare, it sighs for the rare simplicity of the desert. I have never felt this sadness in Egypt and Arabia, and was never without it in India and Zanzibar." So it is no surprise that he explained that the delay in publishing the record of his personal adventure in the Lake Regions was on account of "impaired health, the depression of spirits, and worse still the annoyance of official correspondence, which to me have been the sole results of African Exploration."[112]

Such were the conditions that exposed the contradictions of empire, contradictions lurking in the depths of Burton's own psyche. Europeans could only survive in the tropics if they constructed residences in the hills, avoided pathological sites, and reproduced European infrastructure. But how could these be achieved? Again and again, climate and disease conspired to undermine imperialism's best intentions. "The great gift of Malaria is utter apathy, at once its evil and its cure, its bane and its blessing," Burton reflected. "Men come out from Europe with the fairest prospect, if beyond middle age, of dying soon. Insurance offices object to insure. No one intends to stay longer than two years, and even these two are one long misery. Consequently men will not take the trouble to make roads, nor think of buying a farm, or of building a house upon a hill." How could they, and at the same time, take some of the excellent advice about maintaining a healthy lifestyle in the tropics from Rev. Jones, a West Indian preacher in Sierra Leone, who delivered "the valuable motto 'Take it easy'"?[113]

Plainly, successful imperial careering in the tropical world required attentive, obsessively attentive, oversight to resist the bullish forces of climate's empire. Residence at higher altitudes, or at least frequent flight to hill resorts, was fundamental. And no wonder. At a station called Mzizi Mdogo in the East African Ghauts, Burton spoke of the pleasures of escaping "from the cruel climate of the river-valley, to the pure sweet mountain-air, alternately soft and balmy, cool and reviving." Here, with adjectival exuberance, "dull mangrove, dismal jungle, and monotonous grass, were supplanted by tall solitary trees, amongst which the lofty tamarind rose conspicuously graceful, and a card-table-like swamp, cut by a network of streams, nullahs, and stagnant pools, gave way to dry healthy slopes, with short steep pitches, and gently shelving hills." It was here that the wondrous exchange of climates meant that strength and health had returned to him "as if by magic" after the ghastly experience of being "martyred by miasma" as he left Zungomero.[114] The implications for imperial residence were unmistakable. Burton had documented spaces of sickness as a prelude to mapping places of protection so as to guide future colonization. Medical and moral well-being literally followed the contour lines of the material landscape—a kind of altitudinal metaphysics. What also made the colonizers' dream realizable, moreover, were the material and medical benefits of modernity. In many places, monthly steamers and circulating libraries were already breaking the monotony of colonial existence and offering a much richer way of life to inhabitants whose energies had hitherto been sapped by a punishing sun. These boons, together with "quinine, liberty, and constant occupation," were daily robbing "the most dangerous climate of half its risk."[115]

Torrid Traces

Images of the "torrid zone" as a site of medical and moral peril survived long into the twentieth century. For the triumph of the germ theory of disease, liberating medicine from anxieties over miasmas, exhalations, and the like, did little to change the tropical imaginary long installed in the Western psyche. The writings of several students of climate during the early decades of the twentieth century amply reveal the lingering legacy of tropicality.

The Harvard climatologist and activist for immigration restriction, Robert DeCourcy Ward (1867–1931), to whom we will periodically return, is a case in point. Shortly after his appointment as an assistant in meteorology at Harvard University, he played a major role in the founding of the Immigration Restriction League during the early 1890s with a group of fellow Boston Brahmins who were deeply exercised over what they considered a marked deterioration in the eugenic quality of newcomers to the United States.[116] Ward was a prolific author. Besides a number of textbooks on climate and elementary meteorology, he carried out research on land and sea breeze circulation, economic climatology, atmospheric energy, and climatic classification.[117] Among the specific topics he took up for investigation were the human impacts of fog, the science of local thunderstorms, the physiological effects of diminished air pressure, and the relevance of meteorological conditions for military operations.[118] At the same time, he devoted much energy to such subjects as eugenics, race, and immigration.[119] But Ward did not consider these two passions as separate enterprises. To the contrary, his brand of climatology was a fusion of meteorological science and nativist sentiment. And nowhere, perhaps, is this more powerfully exhibited than in his popular text *Climate, Considered Especially in Relation to Man*, which first appeared in 1908. Within its pages, enduring traces of tropicality remained conspicuously on display.

In his commentary on what he called the "torrid zone"—a "great belt" characterized by "remarkable simplicity and uniformity of its climatic features"—Ward elaborated on the physiological effects of heat and humidity. Here we focus on their perceived implications for health; later we will witness how he portrayed their economic significance. The "hot-house air" of the torrid regions, he told his students, had a debilitating effect, making energetic physical and mental activity difficult if not impossible. In particular, the wet tropics were the most challenging to visitors coming from the higher latitudes. The driest zones were the most hospitable. And the "most energetic natives" were "the desert-dwellers." Not surprisingly, Ward repeated the now familiar recommendation that the highlands and mountainous locations within the tropics offered "more agreeable and more healthful conditions for white settlement." Plainly, atmospheric conditions and the occurrence of disease were causally related. Compared with cold latitudes, he pointed out, "tropical death-rates average high," ranging from "the appalling rate of 483 per 1000 among European troops on the Gold Coast in 1829–1836, through 121 per 1000 for European troops in Jamaica in 1820–1836."[120] For reasons such as these, he devoted considerable space in his text to what he called tropical hygiene. What is notable, however, is that alongside his cataloguing of the deleterious effects of the tropics on the human body, Ward brought within the sweep of his "hygiene of the zones" a political ecology of imperial rule that naturalized the use of indigenous labor on account of the intolerable mental and moral tax burden that "true" colonization would inevitably impose.

To Ward, the acclimatization of the white race in the tropics was a matter of the greatest importance. "Upon it," he declared, "depend the control, government, and utilisation of the tropics." And indeed, "control," "government," and "utilisation" were the precise terms of engagement for imperial dealings with the tropical world. Why? Ward explained that while whites could not with impunity engage in hard manual labor under a tropical sun, they could nevertheless enjoy reasonable health as overseers if they took appropriate precautions. But acclimatization "in the full sense of having white men and women living for successive generations in the tropics, and reproducing their kind without physical, mental, and moral degeneration,—i.e., colonisation in the true sense,—is impossible." Human imperial power was manifestly impotent before the mighty empire of climate. As he concluded,

> It has been well said that the white soldier in the tropics is 'always in campaign; if not against the enemy, at least against the climate.' This sentence may be made to fit the case of the white civilian in the tropics by making it read: the white race in the tropics is always in campaign against its enemy, the climate.[121]

Ellsworth Huntington (1876–1947), the prolific geographical polemicist at Yale, was even more enchanted by the art of tropical moralizing. An ardent advocate of environmental determinism, Huntington's work, according to James Rodger Fleming, gave every appearance of being "rigorous and authoritative"; in fact, "it was neither." His teeming pen was such that Isaiah Bowman once quipped that if Ellsworth forgot to shave one morning, he could produce an article in the time saved. That prompted him to suggest that Huntington would be better advised to "write less and think more."[122] Prolix Huntington surely was, and if only for that reason, his name will periodically feature in our ongoing travels through the empire of climate.

In 1914, Huntington turned his thoughts to the capacity of the white race to adapt to tropical America in a piece he put together for the readers of the *Journal of Race Development*. Here he displayed all the traits of enduring tropicalism by pondering whether what he believed to be the shortcomings of the tropical realm gave any grounds for thinking that races of European origin could "dwell permanently within the tropics and retain not only their health, but the physical energy and mental and moral vigor which have enabled them to dominate the world." He was pessimistic. To him, it was clear that blaming the native peoples of the tropical zone for their "backward condition" was mistaken; they were simply victims of the climate. The reason was plain. Life was too easy there. Tropical abundance meant that there was no need for forethought to plan and provide for a cold or dry season. Add to this climate-induced laxity the ravages of tropical disease, and the consequences for the indigenous inhabitants were obvious. The operations of the mind, no less than bodily activity, were irreparably dulled. And as long as any society remained constantly afflicted with such maladies, there was little chance that it could "rise high in the scale of civilization."[123]

Given these dire circumstances, the dangers for white colonists in tropical environments were palpable. And the challenges were not simply medical and material; they were also moral and mental. Races from the northern latitudes, Huntington insisted, just could not be as efficient in hot climates as they were in their own habitats. Rather, they perpetually found themselves handicapped by a dramatic depletion of their energy levels as well as a diminution in their mental power and moral vigor. Tropical colonists were afflicted by what he called "climatic inhibition," a condition characterized by a "lack of will power," which manifested itself in "lack of industry, in an irascible temper, in drunkenness, and in sexual indulgence." Of all these, perhaps the greatest danger arose from what might be called the cultural ecology of tropical sex. To Huntington, tropical meteorology and sexual morality were causally connected:

> Upon this rock a large number of northerners are wrecked. It is due partly to the low standards of the natives themselves, partly to the mode of dress among the women, which constantly calls attention to their sex, and partly to the free open life which naturally prevails in warm countries. In addition to this there seems to be another reason. Either the actual temptation to

> sexual excess is greater than elsewhere, or else the inhibitory forces are weakened by the same effects which cause people to drink, to become angry, and to work slowly.[124]

The following year, Huntington further pursued this theme in his *Civilization and Climate*. We will return to it at several junctures. For now, we note that the volume was at once influential and controversial as reviews at the time disclose. The University of Chicago's Walter S. Tower, for example, while applauding the project of establishing "climatic laws of control over many important aspects of human life," remained cautious, finding much to commend but recommending "a conservative position" in regard to the climatic thesis.[125] At the same time, R. H. Whitbeck, writing in the *American Historical Review*, judged that the volume's main hypotheses were of "unquestionable importance" pointing "towards truths of great significance" even if some of "Huntington's earlier conclusions regarding the relation of climatic changes to historical events have been vigorously opposed."[126] In particular, Whitbeck found compelling Huntington's insistence that intellectual and physical dynamism were definitely connected to atmospheric temperature. The Columbia University sociologist Alvan A. Tenney, too, reviewing the work for *Political Science Quarterly*, found it "brilliant, stimulating and provocative of constructive thinking" and lauded both its scientific aspirations and its wealth of "detailed information and theories."[127]

It was in the third chapter of *Civilization and Climate* that Huntington turned again to the subject of "the white man in the tropics." Still typecasting topical peoples as dull and unprogressive, he lamented the deteriorating influences that natives had on white colonialists. Once again, the subject of how the "inferior mental ability" of such "despised" races was "almost certain to lead to low sexual morality" came to the fore. It was a subject that evidently fascinated Huntington. To him, the whole matter of sexual relations in tropical regimes was of such importance that it could scarcely be overestimated. Had not missionaries themselves reported that everywhere in the tropics, even the strongest converts to Christianity found it "almost impossible to resist the temptations of sex"? They could be taught honesty, diligence, and many other virtues. But sexual restraint was another matter altogether. "I believe I am speaking within bounds," Huntington insisted, "when I say that any young man of European race with red blood in his veins is in more danger of deteriorating in character and efficiency because of the women of the tropics than from any other single cause." Such evil effects were among the most "potent factors in rendering it difficult for the white man to attain as much success in tropical regions as in those farther to the north or south." A clearer case of transferring moral accountability from the perpetrator to the victim would be hard to find. In short, Huntington portrayed tropical colonists as helpless hostages to tropical inertia, a state of nature that reduced minds and bodies to a nervous and enfeebled condition. It was pernicious. For those who did "spur themselves to work within the tropics as hard as at home" only found themselves in even greater "danger of breaking down in health." And the consequences were long term. Not only were they "almost sure to die before their time," but they were also "not likely to leave many children."[128]

The sentiments of Robert De Courcy Ward and Ellsworth Huntington echoed down through the years. And perhaps no better illustration of their legacy can be found than in a college textbook on climatology by the British physical geographer Austin Miller (1900–1968), which made its first appearance in 1931. By the time of its ninth edition, reprinted in 1969, it was still declaring the sentiment, unchanged in nearly four decades, that direct behavioral correlates of climatic governance were readily discernible across the globe. "The enervating monotonous climates of much of the tropical zone, together with the abundant and easily obtained food-supply," Miller declared, "produce a lazy and indolent people, indisposed to labour for hire and therefore in the past subjected to coercion culminating in slavery."[129] What is notable here is the way in which moralistic descriptions of climate—enervating, monotonous, lazy, indolent—were still being presented as settled scientific maxims and the ease with which intellectual potential and moral behavior were thoroughly naturalized courtesy of the inexorable operation of climate's moral imperatives.[130]

In the early decades of the twentieth century, the tyranny of tropicality continued to be displayed in all its fullness. And its traces persisted among advocates of eugenics, who struggled to inject their climatic fixations into a movement obsessed with genetics and heredity. In the next chapter, we will visit some of these endeavors, not least among the champions of the new science of biometeorology.

The power that the empire of climate was believed to exert on the human frame has nowhere been more prominently on display than in works of tropical medicine in its various guises. Again and again since the mid-eighteenth century, those preoccupied with the connections between climate and health turned their gaze toward the sites of imperial careering. No doubt what further facilitated this long-standing obsession were the health data and mortality statistics available from the ranks of the military on the medical circumstances of troops in colonial settings. From these sources, a picture increasingly emerged, gathering ever greater momentum during the nineteenth century, of the tropical realm as dangerous and debilitating, pestilential and tyrannical. Such bleak adjectives, moreover, were as applicable to the mental and moral state of the white race in hot climates as to their physical condition. And as often as not, these diagnoses went hand-in-hand with a racialist outlook that, at once, denigrated native peoples yet depended on them for labor, service, pathfinding, and protection. Local populations were frequently demonized as the source of medical and moral peril, even though they were the bedrock on which imperial ambitions were erected.

The image of tropical toxicity of course posed acute problems for nations with colonial desires. And for that reason, the question about whether white Europeans could acclimatize to tropical conditions profoundly exercised medical and political minds alike. Here was a scientific question that had immediate implications for imperial strategy. Different stances were adopted on a subject that ranged far and wide. Some were sure that Europeans could never adjust to the tropical world, while others were just as certain that complete acclimatization was entirely feasible even if there was no consensus on the length of time that might be required. Some focused on the long-term consequences of exposure to tropical conditions for the empire's children and for their children's children. Others brooded on the psychological repercussions of monotonous tropical heat, particularly for women. Yet others worried over the specter of gradual but inevitable degeneration of the white race in situations where interracial liaisons were frequent even if frowned upon. Nevertheless, whatever position was adopted, writers of medical treatises laid stress on the urgent need for moral circumspection, hygienic discipline, and a rigorous dietary regimen if the agents of empire were to manage meteorology and if imperial dreams were to be realized.

Whether or not acclimatization was feasible, questions remained about what practical measures could be taken to resist the forces of climate's tropical empire. One strategy was to ferret out and map the location of safe spaces to which imperial officialdom and its entourage could resort for respite and recovery. Hill stations, at once simulating conditions back home and stimulating landscape nostalgia for the old country, were constructed far from lowlands infested with pests and people that colonials feared. In such settings, architectural forms symbolized the distance between inhabitants and intruders, each viewing the alien other with foreboding. Specially designed headgear and clothing to protect settlers from the cruelty of an angry sun only served to reinforce the gulf between colonizer and colonized. The dress codes adopted by Europeans at empire's edge revealed an expatriate community precariously poised between heroism and helplessness.

At the same time, the political ecology of medicine meant that tropical climate could be mobilized as an ideological weapon both for and against colonial enterprises. While imperialists insisted that resources to overcome these health challenges were at hand, others marshaled climate's devastating effects on settlers in the cause of incipient nationalism. The heroic image of Europeans on the pioneer fringe extending the graces and glories of empire to benighted natives was thus regularly subverted by reports recording their frailties no less than their follies under the ruthless heat of a tropical sun. These latter images indeed were sufficient for some medical practitioners to conclude that permanent settlement by white colonizers was impossible and that some means of managing tropical empires from afar had to be devised. As we will later see, this geopolitical strategy was adopted by some keenly aware of the rich natural resources in places like Africa that could be tapped for the benefit of imperial powers.

In the early years of the twentieth century, traces of tropicality lingered long in the imagination of many commentators.[131] Writers of climatology texts, for instance, would incorporate comments, taken to be conventional wisdom, on the physical, mental, moral, and social effects of climatic conditions. Sometimes in the form of short sharp assertions, these gave every impression that they encapsulated the settled certainties of scientific inquiry. Now too some began pushing the medico-meteorological thesis in other directions. Not least of these was Ellsworth Huntington, who, along with several other key figures in the development of what was being called the science of biometeorology, sought to align climatic determinism with a new fascination: eugenics.

4

Climate, Eugenics, and the Biometeorological Body

DURING THE early decades of the twentieth century, Hippocratic medical meteorology found itself mobilized in the cause of eugenics. This turn of events owed a good deal to the writings of the influential, if maverick, Yale geographer Ellsworth Huntington (1876–1947), whose love affair with climatic determinism has already come within our purview, and such medical practitioners as William Ferdinand Petersen (1887–1950) and Clarence Alonzo Mills (1891–1974), who were closely involved with the developing science of biometeorology. Here I want to dwell on how they conceived of the human body and human health in relation to climatic conditions and the way in which they sought to marshal a synthesis of Darwinian evolution and neo-Hippocratic medicine for eugenic ends. In this way, I hope to show the ease with which inquiries into the health effects of different climatic regimes could glide toward the pathologization of particular places and people, as well as allow space for the expression of sentiments shaped by the desire for political biopower. Dividing the world's climates into the sickly and the salubrious provided a rich resource for those with eugenic aspirations who sought to control human populations through managing national immigration and human breeding.

Neo-Hippocratic Health and the Eugenic Imperative

Huntington's interest in climate as an active health force came initially into view during the writing of *Civilization and Climate*, which came out in 1915. A few years later, when reminiscing on his turn to medical climatology, he remarked that it was only when he began studying the civilizations of Central America and Greece that it became clear to him just how formative "the direct effect of climatic conditions upon man's energy and will power, as well as upon his health" really was in directing the course of global history.[1] Of course, he had long been interested in the power that climate exerted on human life more generally. In *The Pulse of Asia* (1907), for example, a work that specifically set out to show that "whatever the motive power of history may be, one of the chief factors in determining its course has been geography; and among geographic forces, changes of climate have been the most potent for both good and bad." In passing, he did characterize certain climates as debilitating or deadening and casually observed that in some places, inhabitants had "the disadvantage of a very unhealthful climate."[2] But, as with his *The Climatic Factor as Illustrated in Arid America* (1914), matters of health were scarcely discernible on Huntington's intellectual horizon. By the time *Civilization and Climate* appeared in 1915, things had changed.

A product of what he called the new science of geography, *Civilization and Climate* stood self-consciously in a tradition reaching back to classical antiquity through Montesquieu and Buckle, who, as Huntington reminded his readers, "believed that climate is the most important factor in determining the status of civilization." His interest was in the sources of civilization more generally, of course, but now he inserted matters of health into his diagnosis using the neo-Hippocratic-sounding language of climate, blood, and circulation. Noting that at higher altitudes, the red corpuscle count in the blood increased enormously and that the capacity to absorb oxygen and to emit carbon dioxide was likewise modified, he went on to observe that while the precise anatomical processes through which frequent temperature changes influenced the body remained obscure, the most likely explanation lay in its influence on the circulatory system. "Provided it does not impose an undue strain on the heart or arteries," he added, "anything that stimulates the circulation appears to be helpful." And drawing upon physiological research on the influence of temperature on animal metabolism, he further observed that "in the driest weather which England enjoys, metabolism is more active than in wet weather."[3]

What drew Huntington to ponder connections between weather and health were the mortality statistics of various states, which he believed were correlated with seasonal variation. And so he reported that for New York State, while mortality was high during the winter months, the maximum number of daily deaths was not recorded until March. This was "because people become sick in January and February, especially those who are elderly, and finally die after lingering illnesses quite unlike those of children." Seasonal patterns of weight gain among patients with tuberculosis also caught his eye. All this led him to seek out comparable statistics for other states and regions of the globe. Soon he was characterizing whole geographical zones in the language of medical meteorology: "We have now seen that from New England to Florida physical strength and health vary in accordance with the seasons. Extremes seem to produce the same effect everywhere."[4]

Just why Huntington embarked on the task of scrutinizing links between weather and well-being is not difficult to discern. His interests in the rise and fall of civilization, presumed to fluctuate in response to the imperatives of climate, had led him to develop a global geography of human "efficiency." The idea seems to have been that some individuals and groups display much greater productive capacity than others. What was lacking in developing what might be referred to as a cartographics of efficacy, and what he hoped to provide, was a "large series of measurements of the actual efficiency of ordinary people under different conditions of climate." To do so, he proposed taking "a group of people who live in a variable climate, and test them at all seasons" using records of their daily work. One group comprised factory workers from a range of U.S. cities located in different climatic regimes—New Haven, Bridgeport, and so on; another was composed of naval and military students. As Huntington pondered the records of these individuals and correlated them with the changing weather conditions, he came across medical data of one sort or another that he believed displayed seasonal variation.[5]

Huntington's turn to medical climatology, it seems, was born of his interest in energetics, a love of hierarchy, and a passion to place human subjects on a graduated scale from high to low—an outlook that fitted snugly with his long-standing commitment to eugenics. So when *Health and Social Progress* by the eugenicist sociologist Rudolph M. Binder appeared in 1920, Huntington welcomed it for the way it repeatedly showed, to his own satisfaction at least, how "the deadening effect of poor health on man's capacity is a prime factor not only in backward regions, but even in our own country among people of high intelligence."[6] Still, whatever his motives, Huntington continued to direct his energies toward elucidating causal connections between weather and health, for which he received the commendation of Charles-Edward Amory Winslow, founder of the Yale School of Public Health.[7] In 1920, in the wake the Spanish flu pandemic, one of the greatest medical catastrophes of the twentieth century, Huntington produced a statistical analysis intending to demonstrate how weather controlled the incidence of pneumonia and influenza.[8] A couple of years later, in 1923, in a separate analysis of the U.S. experience, he concluded that "the favorable conditions of the air were the greatest factor yet detected in helping the people of the cities of the United States to ward off the influenza in the fall of 1918."[9] In this self-same study, though, he took the opportunity to generalize more widely by commenting on how the warm and relaxing climatic conditions at the time of the epidemic in India, Mexico, and South Africa meant that they experienced huge death rates compared with the United States, while the more monotonous weather in eastern and southern Europe meant that the death toll was far higher there than in western Europe. And in a much longer assessment, written in his capacity as chair of the National Research Council's Committee on the Atmosphere and Man, he provided an analysis of the role of weather in explaining patterns of daily mortality in New York City. Here he emphasized the importance of diurnal temperature changes, insisting that a significant drop in temperature was consistently characterized by "a relatively low death rate; a rise by a high death rate." Unsettled weather, it seemed, was systematically associated with a much lower mortality count than was the case when the weather remained unchanged. Of crucial significance in this investigation was his idea that humankind flourished in what he called "optimum" weather conditions—a biometeorological state that, as we will presently see, he explained on evolutionary principles. It formed the conceptual foundation on which his whole study was constructed. While various atmospheric components were at play in determining health, variations in temperature held pride of place in the scheme. As he put it, "Any appreciable departure from the optimum temperature either upward or downward is accompanied by an increased death rate."[10]

Beyond these relatively specialist inquiries, Huntington was keen to disseminate his climatic philosophy of health much more widely. Thus, in his 1933 text *Economic and Social Geography*, cowritten with Frank E. Williams and Samuel van Valkenburg, he devoted a whole chapter to "Climate, Health, and the Distribution of Human Progress." Right upfront, he announced, in a confidently determinist mode, that a comparison of various maps clearly demonstrated "that climate is directly or indirectly the main determinant of the geographical distribution of economic activity and civilization. This happens . . . primarily through its effects on man's health, energy, and mental alertness."[11] His technique, though simple if not simplistic, was visually striking: he used maps to correlate, for example, the distribution of what he called "climatic energy" in the United States with data on health derived from life insurance companies (see figure 4.1).

The take-home message was unmistakable. Huntington wanted to segregate the globe into a hierarchy of spaces that he typecast on a spectrum from

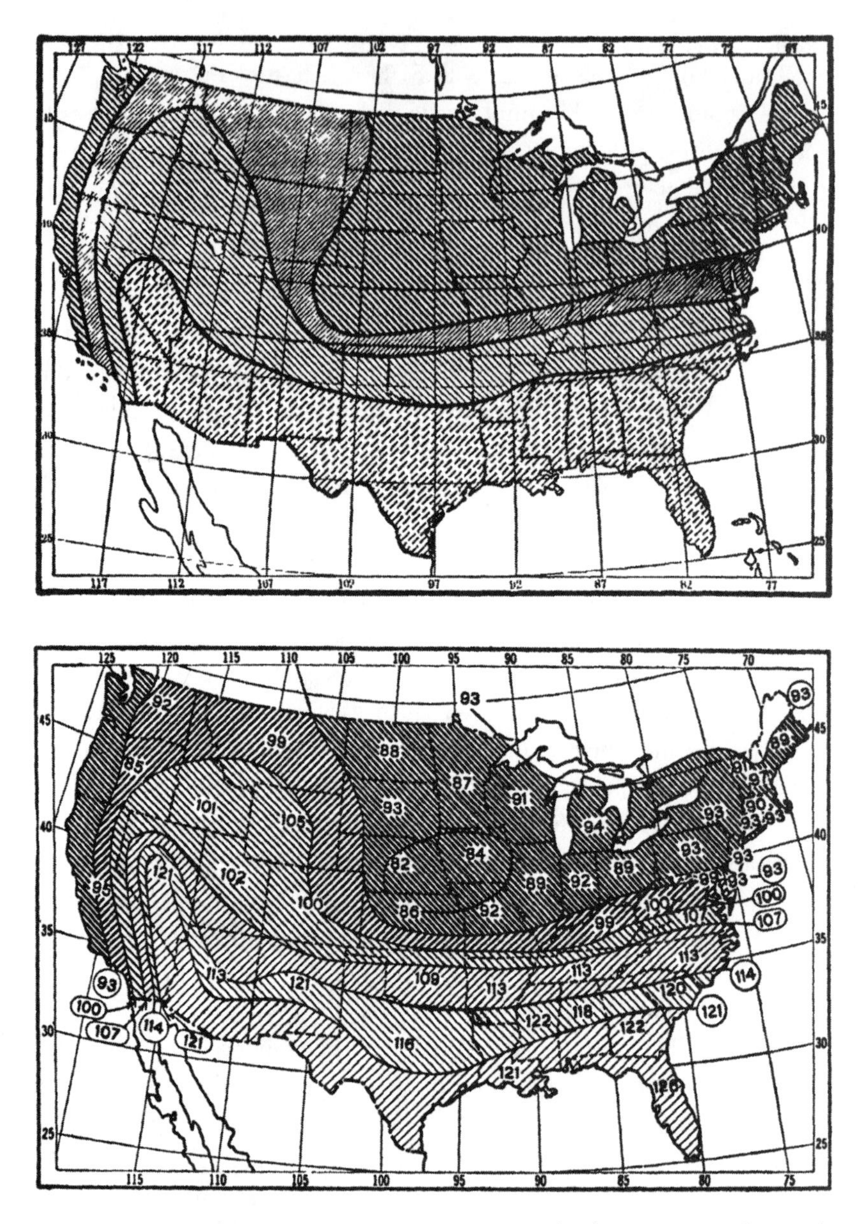

FIGURE 4.1. Huntington's Maps of "Climatic Energy" (top) and "Health" (bottom) in the United States

Source: Ellsworth Huntington et al., *Economic and Social Geography* (New York: John Wiley, 1933)

low to high, from inefficient to efficient, from unfavorable to "nearly perfect."[12] Thereby, he could chart the world's climatic regimes and typecast them as privileged or ill-favored, healthful or disease ridden, stimulating or crippling. It was an imagined geography of enormous scope and power. For not only did he believe he could empirically map global biopower, but he hoped his cartographic energetics could in turn be mobilized to manage the future quality of human reproduction.

Such inclinations, of course, were precisely those of an enthusiastic eugenicist. For indeed, Huntington was long captivated by the eugenic vision, an interest that only widened and deepened as the years went by. As early as 1910, he was in touch with a number of intellectuals whose minds ran along similar eugenic lines. He was in communication with John H. Kellogg (1852–1943) of the Health and Efficiency League of America; William R. Inge (1860–1954), Cambridge University professor of divinity and dean of St. Paul's Cathedral, wrote to him expressing his pleasure at the implementation of new immigration policies since they would raise the birth rate among old-stock American families; Henry Fairfield Osborn (1857–1935), celebrated paleontologist and president of the American Museum of Natural History, invited him to speak at the Half Moon Club in 1910—a leading Progressive Era social group.[13] Besides this, his concerns over unrestricted immigration merged with the eugenic outlook of publicists like Edward Ross, Madison Grant, and Lothrop Stoddart, and he willingly participated in the Second International Congress of Eugenics, which Osborn orchestrated in 1921 at the American Museum of Natural History. Later, when the American Eugenics Society was formed in 1922, he closely associated himself with it, eventually becoming president in 1934. During these years, eugenic concerns increasingly dominated Huntington's output, but two themes in particular warrant comment in the context of his merging of population politics with medical climatology.

First, matters of birth—birth rate, birth control, birth season, and the like—came to dominate Huntington's horizon. Reproduction differentials between the so-called upper and lower classes had become an obsession, one that he readily Darwinized to support his own predilections and policies. So, in a piece he prepared for the American Birth Control League in 1926, at the invitation of its founder Margaret Sanger, on the effects of population growth on Chinese character, he urged that overpopulation stimulated "unfavourable natural selection," which was eugenically detrimental.[14] The biopolitics of birth also dominated *Builders of America*, which he coauthored with the then secretary of the American Eugenics Society, Leon Whitney, in 1927. Here he warned that failure to engage in the discipline of birth control, particularly by the "lower" classes, would prove detrimental to the future of the American people.[15] Darwinian evolution had ensured that the first Europeans to settle New England—the original "builders of America"—were strong in body and mind and possessed "fine temperament, fine intelligence, and fine health"; for they were "one of the most highly selected groups the earth has ever seen." The only way to preserve this rich bioheritage, he believed, was to "diminish the birth rate among the less valuable parts of our society and increase the birth rate among the more valuable parts."[16] For the future, birth rate and biopower had to be intimately intertwined.

Huntington, moreover, did all in his power to disseminate this eugenic vision far and wide. Thus, in 1935, he produced a volume originally intended as an expanded and revised version of Whitney's 1923 *Eugenics Catechism*, a manual widely distributed to learned societies, schools, and various church groups. Entitled *Tomorrow's Children: The Goal of Eugenics* and published in conjunction with the directors of the American Eugenics Society, Huntington's replacement catechism prominently addressed issues congregating around differential birth rates, population policy, and eugenic legislation. Right from the outset, he was concerned that his readers would appreciate that birth rate differentials showed that things were rapidly going "in the wrong direction—low among the more competent and self-controlled types . . . and high among those of less intelligence and foresight." What a contrast that was to colonial America, where the "differential rate of survival" meant that "the more competent and intelligent people increased in number more rapidly than the rest of the population." Huntington's handbook continued in the same vein, catechizing readers on the "causes of the unfavourable differential birthrate," the nature of "eugenic selection," the methods of birth control, sterilization, and the like.[17]

With his mind running along these tracks, it is not surprising that Huntington would soon turn to the whole topic of the season of birth in order to ascertain the most favorable months of the year for conception and parturition. And so in 1938, he produced a full-length study of this very topic. Grounded in Hippocratic thought-forms, *Season of Birth* was no less suffused with Darwinian idioms. Intrigued by the "effect of the seasons" as revealed in "the birthdays of men of eminence," he insisted right from the start that the most propitious temperatures for physical and mental development were due "to the selective effect of the climate in which the species *Homo sapiens* originated during the glacial period."[18] In the 450 or so pages that followed, he elaborated on the geography of birth season by surveying conditions in Belgium, Russia, Japan, and the United States, while at the same time exploring the impact of the seasons on such eugenic themes as sex ratios, race, social rank, longevity, leadership qualities, insanity, genius, criminal behavior, and the like. In presenting his findings, Huntington drew on the work of such figures as the British reproductive biologist Walter Heape FRS; the Finnish philosophical anthropologist Edvard Westermarck, author of *The History of Human Marriage*; the Italian statistician and demographer Corrado Gini, who held the presidency of several eugenics societies; the British travel writer Alleyne Ireland; the Italian physician and criminologist Cesare Lombroso; the American biometeorologist William F. Petersen; and Charles Kassel, the Fort Worth writer and founding partner in the law firm of Ledgerwood & Kassel.

Several things stand out about this investigation. For a start, Huntington's analysis was a conscious Darwinizing of the Hippocratic vision.[19] Right from the outset, he located his interest in birth season in the framework of early human evolution by seeking to ascertain the lingering significance of the climatic regime within which the human species had emerged. In doing so, he called attention to the fundamental role played by the basic animal rhythm of reproduction in human affairs. The seasonal patterns of human birth that he believed he had ascertained were thus rooted in evolutionary biology. The conspicuous "truncation of the mortality curves of both men and women throughout the reproductive period during the months when the basic animal rhythm of reproduction is at its height," he observed, "represents a deep-seated adaptation to seasonal fluctuations of the weather." The reason was to do with the links between seasonality and spermatozoa that the Massachusetts State biologist, the parasitologist David Lawrence Belding, MD (with whom he was in direct communication), had discerned in his research on fisheries. The seasonal fluctuation in the birth rate, Huntington suggested, appeared to be the direct consequence "of a physiological swing whereby the number of spermatozoa ebbs and flows with the seasons." All of this was swept into a synthetic Hippocratic evolutionism, the dimensions of which he spelled out in a chapter devoted to "The evolution of *Homo Sapiens*." His principal conclusion was that in humans, just as in other animals, an annual reproductive rhythm was of primary importance in evolutionary biology. This biometeorological vision welded together climatic optima, seasonal reproductive patterns, food supply, and the survival of offspring:

> One of the most interesting problems . . . is the reason why a temperature in the neighbourhood of 62° is the optimum for reproduction. Why not 40°

> or 80°? We have come to the conclusion that the reason is found in a selective process that occurred among our primitive ancestors. In climates of an intermediate type, neither very cold nor so warm that they are monotonous, the children who are conceived when the average temperature rises to about 62°F. are born in the early spring when reasonably warm weather first arrives. That is the time when the supply of food begins to increase after the scarcity of winter. In other words, birth occurs at the season which is best for the survival of the infant. . . . A major feature of this book is the discovery that the optimum temperature for reproduction is also the optimum for physical health in general.[20]

A second prominent fixture in Huntington's climatic edifice manifested itself in the opening pages of *Season of Birth*. Huntington announced it as a novel discovery, a further extension of the Hippocratic-evolutionary credo. He was sure that the selective operations of climate in the evolution of *Homo sapiens* during the glacial period constituted a persuasive explanation for "the highly puzzling fact that man has two different optima of temperature—physical and mental." He further elaborated, "The physical optimum apparently represents the season when conceptions were most likely to result in the birth of infants that were able to survive. The mental optimum represented the cooler season when births of this kind actually occurred." The physical optimum occurred at 60 to 65°F, the mental optimum at 39 to 54°F. Why? And more particularly, why should the body's metabolic and glandular system function most effectively at that temperature? The story of early human evolution provided the answer. "Our bodies apparently function best," he explained, "at the temperature which prevailed in primitive times at the mating season, for children conceived at that season were best able to survive. This gives us a physical optimum." By contrast, human "minds function best at the temperature which prevailed in primitive times at the normal season of birth. At that time, more than any other, the survival of the new generation depended upon the alertness of the parents. Thus a mental optimum became established at a temperature lower than the physical optimum."[21]

These refinements to his thinking about climatic optima supplemented a theme that Huntington had been toying with for well over twenty years. In his 1915 *Civilization and Climate*, for example, he had developed that idea of what he called "The Ideal Climate." There he had argued for the existence of a distinctive global geography of climatic excellence. The propitious conditions he identified turned out to be concentrated in just a few "chief portions of the globe"—England, North America's Pacific northwestern coast, and New Zealand, with parts of South America being compromised on account of rainfall deficiency.[22] Elsewhere, the absence of annual and diurnal variability of atmospheric conditions condemned inhabitants to physical danger, mental lethargy, and moral peril. Later, in his *World-Power and Evolution* of 1919, Huntington repeatedly deployed the idea of climatic optima in his discussion of such themes as "climate and health" and "the environment of mental evolution."[23] Thereafter, the idea periodically resurfaced. In his 1920 analysis of how weather controlled pneumonia and influenza, he characterized "ideal" atmospheric conditions for the "best treatment for all kinds of respiratory diseases" as air that is "variable in temperature," "fairly moist," and "not too warm."[24] Now, he could weld together his speculations on season of birth with the idea of physical and mental climatic optima to further bolster his eugenic intuitions.

Huntington's mobilization of the idea of climate optima and birth season proved attractive to those interested in cultivating a science of biometeorology—a medical science to which we will presently turn. His priority in identifying the significance of birth season was noted, for example, by William Petersen, a founding father of the movement, in his own reflections on seasonality and the "conception of genius."[25] Similarly, Frederick Sargent, another key figure in the evolution of biometeorology, paused in his survey of the Hippocratic heritage to expound Huntington's concept of the climatic optimum as a critically significant biometeorological concept.[26] It was picked up too in Per Dalén's 1975 study of connections between birth season and schizophrenia and has continued to feature as a landmark intervention in a variety of medical arenas.[27]

For Huntington, it is clear, a synthesis of Darwin and Hippocrates enabled a valorization of the temperate zone, afforded a means of naturalizing human excellence, and delivered a global climatology of health and well-being. But it supplied something else as well—a tool that could be wielded in the interests of eugenics. So he did not hesitate to direct his gaze toward subjects—genius, insanity, criminality, and the like—that were the pet peeves of fellow eugenicists. Early in his investigation, for instance, he made it clear that the season of birth bore a close relationship to the occurrence of genius and eminence in human populations.[28] "Genius," he explained, "apparently arises from a fortunate combination of the genes within the chromosomes at the time of conception . . . births of persons of unusual genius conform to the animal rhythm and to the temperature much more closely than do births in general." To Huntington, moreover, the cognate geographical distribution of optimum weather conditions perfectly well explained what he described as the "scientific tendencies" of the

Swedish and the "philosophical tendencies" of the Scottish. It also made sense of global patterns of civilization. The relationship between low temperatures and intellectual activity that he was sure he had discerned in his study of births encouraged him to "suspect that cultural inventions which render people comfortable in cooler and more invigorating parts of the world" had "enabled civilization to move northwestward from ancient Egypt and Babylonia."[29]

Of course, intellectual capacity, or more especially the lack of it, had yet more direct eugenic significance. For it was the "intellectually weak," alongside the emotionally impoverished, who were "especially likely to yield to the sexual stimulation which marks the chief season of reproduction. Such parents presumably are responsible for a large percentage of the persons who become criminals or suffer insanity." The need for reproductive management was urgent. With the advent, and indeed the sexual incitement, of "the primitive breeding season," he went on, "physically weak people who usually are not able to produce children may become parents"—with ominous consequences. "Their children," Huntington suspected, were "probably especially susceptible to tuberculosis as well as to the influences which lead to crime and insanity," and with "insanity" came dramatically increased incidences of "suicide and sexual crimes."[30]

Just how birth season might be connected with leadership, genius, insanity, and criminality now dominated Huntington's outlook. In a chapter devoted to leadership, for example, he reported that he had detected a distinctive seasonal distribution in the births of distinguished Americans. Those who achieved marked distinction apparently showed "an exceptional tendency to be born in winter." Similarly, in his scrutiny of genius and birth season, he came to the conclusion that prodigies were more likely to be born during the winter months, suggesting that "intellectuality may be fostered by temperatures below those best for bodily vigor alone." In similar vein, he scrutinized the seasonal distribution of births for what he called "mental defectives," "idiots," "imbeciles," "morons," and the "insane"—a finely tuned pathological taxonomy if ever there was one. His findings brought him to the conclusion that these "weak types of people react to the basic animal rhythm of reproduction in a different way from more vigorous types" for "none of these weaker types shows the normal tendency toward increased conceptions when the temperature falls to the physical and then to the mental optimum in the autumn." Further elaboration of Huntington's sometimes tortuous explications of his empirical data is not necessary. Long driven by the febrile yearning to find in climate the mainsprings of human worth, his latest nativist motivation was abundantly clear. It was to bring climatic conditions and season of birth within the arc of eugenic management. As he put it in his closing remarks,

> It has been supposed in the past that eugenics is concerned only with hereditary qualities, and such is undoubtedly the case. Nevertheless, if eugenics means the applied science in which the objective is that children shall be well born, the prenatal and preconceptual influences of climate and weather come close to being included. This does not mean that there is any confusion between heredity and environment. What it does mean is that an unfavourable physical environment may depress a given group of people for generation after generation regardless of the people's genetic constitution.[31]

The implications of Huntington's research program were far-reaching. It meant that the people of the tropics were forever condemned to a low position on the hierarchy of human value. As he had already explained in the textbook on social and economic geography, "No matter what races we deal with, we can scarcely expect the same degree of activity and progress among tropical people as among people in more favorable climates. This same reasoning applies with diminished force to other relatively unfavorable climates such as are found in China, Central Asia, and the far north."[32] Of course, if climate stamped its indelible mark of inferiority on the peoples of certain climatic zones, this had major eugenic repercussions for immigration policy, and Huntington did all in his power to advance a restrictionist agenda. As he put it in 1925 in a letter to Madison Grant, eugenicist, lawyer, and author of *The Passing of the Great Race*, "I believe that by rigid restriction of immigration and the application of the best eugenic practices, America will not only benefit enormously but will do infinitely more good for the world than it can do in any other way."[33] Later, in 1935, that sentiment was enshrined in his catechetical *Tomorrow's Children: The Goal of Eugenics*, where his Hippocratic evolutionism came clearly through. Immigration, his readers learned, was one of the major agencies in shaping the innate quality of any population. "Most authorities agree that it is a selective process," he continued, inasmuch as those "who migrate far face great difficulties and dangers, and for the sake of high ideals, are generally of unusual value eugenically." At the same time, Huntington was sure that what he called "adverse or dysgenic selection" frequently occurred when migration remained unchecked and where the excellence of settlers was not monitored. "The eugenist believes that if we permit immigration, we ought to take stringent measures to make sure of the quality of the immigrants."[34] Natural selection, social policy, and eugenic outcomes all needed careful supervision.

Darwinian Hippocratics and the Biometeorological Body

Huntington's efforts to cultivate a Darwinized form of medical climatology, one that had the potential for reshaping eugenic policy, attracted the eye of a number of medical practitioners during the middle decades of the twentieth century who spearheaded the idea of the human body as a fundamentally biometeorological space. While advocates routinely insist that the science of human biometeorology, when viewed in the longue durée, finds its primary coordinate in Hippocratic philosophy,[35] its modern revitalization is frequently traced to the work of William Ferdinand Petersen (1887–1950), a Chicago physician and a professor of pathology and bacteriology at the University of Illinois College of Medicine, who interrupted his early teaching career to serve with the U.S. Army Reserve Medical Corps at the end of World War I.[36] To him, the human organism was, as he put it on various occasions, a "cosmic resonator."[37] The "individual is weather conditioned," he declared; "we respond to every whim of the air mass in which we exist." Self-consciously following in the wake of Hippocrates and Aristotle and their successors, he could assert, "The cosmos (as Bodin suggested) lives in the organism as truly as the organism lives in the cosmos."[38] Petersen's medical climatology thus sought to connect the intimate spaces of the human body with the far-flung spaces of the cosmos.

While Petersen's early work on the etiology of disease was dominated by a bacteriological perspective,[39] he increasingly moved toward a more holistic conception of illness and produced a number of works on the health effects of climate, notably, the multivolume *The Patient and the Weather* (with Margaret E. Milliken, 1934–1938), *Hippocratic Wisdom* (1945), and *Man–Weather–Sun* (1947). As Frederick Sargent observed of him, "Petersen's investigations demonstrated that man was not free and independent of his external environment. His thoughts and actions, bodily functions, and susceptibility to disease were all conditioned by the meteorological environment."[40]

The more than four thousand pages that constituted the seven-volume *Patient and the Weather* represents the most comprehensive statement of Petersen's medical-meteorological thinking. In many ways, he reconstrued the human body in the language of bioclimatology. Thus, he reconceptualized disease *tout court* as air hunger, insisting that "dysfunction and inadequacy of the mechanism that has to do with oxygen supply is probably the fundamental cause of all disease."[41] To him, the principal agents inducing such vascular failure were meteorological. And so the countless patient measurements that Petersen, along with his assistant Margaret E. Milliken, assembled were designed to demonstrate how the health of individuals fluctuated with changing meteorological conditions, particularly the passing of frontal systems—an idea remarkably consonant with Huntington's obsession with cyclonic civilizations. As Sargent, expounding Petersen's contributions, remarked, "In the northern hemisphere, the air environment exhibited particularly rapid shifts associated with the passage of frequent cyclonic systems and there meteoro-pathology should be most evident."[42] Seasonal variations in storminess, Petersen believed, accounted for annual trends in various physiological processes, and he thus set out to chart the seasonal patterns of inflammatory and vascular diseases. Barometric instability became to him the key to unlocking the source of the biochemical and physiological swings he perceived in the human body. What facilitated his passion for connecting variable weather and health was his assurance that disease could be understood as "energy deficiencies in organs or organ systems" that were the consequences of "meteorological stimulation."[43] As Petersen himself put it in a letter to Ellsworth Huntington in 1933, "I can definitely say that the entire physiological mechanism of the body followed the barometric fluctuations."[44]

The Patient and the Weather thus elaborated in painstaking, not to say mind-numbing, detail numerous other correlations between meteorological and medical conditions. A sampling will suffice to give a flavor of the whole. Attacks of coronary thrombosis peaked during the winter; asthma rose to a climax during late fall and early winter; gastric and duodenal ulcer flareup was associated with the passage of cold fronts; appendicitis struck one to two days after a cold front had passed. The list could go on. Arthritis, thyroid disease, gallbladder disorder, tuberculosis, coronary thrombosis, cerebral hemorrhage: all these and many more were examined as disorders that were meteorologically conditioned. Because of the connections Petersen believed he had identified between weather and well-being, between seasons and sickness, his understanding of pathology bore a strongly spatial imprint. What might be described as the geography of death frequently featured in his analyses. Mortalities from angina pectoris, for example, clustered in the northeastern United States, where there was greatest meteorological instability; fatalities from epilepsy were commonest in the spring, with the highest mortality rate for whites again occurring in the Northeast, while among blacks, the southern states showed the lowest rates; in New England, spring deaths from eclampsia spiked in the spring, whereas in Illinois, this occurred during winter and fall; death rates from appendicitis, closely associated with cyclonic events, were maximal in the Rockies and on the Great Plains.

FIGURE 4.2. Man as a Cosmic Resonator
Source: William F. Peterson, *Lincoln–Douglas: The Weather as Destiny* (Springfield, Ill.: Charles C. Thomas, 1943)

All of this was marshaled in support of Petersen's conviction that the human body was indeed a cosmic resonator (see figure 4.2). As he put it, "I have used the term 'Cosmic Resonator' to indicate the very close integration of the autonomic apparatus of the body and the influences of the cosmic—and particularly meteorological—environment in which we exist." The organic rhythms that governed nearly every aspect of human health, he insisted, were "largely meteorologically conditioned and reflected even minor change to a surprising degree."[45]

Not everyone was convinced, however. Many demurred. And Petersen was forced to concede in the introduction to *The Patient and the Weather* that he had received no support for his research on the subject from his own university. One source of criticism was the fact that his inquiries resolutely focused on the individual. This case-study approach raised troublesome questions for statistically minded investigators who sought to discriminate between correlation and causation and to distinguish between associations of weather and disease that were greater than would arise from simple chance. As Sargent put it, "Since the weather changed continually, the association between clinical event and frontal passage could just as well be random as causal."[46] In response, Petersen began to experiment with a range of statistical procedures, though he always remained more at home with the case histories of individual patients. And this conspicuously manifested itself in his extended study of male triplets, *Man–Weather–Sun,* which appeared in 1947.

Petersen's triplet inquiries were based on investigations he had carried out in the summer of 1940 on male triplets, two of whom were identical, with the third displaying heterozygous variations. This venture was intended to allow him to make comparisons between biological variability and weather changes over a period of six weeks. The triplets were exposed to a string of tests, the results of which were subjected to a battery of statistical procedures such as the use of correlation coefficients and analysis of variance. In the end, however, Petersen resorted to visual representations—graphs in particular—of specific episodes as these served his needs better than complex statistizing. Either way, his claims were obvious: there were causal links between meteorological activity and such vital processes as physiological function, morbidity, and mortality.

Despite the small number of subjects and the short period over which they were studied, Petersen elaborated on a large range of clinical conditions over which the weather reportedly exercised its imperial power. The passage of cold fronts played a dominating role in Petersen's mind-set, and he recorded the time and date of every passing front between June 18 and August 4. His self-appointed task was to connect these data with a wide range of clinical measurements of the patient—changing pulse rate, blood pressure, body temperature, bacterial count, and the like. Petersen then moved on to identify connections between the passage of cold air masses and the advent of appendicitis, an inquiry that he described as "an excursion into Hippocratic medicine." His conclusion, based on the experience of one of the individuals, was that an acute attack of appendicitis that this patient experienced was directly related "to major environmental disturbances"—extreme heat and the sudden onset of cold conditions. Petersen went on to suggest that similar meteorological conditions underlay the occurrence of migraine, various psychotic conditions, arthritic flareups, and visceral disturbances connected with gallbladder, ulcer, and colitis. Epileptic episodes were likewise associated with seasonal weather patterns, as were cardiovascular-related deaths. In the case of epileptic seizure, characterized as "a biochemical crisis," Petersen was convinced that he had found significant correlations of attacks with lunar phases and periods of "unusually high temperature."[47] And it was the same for a range of psychotic conditions. Admissions to "psychopathic"

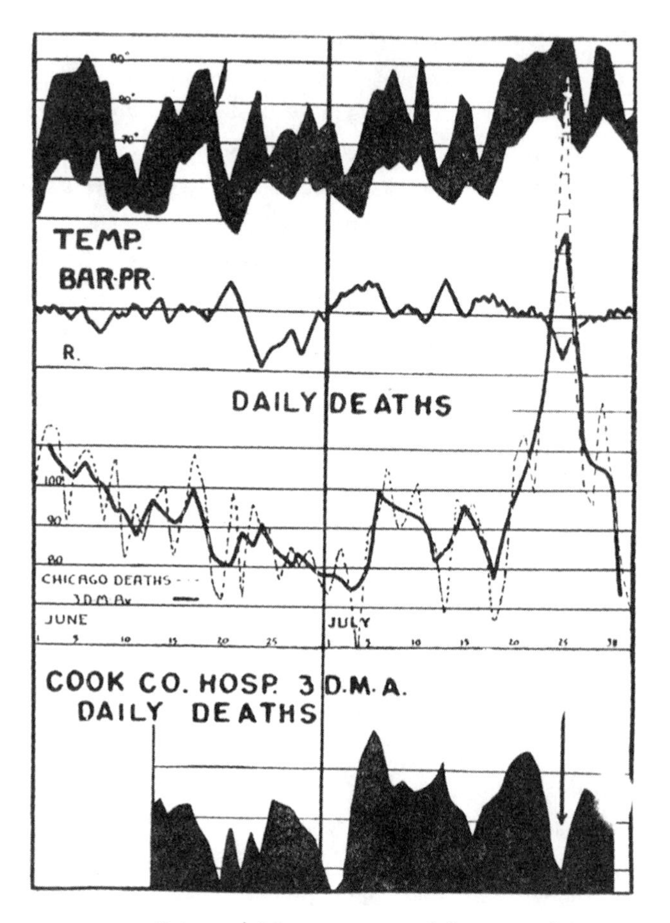

FIGURE 4.3. Peterson's Meteorogram with Interposed Curve of Daily Deaths in Chicago
Source: William F. Peterson, *Man, Weather, Sun* (Springfield Ill.: Charles C. Thomas, 1947)

hospitals, he reported, were directly correlated with environmental temperature, as were deaths from tuberculosis.

Throughout this venture, Petersen was engaged in a project to reenvision the human organism as a biometeorological entity. We have already noted his portrayal of the human body as a cosmic resonator. In *Man–Weather–Sun*, he liberally deployed what he called "meteorograms" to highlight precisely the sorts of association he divined between atmospheric and clinical conditions (see figure 4.3). These rhetorical devices of persuasion gave visual expression to how organic tissue was conditioned by the inorganic environment. It was a comprehensive vision. For it was not just the body that came within the span of climate's imperial reign; mind and emotions were no less governed by its imperatives. Petersen thus devoted chapter 22 of his book to the subject of "Genius"—a theme beloved of eugenicists since the time of Francis Galton. Here, while acknowledging the dominant influence of genetic background, he drew on the writings of figures like the Italian physician and criminologist Cesare Lombroso (1835–1909), as well as Huntington, in support of the claim that there were direct ties between birth month and intellectual accomplishment. Petersen's own feeling was that the season of conception was of even greater significance owing to the plasticity of the embryo's cerebral tissues and thus their susceptibility to climatic stimuli. His own inquiries suggested that the conception of what he called "unusual human types" took place during late winter and spring when meteorological variability was at its height. The effects, moreover, could be generalized. What he referred to as "our western culture, primarily Hellenic in origin," for example, revealed "the interesting phenomenon of the sudden appearance on the world stage of the largest group of unusual individuals that ever appeared in a short time"—a circumstance that coincided with "the longest recorded period of the Aurora Borealis." This association encouraged him to direct his attention to more recent European and American data on birth years and genius using information contained in biographical dictionaries and connecting these with sunspot occurrence. It all confirmed his judgment that there was a significant relationship between mental capacity and the meteorological conditions at the time of an individual's conception. Why did such a relationship exist? Darwinizing climate provided the answer. Prenatal and infant environmental stress arising from climatic conditions introduced a more rigorous "selection" of certain "human types at certain periods." Climate, to Petersen, was thus an active agent of natural selection, and this meant that organic evolution should not be thought of "as bound by the (relatively) immutable laws of genetics" but rather driven by the "far greater organic plasticity arising from the interaction of the environment, the parental organisms, and the egg or the very young embryo."[48] Evolutionary theory provided a vocabulary—variation, selection, adaptation—by which to make sense of the body as a biometeorological phenomenon.

By now, Petersen was operating at quite some distance from clinical medicine. But this was entirely in keeping with the spirit of *Man–Weather–Sun*. As he had already made clear in the preface, the way in which the human body registered organic reactions to atmospheric circumstances was "reflected in cultural cycles, in the waxing and waning of religious and ethical concepts, in social revolution, in mass migrations and in wars."[49] So it is not at all surprising that in the final chapters, he turned his attention to such works as Huntington's *Season of Birth: Its Relation to Human Abilities* and Lombroso's *The Man of Genius* in support of his belief that mental ability was causally linked to climatic variability;[50] to *The Revolutions of Civilisation* by Flinders Petrie (1853–1942), the English Egyptologist, who used meteorological data in archaeology to explain the rise and fall of civilizations;[51] and to the writings of the Russian-German climatologist Wladimir Köppen (1846–1940), who, Petersen observed, "predicted the Russian Revolution of 1917 on the basis of solar disturbance."[52]

This was not a new departure for Petersen. A few years earlier, in 1943, he published a study entitled *Lincoln–Douglas: The Weather as Destiny*. Ostensibly dealing with the lives of Abraham Lincoln and his political opponent Stephen Douglas, its "true subject," as one reviewer put it, was "everyman." The fundamental argument was simple: while "willingly" granting the "importance of hereditary influences and of environmental factors other than the weather," this same reviewer noted, Petersen "contends that the relationship between the human body and the atmosphere in which it functions is of primary importance."[53] And indeed in his preface, Petersen eagerly universalized his method. Because the human body was so profoundly conditioned by the broad array of atmospheric forces, nothing seemed insulated from the long reach of the climate. As he reflected, "These reactions to the rays of the sun, to the cold breath of the north wind and the warm caress of the south, to rain and to drought, to quiet and to turbulence, to spring and to autumn—to these basic effectors we all must defer."[54]

Right from the start, two things stood out. First, Petersen's purpose was to connect in causally compelling ways climatic forces with the life paths of Lincoln and Douglas. And second, he was engaged in a project of literal typecasting. On the first count, Petersen focused on critical moments in the experiences of both figures to uncover their susceptibility to atmospheric circumstances. Thus, the prevailing climatic conditions when each was conceived (in May 1808 and July 1812, respectively), the pattern of storm tracks in central Illinois, the effects of seasonality, and the like, were called upon to explain such things as the "complexity" of Lincoln's personality and the "simplicity" of Douglas's, Lincoln's moodiness and depressive tendencies, and Douglas's easygoing temperament and indifference to philosophy. And so, in recounting the presidential election campaign in 1860, he noted that during the winter months, Lincoln "would talk of suicide," "but when the heat of summer came, then no

FIGURE 4.4. Weather and Body Type
Source: William F. Peterson, *Lincoln–Douglas: The Weather as Destiny* (Springfield, Ill.: Charles C. Thomas, 1943)

campaign could be too strenuous—then 'he came out of the campaign better than ever'—while Douglas was a wreck." The reason? Lincoln "reflected the earth and the atmosphere in every cell and in every fiber, was disturbed when the environment was a bit too strenuous, when an emotional trauma impinged on a cerebral mechanism that was conditioned by an unusual storm, then fatigue and dysfunction followed."[55]

Second, the particulars of Lincoln and Douglas were intended to serve as proxies for far greater generalizations. For example, these two figures were staged as instantiations of more general body types. Lincoln was a case of what he called the "Slender Type," Douglas of the "Broad Type" (see figure 4.4). The former, "tough and wiry" but more likely to be "deficient in muscle and connective tissue," was "distinctly more sensitive to the environment" than the "heavy-boned" and "well-muscled" Broad Type. They thus reacted in different ways to the weather, especially so for those inhabiting the northern latitudes—"those regions of the world that are atmospherically unstable." What was true of Lincoln and Douglas was obviously true of the types they represented. Because in turbulent meteorological environments, "the oxygenation of the tissues" of slender types was "not perfect," gastric ulcers, asthma, nervous disorders, and endocrine dysfunctions were widespread among them. As slender organisms grew older, however, they acquired greater stability and achieved a "better equilibrium with the environment." Broader types followed a different pattern. As he summarized it,

> The broad individual passes from the stage of relative stability (and perfect health) to the stage of relative fixation. Then these types can no longer properly adjust to environmental swings. As they get older we say they 'wear out.' This process of 'wearing out' finds its medical expression in the development of Bright's disease [nephritis], heart disease, diabetes, arthritis, ulcer, manic-depressive insanity, arteriosclerosis, etc. Such types die at an earlier age than the slender individuals who have lived to adult years.[56]

Petersen's meteorological biography of Lincoln and Douglas was self-consciously tethered to Hippocratic moorings. Chapter after chapter was inaugurated with an epigraph from the classical father of medicine. And in Petersen's hands, this Hippocratic impulse had the effect of reducing human agency to the forces of natural law. The "tragedy" of Lincoln and Douglas (and indeed Mary Todd, Lincoln's wife), he mused, was "valid for the rest of us." "All three were Prometheans—who dared; but all had to bow to the laws of the Universe." For "all that we are, think, will, or do," he insisted, "is conditioned by these forces of the environment."[57]

All of this, as I have said, was a reenvisioning of the classical tradition going back to the Hippocratic corpus. From the outset of *Man–Weather–Sun*, Petersen made it clear that his entire outlook on what he called meteoropathology was rooted in earlier ways of thinking. His book provided "no new gospel"; on the contrary, it reverted, he confessed, albeit with "trepidation, to the Stone Age." His conclusions simply substantiated "intuitive folk perception and folk transmitted interpretation of the oneness of the organic and inorganic world."[58] Indeed, Petersen had already been devoting his energies to retooling the Hippocratic corpus for the twentieth century. In 1938, he conceived the idea of producing a collection of Hippocrates' writings translated into the idioms of the modern medical lexicon. Initially, he worked on the project with a physician friend, Edmund Andrews, who could read classical Greek.[59] When Andrews died in 1941, Petersen continued with the task, eventually publishing *Hippocratic Wisdom* in 1946. Petersen was clear from the start just what he hoped to achieve. As he had explained in a 1938 letter to Andrews, he wanted to get out what he called "a *modern* Hippocrates," and by that he meant "changing archaic terms into what Hippocrates actually meant, for instance fire to catabolism water to anabolism, etc."[60]

In Petersen's judgment, Hippocrates' evident appreciation of the influences the meteorological environment exerted on the organic world was of enduring importance, and his declared aim was to bring the text up to date by "substituting certain modern equivalents for words used in a symbolic sense by the Greek physician."[61] Calling on the legitimating authority of the polymathic biological chemist, Lawrence Henderson, who likewise enthused over Hippocratic methodology, Petersen went about the task of translation.[62] But the work was more than a labor of lexical modernization. It was also a kind of self-apologia as Petersen repeatedly recruited Hippocrates for his own cherished theories. Right from the outset, for instance, he valorized the attention Hippocrates devoted to particular medical cases so as to vindicate his own preference for the study of individuals over the increasingly fashionable use of depersonalized statistical data. Similarly, Hippocrates' comment that "the greatest need of a body is air" was harnessed to serve Petersen's interests by being translated into the vocabulary of "anoxia" or what he called "air hunger"—a concept, as we have seen, he assiduously promoted as fundamental to the understanding of disease. As he interjected into the Hippocratic text, "Proper air supply is basic and Hippocrates not only recognized that, but also that local inadequacy would be followed by local symptomatology." Then again, he marshaled the Hippocratic corpus in support of his own thinking about the impact that the passage of cold fronts exerted on human health, epilepsy in particular; Hippocrates' observation on "corruption in the coagulation of the seed" was rendered into "disturbance in the early development of the embryo"; and the Hippocratic reflections on early human diets were made to confirm Petersen's Darwinian-sounding judgment that "Hippocrates considered man as having evolved by the selection of survival types."[63] Symbolic of the whole enterprise was the final chapter of the work. Entitled "The Human as a Cosmic Resonator," it exemplified how Hippocrates was made to ventriloquize Petersen.

In the last analysis, *Hippocratic Wisdom* amounted to little more than a sustained effort on the part of Petersen to insert himself into Hippocrates' text. It was therefore less a work of exposition than an exercise in expropriation. As Sargent noted, "The physiology that Petersen found in the Hippocratic *Corpus* reflected his own knowledge and theories. Petersen could not think in the context of Greek medicine. He was thinking in the context of medicine in the twentieth century and read into the words of Hippocrates his own modern insights."[64]

The comprehensive climatic philosophy that Huntington and Petersen cultivated also found expression in the thinking of another key member of the nascent biometeorological circle, Clarence Alonzo Mills (1891–1974), professor of experimental medicine at the University of Cincinnati and attending physician at the Cincinnati General Hospital.[65] His interest in medical climatology had been stimulated during a visit to China in 1926, where he observed among Chinese diabetics a curious sensitivity to insulin and, among a number of patients, a functional insufficiency of the adrenal glands. As he pondered on such problems, Mills came to ascribe these syndromes to peaks of humidity and heat.[66] This led to a lifetime's work on the effects of climate on the body and resulted in the publication of *Medical Climatology: Climatic and Weather Influences in Health and Disease* in 1939. Here, and in a host of particular case studies, Mills scrutinized the influence of climatic conditions on metabolic, cardiovascular, and infectious diseases; later, he also devoted his attention to matters of air pollution and air conditioning.[67] A more or less random listing of the medical disorders coming within his purview will give a flavor of the scope of his climatic diagnostics: links between climatic conditions and thiamine deficiency;[68] the causal connections between storminess and high blood pressure, resistance to infection, and respiratory problems;[69] atmospheric temperature, sexual cycles, and the onset of puberty;[70] the consumption of caffeine and alcohol in different climatic regimes;[71] mortality rates during periods of extreme heat;[72] and the influence of weather on such specific conditions as diabetes, pernicious anemia, leukemia, tuberculosis, pneumonia, sclerosis, and failures of the vascular system.[73]

Perhaps not surprisingly, Huntington, in the main, liked Mills' effort. In a review for the American Association for the Advancement of Science in 1939, he described it as "a perfect mine of valuable suggestions" and as work of "high value" that "blazes a pioneer trail in its suggestion that the study of climate may be the clue which will lead ultimately to a reduction in metabolic or degenerative diseases."[74] Joseph J. Spengler, the distinguished American economic historian, was also enthusiastic. Writing in *Social Forces*, he was convinced that sociologists would have to revisit their modes of explanation in light of Mills' work, which, to him, demonstrated "at length the importance of the effects of climate and weather upon the fundamentals of human existence."[75] By contrast, the reviewer for the American Medical Association was troubled by Mills' capacious claims and observed, "In spite of the author's statement in the preface that he does not desire to hold climate and weather responsible for all human reactions, the reading of the book demands the constant exercise of the critical faculties to avoid this conclusion."[76]

Like both Huntington and Petersen, Mills increasingly cultivated a more general climatic philosophy of history that he happily deployed for causes dear to the hearts of eugenicists. Perhaps the most prominent manifestation of his climatic historicism surfaced in a report that appeared in March 1941 in Iowa's

Mason City Globe-Gazette. "Dr Mills believes," the article announced, "that the rise to power of Adolf Hitler in Germany and Benito Mussolini in Italy may be due in part to the gradually warming temperatures of the world. People are more docile and easily led in warm weather than in cold, Dr. Mills insists."[77] In Mills' vision, the scope of climate's empire was vast indeed. So comprehensive was its power that he told the readers of his 1942 volume, *Climate Makes the Man*, that "climatic influences . . . affect man's rate of growth, speed of development, resistance to infection, fertility of mind and body and the amount of energy available for thought of action." With climate's dominion extending over such an expansive territory, it is not surprising that Mills should conclude that the human race was nothing less than "a veritable pawn of the universe."[78] A mere sampling of the sphere of political influence over which climate held sway reveals something of the scope Mills attributed to it. In cultural history, he ascribed the triumphs of ancient Chinese civilization, the rise of early Greece, the advent of the "Dark Ages," the French Revolution, the modern imperial expansion of Japan, and the like to the vicissitudes of climate. In terms of economics, Mills connected business cycles with what he called "weather stimulation" and "human energy."[79] So convinced was he of his "energy theory of economic and social trends" and that the weather was "the basic dictator of man's energy and business activity" that he held out the hope that economists would "seek the help and advice of scientists in the fields of meteorology and physiology" in their aim of keeping "business on an even keel."[80] There were political implications too, as the previous five years of remarkably "depressive weather" had shown. The decline in human energy that the weather had induced not only brought economic stagnation to 1930s America but had also "induced a marked decline in the desire for Liberty and a willingness to shift responsibility for personal welfare to the shoulders of the state." Great concern, he assured his readers, "is being exhibited by the more dynamic, intelligent portion of the population toward the growing paternalism of the Washington Administration."[81] It seems that Mills, obviously himself untouched by the hidden hand of climate, could discern in Washington statecraft the effects of dismal weather.

Underlying Mills' climatic faith was the foundational conviction that weather and climate "determine the level of activity or 'pep,' the restlessness or complacency, bodily vigor or sluggishness, progressiveness or contentment with mere existence." Mills believed he had empirically established this claim from laboratory experiments he had carried out on rats—two in particular, Ivan and Hilda—kept in temperature-controlled environments, as well as from observations of the behavior of domestic animals, particularly in the tropics. He was eager to directly apply his findings to the human species. His work on hot-room rats fed on diets deficient in thiamine, for example, apparently demonstrated "the tropical resignation and lack of initiative which keeps large sections of the human race content with a low level of existence."[82] And he found it remarkably easy to inflate the behavior of rats in a hot box into a global geography of what he called climatic drive—a concept akin to Huntington's notion of climatic energy. As Mills explained to readers of his *Medical Climatology*, the most significant aspect of the climatic environment was the way in which it exerted "its major and most direct effect on the ability of the body to produce energy." These effects struck at the "very dynamics of all existence." And nowhere was this more clearly evident than in the fact that "without the vitalizing action of energy and vigor to transform it into ability to do and to accomplish," both inherited and native "intelligence is left sterile."[83]

Mills' fascination with climatological energetics became a dominating preoccupation and shaped his global vision. And so, on a number of occasions, he eagerly embarked on the project of drawing up a hierarchical catalogue of the world's regions arranged in "descending order of climatic energy." The results were both dramatic . . . and dramatically self-serving. First place, happily, went to "the storm belt of North America," in particular the region "around and just to the west of the Great Lakes."[84] This zone possessed "one of the most invigorating climates the earth has to offer, and there human physical development shows clearly the effects of the intense climatic drive."[85] By contrast, his zone of lowest climatic drive incorporated Africa, India, South China, the Malay Peninsula, the East Indies, Philippines, northern Australia, much of South America east of the Andes, and low-lying parts of Mexico and Central America. Here things were pretty grim. "Africa—'darkest Africa'—has little to offer for human development," he announced, while "Asia . . . is likewise damned by climate."[86] Climate had consigned the tropical world to the fringes of human significance. Medically, its depressive heat brought about a lowering of bodily energy with the result that "people there cannot meet physical emergencies with the vigor that temperate zone residents show."[87] Politically, it was impotent to govern itself. As he starkly put it, "Only under the most rigid dictatorship can tropical low-energy nations achieve and maintain today's ideal of an industrialized society."[88]

In portraying the tropical psyche, gloomy verb was piled upon gloomy adjective: "The heat of the tropics lulls people into a passive complacency and saps their vitality," he proclaimed.[89] Passivity, resignation, sluggishness, lassitude, monotony: these were the tropical tropes of Mills' biometeorology. Correspondingly, the temperate world, with its storminess and variability, was home to cyclonic civilizations where the greatest wealth and industry were to be found, as well as "the most advanced procedures in public health and the lowest death rates." Why? Stormy weather delivered to the occupants of the mid-latitude cyclonic zone "a wholesome and stimulating variety of life" that induced a kind of restlessness among its inhabitants that urged them on "to build skyscrapers, set up great factories, and pursue other energetic activities" (see figure 4.5). Because life in the temperate belt was blessed with "little monotony, either climatic, mental, or physical," the state of the population's health was "most buoyant," and life spans were full of interest and stimulation. Here, in short compass, Mills traversed the space between weather and well-being. It was easy to take the next if altogether gigantic step. "The men dwelling amid such influences," he casually remarked, "are the ones who have dominated the world in the past."[90] The consequences were of epic proportions. The global dominion exercised by the imperial powers of the temperate North could simply be reduced to the propelling force of the weather. For the wealth and high living standards that the nations of the North Atlantic enjoyed were nothing but the "long-term dividends of centuries of residence under the world's most driving climates."[91]

The continuities between Mills and Huntington are striking. And indeed, on such subjects as climate change, the cyclical character of climatic pulsations, the use of tree-ring data to chart long-term fluctuations, the role of climate in determining "the rise and fall of human development," patterns of storm pathways, and the mapping of climatic energy, Mills specifically called on Huntington's authority.[92] Even more significant, though, were a number of concepts he derived from Huntington that were central to his eugenic aspirations. His conviction that human excellence only truly flourished in the temperate storm belt mirrored Huntington's thoughts on cyclonic civilizations. "Most stimulating of all for man," Mills declared, "is a cool climate, with wide and frequent storm change in temperature. Under such conditions greatest bodily vigor develops—a vigor that drives people into action."[93] Mills was no less captivated by Huntington's ideas about climatic optima and mobilized them to calculate his own index of climatic stimulation.[94] Like Huntington, he insisted that "man functions best along physical lines when the mean temperature is near 64°F," while "for mental effort a lower temperature (38°F) seems best."[95] Later, during the Cold War, when he kept a close eye on Russia, he would connect this idea with recent work on climate change, not least associated with polar ice melting, to argue that the "rise and fall of nations" were "closely keyed

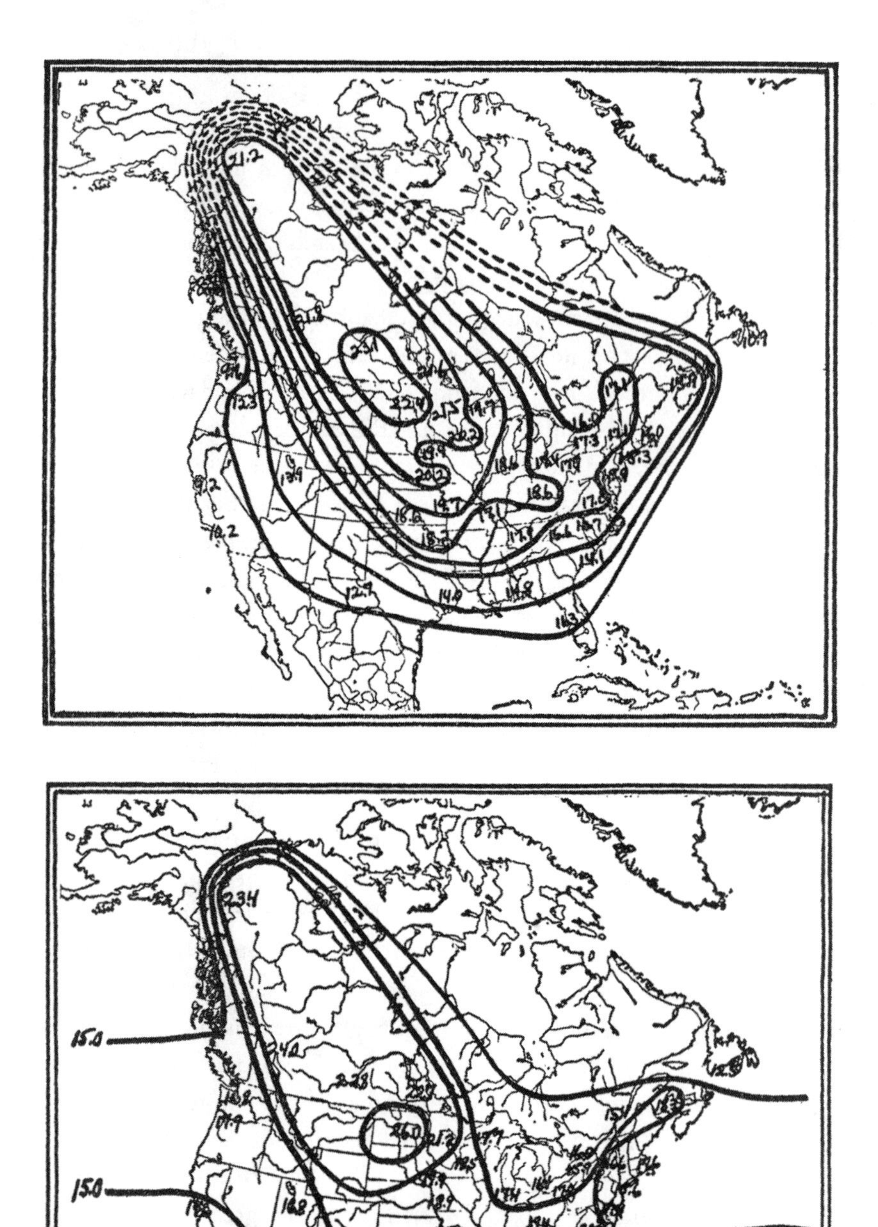

FIGURE 4.5. Mills' Maps of Storminess (top) and Climatic Stimulation (bottom) *Source*: Clarence A. Mills, *Medical Climatology: Climatic and Weather Influences in Health and Disease* (London: Bailliére, Tindall & Cox, 1939)

to major shifts in climatic optima."[96] At the same time, he was entirely persuaded by Huntington's thinking on the significance of the season of conception, urging that "temperatures not only affect the number of progeny, but also their vitality and ability to survive."[97]

These tenets of the biometeorological creed readily translated into two related eugenic fixations: immigration and reproduction. In 1939, Mills warned the users of his *Medical Climatology* textbook that it was impossible to overemphasize the dangers that migrants from "regions of low climatic vigor" faced when they settled in the northern storm belt where "climatic stimulation" was more intense than any other meteorological zone on the face of the earth."[98] A few years earlier, he devoted a whole chapter in *Living with the Weather* to the subject of "Climatic Stimulation and the Immigration Question." The banal clichés of tropical denigration and temperate veneration to which Mills resorted here were underwritten by the presumption that "the energy level and vitality of a people depend greatly on the character of the climate in which they live." This gave him grounds for pronouncing on the indissoluble links between regional climate, immigration restriction, and the need for the eugenic management of national efficiency:

> The day of unrestricted migration of large masses of people should not be permitted to return. With the knowledge we now have of climatic effects, due foresight would prevent a repetition of past errors. Migration from a more to a less stimulating region is not so likely to lead to trouble as is the reverse. . . . With migration from a less to a more stimulating climate, however, it is the new-found energy of the new arrival which causes him to be such a misfit in his surroundings, affecting not only himself but others about him. This idea of climatic control over man offers interesting ground for further study and speculation. If by such study we can in future avoid past blunders in dealing with world problems of migration, any effort expended will be well worth while. Far too long has immigration been considered from the economic aspect,—what it now needs is a concern for its biologic results and possibilities.[99]

In this same volume, when reflecting more specifically on the problems of life in tropical regions with moist summer heat, he encapsulated his immigration stance yet more pointedly: "Hand in hand with the low energy existence goes a lack of inhibition, a free and easy life. . . . On such a basis as this probably lies much of our immigration troubles with peoples from less stimulating lands."[100] In Mills' eyes, declining biopower was the net result of immigration from undesirable regions of the world. And at the same time, he seemed troubled by the thought that people from regions of low climatic vigor would acquire newfound energy when they arrived in a more stimulating climate.

A second prong of Mills' eugenic crusade focused on matters of sex and reproduction. His combination of disconsolate tropicality and procreative seasonality enabled him to offer eugenic advice to colonial settlers. In regions experiencing "real tropical heat," he insisted that there was no such thing as an "optimal period" and therefore that human energy and vitality was consistently low throughout the entire year. As a consequence, prospective parents living in the tropics who desired "to practice the highest type of eugenics and give their children all possible benefits should spend several months in northern cold before conception takes place."[101] Not surprisingly, Mills held out the hope that such advice would meld with new marriage laws compelling young people planning matrimony to undergo a medical examination before a wedding license was issued. The message that there were decided advantages in "cold-weather conception" needed widespread circulation, and he hoped that the "facts" he had accumulated would soon "find expression in high school or college eugenics courses."[102] This concern for greater oversight of reproduction was only one manifestation of Mills' fascination with the biopolitics of climate and sex. In his *Medical Climatology*, for example, he devoted a whole chapter to "Variation in Fertility and Sexual Functions," dwelling on the onset of menarche and sexual maturity and the relations these supposedly bore to the different degrees of climatic stimulation experienced in different regions of the globe. According to Mills, tropical children lagged so far behind those of the stormy temperate zone that the average tropical female of fifteen years of age appeared "no nearer maturity than the 12-year old girl of cool temperate regions."[103] Elsewhere, he used the authority of Hippocrates to counter "the fallacy of early tropical maturity."[104] Such perceptions, of course, had immediate implications for immigration too. In a chapter arrestingly titled "Flaming Youth and Early Aging," he declared that among the countless millions of immigrants who had come to the United States—"this stormy stimulating continent," as he called it—much earlier sexual maturity was evident in their children born in the New World. This climatically determined change in the onset of puberty would create long-term problems on account of the "more volatile nature of sub-tropical natives" compared with "more reserved northerners."[105] That meant that the management of sex was urgent as these newcomers were exposed to a degree of climatic and sexual stimulation for which they were entirely unfitted.

To Mills, research on the season of conception, racial reproduction differentials, and the role of tropical-to-temperate migration in lowering the age of sexual maturity were of massive eugenic significance. And so, calling on the support of Huntington, he concluded his analysis of birth season:

> The days of leaving human reproduction to blind chance in its timing should no longer exist for intelligent couples. Optimal parental health and proper season of conception mean much to the existence of the next generation, and no parent should wish his children in future years to look back and feel that perhaps he did not do all he might have done for their welfare. To appreciate in full measure the importance of these matters, every prospective parent should read Huntington's *Season of Birth: Its Relation to Human Abilities.*[106]

During the mid-twentieth century, Petersen's and Mills' perspective persisted in the endeavors of the International Society of Biometeorology, which came into being in 1956. The Society itself was remarkably heterogeneous, with members representing a wide range of scientific backgrounds—geology, geophysics, botany, meteorology, medicine—and this was reflected in a range of contesting definitions of, and indeed uneasiness about, just what biometeorology actually named.[107] Still, with interest in the influence of meteorology on organic life, a good deal of effort rotated around medical matters and the relationship between atmospheric conditions and human health.

The chief architect behind the new venture was Solco Walle Tromp (1909–1983), a Dutch geologist and geophysicist based at the Biometeorological Research Centre at Leiden. According to David Tout, Tromp "probably did more than any other person to establish biometeorology as a distinct discipline."[108] Besides instigating a series of books dealing with the progress of biometeorology in related fields of endeavor, he also brought out the monumental *Medical Biometeorology: Weather, Climate and the Living Organism* in 1963, which provided an interdisciplinary survey of the entire field, drawing on the expertise of contributors from many disciplines. Later he provided his own synoptic overview in *Biometeorology: The Impact of the Weather and Climate on Humans and Their Environment,* in which he portrayed the new science as "the study of the direct and indirect effects . . . of both the earth's atmosphere and of similar extra-terrestrial environments, on . . . living organisms."[109] Again tracing the enterprise back to the writings of Hippocrates, Tromp reaffirmed the Hippocratic dictum that students of medicine ought first to ascertain the health effects of each season of the year. While he was convinced that nearly two and a half millennia had to elapse before the first properly scientific investigations were undertaken on the effects of the weather and climate on human health and disease, the Hippocratic proclivity for looking toward seasonal influence remained foundational. Accordingly, he devoted much space to identifying what he described as seasonal diseases such as angina tonsillaris (peaking in December–April in Europe), cholera (August in Europe), lobar pneumonia (December–February in the Northern Hemisphere), glaucoma (maximum in November), and so forth. Such "meteorotropic diseases," of course, were not uniformly calendrical. Correlating cardiac arrest with weather conditions, he reported, revealed a large number of cases where sudden heart failure "occurred on days of considerable meteorological disturbance characterized by the passage of active weather fronts, either cold or warm." There were too, as Hippocrates had noted, important ways in which "the developing embryo" was "influenced by the condition of the mother and by the climatic conditions in which she lives."[110] So he sought to identify correspondences between the incidence of certain congenital malformations such as anencephaly (absence of a major portion of brain and skull) and pyloric stenosis (blockage of food at the stomach outlet) and seasonal variation. Tromp's biometeorological conspectus, moreover, did not simply consist in cataloguing climate's human health toll. He also addressed such subjects as the influence of climate on plants and animals, the connections between external weather conditions and the microclimates of buildings, the effects of meteorological stimuli on basic physiological functions, and the therapeutic applications of biometeorological effects—what he called "climatotherapy."

The entire enterprise, it is clear, was grounded in the conviction that the human body comprised a series of meteorotropic zones or centers that registered atmospheric stimuli—the skin, the nose, the lungs, the nervous system, and so on. Of crucial importance was the hypothalamus, a portion of the brain, through which he insisted changes in weather and climate affected the body on account of its thermoregulatory function.[111] The biometeorological body was thus envisioned as a suite of somatic zones, a landscape, responsive to the imperatives of weather and climate.

Given the symbiotic relationship biometeorologists perceived between climatic events and corporeal regions, it is perhaps not surprising that advocates found themselves drawn to ecological modes of thought and to Darwinian adaptationism in particular. Two cases are illustrative. Frederick Sargent II (1920–1980), who took an undergraduate degree in meteorology before studying medicine, was elected first president of the International Society of Biometeorology when it was formed in 1956.[112] According to Wolf Weihe, while Tromp was an enthusiast, Sargent was the scientific expert among the founders. His early work at the Harvard Fatigue Laboratory and the U.S. Army Medical Nutrition Laboratory in Chicago had confirmed his expertise in physiology, and he later held academic positions at the University of Illinois, University of Wisconsin, and the University of Texas School of Public Health. Besides his scientific labors, Sargent produced a lengthy historical treatise—the first history of human biometeorology, according to Helmut Landsberg—under the title *Hippocratic Heritage: A History of Weather and Public Health,* which appeared posthumously. In fact, it was a work of apologetic dedicated to constructing an extended historical tradition for contemporary biometeorological science and was thus presentist through and through. As he himself put it, "Although the first systematic record of meteorological influences on human health was made by Hippocrates some 2,400 years ago, it was not until the present century that this linkage was verified."[113]

If this work displayed the Hippocratic roots of Sargent's biometeorological vision, his most influential scientific publication, the collection of essays he edited in 1974, entitled *Human Ecology,* displayed to the full the ecological cast of his thinking. In his own introductory chapter, he emphasized the reciprocal bonds between "life-process and environment" but was insistent that the emergence of the human species in a tropical environment had so impressed itself on the human constitution that it limited its "capacity to cope with cold environments."[114] Thereby something of the human race's climatic destiny was sealed. Here the critical role of evolutionary adaptationism asserted itself. Elsewhere in the collection, other contributors drew on Dubos' claim that "states of health or disease were the expressions of the success or failure experienced by the organism in its efforts to respond adaptively to environmental challenge" to confirm that "disease" was fundamentally synonymous with "maladjustment."[115] The observation made by these authors that those with diseases were "unfit people" raises the specter of eugenics—an issue that Sargent hinted at when reflecting on "Malthusian limits." "In a world of prejudice," he mused, "the problems of eugenics, except for overwhelming diseases, can scarcely be broached."[116]

The Darwinian motif also featured conspicuously in the work of the influential German-born climatologist Helmut E. Landsberg (1906–1985), who achieved distinction for his pioneering work on the statistical analysis of climate, the biometeorological consequences of urbanization, and the effects of weather on human health and behavior. During his long career, Landsberg held many posts: he worked as an operations analyst with the U.S. Army Air Corps, occupied the position of director of the U.S. Weather Bureau's Office of Climatology, and was appointed to the faculties of the Universities of Chicago and Maryland.[117]

Among his numerous contributions was his 1969 text, *Weather and Health: An Introduction to Biometeorology*. Here he introduced his readers to a range of key topics such as the science of the sun's radiation and its effects on human skin, the impact on the body of air and altitude, and the medical effects of urban temperature, contaminants, air pollution, and sulfur dioxide. Landsberg's familiarity with contemporary meteorological science, moreover, did not prevent him from resorting to the ancient writings of Hippocrates. And so in his reflections on what he called "Weather Suffering," he found space to reprise Hippocratic wisdom on the influence of wet years on "fevers, gangrene, epilepsy, apoplexy, and quinsies," adding that some of these ideas had stood both the test of time and the rigors of scientific scrutiny. The experience of pain, the flareup of rheumatic conditions, the onset of asthma, and the occurrence of heart attacks all displayed seasonal patterns. At the same time, Landsberg's Hippocratic sympathies were suffused with Darwinian sentiments. His reflections on human responses to cold conditions, for instance, were couched in the language of migration and "the race for survival in a cold environment" and were buttressed by Darwin's own observations drawn from the *Beagle* voyage. Similarly, his comments on hot environments and heat stress were grounded in the human biology of evolution in a tropical climate. The close relationship the human species sustained with its climate thus stemmed from the operations of natural selection, which had ensured that early humans "were obviously adapted to an environment in which they could easily survive."[118] This entire conception of the human body as a biometeorological entity was thus rooted in a synthesis of the Hippocratic and Darwinian visions.

What is just as notable, however, is that despite the ease with which certain neo-Hippocratics and ultra-Darwinians alike lapsed into deterministic modes of explanation, Landsberg, on different occasions, issued cautionary warnings about the fashion of casually attributing causal powers to climate. In *Weather and Health*, for example, he warned that students concerned with the interface between climate and well-being were dealing with entities at the conjunction of several very complex systems. As he further explained, "On one side are atmospheric fluctuations; on the other, the complex responses of the human body and the even more complex responses of a large group of individuals. Therefore, many responses are established only in a statistical sense. They apply validly to a large group of persons and yet some individuals in that group may show very divergent responses. All that follows has to be read with this important reservation in mind." More particularly, when reflecting on weather and heart disease, he reminded his readers that influences were both direct and indirect. Research on malaria, for example, had revealed that it was only possible in a very few cases "to trace the weather effects through the whole chain of events" and that in many tropical diseases, such as yellow fever, there was simply no evidence that the influence of climate on the human body had any role to play whatsoever.[119] Indeed, the previous year, Landsberg had issued warnings about the dangers of naive determinism among those tracking human disease, mental excellence, and cultural development to the influence of atmospheric variability. Here he focused on the statistical difficulties in establishing connections between the human subject and weather-related conditions. "Earlier analyses," he elsewhere declared, "simply cannot meet tests that would show statistical relations, let alone cause and effect." In particular, he was skeptical of those "vague associations" that were "supposed to relate climate to the energy of populations, their state of civilization, and their mental productivity." Huntington in particular was guilty not least when he "reduced" these phenomena "to a number by a poll of opinions." Landsberg dismissed these findings as based on "a most dubious procedure."[120]

Taken in the round, though, biometeorology was fundamentally a revisioning of the Hippocratic project in which the human body as a "cosmic resonator," to use Petersen's term, was foregrounded. While construed more or less deterministically and couched in the language of Darwinian adaptation, it attributed significant causal powers to the climate in conditioning human well-being. Championed by a number of university researchers and medical practitioners, it has remained something of a minority interest within the scientific community. Nonetheless, in more popular settings, the concern to identify the interactions between the biosphere and the atmosphere has continued to attract writers fired with a passion to bring its message to a wider public.

Lingering Resonances

Whatever his scientific reputation, Petersen's own construal of human biometeorology has continued to captivate the minds of assiduous weather watchers. The prize-winning essayist Sallie Tisdale, a self-designated "weather-sensitive" soul, is a case in point. Writing in 1995, in an appropriately entitled op-ed "Confessions of a Cosmic Resonator," she paused to assess Petersen's reputational legacy—a legacy marooned somewhere between crank and sage. Her revisiting "the *magnum opus*" of Petersen, "a fellow-traveler" in the world of weather watching, was in the context of pondering on what the "Germans call . . . *Wetterschmerz*, weather pain." Her enthusiasm was palpable even if tempered by the denigration of several scientific professionals whom she consulted. "Petersen was the consummate crackpot scientist, a zealot who was either way ahead of or way behind his times, depending on your point of view," she mused. "He devoted the last 20-some years of his life to a collection of anecdotes, statistical minutiae, and page upon page of graphs and tables analyzing everything from nasal pH to leucocyte counts in relation to changes in the weather."[121] Yet whatever the ambivalence of later judgments, including Tisdale's, during the mid-twentieth century, Petersen's perspective persisted in the endeavors of the International Society of Biometeorology. The links to Petersen are readily discernible not only in the new Society's entire focus but also in the fact that it established "the William F. Petersen Foundation" in 1963 "to commemorate the great biometeorologist."[122]

For a wider public still, biometeorology in the Petersen mode flourished in a lengthy work by Stephen Rosen, MD, entitled *Weathering: How the Atmosphere Conditions Your Body, Your Mind, Your Moods—and Your Health*, which first appeared in 1979. A kind of biometeorology primer for a popular audience, it contained more references to Petersen than to any other source. Rosen, born in 1934, enjoyed careers as a physicist and in career management, publishing on cosmic ray origin theory, career rejuvenation, and, in 2013, a rather curious autobiographical memoir.[123] That same year, 2013, in a piece for the *East Hampton Star*, he revisited his earlier weather passion and announced that "Landmark investigations by William F. Petersen, a pathologist," had "demonstrated that clinical symptoms, mental reactions, and abnormal behavior are conditioned by changing seasons and weather."[124] And indeed, three and a half decades earlier, in his 1979 volume, Rosen devoted much space to elucidating Petersen's triplet study, reviewing his ideas on oxidation, rehearsing his climatic biography of Lincoln and Douglas, and highlighting the significance of his research on weather sensitivity, mental types, and genius.

Right from the first page, it was clear that Rosen was intent, as was Petersen before him, on cultivating a comprehensive meteorological metaphysics. "The atmosphere energizes us," he began. "It has engineered the appearance, the development, the evolution, the racial differentiation of *Homo sapiens*—perhaps even the creation of human genius." Later he declared that the average temperature in any locality conditioned its religious sentiments, its ethical standards, and the size of its standing armies and that climate "not only determines achievement but also anatomy." Given the sprawling cultural terrain over which the climate held sway, it is hardly surprising that Rosen should announce that "human biometeorology *is* a majestic subject." In support of this credo, he called for an alliance "between *med*-men and *met*-men" to counter "weather-induced disorders" and produced a multitude of charts connecting weather conditions with such ailments as angina pectoris, tonsillitis, apoplectic attacks, asthma, and epilepsy. For all his declarations about the need for interpreting cause and effect with great care, he happily announced that "we are inescapably yoked to and enslaved by the atmosphere."[125]

All of this, of course, was biometeorological convention, and it is entirely in keeping with that project that Rosen should approvingly cite the publications of Huntington and Mills, alongside Petersen. Huntington, for example, was still being marshaled in support of a climatic determinist reading of civilization and speculations on the social manifestations of climatic energy. "Ellsworth Huntington," Rosen observed, "has made the persuasive case that climate stimulates achievement. No great civilization, he argued, ever flourished in polar or tropical regions. He mapped the world's climates according to how much energy they induce in man, what he called *climatic efficiency*."[126] In further support of this brand of reductionist geophilosophy, Rosen called on the testimony of the American psychologist David McClelland's *The Achieving Society* to demonstrate that "low-achievement cultures are found in rainy areas and high-achievement cultures in relatively dry climates that also tend to have poorer soils." No less inspirational for Rosen were the writings of Mills. He thus began his reflections on the theme "Tropical or Polar Sex?" with the observation that the "rate of human development, fertility, and sexual motivation depends upon climatic stimulation, according to studies by Clarence Mills," and went on to reprise Mills' assurance that temperature fluctuations, changes in atmospheric pressure, and the prevalence of humid conditions were powerful stimuli for human growth and mental development.[127]

Tellingly, many of these extracts made their appearance in a chapter with the daring, not to say foolhardy, title "The Elegant Science of Racism." While no doubt eschewing scientific racism, Rosen nonetheless warmed to themes dear to the heart of eugenicists and racialist thinkers. Climatic efficiency (what he called "energetic skies") together with the links he divined between climate and mental activity, weather and genius—all these were in keeping with the eugenic spirit of the Huntington–Mills biometeorology project touched with the spirit of Darwinism. So too was his exposition of "Tropical Neurasthenia," that remarkably flexible piece of medical nomenclature naming a range of pathological conditions that were a colonial obsession among those concerned at the purported effects of a tropical climate on the nervous system of imperial travelers, particularly but not exclusively women. We will later examine this purported condition in more detail, but for the meantime, Dane Kennedy's inventory of symptoms that neurasthenia supposedly named is illuminating. It included "fatigue, irritation, loss of concentration, loss of memory, hypochondria, loss of appetite, diarrhoeas and digestive disorders, insomnia, headaches, depression, palpitations, ulcers, alcoholism, anemia, sexual profligacy, sexual debility, premature and prolonged menstruation, insanity, and suicide."[128] A remarkable medical miscellany!

Rosen's own account of neurasthenia built on the diagnosis provided by Charles E. Woodruff, whose 1905 work, *The Effects of Tropical Light on White Men,* will later come under scrutiny. He supplemented this source with the thinking of figures like Frederick Sargent II, who had "enumerated over seventy symptoms and signs" of the condition, and the American chemist-historian John William Draper, who, in 1868, postulated the lawlike generalization that for "every climate there is an answering type of humanity."[129] Rosen also canvassed the testimony of Raymond H. Wheeler, author of *Climate: The Key to Understanding Business Cycles,* which will later also attract our interest, in support of the view that in "cooler climates man is more vigorous, more aggressive, more persistent, stronger physically, larger, braver in battle, healthier, and less prone to sexual indulgence. In warm climates man is more timid, smaller, physically weaker and less courageous but more inclined to physical pleasures, more effeminate, lazier, and less aggressive." Compared with inhabitants of cooler regimes, Rosen's readers were told that the "warmer races" were considered less emotionally stable and less reliable than their colder counterparts.[130]

In Rosen's telling, then, grasping how the atmosphere conditions "your body, your mind, your moods—and your health" was all of a piece with the inclination among the early twentieth-century champions of biometeorology to cultivate a racial economy of climate. That Rosen perpetuated a much longer tradition of medical climatology grounded in environmental reductionism is no less obvious. And while his work in *Weathering* stands as a popular statement of the lure that climatic determinism exerted on some students of health, it also incorporates something of the darker side of climatic thinking that manifested itself in eugenic biometeorology. Moreover, its popularity meant that its influence, one way or another, has persisted. *Weathering* is cited in support of links between meteorological conditions and glaucoma, stroke and gallstone trouble, for instance, in Marsha Baum's 2007 *When Nature Strikes Back.*[131] Nan Moss' distinctly new-agey *Weather Shamanism* turns to Rosen for his statements on the nature of biometeorology and the effects of the passage of cold and warm fronts on blood pressure, alkaline levels, joint ache, and headaches,[132] and Jan DeBlieu, writer and journalist, quotes Rosen on weather-related mortality in her *Wind: How the Flow of Air Has Shaped Life, Myth, and the Land.*[133] Through outlets like these, and in magazine articles penned by writers sympathetic to Petersen's and Rosen's idea of weather sensitivity, neo-Hippocratic biometeorology continues to flourish in popular consciousness.

The conception of the human body as a biometeorological entity has been a persistent refrain over the past century or so. Indeed, if the declarations of the figures whose contributions we have inspected are to be believed, its paternity has an even greater historical reach, stretching back to the earliest days of the Hippocratic tradition. And, to be sure, the impulse to connect the human body in the most profound of ways to the elemental forces of the climate owes much to those Hippocratic moorings. Garnering support from a range of classical writers, the medical philosophy enshrined in *On Airs, Waters and Places* found itself rejuvenated again and again by a range of physicians in the centuries that followed. Sometimes in the ascendancy, sometimes relegated to the fringes of the medical establishment, Hippocrates has been installed in our own time as an icon for a medical community attuned to the health challenges of climate change.

Because of the Hippocratic inclination to evaluate climatic regimes as much as to diagnose patients, the history of medical climatology has been troubled by those determined to divide the world into sickly and sanitary spaces. That was a species of global taxonomy that easily drifted into portrayals of regions of the world in the bipolar language of stagnation or stimulation. This proclivity manifested itself with particular force in colonial contexts where the need to manage human bodies, especially in tropical conditions, presented imperial regimes with seemingly insurmountable problems. Some turned to the science of acclimatization, others to architectural innovations, still others to what I call moral medicine, in the attempt to figure out ways of bringing tropical turmoil under the control of temperate regimens. These ambitions persisted into the twentieth century, when, even in the wake of the germ theory and the development of parasitology, the tropical imaginary that brought both climates and communities under judgment continued to thrive.

Perhaps not surprisingly, the tropes of tropicality readily found a home in the writings of a number of figures intent on constructing a new science of biometeorology. What was conspicuous about that revivification of neo-Hippocratic dogmata was the effort of its advocates to integrate medical climatology with Darwinian evolution. What facilitated this synthesis was the way in which natural selection and adaptation to environment could be fused with a growing awareness of the climatic conditions that prevailed during the period of the emergence of *Homo sapiens*. This Darwinizing impulse facilitated a range of moves in newer directions. Combining ideas of struggle, selection, and fitness with the deliverances of medical meteorology suited those with eugenic sympathies, for the Hippocratic–Darwinian synthesis enabled the identification of those deemed "unfit" for the environment in which they found themselves or from which they came. Careful oversight of the population was needed, they urged, to eliminate the reproduction of the maladapted while encouraging the breeding of more vigorous types. At the same time, those with vested interests in immigration restriction could find in Hippocratic Darwinism a new weapon to wield in support of nativist policies. Again and again, research on the effects of various atmospheric conditions—seasonal temperature, barometric pressure, cyclonic systems, and suchlike—on human health easily spilled over into fixations with such subjects as birth rate, sexual behavior, racial quality, criminality, insanity, and the origins of genius, all of which were the *idées fixes* of eugenicists. Using maps associating climatic conditions with various medical distributions enabled correlation to be taken or, better, mistaken as cause. And this, together with the deployment of ideas like "climatic energy" and "human efficiency," allowed advocates to restage older ideas of racial and regional hierarchy and to pathologize people and places by segregating the globe into elevating or debilitating, nourishing or noxious environments.

Whether in popular consciousness or in the specialist publications of the medical profession, concerns over the health costs of climate change have returned the Hippocratic project to prominence. There is surely little doubt that technical research on biometeorology or that the holistic therapeutics of the Hippocratics over against atomistic conceptions of illness are to be welcomed. But a greater awareness of the sometimes ominous paths into which this whole tradition has deviated should caution enthusiasts about the perils of climatic reductionism. The warning signals manifest in the denigrations of tropicality, the lure of exerting biopower through eugenic management, and the stereotyping of races and spaces ought to be recalled. The need to inform citizens about the health dangers of global warming, greenhouse gases, rising sea levels, and the like is certainly urgent. But for those tempted by political exigency to indulge in rhetorical overstatement or explanatory imperialism, a greater awareness of the troubling ways in which climatic explanations could foster policies that were politically manipulative, socially disempowering, and racially denigrating should serve as a reminder of the ease with which prejudice can all too easily masquerade as objective scientific inquiry.

PART TWO

Mind

5

Climate, Cognition, and Human Evolution

"DID CLIMATE Change Make Us Intelligent?" The BBC Teach website, which hosts a range of educational resources, features a piece addressing that very question in its section on "The Amazing Human Body." Here it reports on the scientific claim that the evolutionary leap in "brain power" between our ape ancestors and *Homo sapiens* was "driven by the impact of climate change." Rapid and violent climate oscillations have occurred frequently in the earth's history, and herein lies the clue for "how apes got smarter." In the accompanying video clip from his 2014 "Human Universe" documentary series for BBC Two, Brian Cox, the English physicist and popular science media personality, described the "explosion" in brain volume between *Australopithecus*, *Homo erectus*, and lastly Omo II. Speaking from East Africa's Great Rift Valley, long host to dramatic climate shifts, Cox observed, "It's thought that it was this rapidly changing environment that drove our transformation from ape to man."[1]

Finding connections between climate, cognition, and human evolution has become increasingly common.[2] Pondering on the processes of hominization, an event routinely associated with massive brain enlargement relative to body size—what is often called encephalization—researchers have turned to the effects of rapid climate change, and in particular to the onset of ice ages, in their search for a compelling explanation. Surveying something of this terrain is my concern in this chapter. Along the way, we will pause to note how students of climate and human evolution often, but not invariably, connected their findings with racial projects intended to map the cognitive capacities of different ethnic groups and to derive a taxonomic hierarchy of racial excellence. Such preoccupations were increasingly disapproved during the middle decades of the twentieth century. But the impulse to find in climatic conditions the cause of human brain development remained and continues to flourish, as indeed has the urge to connect brain expansion with the history of civilizational accomplishment. Placing climate's influence on the human brain and its cognitive powers in the wider context of evolutionary history is no new preoccupation, however. And charting this history, in broad outline, will highlight a number of directions, some more troubling than others, in which the climate-and-evolution narrative has moved. While later repudiated, an early concentration on presumed racial correlates of brain size and cognitive capacity dominated proposals. Subsequently, the fascination with the role of climate in shaping mind and brain was resurrected albeit mostly bereft of racial typecasting. In an era of anxiety about anthropogenic climate change, the implications of the role of climate in early human evolution have taken on contemporary political significance.

Climate, Human Evolution, and Racial Cartography

William Diller Matthew (1871–1930) spent the vast bulk of his career at the American Museum of Natural History and continues to occupy an honored position in the pantheon of historical biogeography.[3] For him, climate was the driver of organic evolution. Drawing on Thomas C. Chamberlin's analysis of the geological consequences of changes from extreme warm, moist tropical climates to intensely cold, arid conditions, Matthew set out to apply these insights to the evolution of land vertebrates.[4] Of critical importance in his scenario was faunal adaptation to prevailing climatic conditions. With this assumption, Matthew expected "to find in the land life adapted to the arid climatic phase a greater activity and higher development of life." At this point, "special adaptations" designed to meet the challenges of "violent changes in temperature" would emerge along with specializations fitted to life in "open grassy plains and desert." By contrast, in "the moist tropical phase of land life, we should expect to find adaptations to abundant food, to relatively sluggish life and to the great expanse of swamp and forest vegetation." The onset of chilly, dry conditions induced outward migrations from northerly landmasses. In these circumstances, the tropical and southern continents became refuges for "the less adaptable and progressive types." In consequence, this phase of climate fluctuation favored "a higher development and greater activity of land life" and "cosmopolitan faunae." Subsequently, "when the climatic pendulum began to reverse its swing, the continents became isolated and their faunae developed independently."[5]

Climate and Evolution was Matthew's most significant scientific contribution, and it played a critical role in bringing the influence of climate within the sweep of the Darwinian empire. Indeed, in May 1915, a year or so after he returned from the Brazilian jungle, Teddy Roosevelt wrote to Matthew as curator of vertebrate paleontology at the American Museum of Natural History, which had sponsored his South American expedition. Roosevelt was "immensely interested" in Matthew's monograph and judged it to be "one of the best pieces of work that has recently been done."[6] Others were just as enthusiastic, and Matthew's theory, along with the cartographic apparatus that sustained it, were mobilized by a number of other contemporary publicists who are routinely taken as paradigmatic advocates of climatic determinism.

To Matthew, climatic change was the principal cause of land vertebrate geography. The migration of organisms over time had depended not on hypothetical land bridges but rather on minor changes in the elevation of landmasses through isostatic readjustments. Periods of land uplift were associated with particular climatic circumstances, notably, a reduction in amounts of carbon dioxide and the advent of arid conditions, which stimulated evolutionary change through migration. Vertebrate relocation took place from a series of distinct centers of dispersal, the vast bulk of which were in the Holarctic ecozone, namely, the habitats of the globe's northern continents.[7] What drove the entire evolutionary system was adaptation to climate.

Central to the communication of Matthew's theory was a sequence of north polar projection maps exposing linkages between the continents. The base chart, on which the others were modeled, showed the configuration of major zoological regions and established global unity when viewed from the north polar perspective. The map turned out to be exceptionally influential not least among a number of early twentieth-century students of human evolution enamored of climatic determinism. For them, it made manifest the intimate connections among evolution, race, and migration and how these were regulated by climate change. As a rhetorical contrivance, the projection unified the earth's surface into a coherently connected whole and gave striking visual expression to Matthew's argument that land vertebrates had radiated outward in successive waves from specific centers. This implied that "the most advanced stages should be nearest the centre of dispersal, the most conservative stages farther from it."[8] For more advanced types, developed in response to stimulating climatic conditions, inevitably forced primitive varieties into more peripheral habitats.[9]

The human species was not exempt from these generalizations: "It is not in Australia that we should look for the ancestry of man, but in Asia," he insisted.

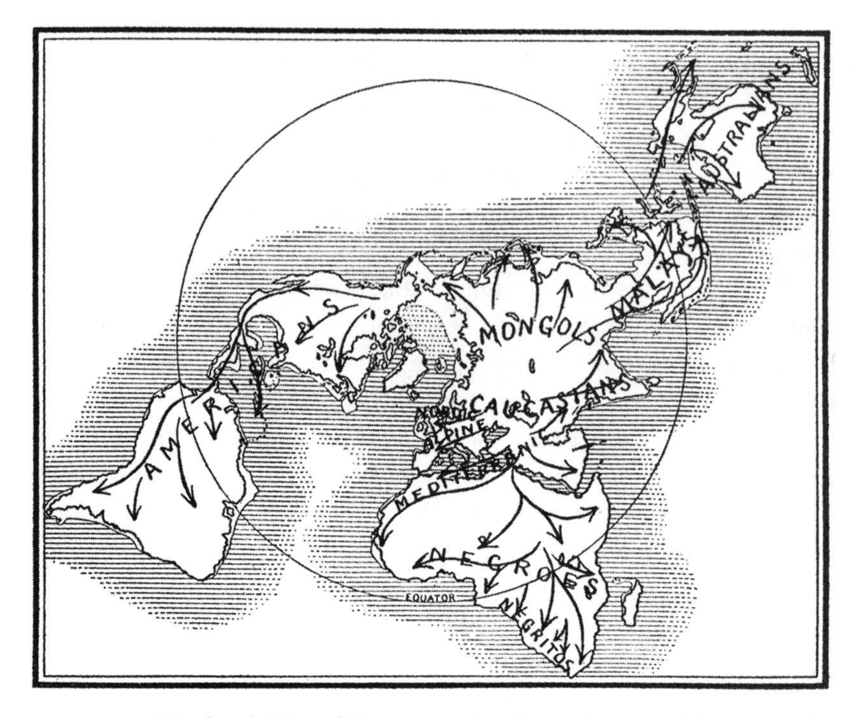

FIGURE 5.1. Matthew's Map of the Dispersal and Distribution of the Principal Races of Man
Source: W.D. Matthew, "Climate and Evolution." *Annals of the New York Academy of Sciences* 24, no. 1 (1914)

Certainly his treatise covered primates, carnivores, rodents, tapirs, horses, and many more, but the "Dispersal and Distribution of the Principal Races of Man" was his very first port of call. And the accompanying map charted "the broader lines of dispersal" of such human groups as "Negritos," "Caucasians," "Mongols," "Negroes," and "Australians"—contested descriptors that still make an appearance in writings on paleoanthropology (see figure 5.1).[10]

Assuming humanity's Asian origin, Matthew used his cartographic craft to depict how the human species had radiated outward from an original hub toward the fringes of the map and thus toward the edges of global space. This meant that the "lowest and most primitive races of men" were to be found in such places as Australia, the more remote regions of south India and Ceylon, the Andaman Islands, central Africa, and northern Brazil. These locations were "so far as practicable travel-routes are concerned," the most distant from humanity's central Asian homeland. Presupposing a rigid racial hierarchy, he reported that "successively higher types from the east" had persistently overrun African regions north of the Sahara. In Matthew's scenario, a long-standing celebration of the temperate zone and a corresponding disparagement of tropical latitudes also reasserted itself: "The higher races of man are adapted to a cool-temperate climate [whereas] in the moist tropical environment, the physique is poor, the death rate is high, it is difficult to work vigorously or continuously." Nor indeed was the tropical world suited to "the lower races," who likewise reached their "highest physical development not in the great equatorial forests" but in drier, cooler environments. Such circumstances confirmed to him that "the center of dispersal of mankind in prehistoric times was central Asia north of the great Himalayan range."[11] Matthew's map gave dramatic expression to this evolutionary epic by consigning the tropical world to the margins and depicting the human races branching outward from a central cradle. It delivered the message that evolution in general and human evolution in particular were intimately connected with climatic conditions, which acted as a forcing agent for adaptation.

The Matthew Cartographic Circle

The narrative power of Matthew's cartographics soon attracted the eye of others concerned with human evolution and race history. And it incited their desires to connect this racial scenario to old cephalic indices and similar anthropometric measurements. No sooner did it appear in 1915 than it was fastened upon by the Australian geographer Griffith Taylor, who found it inspirational and readily turned to it as a tool to disentangle human ancestry. He had read Matthew's monograph that April and scribbled in his diary his shorthand verdict: "Poor animals flee from bad land: Stronger stay and adapt."[12] Just how Matthew's theory of climate-driven evolution could shed light on humanity's racial history gripped Taylor's imagination. What absorbed him was the thought that the most "primitive" human types ended up at the farthest distance from the center of dispersal, having been thrust to the margins by superior groups under pressure from climatic changes and the advent of ice ages in particular. Matthew's maps tantalized him. They were the representational vehicle that allowed him to construct connections between climate change, evolutionary dispersals, and cultural zones.

By 1919, he had put his thoughts on the subject in print. In "Climatic Cycles and Evolution," he urged that climate was a chief factor "in determining evolution" and that a sequence of climatically induced migrations had forced a range of human types out from their point of origin in Asia across the surface of the earth. In this piece, head shape and cranial capacity dominated his analysis. The cephalic index was very largely the foundation on which he erected his racial hierarchy, an emphasis he justified by calling on the work of the Franco-Russian anthropologist Joseph Deniker (1852–1918). And it remained an obsession for the rest of his life. Two forces dominated his race history horizon: migration spurred by climate change and the progressive encephalization of the human species. With each migration cycle, Taylor insisted he could discern the same general trend in evolution: "The head becomes more compact, e.g. spherical, as the brain increases in size. Hence the cephalic index usually increases with the growth of culture." The Chellean Migration, now considered to have taken place during the Lower Paleolithic, brought into being a "race of taller dark dolichocephalic people" during "the first of the great ice ages." The Mousterian Migration, during the Second Ice Age, delivered peoples who had "brow ridges" that were decidedly more prominent than their predecessors, "were much more expert workers in stone and had developed some primitive industries." Subsequent ice ages—cooler and more moist—brought further advancement. That "healthful energy-promoting climate led to a growth of civilization and intellect, which reacted on the cranium" developing a cephalic form "much less oval and more compact." Plainly for Taylor, an increase in brain size and cultural advancement went hand in hand. The more ambiguous Aryan Migration, stimulated by the Riss glaciation, "included the earliest 'Mediterranean' peoples, whose cephalic index was below 78." A sequence of later migrations from Asia, Taylor reported, took place during the fourth interglacial era. With the end of the glacial period, of course, climatic control did not cease to cause migrations. Drought had "caused the later people of Asia to swarm forth from time to time." Nor did climate simply sculpt the shape of global migration. It had also cultivated "the childlike behavior of the negro"; the "versatile, gay, and inventive" white races; and the meditative and melancholic "yellow people."[13]

For this entire venture, Matthew's cartographics proved inspirational, and Taylor resorted to them for his own map of "Man's Precursors in Late Tertiary Times." Explicitly constructed on Matthew's principles, it portrayed humanity's anthropoid predecessors moving along various migration corridors. His own chart of the "Zones of Migration Showing the Evolution of the Races" was modeled on Matthew's iconic image, designated by Taylor as the "polar zenithal equidistant projection." Taylor's strategy was to build his version of the map on "the cephalic indices of the most primitive tribes in each region" in support of a scientific narrative that, he claimed, "clearly indicates, as Matthew suggests, that the biological center is in Asia and that all the races have migrated thence as a result of climatic thrusts."[14] Its positioning in Taylor's lengthy article is also significant, coming as it did after an extended excursus on the migrations of various racial types and what he called the genetic basis of ethnology. By this Taylor intended to depict the various time periods when successive racial migrations—Late and Early Mongolian, Aryan, Hamitic, Negro, and Negrito—had branched off from the Central Asian stem during the previous million years or so. It served to suggest that these different families had evolved phylogenetically over lengthy periods of time. This "crypto-polygenist position," as Nancy Christie calls it, suited his politics perfectly.[15]

In the years to come, Taylor would resort again and again to Matthew-inspired cartography. In his 1934 article for *Ecology* on "The Ecological Basis of Anthropology," he used the self-same projection to convey global patterns of head indices and hair type and later, in 1936, to depict racial migration corridors for the readers of *Human Biology*.[16] The colorful migration map, based on cephalic counts, itself appeared in a slightly modified form as the frontispiece to his 1927 *Environment and Race*—considered by many at the time to be his magnum opus. Here he elaborated in further detail just why the "most satisfactory criterion in ethnology . . . is based on *skull-measurement*," calling on the witness of the American economist William Z. Ripley (1867–1941) and the anthropologist Roland C. Dixon (1875–1934). As he explained, "The chief difference between man and the lower animals is in the development of the reasoning faculties. This is correlated with a growth of the brain." Later, he found it necessary to contest the judgment of the anthropologist Franz Boas (1858–1942), who claimed that head forms were of little value in racial taxonomy and that, in any case, they were subject to hereditary modification in different environments. Predictably, Taylor assigned the black races, with their "low brain capacity," to the bottom rung of the racial ladder; Mediterranean skulls indicated "medium brain-contents"; European and Chinese came out on top with "the largest capacity." In his portrayal of the peoples of Africa, he remarked on their dolichocephalic head form and on their "retreating foreheads." At the same time, he further elaborated on his "migration-zone theory of race evolution," arguing that the most "primitive races" were to be found in regions farthest away from the center of the origin of the human species.[17] Matthew's base map provided a perfect projection template to carry this racialist message (see figure 5.2).

All these self-same fixations likewise recurred in the augmented 1937 version of this text—and in later editions—under the revised title *Environment,*

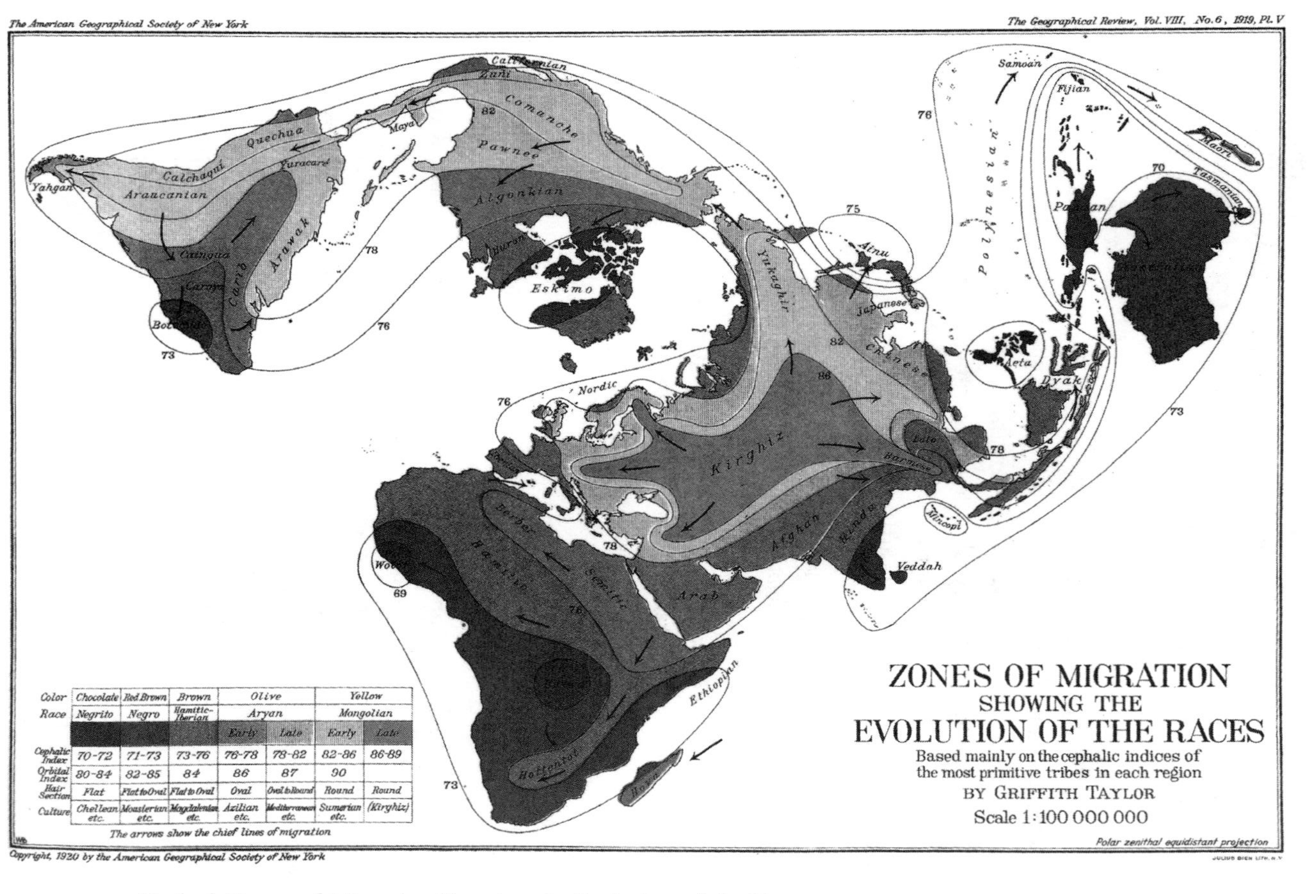

FIGURE 5.2. Taylor's Zones of Migration Showing the Evolution of the Races
Source: Griffith Taylor, "Climate and Evolution." *Geographical Review* 8 (1919): 289–328

Race, and Migration and reemerged in his essay on "Racial Geography" during the 1950s. To Taylor, Matthew's polar projection was far and away the best cartographic device for conveying an immediate sense of what he called the "general world plan," which incorporated the globe's "topographic plan" and "climatic plan."[18] And the critical role Matthew had assigned to climate in shaping the course of evolution was the leitmotif that wound its way through the length and breadth of Taylor's geographical vision of global evolutionary history. To be sure, he affected to depart from what he insisted was the too-rigid determinism of earlier figures like Buckle and Demolins and claimed to advance a scientific version—what he called "modern scientific determinism"—that was empirically testable.[19] Yet for all that, it was still "*variation in the environment*" that he considered "the most potent factor of all in influencing human evolution, whether biological or social"—and more particularly "climatic change" that had "affected man most directly."[20]

Two figures on whom Taylor relied—Ellsworth Huntington and Henry Fairfield Osborn—were similarly drawn to Matthew's visual apparatus and the way in which it could serve their racial ideologies. It was Taylor's reworking of Matthew's map that caught Huntington's eye, and it showed up in his 1924 volume on *The Character of Races*. Though the map was presented as a "rough generalization," Huntington believed it was highly valuable since it presented a graphic illustration of what he claimed was the standard anthropological consensus on human evolution. His turning to Taylor's cartography was entirely understandable, of course, since he supported Matthew's theory of humanity's Asian origins and approvingly cited his observations on the inferiority of the tropical world as well as his conviction that "the higher races of man are adapted to a cool temperate climate."[21] That fitted Huntington's mind-set perfectly. In a typescript on "The Role of Deserts in Evolution," he sided with the view that the human species' remote ancestors would never have "come down from the trees" or adopted an erect posture if their habitats had not become so arid that tree cover largely disappeared to be replaced by grasslands. For that reason, he was sure that human origins should not be sought in tropical zones and that "parts of Central Asia or North Africa that are now deserts are the region where the search for the earliest traces of man is most likely to be successful."[22]

The Character of Races also repeatedly connected the pace of evolution with climatic circumstances: climatic variability was the reason why "evolution is so rapid in the interior of large continents," while those early hominids who migrated into more uniform tropical climates evolved more slowly or simply stagnated. The evolutionary modifications that climate induced, moreover, reached into the interior realm of mind and brain. Climate fluctuations, migration, and the evolution of the human brain were intimately connected. As Huntington observed, "During those half-million years human evolution has been chiefly guided by two sets of facts, namely, the mutations that have occurred in man, especially in his skull and brain, and migration and natural selection under the influence of great pulsations of climate." Along with both Matthew and Taylor, Huntington was convinced that this evolutionary narrative told the story of racial quality and thus, for him, racial inequality. "The form of the head," he explained, "is the most permanent and distinctive of racial traits." This meant that the "most primitive heads" were "long, narrow, and low, with small brain capacity." During the course of evolution, the human head had elongated, increased in height, and expanded in breadth. Climate-induced migration history mapped the course of this encephalization pathway. Brachycephalic skulls were thus the biologically "highest and most specialized" crania since they could house "the largest brain in proportion to its surface and weight." And that, of course, produced a global cartography of intellectual power. Patterns of human "intelligence," Huntington pronounced, were a telling example of the way in which evolution's center of gravity lay in "regions of physical extremes" and was "accelerated by climatic variations." A clear consequence of that principle was that in "equatorial regions the mental type of specialization" had been conspicuously slow simply because there had been "no really great changes throughout man's history, not even during the severest glacial epochs."[23] It was different in the cooler northern reaches. But at the same time, conditions could be too severe. After all, it was the punishing iciness of the Far North that had induced, particularly in women, what he described as Arctic or Siberian hysteria—a psychopathology that continued to afflict large portions of North America's native peoples whose ancestors had crossed the Bering Strait.

Even before Taylor's cartographic rhetoric came within his sights, Huntington had been importing evolutionary theory directly into his climatic credo. In 1919, his tellingly entitled *World-Power and Evolution* had laid out his thinking on the role played by climatic variability in what he dubbed "the voyage of evolution." Each of the great "crises" in evolutionary history—the emergence of the vertebrates, the development of warm-blooded forms of life, and the coming of "man"—that "two-handed, two-footed, big-brained creature"—was associated with profound and sudden climatic changes. In the case of *Homo sapiens*, it was "the Glacial Period which chiefly stimulated man's mental development and caused his intelligence to dominate the earth." The speedy brain enlargement that took place was on account of the fact that it was "the most plastic part of the human organism." Indeed, under the stress of rapid climatic fluctuations during the Ice Age, it developed so quickly that the human species "changed a thousand times faster than the animals had changed during the vast periods of relatively uniform climate in earlier geological times." In summary,

> Man's brain, the most sensitive of all his organs, made by far its most rapid evolution under the stress of great climatic extremes. The greater the climatic changes, the more rapidly new types were evolved not only among plants and animals, but in the human species. Whatever may be the inner causes of the rise and extinction of new forms of life, there can be no question that these have occurred most rapidly at times when the climate of the world swung rapidly from one extreme to another.[24]

Huntington happily connected all of this to racial temperament and to what he regarded as the variability of mental acumen from race to race.[25] In the *Pulse of Progress*, which came out in 1926, he explained how the development of racial character was illuminated by early hominid evolution and directly linked the climatic stresses of the glacial period—"the greatest of all selective factors in hastening the development of the human brain"—with the differentiation of "the relatively inactive races of low latitudes, or of very high latitudes, from the highly active races of middle latitudes."[26] Like Taylor's, Huntington's climatic evolutionism was thus suffused throughout with racial sentiments. So it is not surprising that, as we have seen, he threw his weight behind the eugenics movement and found himself drawn, again and again, to the writings of the "leading scientist," Henry Fairfield Osborn, vertebrate paleontologist, eugenicist, and, from 1908, president of the American Museum of Natural History in New York.[27]

Matthew's cartographic saga depicting the diffusion of human types under pressure from climate changes also found its way into Osborn's writings. That he should be drawn to Matthew's iconic chart is entirely understandable, for Osborn, "an artist at heart," always had a taste for eye-catching design, and as a museum curator, he routinely displayed colorful images, paintings, and sculptures to illustrate the epic of human evolution.[28] Osborn indeed was a convinced, albeit idiosyncratic, evolutionist. He robustly opposed William Jennings Bryan, the prosecutor in the notorious Scopes Monkey Trial,[29] but rejected standard Darwinism, the chromosomal theory of inheritance, and the mutation theory of Hugo de Vries, opting instead for a kind of orthogenetic version that straddled neo-Lamarckism and Weismann-style neo-Darwinism.[30] To Osborn, evolution ran in certain definite pathways and impelled organic change along a number of separate, parallel lines of development.[31] One consequence of this extreme form of polyphyleticism was an inclination toward excessive taxonomic splitting. As for questions of human evolution, he proposed that an ancient humanlike ancestor, a hypothetical "Dawn Man," had arisen in Asia some thirty million years before the appearance of modern humans. This line constituted a completely separate evolutionary lineage from that of the anthropoid apes.

Man Rises to Parnassus—Osborn's infamous 1927 manifesto that had the misfortune to appear after reports about the discovery of Piltdown Man but before the find was exposed as a fraud—laid out the "Dawn Man" hypothesis. And it was here that Matthew's evocative map, with its critical Asian fulcrum point, proved invaluable for it could be used to confirm his own settled conviction that the as-yet-undiscovered Dawn Man "*will be found in the high Asiatic plateau region and not in the forested lowlands of Asia*."[32] As with Matthew, Taylor, and Huntington, Osborn was sure that human evolution took place in challenging climatic regimes. As he explained in a 1926 article titled "Why Central Asia?" human evolution was "arrested or retrogressive in every region where the natural food supply is abundant and accessible without effort." By contrast, "the Dawn Men were evolving in the invigorating atmosphere of the relatively dry uplands."[33] Environmental compulsion is prominently displayed here. Tough climatic conditions were midwife to the birth of humanity. In Osborn's view, the Dawn Men were derived from the more progressive, alert variety that displayed mental superiority by adapting to the new environmental circumstances.

Osborn's paleoanthropology, of course, was just as racially charged as Huntington's. As Ronald Rainger explains, he had long "expressed the views of a typologist and polygenist, one who maintained that human races were in fact separate species and the product of separate lines of evolutionary descent."[34] Indeed, in his 1915 *Men of the Old Stone Age*, he had insisted that the "lines of descent of the races of the Old Stone Age consisted of a number of entirely separate branches."[35] This outlook, for all its evolutionary apparatus, resonated with the pre-Darwinian writings of Joseph Leidy, another advocate of a Central Asian origin for humanity.[36] So it was entirely in keeping with Osborn's frame of mind that no sooner had he taken advantage of Matthew's cartographic device than he embarked on a racial diagnosis that would, within a few pages, issue in a eugenic jeremiad.

Matthew's map, he informed his readers, plainly illustrated the principle "that primitive races of man, as well as primitive races of mammals, are constantly being thrust out from the center of dispersal into the most remote terminal regions of the earth's surface, whereby, viewing the earth from the North Pole, we see a fringe of primitive peoples—Australians, Bushmen, Negritos, Tierra del Fuegians—thrust into peripheral regions as companions of primitive mammals like the monotremes, marsupials, and insectivores." The cartographic power of the map was so enchanting that he presently unveiled one of his own making—a Matthew-inspired chart of the "Dispersal of Mankind from Central Asia"—to further advance the political imperatives flowing from the cartography of human origins. He immediately plunged into racial diagnostics. The study of racial origins, he declared, will become "a matter of political importance, a matter to be taken into consideration by the State. Indeed it is already being considered in this way in the United States as we begin to realize that different races respond very differently to our political institutions."[37]

In opposition to the French zoologist and anthropologist Armand de Quatrefages, Osborn denied the specific unity of the human race and elaborated on what he called the three "Primary Human Stocks, or Super-Races"—namely, "the Caucasian, the Mongolian, and the Negroid." In zoology, he claimed, each "would be given the rank of species, if not genera." Their respective "spiritual, intellectual, moral, and physical characteristics," Osborn insisted, were "very profound, and ancient" since they had "diverged from each other during the Age of Mammals, even before the beginning of the Pleistocene or Ice Age." In his opinion, the "Negroid stock is even more ancient than the Caucasian and Mongolian, as may be proved by an examination not only of the brain, of the hair, of the bodily characters . . . but also of the instincts and the intelligence."[38] Climate-driven evolution had sculpted the brain, body, and psyche of the different races. All of these scientific-sounding claims were intended to cement in his readers' minds the critical significance of evolutionary history for contemporary race politics. This was only to be expected from a writer who had provided the preface to Madison Grant's *The Passing of the Great Race* (1916), as well as various pieces on immigration, and who lauded the writing of history along racial lines.[39] To him, the "moral tendency of the hereditary interpretation of history . . . is in strong accord with the true spirit of the modern eugenics movement in relation to patriotism, namely, the conservation and multiplication for our country of the best spiritual, moral, intellectual and physical forces of heredity."[40] Osborn's eugenic zeal was not simply a personal predilection. In 1921, he hosted the Second International Congress of Eugenics at the American Museum of Natural History; the exhibits in his Hall of the Age of Man were suffused with eugenic themes, and he himself "proselytized for immigration restriction," or, in more Darwinian-like idiom, "immigration selection," as he dubbed it.[41]

With Osborn, as with Matthew, Taylor, and Huntington, questions of human origins were intensely political. To them, the project to excavate the paleogeography of human evolution, guided by the imperatives of climatic change, was an inherently racial undertaking, and the north polar map projection served as a critical accomplice in their mission. Together, they constituted a cartographic quadrilateral united in their appeal to climate variability as scripting the master narrative of humanity's evolutionary story and of the evolution of cranial capacity in particular. The possibilities for using the science of climate and human evolution as a vehicle to carry racial cargo offered wide appeal in the early years of the twentieth century. Indeed, climatic zones were readily adopted as a proxy for racial hierarchy at the time. Those with eugenic tastes could certainly find here scientific-sounding idioms to advance their crusade. And it crystallized in other arenas too. In an examination of the way in which climate was mobilized as a racial category, Ashwini Tambe remarks on the problematic tensions inherent in the ways in which liberalism—exemplified in the League of Nations—could espouse universal political equality even while naturalizing hierarchies between nations to justify various forms of tutelage.[42] As maneuvers of this stripe demonstrate, climatic readings of human mental evolution were often freighted with the politics of race representation.

Crania, Climate, and Cleverness: Environmental Determinism Reborn

In the decades that followed, climatic determinism fell into disrepute. Critics insisted that its crude fatalism rode roughshod over the role of human agency as a historical force. Commentators also identified what they considered empirical evidence refuting extravagantly determinist claims, even if they were willing to concede that climate and other environmental features might act as constraints on human action.[43] During and immediately after World War II, the crudely racial and imperialist cast of much of this tradition also came under fire from a variety of perspectives that served to cast doubts on the climatic determinists' entire credo. Franz Boas, for example, who turned from his geographical heritage to cultural anthropology, came to believe that culture and human behavior had to be understood sui generis rather than as an epiphenomenon of environmental conditions. By midcentury, the intellectual mood had shifted decisively in his favor. So when Stephen Schneider and Randi Londer brought out their wide-ranging volume, *The Coevolution of Climate and Life*, in 1984, they remarked that while "the generally benign climate that replaced the ice age coincided with the rise of civilizations," they had no intention of suggesting that "climate was the only—or even the dominant—factor" in the advent of Holocene civilization. Indeed, they allocated space to the importance of "nonclimatic factors" in explaining historical events and consciously distanced themselves from the "simplistic" climatic determinism and scientific racism of the likes of Huntington.[44]

This does not mean, of course, that connecting human brain size and head form with climatic conditions disappeared from the scientific horizon. In 1972, Kenneth Beals at Oregon State University addressed the question of head form and climatic stress, recalling the earlier insistence of Carleton Coon, whose controversial racial theories were adopted by segregationists and white supremacists, that head size and shape had adaptive significance for the regulation of body temperature.[45] Focusing on matters of heat exchange, Beals proposed that "under conditions of cold stress, the most advantageous head shape would be rounded since this most closely approximates the spherical ideal." Occupying a cold climate was thus a circumstance that increased the frequency of brachycephaly (short, broad-headedness), a proposal that correlated with his map of zones of climatic stress. By the same token, Beals insisted that adaptation to the thermal atmosphere was not the only factor determining the human head form. Nonetheless, because long-headedness was suited to populations living in hot climates and represented the "retention of a primitive characteristic," the evolutionary trend toward "brachycephalization" was closely connected with human exploitation of the resources, and adaptation to the rigors, of the globe's colder climates.[46]

In a subsequent investigation of broad-headedness a decade or so later, climate again prominently featured.[47] Still dwelling on the thermoregulatory significance of cranial architecture,[48] disavowing earlier racial associations, and building on a large computerized data set, Beals and his coworkers continued to push the claim that, of all the variables in question, climatic zone enjoyed the highest correlation with the cephalic index, even if they conceded that climate could not be the only explanation and might "not even be the most important."[49] Still, because it limited heat loss, and thus death by exposure, brachycephaly delivered an adaptive trait in cold environments on which natural selection could exercise its powers. All of this meant that their bioclimatic model of head-form evolution predicted that cranial capacity would "increase with distance from the equator—latitude being correlated with a decrease in solar radiation." Their empirical findings, they reported, supported their inference from conditions of solar energy that "brain container volume and latitude are highly correlated." What confirmed to them the rightness of this finding was their judgment that "latitude associations are supported by the culture history of each continental area."[50]

This body of work elicited responses from a range of scientists from different disciplines—among them anthropology, anatomy, ecology, and cell biology. Most were supportive. But one, Kathleen Gibson, of the University of Texas McGovern Medical School, expressed the view that the project dressed up "old-fashioned physical anthropology with new-fashioned computer techniques" and that the authors made the mistake of assuming that "statistical correlation implies causation via natural selection." Nor was she persuaded that brain enlargement took place in order to conserve body energy. Surely, she argued, there were more metabolically effective ways of conserving heat such as "the evolution of insulating layers of hair, fat or clothing." For herself, she preferred to consider brain expansion to be the consequence of neural functions that delivered selective advantage.[51] Around the same time, an extensive review of proposals advanced to explain the enlargement and reorganization of the hominid brain canvassed a wide range of candidates, including Pliocene and Pleistocene climate change, but concluded that none of them was entirely persuasive.[52]

Despite the qualifications expressed by Beals and his coworkers, as well as reservations of the kind just mentioned, the final decades of the twentieth century bore witness to an intensification of interest in causally connecting climate, evolution, and the development of the human brain—a trend evident both in specialist literature and in works intended for consumption by the general public. Pronounced dead at midcentury, environmental determinism of one stripe or another had undergone a resurrection by century's end.[53] Concerns about global warming, and then global cooling and the possible onset of another ice age, had been growing in public consciousness during the middle decades of the twentieth century. These anxieties were dramatically heightened during the 1980s, when rising concentrations of carbon dioxide and other greenhouse gases made the headlines.[54] As the century came to an end, climate change was more firmly on the agenda than ever before. And disciplines like paleoanthropology did not remain untouched—certainly if more recent pronouncements are anything to go by. What, at least in part, facilitated this resurgence of climatic explanations was the rapid accumulation of fresh paleoenvironmental data.[55] By the mid-1980s, new techniques had produced a dated isotopic record of changing environmental conditions from marine sediments, providing a chronology for late Cenozoic climate change and supporting the view that such changes could take place very rapidly. These findings enabled researchers to use remarkable and increasing amounts of information to begin searching for connections between climate-induced environmental change and key human-evolutionary events. The possibilities for moving casually between correlation and causation were rich indeed.

By 1990, scholarship on the whole subject was sufficiently extensive for the science writers John and Mary Gribbin to bring out their *Children of the Ice.* Subtitled *Climate and Human Origins,* it sought to trace the deep history of human-type traits back to the end times of the dinosaurs. Of critical importance was the shift in primate behavior from living in the trees to dwelling on terra firma. Climate change was the driver of these developments. The "cooling and drying of the globe in the Oligocene" brought about conditions that the Gribbins regarded "as ideal, an Eden-like paradise on earth." For it was "the invention of winter that made Earth an Eden for our family." As the forests retreated, they went on, some apes further developed apelike features for arboreal living, while others evolved traits that enabled them to thrive on the more open grasslands. Eventually, the latter "developed large brains, and became human." This transition took place with remarkable speed during a "breathtakingly fast spurt of evolutionary change" stimulated by responses to rapid climate fluctuations. Of critical importance was adaptability to the new ecological circumstances. The authors laid out the story with an air of teleological compulsion. It went something like this. During ice ages, many die. Some respond to the challenging climate by becoming better forest apes. On the plains, "only the most intelligent and adaptable" apes survived. During interglacial times, their descendants fanned out and flourished because they had been "selected for adaptation and cunning." In sum, as they put it, "Our ancestors survived, and we are here today, because of the unusual pattern of climatic change they experienced. We are, indeed, children of the ice." The advent of the big-brained *Homo sapiens* was the outcome of these "climatic rhythms."[56]

Since then, experts have further extended the scope of climate's imperial rule over humanity's evolutionary past. Richard Potts of the Smithsonian National Museum of Natural History, for example, could announce in 1998 that the "origin of bipedality, stone toolmaking, increased brain size, other adaptive trends in *Homo* and key turnover events" were all "linked specifically to cooling, drying, and savannah expansion."[57] Taken as a touchstone, something of the range of these purported impacts can be gleaned from a major collection of essays that appeared in 1995 under the title *Paleoclimate and Evolution*. Because the "key concern" of the volume was "to identify external forces that drive evolution," the climate record was fastened upon as the most likely suspect.[58] The editors' aim was to bring together findings from two disparate traditions of inquiry dealing respectively with proxy climate records derived from deep-sea sediments and patterns of evolutionary change.

At the more generalized level, climate fluctuations were called upon as the critical determinant of evolutionary pathways and the fragmentation of species into isolated populations.[59] In these accounts, climatic variation crystallized as the explanation for features such as changes in teeth—an adaptive response to dietary modifications—and associated alterations in facial and cranial architecture.[60] A falling predictability in climatic patterns and food supply in certain ecosystems was put forward as the reason for the emergence of more plastic, less rigid behaviors that conferred selective advantage. And the manufacture of stone tools was interpreted as a strategy to secure greater reserves of fat during prolonged dry seasons.[61] The changing character of habitats in direct response to climatic triggers, the paleontologists argued, confronted early hominids with the possibility of pursuing what has been called "a generalist strategy" that facilitated their survival in a range of diverse climate regimes compared with the much more restrictive response of australopithecines who became specialist frugivores.[62]

The deterministic tenor of claims like these was plain. For much of the empirical work recorded in this expansive volume engaged with what Kimbel called the "strong environmentalist argument" of Elisabeth Vrba—one of the collection's editors and a prominent Yale paleontologist.[63] And indeed, the necessitarian cast of her vision had long been conspicuous. In 1985, for example, she had announced that "environmental changes *caused* the evolutionary changes in hominids that are observed in eastern and southern Africa" and added that "widespread change in temperature, rainfall, and vegetation cover *caused* the evolution of the hominid phenotypes currently included in the species *Homo habilis, Australopithecus robustus,* and *A. Boisei*." Among these climate-induced effects were modifications to musculature and changes in social behavior that provided selective advantage for "life in more vegetationally open and arid environments." All of this was part and parcel of "a wave of evolutionary activity that was forced upon the biota by a common environmental cause."[64]

Subsequently, Vrba extended her project in additional directions, notably with regard to hominid brain size and accompanying behavioral traits.[65] Evolutionary changes in the onset of particular structural developments, she suggested, were climatically initiated, especially the preservation for longer periods of ancestral juvenile features in adult descendants. Most critically, Vrba noted, human brain growth is prolonged after birth for much longer periods, and this progressive encephalization had dramatic implications.[66] Vrba was convinced, moreover, that the "the climate-forcing hypothesis" could be extended further. Believing that "paleoclimate changes could trigger hominid behavioral evolution," she maintained that episodes "of global cooling have not only been a major causal influence in human origins, but also have been instrumental in the evolution of human culture and social behavior." Because cold temperatures "had a strong causal influence on progressive neotenization"—the prolongation of earlier juvenile features—traits like investigative curiosity, pair-bonding, improved bipedality, monogamy, and the origin of the nuclear family were all presented as the products of climatic control.[67] In progressive cooling too, she found the genesis of changed feeding habits, innovative tools for digging out below-ground foods, and less specialized dietary behaviors. All in all, the causal scope of climate's evolutionary domain was extensive indeed.

The correlation between hominization and cooling climatic conditions, of course, left open the issue of precisely what mechanisms might have been operative in delivering the big-brained human. No doubt many candidates have been canvassed. But three in particular seem conspicuous: the move from tree-dwelling to life on the plains, adaptability to a diverse range of climatic conditions, and the significance of thermoregulatory mechanisms in human physiology. To advocates of these climatic causes we now turn.

Down to Earth

In 1996, award-winning paleontologist and evolutionary biologist Steven M. Stanley, well known for his contributions to the theory of punctuated equilibrium, brought out an interdisciplinary account of the emergence of the human species under the title *Children of the Ice Age*. It was lauded on the back cover by his lifelong colleague, Stephen Jay Gould, as a "powerful and compelling hypothesis for the most crucial step in human evolution—our descent from the trees to life on the ground." Coming down to earth from the trees on account of changed climatic conditions, in Stanley's scenario, turned out to be of critical importance for the development of the human brain.

It is a complex narrative with an intricate suite of interrelated events and operations. The pattern of brain growth among humans, Stanley explained, "is unique among primates." Compared with lower primates, the human species retains an extremely high rate of brain growth both prior to, and during, the first year after birth. "During this brief period an average infant adds slightly more tissue to its brain than it will add throughout the remainder of its life." This meant that the human infant is typically "top-heavy" by the age of one. While there is no direct connection among humans between brain size per se and intellectual aptitude, Stanley was sure that on average, "big brains contain more neurons and more connections between them than do small brains." For Stanley, this meant that any species that possessed a much larger brain for its body size than others would "be a quantum leap ahead in brain-power."[68] All of this posed the question, what were the conditions necessary for the evolution of such a big-brained creature to emerge? Stanley's answer was multifaceted.

First, there was what Stanley called "the terrestrial imperative." Because of the time requirement needed for newly born offspring to develop the brainpower that distinguishes humans from other creatures, a much longer period of maturation was a necessary prior condition for encephalization. The "highly immature offspring" that uniquely characterize the human species, he affirmed, was "a necessary concomitant of the way in which we grow our large brain." During the first year after birth, humans "move far beyond apes in brain size." Only when tree-dwelling apes came down to earth and dwelt on terra firma could prolonged infancy take hold. Before the momentous transition to *Homo sapiens* could occur, Stanley affirmed, "our ancestors had to have abandoned the simian habit of regularly climbing trees." "The essence of this restriction, which I call the *terrestrial imperative*," he continued, "is that before large brains could evolve the transitional animals needed to have their hands free to carry and tend helpless infants." The forelimbs had to be freed from the obligations of climbing in order to leave them available for the tasks of parenting feeble infants. Long immaturity, terrestrial living, and the massive amplification of the human brain were intimately interrelated. As Stanley summarized it: "First came life on the ground and, much later, the large brain."[69]

A further requirement for this scenario to unfold was some agent introducing dramatic landscape change, an environmental-forcing driver that fostered down-to-earth living. The answer was ready to hand. It was the Ice Age, which "began with what amounted to a flip of the climatic switch" that transformed the African landscape by bringing about an extraordinary shrinkage of its previously dense forest cover.[70] A novel feature of Stanley's storyline was that this dramatic change in global climate could be traced to movements in the crust of the earth that formed the isthmus of Panama two and a half million years ago. This narrow neck of land suddenly became a barrier between the Atlantic and Pacific, preventing the flow of water between the oceans, thereby increasing the salinity of the Atlantic. These changed conditions triggered actions that resulted in the Ice Age.

A critical consequence of these events for human evolution was the fate of the Australopithecines, a transitional hominin between tree-inhabiting apes and ground-dwelling humans. "The wrenching contraction of the woodlands" that the changed climate stimulated and that had long offered the Australopithecines "safe haven above the ground" now delivered "an entirely new opportunity to any population of *Australopithecus* that straggled into the Ice Age—the chance to evolve a much larger brain." The inference for Stanley was that because of the way human brains develop, "a population of *Australopithecus* could only have evolved such a structure after abandoning the habit of climbing trees every day." The changes that the Ice Age brought about in Africa, the continent to which *Australopithecus* was confined, "were exactly the ones needed to close the door on this apelike creature and to open the way for the genesis of *Homo*." It was thus "the onset of the modern Ice Age" that brought about the transition from *Australopithecus* to *Homo* by rapidly transforming the African landscape. The irony was plain. The changed environment that wiped out the previously successful *Australopithecus* removed the very barrier that had hindered this creature from developing a much larger brain. In these conditions, "one small surviving group soon emerged from the environmental crisis as an entirely new, big-brained form that had the wherewithal to make its way in the Ice Age world of Africa: this was early *Homo*." While nothing short of catastrophic for our ancestors, the Ice Age was exactly what allowed our own species to emerge. A case of *felix culpa* for us, disaster for them. As Stanley put it, "The idea of a catastrophic origin for *Homo* carries with it a sobering implication: global change is not all bad." While the human race now lives in fear of the ecological consequences of global warming, he pointed out, it owed its "very existence to a profound, climatically driven transformation of the earth's ecosystems—one that wiped out many members of the human family but that allowed at least a few to survive by what amounted to an evolutionary rebirth."[71]

Besides facilitating the physiological conditions that permitted encephalization to take place, what was it about life on the plains that made superior mental agility such an adaptive trait? Stanley's suggestion was that while language acquisition was undoubtedly important, yet more significant was the advantage it gave in surviving the brutal hardships and frequent carnivore attacks that must have decimated the population. Those with greater intellectual acuity enjoyed marked selective advantage. "I suggest," he remarked, "that the need for self-defense while living freely on the ground was the primary driving force behind the natural selection that created the large brain of *Homo*." And it was from this starting point that *Homo sapiens* acquired the capacity to construct language "using its newly expanded frontal lobes."[72]

Piecing these component parts together, a number of diagnostic features of *Homo*, what Stanley called the "adaptive complex," came into clear view. Among these were the presence of an extremely large brain with a greatly enlarged prefrontal cortex, a much-delayed development postbirth compared with the apes, and a set of anatomical adaptations selected for terrestrial life. Attributing these developments to the effects of an abrupt cooling of the climate meant that early *Homo*'s large brain should no longer be thought of as the product of an extended evolutionary trend but rather the result of a geologically abrupt event. Moreover, the large human brain emerged not because it became *useful* for the first time but because the environmental setting made it *possible* for the first time. In sum, the big-brained human species was the outcome of a dramatic climatic crisis: an Ice Age that literally brought us down to earth. To climb up the tree of knowledge, it seems, we had to climb down from the trees.

A Mind for All Seasons

In January 1998, *The Atlantic Monthly* ran as its cover story a piece by the neurophysiologist William H. Calvin, well known as a popularizer of what is called neural Darwinism—a synthesis of evolutionary biology and neuroscience. On this occasion, his subject was "The Great Climate Flip-Flop." Here he reported on one of the "most shocking" scientific revelations "of all time"—namely, that the "earth's climate does great flip-flops every few thousand years, and with breathtaking speed." His argument was that within a decade or so, the modern world could revert to ice-age temperatures—a sudden chilling that, paradoxically, could be triggered by global warming. That would be a culturally devastating turn of events, he urged, as there would be no time for agricultural adaptations to be made. The article attracted a good deal of attention, and not surprisingly. For Calvin noted a remarkable concurrence between the time when the "back and forth of the ice started 2.5 million years ago" and the period when the "ape-sized hominid brain began to develop into a fully human one"—an organ "reorganized" for using language, cultivating music, and making inferences. Civilizations, he further noted, began to develop immediately after the continental ice sheets melted around 11,000 years ago. With an eye to the future, he warned that an abrupt cooling phase could be triggered by global warming with such "breath-taking speed" that adjustments to production and supply simply could not be made in time to avert the "civilization-shattering" consequences of the climatic "whiplash." With determinative certainty, he declared, "The effects of an abrupt cold last for centuries. They might not be the end of Homo sapiens . . . but the world after such a population crash would certainly be full of despotic governments that hated their neighbors because of recent atrocities."[73]

In many ways, Calvin's article was something of a promissory note. For the piece came to constitute several chapters of the book he subsequently published in 2002 under the title *A Brain for All Seasons*. As its subtitle indicated, Calvin wanted to place the development of the human brain in the twin contexts of "human evolution and abrupt climate change." Staged as an e-seminar on human evolution, the book presented itself as a kind of scientific travelogue—a series of communiqués issued by the author from various locations around the world on the relationships among rapid climate change, the human brain, and hominid evolution. Climate flip-flops featured as the key driver of evolutionary transformism. Humanity's ancestors, Calvin explained, lived through hundreds of such episodes, each jolt eliminating the vast bulk of the population. Those who made it through were the ones able to adjust to the fast-changing climatic conditions. To survive was to improvise. They had to amend their diets, alter their methods of hunting, modify their language, invent new tools, and much else besides. Those who responded to the challenges with such adaptations were picked out by natural selection for survival, and those traits spurred further evolutionary change. In language reminiscent of Daniel Dennett, the see-sawing climate acted as a pumping device inducing physiological and cultural transformation.[74] As Calvin himself put it, the Phoenix-like "paleoclimate pump is the longest-running rags-to-riches play in humanity's history."[75]

As for the human brain, even while admitting that brain size per se is typically a function of body size and not any simple indicator of higher cognitive abilities, Calvin nonetheless remained impressed with the coincidence between the advent of the ice ages, the spread of tool-making skills, and the progressive enlargement and reorganization of the human brain. What linked encephalization and the restructuring of the neurological machinery that he deemed fundamental to human evolution was the simple thought that "brain reorganization could proceed more easily in the bigger-brained variants, just because of more room to manoeuvre—but that size per se wasn't the name of the game." It was not average temperature or the extent of ice coverage that directly spurred the enlargement of the brain. It was rather that a raft of whip-lash climate changes gave selective advantage to those hominids who had developed the cognitive repertoire to adaptively respond to fast-changing environments. During those climatic oscillations, it might be said, they acquired flexible brains—brains "for all seasons." And that feature would, courtesy of natural selection, rapidly develop from generation to generation. Again, as with Dennett, Calvin described this process as a bootstrapping operation. "Nature can be seen to pull itself up by its own bootstraps," he declared, and that process was what delivered to the human species everything from concealed ovulation, tool-making skills, and linguistic complexity to logical inference, ethics, and music.[76] Add to that the influence of climate change on the prolongation of infancy, differential reproductive rates, early maturation, tooth size reduction, and much else besides. An expansive empire for sure.

Calvin's whole story was rooted in Darwin's theory of evolution, and he happily pointed out that Darwin himself had acknowledged the role of changing climate in the evolution of species. Certainly he felt that Darwin had subscribed a little too fully to gradualism while he himself was attracted to a form of catastrophism, insisting that climate "catastrophes are often mixed up with evolutionary jumps" and that "the need to discover a new way of making a living within a single generation" showed "how jack-of-all-trades variants could survive better."[77] But, for all that, his entire project remained firmly grounded in natural selection. Indeed, Calvin had long been invested in what is best described as climate-honed evolution as his 1991 book, *The Ascent of Mind*, amply revealed. Many of the arguments in *A Brain for all Seasons* were prefigured in that volume even though its title foregrounded climate and the evolution of intelligence rather than the anatomical enlargement of the brain itself. The first sentence of the first chapter summarized the book's message: "Matching wits with the fickle climate is how we became human." The Darwin Machine, as he called it, had gifted humanity with "a brain that can function in various different climates" and that gave enormous survival advantage over minds that were merely "efficient in a single climate." Climate's evolutionary largesse was extensive. In the transition from ape to ex-ape (to use Jon Marks' felicitous expression[78]), climate changes fostered "some beyond-the-apes abilities that we value most highly: a versatile language and a plan-ahead consciousness that enables us to feel dismay when seeing a tragedy unfold, enables us to develop ethics."[79]

The Ascent of Mind traveled in many directions. Speculations on the evolution of the hand axe, the significance of juvenilization in human offspring, tooth size reduction, and much else besides were all swept into Calvin's narrative. But the upshot of his weather-and-wits saga remained clear: *Homo sapiens* is an animal adapted by evolution to perform well in a diverse array of climates, a creature with sufficient flexibility to survive in locations everywhere from the frost-bound to the sweltering. Perhaps predictably, the valorization of the temperate zone, which has routinely surfaced among those attributing causal powers to the climate, reasserted itself. The most suitable locations to facilitate speedy hominid encephalization, Calvin judged, would most likely be "in the temperate zone, where every year the winter speeds up natural selection." Temperate conditions exposed its populations to "the ice age's tendencies to create islands on which evolutionary change is faster—and incidentally enhance speciation opportunities and reduce backsliding." This meant that ice ages caused "temperate-zone traits to become far more important than one might initially think." Indeed, he reckoned that "African models of hominid" had actually developed in the temperate regions during migration from an African center to which they subsequently returned. That suggested to him that "such 'African' hominids had some temperate-zone specializations that weren't really essential in the tropics. And the big brain may be one of them."[80]

Facilitating Calvin's portrayal of all these evolutionary-climatic maneuvers was a fertile imagination liberally trading in anthropomorphic metaphors to explain early human development. Concealed within our subconscious, he remarked, "are some instructions for how to behave in an ice age, a suite of behaviours (what used to be called a 'racial memory') for times past." Or again: "Ice-age climates that switch back and forth illustrate one reason why 'backing up' might be a good thing: suppress the current genes for big bodies when the climate warms up, and go back to the old set of genes that emphasized rapid reproduction instead." With a penchant for future planning, Calvin was sure that Nature had learned not to "throw away the genes for big bodies as maybe they'll be needed again, many thousands of generations later when the unstable climate cools." Whether with or without the calculated quotation marks, his metaphorical enthusiasm shone through not least when he told his readers that "some animals 'forecast' future conditions and 'plan' accordingly" and that in "an improving environment, the calculating gene might well decide to decrease birth spacing."[81] However these operations precisely worked, what is clear is that Calvin intimately connected the evolution of the human brain and mind with the contingencies of rapid climate change.[82]

Keeping a Cool Head

The connection between thermal regulation and brain size is another avenue down which a number of students of human evolution have traveled in trying to make sense of brain development. In August 2009, the science writer Bob Holmes put the question, "Did an Ice Age Allow the Boom in Brain Size That Made Humans What They Are Today?" to readers of the *New Scientist*. What prompted this query was the research carried out by David Swartzman and George Middendorf, both biologists at Howard University in Washington D.C., on the effects of a high temperature on the physiology of the brain. Under the emotive headline "Cool Times, Smart Times," Holmes recast his original question by turning a historical supposition into a future speculation: "If global cooling allowed humans to evolve their big brains, will today's global warming take them away again?" His answer was a hopeful "no." Human culture surely had the technological know-how to "buffer" the worst effects of a hot climate. But he did nonetheless foreground the dread of Middendorf, who commented: "I'd hate to think that a difference of 1.5°C might mean the end of humans because our brains cook . . . but I guess it's a scenario that might play out."[83]

It was a decade earlier at a 1999 conference entitled "A New Era in Bioastronomy," organized by the Astronomical Society of the Pacific, that they presented their research on the emergence of human intelligence and its connection to atmospheric cooling. They had noticed the concurrence between "two periods of pronounced Phanerozoic cooling" in the earth's history and "the emergence of mammal-like reptiles and a burst of encephalization in homeotherms"—organisms maintaining their body temperature at a constant level by metabolic activity. The pattern they discerned pressed them to the conclusion that a cooler environment was "a necessary condition for rapid encephalization." Rather than having anything to do with life on the ground, cognitive versatility, or flexible intelligence, their emphasis was rather on thermoregulation and the claim that the modern human brain could not have evolved until the Quaternary Ice Age around two and a half million years ago. Because of the heat generated by the human brain and the need for that heat to dissipate, the argument was that cooler periods, such as ice ages, facilitated cerebral enlargement. The need to literally keep a "cool" head and to prevent cerebral overheating was critical for increased encephalization. The high energy requirement of larger brains generated the principle that the "temperature differential between an animal and a cooler environment . . . will increase the efficiency of heat loss from the brain." Hominization coincided with a drop in mean temperature and with increased fluctuations in diurnal temperature on account of the more open habitats early hominids colonized as they moved from forests to grasslands. For this reason, Swartzman and Middendorf urged that the greater differential in temperature between homeotherms and environment was likely "a necessary condition for critical levels of encephalization."[84] Cooler conditions evidently generated bigger brains.

Ten years on, and now joined by another Howard University colleague, Miranda Armour-Chelu, Swartzman and Middendorf further elaborated on their theory. Updating their case, they presented what they called a "first approximation estimate of the cooling required for hominin brain size increase." The approximate estimate, in fact, turned out to have remarkable precision. "For an assumed brain temperature of 37°C and a local mean climatic temperature of 25°C at the time of both transitions," they declared, "the required cooling (drop in mean local environmental temperature) for emergence of *H. habilis* and *H. erectus* is 1.5°C, a plausible drop in the sub Saharan African venue for a Pleistocene interglacial to glacial shift." That was plainly where the 1.5°C figure came from in Holmes' *New Scientist* report. What bolstered the authors in their conclusion that "Pleistocene glacial episodes were likely sufficient to serve as prime releasers for emergence of *Homo habilis* and *Homo erectus*" was the just-published research of Axel Kleidon, at the Max Planck Institute for Biogeochemistry in Jena. Kleidon, they reported, had concluded that "warm climates impose important constraints on the evolution of large brains relative to body size confirming our previous hypothesis."[85]

The publication to which Swartzman and his colleagues referred was centered on the links between metabolic activity and climatic conditions. Of critical importance in Kleidon's analysis was the connection between two operating forces: the free chemical energy arising from the terrestrial biosphere and the generation of heat within the human body. Both of these processes were directly linked with climatic conditions. Climate constrained both the productivity of the biosphere and the release of human body heat. The trade-off between these operations determined the maximum level of human metabolic activity in particular climatic zones. For present-day conditions, Kleidon noted, this meant that while tropical regions were extremely productive and sustained a large supply of free energy, their warm and humid conditions resulted "in a low ability to loose [*sic*] heat, especially during daylight." By contrast, polar regions, while less naturally productive, nonetheless allowed for much more efficient heat loss to the external environment. The counterbalancing of these constraints revealed that there was "an optimum latitude (and altitude) at which the climatic environment allows humans to be metabolically active and perform maximum levels of physical work." Mid-latitudes turned out to be best inasmuch as "maximum human metabolic activity shifts towards lower latitudes for colder climates, and to higher latitudes for warmer climates." Metabolic activity followed the contours of the globe's climatic zones. All of this was supported by the paleobiological work of Swartzman and Middendorf, who had shown the relationship between the onset of ice age conditions in the Northern Hemisphere and the emergence of *Homo habilis* and *Homo erectus*. Characterized "by a greater brain biomass in relation to body weight compared to its ancestors," these hominids confirmed Kleidon in the claim that "warm climates impose important constraints on the evolution of large brains in relation to body size."[86]

Culturally, Kleidon insisted that this approach was consistent with "the geographic locations of where higher civilizations first emerged." When applied to climatic changes in the past, he went on, "this perspective can explain why major evolutionary events in human evolutionary history took place at times of global cooling." Such claims strongly resonated with the early twentieth-century figures we have examined. And indeed, referring specifically to Huntington's *Civilization and Climate*, Kleidon noted that while "the strongest form of environmental determinism can easily be rejected," such rebuttals did nothing to "lessen the fundamental role of climate in constraining what humans can—and cannot—do in a given environment."[87] As for the future, Kleidon issued the warning that global warming would inexorably lead to conditions poorly suited to human metabolic activity, certainly in the natural environment, on account of a much reduced capacity for heat loss.[88]

With so much at stake, it is hardly surprising that Bob Holmes at the *New Scientist* should fasten onto the story. And that warning was only one of a number of comparable distress signals. *ScienceDaily* had already announced in March 2007 the results of a research project at the University of Albany. The findings, it reported, had suggested that seasonal variation in climate had acted as a key selective force in the evolution of human cranial capacity and had played a crucial role, alongside migration away from the equator, in promoting human intelligence. The warning sign, recalling nineteenth-century fears about degeneration, was that "the recent trend toward global warming may be reversing a trend that led to brain expansion in humans."[89]

Now, Then, and Again

What united these various interventions was the resort to the role of climate or climate change in the process of hominization in general and in both the enlargement of the brain and the expansion of mental powers in particular.[90] The idea has made good copy, not least for science journalists. A piece appearing in the *Guardian* newspaper in 2007, for example, by the newspaper's science correspondent Ian Sample, described how the "evolution of our earliest human ancestors was driven by wild swings in eastern Africa's ancient climate." The "rapidly changing climate," he observed, placed "enormous pressure on early humans to adapt." Some were pushed to the "brink of extinction, while other better-suited relatives emerged and flourished."[91] Writing for *The Register*, an online enterprise technology news publication, another journalist reported that research had shown that humans "grew bigger brains as the climate they lived in got cooler" and concluded that "humans got brainier because they had to adapt to a more challenging environment."[92]

Such certainties, however, have not been unanimously shared. A review piece for *Science* by the paleoanthropologist Anna K. Behrensmeyer of the Smithsonian's National Museum of Natural History is illustrative. In this assessment, she expanded on a range of hypotheses advancing the claim that climate-driven environmental changes over several million years were responsible for hominid speciation, including such developments as "enlarged cranial capacity, behavioral adaptability, cultural innovations, and intercontinental immigration events." She remained cautious, however, about the deterministic language that frequently insinuated its way into such portrayals. Deciding "what constitutes a strong case for a causal link between climate change and an evolutionary event," she reflected, was challenging, not least since "climate was only one of many factors affecting human evolution." "Synchronous events" in the geological and paleontological record were often assumed to be causal but, "while seductive in their simple explanation of how our species came to be," they failed to address the full complexity of the processes involved.[93] Strong environmentalist theses about the patterns of hominid evolution thus continue to be subjected to empirical scrutiny. But the impression that climatic determinism is a settled fact—whether in stronger or weaker forms—lingers powerfully in some quarters, not least in relatively popular accounts of human evolutionary history by a number of distinguished archeo-anthropologists as they reach out to wider audiences.

In his self-confessedly speculative *Out of Eden*, the geneticist Stephen Oppenheimer assigned much in human history to the influence of climate, particularly the movements of human populations. “Our physical and behavioural adaptations,” he announced, “were focused on surviving the struggle with our greatest enemy and stern teacher, the climate.” Musing on what he dubbed, rather agriculturally, “Growing brains in the big dry,” he noted how various species of *Homo* developed substantially larger brains. For Oppenheimer, the thought that humans somehow first sprouted a big brain and only later decided what it was for was both confused and anti-Darwinian. Instead, he was sure that there “must already have been some aspect of our behavior” that conferred survival value on “large, energy-expensive brains.” Whatever it was, it was “something to do with the way we faced the climatic challenge.” During the earliest period of increased climatic adversity at the end of the Pliocene, and over the Pliocene–Pleistocene climatic transformation, he went on, evolutionary processes selectively advantaged brain enlargement among the various species of new hominid. Around 300,000 years ago, he reported, “the climate-driven brain-growth machine reached a plateau.” And so climatic forces made regular appearances throughout the drama he scripted. The choices that were presented to “the formerly highly successful hunter-gatherers of the North Eurasian steppe,” his readers were told, “were determined by geography and climate.”[94] Everything from bipedality and manual dexterity to the acquisition of language and brain size came within the sweep of climatic influence.

In *Neanderthals and Modern Humans*, Clive Finlayson, director of the Gibraltar Museum, similarly developed the argument that there was a close relationship between climate change, ecological transformation, and biogeographical patterns among communities of *Homo* during the Pleistocene. Thus, the reduction in size of our ancestral African population that experienced “a genetic bottleneck” was “due to climate-driven habitat fragmentation.” The extinction of Neanderthals was likewise the “result of environmental change caused by climate”—a theme he pursued even more fully in *The Humans Who Went Extinct: Why Neanderthals Died Out and We Survived*.[95] While there were differences of emphasis between Finlayson and other students of hominid evolution, he nonetheless resorted to Beals to confirm that, at least in part, neurocranial variations observed in modern humans were “related to climate” for the simple reason that thermoregulation had a much greater effect on “the cranium than upon the body as a whole.” The brain enlargement that reached its maximum in such archaic humans as the Neanderthals and their modern counterparts was an adaptive trait related to the increasing specialization suited to the changed ecology of the tropical African landscape.[96]

Other popular science works pointed in the same direction. Chris Stringer, a research paleontologist at the Natural History Museum in London and champion of the "Out-of-Africa" theory of human origins, for instance, laid out for readers of his 2006 award-winning *Homo Britannicus* the role that climate fluctuations played in the history of human life in Britain.[97] Here he fastened on the advent of successive ice ages to explain the extinction of the Neanderthals and the success of the Cro-Magnons. Crucial at this stage of human evolution were the effects of an oscillating Gulf Stream on the climate of Britain, as well as on the remarkable number of occasions, and the rapidity with which, "this climate switch was turned on and off." In "such unstable times, with severe climate swings happening even within the lifetime of a single Neanderthal or Cro-Magnon, it would have meant survival of the most resourceful and adaptable at a time when environmental change must have been at its most challenging." Evidently, versatility of mind and flexibility of behavior were picked out by natural selection for survival. And climatic effects did not stop with the influence of periodic glaciations on early patterns of migration. Looking at our more recent history, Stringer declared that "from about AD 800 to 1300 Europe experienced conditions slightly warmer than today, and this led Erik the Red to found colonies on the coast of Greenland." Similarly, the colder conditions of the so-called Little Ice Age during the 1400s "shaped Europe irreversibly, influencing social, political and agricultural changes that are still with us today."[98] Brian Fagan too, an anthropologist and prehistorian whose numerous archaeological works will later attract our attention, subtitled his account of the Cro-Magnons, which came out in 2010, *How the Ice Age Gave Birth to the First Modern Humans*. Like others, he reported that what "really came into play with the onset of the cold were the qualities of restless innovation that had marked *Homo sapiens* from the beginning."[99] The qualities that were later exemplified by arctic hunter–gatherer communities—cautious temperaments, an innovative mentality, and behavioral versatility—were birthed among the Cro-Magnon during the Late Glacial Maximum. In environments where temperatures could fluctuate with remarkable rapidity, these were precisely the survival skills Fagan insisted he could infer from the archaeological record.

No doubt the catalogue of works, both popular and specialist, attributing the emergence of *Homo sapiens* and the expansion of the human brain to the influence of climate could be elaborated ad libitum. But even this thumbnail sketch discloses a number of resonances between this corpus of work and the earlier twentieth-century figures who have attracted our attention. Four in particular seem worthy of comment.

First, attributing patterns of human evolution and the progress of encephalization to climatic forces has been a persistent refrain. So it is not surprising that a symposium and exhibition on "Ancestors, Four Million Years of Humanity" that the American Museum of Natural History hosted in 1984 paid tribute to the earlier contributions of William Diller Matthew and Henry Fairfield Osborn. In introducing the collection of published essays from the proceedings, Harry Shapiro paused to comment on the museum's role in the development of the science of paleoanthropology. While his Asian origins theory was no longer widely adopted, Shapiro was certain that Matthew's contributions had a "relevance that has survived."[100] Indeed, Matthew's interventions continued to exert a discernible influence. Even critics who remained unpersuaded by his idea of centers of origin, opting instead for evolution by vicariance, identified him, sometimes alongside Charles Darwin, as a foundational figure and made reference to the "Darwin-Matthew model" of evolution.[101] Not surprisingly, advocates of centrifugal speciation remain enthusiastic. Linking Matthew with Darwin's proposals, John Briggs argued that Matthew gave the theory its modern form in *Climate and Evolution* and that it was later substantiated by Darlington.[102] And, in 2009, E. O. Wilson began some reflections on the state of island biogeography in the 1960s with the comment, "When I was still a graduate student, in the early 1950s, an idea was circulating that I found inspirational. It originated with William Diller Matthew." It was what he termed the "Matthew-Darlington epic view of global territorial biogeography."[103]

Second, the revitalization of the role of climatic swings in paleoanthropological explanations resonated with the place allocated to climate oscillations and ecological change among early twentieth-century students of human paleogeography. Ronald Pearson's *Climate and Evolution*, which appeared in 1978, for instance, drew inspiration from Ellsworth Huntington's *Pulse of Asia*. Pearson lamented the "pall of neglect" Huntington's work had suffered during the middle decades of the century and lauded the innovative nature of his insistence that "climatic changes had been of a pulsating nature."[104] Indeed, in the Liverpool zoologist's estimation, the more reliable knowledge of climate history that had come to light in the late twentieth century undermined many "of the older criticisms of 'climatic determinism'" that had been "inspired by doubts about the validity of climatic data, and more particularly about their possible periodicity."[105] In point of fact, these convictions sparked in Pearson a moment of historiographical contemplation. "Ideas gain acceptance in one generation, become *passé*, and are then rediscovered, often independently," he mused. The reason was "that the ideas are, at various times, unacceptable to certain arbiters of fashion" and "their history clearly says more about those arbiters, and the socio-political environment in which they lived, than about the ideas themselves." Climatic influence was evidently a case in point, and Pearson set about rehabilitating it in the context of contemporary evolutionary theory. Huntington was a critical early port of call, and Pearson portrayed him as "a prime example of the way in which ideas can fall into disrepute, disappear from current knowledge, and then stage a re-emergence."[106]

While not explicitly drawing on Huntington's work, Vrba's description of evolutionary events "as a direct function of environmental change" in terms of "concerted pulses" similarly echoed his repeated resort to both meteorological and cultural pulsations.[107] So too did Stringer's comments on the evolutionary consequences of the oscillations of the polar front in Britain. Only the most flexible and creative could have survived the hardships of the environmental swings that Neanderthals and Cro-Magnons faced.[108] Indeed, in Vrba's case, her "'traffic light' model of interland migration" was closely bound up with climatic fluctuations. In this schema, historic movement between landmasses only took place when a land bridge was present and a "suitable habitat" extended across it "from the ancestral habitat" to new homelands. The length of time available for such conditions made migration responsive to the regulating activities of a "biased traffic light." During periods of global cooling, as she explained, the "traffic light turns green for migrants only when both the habitat and landbridge conditions are met and then stays green for a long time while cooler habitat conditions and lowered sea level coincide throughout most of a glacial period." For Vrba, this traffic light model linked migration history with "climate-induced habitat movements and eustasy."[109] In spirit, if not in design, this conception brings to mind the "stop-and-go determinism" of Griffith Taylor with its emphasis on the episodic environmental triggers that regulated human behavior.[110]

A third echo, and to some more troublesome, has been the way in which matters of race and identity seem to have lingered in the choice of vocabulary adopted by a number of authors on human origins and hominization. While certainly in a different political register from the proclamations of Matthew, Taylor, Huntington, and Osborn, Oppenheimer's *Out of Eden* might be thought vexing in this regard. While eschewing the "racist stereotyping of Europeans as *advanced* and aboriginals as *primitive*," his continued use of what he himself acknowledged to be contested racial categories such as "Negritos," "Caucasoids," "Mongoloids," "Australoids," and the like required a good deal of finessing. He fully acknowledged that "the issue of physical difference is loaded with that ugliest and most destructive form of human group behaviour, racism," yet he declared that "we get nowhere by shutting our eyes." Oppenheimer retained these categories, he insisted, simply "because they are in common usage" and were no more "misleading" or "vague" than "alternative terms for race such as 'population' or 'ethnic group.'" Not everyone will be persuaded by the argument that "other misleading labels are no better." Besides, references to "miscegenation" between archaic and modern *Homo sapiens*, as well as scrutiny of skin hue and facial shape—both beloved measurements of Victorian anthropometry—placed Oppenheimer in the awkward position of recalling the anthropological preoccupations of the heyday of scientific racism.[111]

Reflecting on comparable matters, Paul Graves-Brown observed that the question of Neanderthal extinction was often freighted, by an "implicit association," with "questions of race and racism." Writing on the eve of the 500th anniversary of Columbus' transatlantic enterprise, he suggested that matters of identity, equality, and race can still be found in modern narratives about the fate of the Neanderthals by raising questions congregating around just what it is to be human. "As a metaphor for the issues of race and racism," he contended, "the Neanderthal controversy returns again and again to the related concepts of identity and equality. Were Neanderthals the intellectual equals of their Cro-Magnon counterparts?" In Graves-Brown's opinion, it was "time to seek explanatory frameworks which offer an alternative to those derived from the dichotomy between indigenous development and colonisation/displacement," not least on account of the inclination to portray the relative evolutionary success of Neanderthals and Cro-Magnons in the language of inferior/superior cultures, species difference, and hybrid cross-breeding—conceptualizations all too troubled by the burdens of history.[112] Whether or not his analysis is in the right neighborhood, there is little doubt that aspects of the contemporary debate about human origins have been saddled with the cultural politics of race.[113]

A final point of continuity linking more recent paleoanthropological inquiries with early twentieth-century writings on climate and hominization is the concern to demonstrate the contemporary implications of the role of climate change in human evolution. Twenty-first-century crises are as conspicuously to the fore among today's students of hominid evolution as their counterparts were in the 1920s. Stringer, for instance, declared upfront that by "showing the vulnerability of humans to past climate and environmental changes," his project was "also providing a warning for the future, for the human race will face challenges every bit as serious in the near future." Because "the effects will be every bit as severe as those that caused our predecessors to flee or die out," he devoted the final chapter of *Homo Britannicus* to our own challenging climates.

Reprising warnings issued by Sir David King, the U.K. government's former chief scientific adviser, Stringer referred to the worrying conditions that prevail. But the worst has yet to come, as he foresaw "a runaway greenhouse effect [that] could return us to the scorching heat of the hot-house Earth of 250 million years ago." Concerned to find lessons from our deep history that could equip the human species to meet today's climate threat, he asked "which populations have coped best with rapid climatic or environmental change, and how did they do it?" Inspecting these hominid groups led him to the conclusion that "the importance of cooperation rather than conflict in challenging times is a lesson for Europe and the world today."[114]

In a comparable vein, though with rapid cooling in mind, Calvin concluded his *A Brain for All Seasons* by prognosticating what a climatically determined future for humanity would look like. In the event of a population crash following the next climatic flip-flop, he announced that plummeting crop yields "will cause some powerful counties to try to take over their neighbor or distant lands—if only because their armies, unpaid and lacking food, will go marauding, both at home and across the borders." More efficiently organized nations "will attempt to use their armies, before they fall apart entirely, to take over countries with significant remaining resources." In a classical Malthusian struggle for existence, exterminating competitors for scarce food resources will become "a worldwide problem—and could easily lead to a Third World War."[115] As environmental conditions that steered early human evolution had shown, the effects of abrupt cold lasted for centuries. If *Homo sapiens* survived at all, the species would now inhabit a world of internecine violence, political repression, plunging population, and agricultural decline. Here climate and climate change are staged as the author of humankind's coming doom.

In his account of why the Neanderthals went extinct, Clive Finlayson likewise looked forward from the vantage point of his own time perilously perched on the cusp of hazardous global climatic change. *Homo sapiens*, he remarked, was the "product of marginal people who had to do a lot of improvising to get by" and whose adoption of technology of one sort or another enabled it to respond "much faster than our genes could" to the challenges of climatic and ecological change. That provided the lens through which he peered into a precarious future. When "it all comes tumbling down," he wondered, "who will survive?" Not "those of us in the comfort zone, the auto-domesticated slaves of electricity, motor cars, and cyberspace," he mused, but rather the "children of chance, those poor people who today must scrap for morsels each day . . . will once again be the most capable at survival."[116] Hominid evolution had apparently taught that lesson. The *BBC News* picked up the story, and Finlayson provided a "Viewpoint" piece on "Climate and Humans: The Long View." "We seem preoccupied today by looming predictions of imminent climate change," he began, but if "we were to use the deep history of this planet as our yardstick, the unusual thing would be for our climate to remain immutable." The lessons from Neanderthal extinction, whose fate was "shaped by climate change and luck," were immediate: the Neanderthals "vanished because of too much global cooling." Many strands of *Homo* "also went extinct because they couldn't handle the climate and its effects."[117] Evolutionary paleoanthropology, trading on a moral economy of fear, was being mobilized in the interests of contemporary environmental politics.

These remarks demonstrate something of the ways in which the links between climate oscillations and early hominization have been marshaled in the context of the alarming anthropogenic climate change facing society today. And indeed, this contextual association has suggested to some that recent fears about global warming have directly stimulated paleoanthropologists to inquire into the role of climate variability in the early history of human evolution. For instance, Paul Coombes and Keith Barber argued that during the 1990s, "concern over anthropogenic impacts on global warming and a growing appreciation of the magnitude and rapidity of Holocene climatic change combined to produce an intellectual climate suddenly sympathetic to the idea of environmentally triggered catastrophes."[118] Others have noted that a major change—perhaps akin to a paradigm shift, according to Chambers and Brain—in Holocene research toward climate change and global warming took place in the early 1990s shortly after the establishment of the Intergovernmental Panel on Climate Change in 1988.[119] At the time, Frank Oldfield, a distinguished environmental scientist, observed that the research agenda was "being set not by what has come to light as a result of changes in the past, but by projections of what *may* happen . . . in the future."[120] Whether or not such analyses will bear the weight of concentrated empirical scrutiny remains to be seen. But there can be little doubt that climatic determinism resonates with our present cultural moment as commentators speculate on the dire human consequences of runaway global climatic change.

The whole question of the sociopolitical context of human origins narratives, taken in the *longue durée*, was the subject of a provocative assessment by the French anthropologist Wiktor Stoczkowski in 1992. While his analysis dwelt on proposals dating from 1820 to 1986, his reflections may well have a bearing on more recent interventions on climate change and the evolution of the human mind and brain. Charting a variety of postulated scenarios for the genesis of *Homo sapiens*, he argued that the history of hominization chronicles revealed that they had frequently been undertaken with "the intention of laying down an order for the present," namely, to excavate emblematic instances of matters deemed relevant to "the desirable organization of society in the present."[121] These explorations convinced him that he could discern, in different modes of scripting human origins, echoes of Enlightenment thinking on progressive decadence, Soviet–Marxist theories of labor and collectivism, countercultural romanticism about the noble savage, and feminist emphases on cooperation and compassion rather than bloody combat. Stoczkowski thus argued that the storylines developed by the scriveners of prehistory were shaped by what he called a "conditioned imagination" predisposing them toward what they considered plausible narratives. His inquiry provokes the thought that a contextual interrogation of current hominization theories might well identify the cultural politics of contemporary climate change as a significant motivating force buttressing a return to climate as the prime mover of anthropogenesis.

Over the past century and more, the process of hominization and the emergence of the human mind and brain, more particularly, have frequently been attributed to the role played by climate in human evolution. During the first decades of the twentieth century, this suite of inquiries was routinely domiciled in a framework that placed different human races on a hierarchical ladder of excellence. A number of key figures attracted to some form of climatic determinism resorted to cartographic rhetoric to portray their story of human evolution and involved themselves in the politics of eugenics and race relations. By mapping the worldwide patterns of climate-forced migration, they could relegate what they considered inferior racial types to the cartographic and cultural edges of the globe. A major element in this endeavor was the resort to climate oscillations as a crucial driver of the massively enlarged cranial capacity in *Homo sapiens* and of the intelligence, cunning, and creativity of our species in adjusting to the rigors of changed climatic conditions. By midcentury, however, the troubling racial and deterministic ethos of these endeavors, together with reservations about the complacent and convenient elision of cause and correlation, had, by and large, consigned this whole tradition to the margins of scholarly significance. For some, though, climate reductionism had been socially disapproved rather than scientifically disproved. And so, in one form or another, climate's role in the operations of natural selection in the path to modern humans retained some committed supporters.

As the twentieth century drew to a close, a revivification of climatic environmentalism in the study of human evolution was clearly discernible. A whole spectrum of peculiarly human traits from bipedalism, tool-making, and free-handedness to encephalization, prolonged infancy, and intellectual acuity was being attributed to changing historical climates. For some, finding an explanation for the "big-brained" human became a central concern, and a number of evolutionary scenarios were constructed to account for its emergence. The ecological changes brought about by cooling climatic conditions was fastened on by some as the crucial stimulus for the enlargement of the hominid brain as our apelike ancestors descended from the trees to dwell on solid ground. For others, it was the need to evolve a flexible intelligence that could adjust to rapidly changing weather variations that was decisive in the big brain narrative. For yet others, the anatomical solution to problems of thermoregulation that larger brains generated was a fundamental precondition for the evolution of larger head forms. Either way, the argument was that the human mind and brain were made in the crucible of climatic challenges. We are all the children of the climate.

What is noticeable about this whole tradition of research was the urge to connect the ways in which climate exercised control over the evolution of hominid anatomy and migration to the urgent political questions of the day. While early twentieth-century champions of the climatic thesis practiced their art in an era of racial fantasy and eugenic fixation, their twenty-first-century successors conduct their science at a time of unprecedented challenge from anthropogenic climate change. In both cases, it might be thought, the political context within which the study of hominization was domesticated made certain narratives more or less plausible and thus relevant to contemporary needs. The concurrence between the pronouncements of such bodies as the Intergovernmental Panel on Climate Change, the priority accorded to the challenges of climate change by numerous funding agencies, and the reemergence of climatic explanations in paleoanthropology and the study of hominization have suggested to some that the rebirth of various forms of climate determinism owes much to the prevailing anxieties over global warming. Not surprisingly, while the theories connecting climate and human evolution were advanced by serious scientific practitioners, they attracted the attention of science journalists and more popular media frequently on account of the lessons they believed these inquiries held for the future of our species. Whether here too correlation and causation are being confounded remains to be seen.

6

Mind, Mood, and Meteorology

IN 2017, the American Psychological Association, in conjunction with the "Climate for Health" initiative and its parent nonprofit organization, ecoAmerica, brought out a report entitled *Mental Health and Our Changing Climate*. In a wide-ranging review of the impact of climate change on the human mind and psyche, the authors identified a broad spectrum of ways in which mental health could be adversely affected, in the short and long term, by stressors arising from both gradual changes in the weather and climate-induced natural disasters. The mental health conditions experienced by communities and individuals alike on account of changing climates were both acute and chronic and also manifested themselves in different ways among vulnerable and resilient populations. In 2005, for example, Hurricane Katrina induced posttraumatic stress disorder among people living in areas affected by the calamity with incidences of suicide and suicidal ideation doubling. Comparable findings were reported in the wake of Hurricane Sandy (2012) and after the experience of flood disasters, extreme bushfires, and prolonged drought. Among the chronic impacts of gradually warming weather, the report called attention to increased anger and violence, observing that experiments and surveys had "demonstrated a causal relationship between heat and aggression." The arousal of rage and a diminishing of self-control on account of higher temperatures meant that climatic conditions reached right into the inmost workings of the human mind, provoking "negative and hostile thoughts" and increased levels of suicide. In consequence, the guidebook noted, higher temperature brought with it much greater use of emergency mental health services. More generally, the effects of climate change penetrated the emotional depths of the human soul. A sense of helplessness, depressive disorders, feelings of fear, and surrender to fatalism were just some of the emotional symptoms of what has been termed "ecoanxiety." Besides these individual experiences, the report highlighted the effects that climate change exerted on communities and societies such as interpersonal violence and intergroup aggression, and it stressed too that disadvantaged communities experienced "climate impacts more severely."[1]

Predictably, this report and the body of research it represented, were picked up by a host of popular online sources. Robert S. Eshelman, writing for *Seeker*, a website devoted to science communication, provided a summary informing his readers that the "mental health toll of climate change could be dire." Itemizing the effects of acute weather-derived disasters as well as slow-onset global warming, he warned that climate change was "making us mentally unhinged."[2] In comparable vein, an earlier 2014 post by Kristen Rodman, on the website of the forecasting service AccuWeather, informed visitors to the site that severe weather incidents have had "a negative effect on mental health," which could materially alter rates of anxiety and depression.[3] At the same time, readers of a variety of internet health forums have reported on moods and meteorology. Visitors to Every Day Health, a consumer health website with an editorial team of accredited health journalists, for example, heard from Therese Borchard that pleasant weather induced better memory and "broadened cognitive style," that higher levels of humidity resulted in lower powers of concentration, and that warmer temperatures resulted in increased violence. She noted too that spring and early summer witnessed a rise in the number of suicides compared with autumn and winter in both hemispheres.[4] WebMD, another online medical information service, heard from Julie Taylor that "rainy days really *can* get you down" by generating feelings of sadness, anger, and low self-esteem. Light therapy, she noted, has been offered by psychiatrists to combat such negative emotional effects of high precipitation.[5] Writing in 2013 for *The Conversation*, a forum synthesizing "academic rigour and journalistic flair," as it describes itself, the Australian psychologist Nicholas Haslam reported that sunshine had "repeatedly been found to boost positive moods, dampen negative moods and diminish tiredness." It could also "affect our mental sharpness," he went on, while at the same time increasing "verbal aggression." By way of illustration, he noted that American journalists resorted to more "negative words" during hotter days in their bulletins on the 2008 Beijing Olympics. On the other hand, certainly if Minnesotans were anything to go by, tipping in restaurants was apparently more generous on sunny days.[6] For the same outlet, two agricultural economists told their readers in March 2020 that "hotter weather brings more stress, depression and other mental health problems."[7]

These desultory remarks, sometimes concurring, sometimes conflicting, are the merest sampling of a burgeoning suite of commentaries on climate and consciousness. But as even these snippets indicate, the idea that the weather infiltrates its way into the inner reaches of the human mind seems widespread in our modern culture. Even if, among these commentators, the focus of attention is typically on recent climate change, there is nothing novel about the thought that prevailing weather conditions, directly or indirectly, have a profound influence on the human psyche, whether individual or corporate. Since ancient times, the ways in which this influence has been characterized have been many and diverse. Notwithstanding this assortment, the foundational conviction remains that moods and meteorology, temperature and temperament are intimately connected. However different the metaphysical edifices that have supported such assertions may have been, and however varied the political inferences that have been drawn from them, the same resort to climatic conditions as shaping mental dispositions has persisted. This is the territory over which we initially pass in this chapter as a prelude to inquiring into more recent manifestations of the mind, mood, and meteorology impulse.

Temperature and Temperament: A Deeper History

The Hippocratic roots of environmental medicine have already come within our purview when charting something of the historical sources of the project to connect climate and health in causal ways. The scope of Hippocrates' inquiries in the fifth century BC and the legacy that the Hippocratics established were holistic through and through. Accordingly, it was not just the human body that registered the imprint of the climate; minds, morals, and emotions were alike subject to its conditioning power. Cataloguing physical ailments in different climatic regimes arising from imbalances in the humors dominated the first part of the celebrated *On Airs, Waters and Places*. But in the second half, ethnographic observations came to the fore as the writer—or writers—sought to identify the impact climate exerted on national dispositions, regional temperaments, and the operations of the human mind more generally. In depicting the character of the Scythians in the Far North, for example, the Hippocratics insisted that because seasonal change there was neither "great nor violent," the people were resistant to engaging in "any laborious exertions." This was because "neither body nor mind" in the High North were "capable of enduring fatigue," a state of being generated by the monotonous invariability of the climate. Consequently, the Scythians were portrayed as having what Hippocrates described as "a humid temperament." As for contrasts between the "Asiatics" and "Europeans," differences in seasonal variability explained why the

European temperament was unruly, aggressive, and passionate. While uniformity of climate induced Asiatic indolence, *On Airs, Waters and Places* declared that great seasonal variations brought about "frequent excitement of the mind" among Europeans, which in turn induced wildness and extinguished "sociableness and mildness of disposition." In sum, lethargy and sloth, vigor and vivacity, intellectual acuity and mental dullness were all the outcome of local climatic and environmental conditions.[8]

Doubtless allusions to the Hippocratic doctrine of climate and character were dotted through many of the medical treatises that followed in the wake of *On Airs, Waters and Places*. But the rejuvenation of this impulse in more modern times owes much to the 1748 *De L'Esprit des Lois* by the French jurist and political theorist, Charles-Louis de Secondat, Baron de Montesquieu (1689–1775). Montesquieu's more general climatic philosophy of civilization and governance will come to the fore in a later chapter. Here, we simply note his cultivation of a thermal theory of the physiology of temperament. What sustained Montesquieu's analyses were two discrete bodies of Enlightenment literature: first, the travel writings of a range of authors reporting on overseas voyages and, second, the physiological theories of figures like Boerhaave that enabled Montesquieu to conceive of the nervous system as consisting of tiny tubules that carried nerve juice, or animal spirits as they were called, around the body. It was this conviction that allowed him to claim that the inhabitants of cold climates were more vigorous since the fibers in their cardiovascular systems contracted in cold air and this, in turn, allowed for faster flowing of the blood.

Montesquieu's physiological theory of mind came through with particular clarity in "An Essay on the Causes That May Affect Men's Minds and Characters" that he put together in preparation for his magnum opus.[9] Here he stressed the need to keep physical and moral causes in tandem, going so far as to insist that "moral causes contribute more than do physical causes to the general character of a nation and to the quality of its thinking."[10] Moral causes could override physiological impulses. This meant that legislation and climate often moved in contrary motion. The reason was simple: the "more the physical causes incline men to rest, the more the moral causes should divert them from it." The ethical imperative here was nowhere more conspicuously displayed than in the cultivation of the land. In combating the "bad effects" of hot climates, namely, "natural laziness," Montesquieu insisted that the "more their climate inclines them to flee . . . labor, the more their religion and laws should rouse them to it."[11] In the light of such declarations, Diana Schaub is perhaps correct to depict Montesquieu's overall message as "not one of climatic determinism, but of the need for wise legislators" to take full cognizance of the regional atmospherics of their regimes.[12]

Yet for all that these sentiments may have muted the sharpest edge of climatic necessitarianism, there can nonetheless be little doubt that a deterministic spirit—albeit ambivalent on occasion—was woven into the fabric of Montesquieu's understanding of the influence of atmospheric and environmental conditions on the human mind. What Richter describes as his "materialistic psychology" manifested itself in Montesquieu's ready adoption of the zonal physiology he encountered in Arbuthnot's *Essay Concerning the Effects of Air on Human Bodies.*[13] In different temperature regimes, the surface "fibers" of the human body reacted by relaxing or contracting according to hot or cold conditions, and these responses in turn influenced blood circulation and pulmonary power. That was why human physiology clearly mapped onto climatic zone. By way of empirical corroboration, Montesquieu conducted his own experiments on the papillae of a sheep's tongue. Having frozen the organ, he closely examined, with the use of a microscope, the tiny hairs on the papillae and found them greatly diminished; as it thawed, they reappeared.

The implications turned out to be of epic proportions. On such physiological details hung the destiny of nations. For the despotism of nature was not easily defeated. And the impact on human cognition was dramatic. "An excessive rigidity or coarseness of the fibers may produce slowness of mind," he recorded in his preparatory notes for *The Spirit of the Laws,* but when the "fibers are too flexible," he went on, "this may produce weakness." The operations of fiber physiology in cold climates confirmed him in the judgment that "Northern peoples will not have that quick penetration, that vivacity of conception, that ease of receiving and communicating all sorts of impressions—qualities found in other climates." Still, this deficiency seemed more than compensated for by their dogged persistence in seeing through enterprises that they initiated. Accordingly, the people of Holland, he declared, while "famous for the slowness with which they arrive at new ideas," exhibited a "constancy in their passions which has led them to achieve such great things."[14] Because of the influence of the temperature on national temperament, inhabitants of cold zones also displayed "little sensitivity to pleasures"; dwellers of temperate regions were somewhat more responsive to pleasure, while in hot countries, desire rose to extremes. The geography of human response mirrored atmospheric conditions. Just as climates could be differentiated by degrees of latitude, Montesquieu confirmed that "one can also distinguish them by degrees of sensitivity."[15] The same opera performed in England and Italy elicited markedly different emotional responses from their audiences. Here was evidence, if evidence were needed, that the "spirit and passions of the heart" were "extremely different in the various climates."[16] And it necessarily followed that laws should be relative to these varying circumstances.

Diverse mental and temperamental constitutions also meant that meteorological conditions differentially influenced various national psychologies. Take the effects of wind. The Sirocco, "having passed over the sands of Africa . . . governs Italy" and "exercises its power over everyone's mind," he asserted. In England, it was the gales from the east that made their impact. Plainly, the inhabitants of England and Italy were subject to the strong winds they experienced. "But," Montesquieu continued, "there is a difference between the illnesses that attack the minds of the Italians and those that attack the minds of the English." Different national psyches were evidently conditioned in different ways by the power of the air, thereby generating distinctive geographies of mental health. And the habits of mind that characterized Europe's northern and southern latitudes even shaped their differing tastes in matters of religious identity. Two types of religious sensibility presided over Europe, namely, "the Catholic, which demands submission, and the Protestant, which seeks independence." In the North, Protestant inclinations tended toward intellectualizing; by contrast, the Catholics in the South were preoccupied with mystery—something Montesquieu regarded as "very reasonable." In consequence, he concluded, "the peoples of the south, with ideas that are better founded on the great truths even with minds that are naturally better, are in other respects at a very great disadvantage when compared to the people of the north."[17]

English physician, fellow of the Royal Society, and prolific author William Falconer (1744–1824) also ventured into the mind–meteorology terrain. Falconer was renowned for his inquiries into the efficacy of the Bath waters for a range of chronic medical conditions, and while personally abrasive, he nonetheless counted among his patients England's Lord Chancellor Edward Thurlowe, Prime Minister William Pitt, and Admiral Horatio Nelson. Besides his publications on plague, fever, influenza, and other medical subjects, he wrote on theology, classics, and natural history as well as on the influence of climate on the human psyche.[18] Here he consciously built on the foundations laid by Montesquieu. Later, in chapter 10, we will visit his proposals on how climatic conditions variously predisposed inhabitants of different regions toward courage or cowardice, fortitude or frailty, in military affairs. At this point, we dwell on how the human mind as expressed in national disposition, cultural tradition, intellectual accomplishment, and religious persuasion came under the dominating rule of the climate's powerful empire.

In the introductory remarks to his 1871 volume on the influence of the climate, Falconer explained just how he understood the ways in which climate sculpted, within certain boundaries, the diagnostic features of different ways of life. Atmospheric effects were neither ubiquitous nor uniform, he explained, and were at times counterbalanced by the actions of other powers. But they were real nonetheless. As he put it: "if a considerable majority of the nations, as well as the individuals, that live under a certain climate, are affected in a certain manner, we may pronounce decisively on its influence, even though there may be some exceptions." Thus, while hot climates naturally rendered their populations both "timid and slothful," the "necessity induced by a barren country" was likely to "correct this tendency of the climate, and dispose the manners to a different turn." To Falconer, heat was "perhaps the most universal stimulus" among the components of the climate because it excited the neurological system and "the cutaneous nerves in particular." Indeed, it was because of the impact of heat and cold upon the body that its influence on the mind could be explained. As he put it, the "sensibility of the body is by sympathy communicated to the mind." For this reason, he organized the first book of his treatise, *Remarks on the Influence of Climate,* around the ways in which hot, cold, and moderate climatic regimes governed everything from temperament and morality, customs and laws, to national character and religion.[19]

These preliminary observations paved the way for a comprehensive inventory of climate's moral and political governance. On patterns of national temperament, Falconer ascribed passionate, amorous, and vindictive dispositions, as well as levity, to hot conditions, while cold climates were characterized by a tougher temperament, less disposed "to the tender passions" though displaying a charitable attitude toward the poor. Moderate regimes were best. Friendship, he reported, was "seen to most advantage in temperate latitudes." Why? Because moderate conditions induced moderate conduct—a "proper medium" between excessive "severity and too great forbearance." But meteorology's sphere of influence spilled well beyond national disposition, infiltrating its way into the manners, morals, and mental accomplishments of different peoples. The inhabitants of hot climates were "disposed to be quarrelsome, passionate, litigious, and revengeful," displaying little "probity and honesty in the common dealings of life." In these regimes, the way of life—"the mode of manners" as he put it—had stagnated. What he referred to as "the form and ceremonial of behaviour" in such places as Persia and India, for example, were frozen in time, remaining unchanged over two thousand years. By contrast, the geographical distribution of violence, drunkenness, and gaming showed that these traits were largely the preserve of cold climates. At the same time, the high latitudes were home to decency of conduct, moral excellence, and enduring sexual relationships. Yet again, however, the temperate world exhibited its superiority over more extreme climatic zones, with "politeness and elegance of behaviour" and more dignifying relations between the sexes attaining their "greatest perfection" in the middle latitudes.[20]

Given the mental territory over which climate exercised its power, it was only to be expected that the intellectual faculties and religious sensibilities would likewise fall within its dominion. And here too Falconer's analysis followed the contours of his tripartite climatic scheme. Hot environments stimulated the imagination, and so the literary styles and modes of thinking that "abounded most in the south," embellished as they were by "the fruits of fancy," cultivated mythological symbolism and poetic allusion. He called on the psalms and lyrical poetry of the ancient Hebrews as witness to "the passion of love, great beauty and richness of imagery" they represented. By the same token, the contribution of hot climates to natural philosophy was much inferior to the other zones, even if the invention of the plough, printing, and gunpowder were to be credited to them. In marked contrast, the cold northern reaches of the globe, notably along the shores of the Baltic, had made great advances in mathematics and astronomy as Copernicus, Tycho Brahe, and Kepler illustrated, while moral philosophy and metaphysics remained unsuited to a northern temperament lacking in sensibility. Ultimately, of course, the triumph of the temperate reasserted itself. In literary culture, "the more regular and temperate genius of Europe" showed itself "far superior to the flighty luxuriance of hot climates." And the same was true of scientific disciplines:

> In the more sober sciences . . . the superiority is still more visible. History, geography, chronology, &c. are cultivated with most success in temperate climates. The mechanical part of history, if it may be so termed, has been much studied in the North; but the reasons I have before given, prevent their labours being admitted on a rank with history. In geography, indeed, a study which requires more toil than invention, more labour than genius, the northern nations have been more successful; but still the most considerable improvements and discoveries have been always made from temperate latitudes. Natural philosophy, also, though great additions have accrued to it from cold climates, has still received its most solid improvement in moderate ones.[21]

In Falconer's portrayal of the effects of climate on religion, a kind of thermogenic theory of theology came to the surface. Because of the way in which hot conditions acted on the mind, he noted that the mental disposition of low-latitude peoples was orientated more to "sensible and visible objects" than to "mere spiritual ideas." It was for this reason that "the ancient religion of the Persians was directed to the sun principally." Similar stories could be told for the moon, other celestial bodies, and fire. "Almost all the religions of hot climates," he reported, "were connected with some sensible object." But it was not just what Falconer described as "false religions" that came under the hegemony of climate. It had "also infected the true," as witnessed in Christianity itself. Here Falconer's anti-Catholic sentiments were on full display. The "Romanists," he declared, "pay an extravagant and absurd, not to say an impious, adoration to certain images of saints." From this proclivity flowed the "extraordinary regard and veneration paid by that religion to relics or remains of persons illustrious in religious concerns," adoration of the Virgin, and the "ridiculous farce" of transubstantiation. Cold climates pulled their populations in the opposite direction. There, religion was "a subject of internal contemplation; and its influence is directed more to the reason than to the passions. Hence they have always been averse to representations of the Deity by sensible objects." Still, as with scientific pursuits, the truest religious ideas were always found in temperate conditions. Even though it had "pleased the Almighty to make a warm climate the scene of his particular revelation," Falconer could persuade himself that it was "in temperate latitudes that Christianity has been best understood and practised."[22]

Falconer's climatology of religion spiraled in many directions. Referring to the writings of Montesquieu, "one of the greatest men of the age," Falconer noted how various attributes ascribed to the deity revealed the influence of the climate. Religious regulations, moral prohibitions, dietary requirements, the practices of prayer, doctrinal commitments—all, in one way or another, mirrored climatic conditions. The doctrine of predestination, for example, was apparently the offspring of a hot climate. Those same atmospheric conditions were "likewise better suited to an idle religion, as the necessaries of life are procured with less labour" and to the use of "set forms of prayer" rather than extemporary supplication; both were "adapted to the indolent turn of hot climates." For this reason, he believed he could discern direct connections between meteorology and monasticism. Because hot climates inclined their inhabitants to leisure and repose, the monastic life was "highly suitable to such an indulgence." In this way, lethargy was able to masquerade as piety. How different that was from the principles of religion in moderate climates, where both theology and observance enjoyed a far "greater cultivation of science and literature" and were in consequence "more liberal and enlarged."[23]

Between them, and building on Hippocratic foundations, Montesquieu and Falconer elaborated in extenso on how cold, moderate, and hot climates conditioned both national and local temperaments. The ways in which climate could colonize the human mind were myriad. Cognitive styles, behavioral traits, moral dispositions, literary tastes, and religious affections were all swept into the psychophysical domain over which the climate held sway. But one particular state of mind and emotion that was often attributed to the effect of atmospheric conditions captivated many writers. It went under various names: melancholy, neurasthenia, nervous exhaustion, and indeed many more.

Meteorology and Melancholia

Many of the authors whose treatises on tropical medicine attracted our attention in earlier chapters paused in their catalogue of physical diseases to comment on the mental and psychological impact of a hot climate on European constitutions. James Lind, the Scottish military surgeon and pioneer of naval hygiene, for example, was convinced that the two organs that were chiefly affected by the "impure air" of hot lands were the brain and the stomach, "or in other words, the nervous system, and the organs of alimentary digestion." Psychologically, the consequences for both travelers and settlers were dramatic. In a treatise first published in 1768 and routinely consulted for more than fifty years thereafter in Britain, Lind warned that strangers "unaccustomed to such an air, though seemingly in health, feel an oppression and lowness of spirits;—they become inactive, have a great inclination to rest or sleep, and often complain of a headach [*sic*];—their reasoning faculties are sensibly impaired, particularly the memory. Every kind of study, or long attention of the mind to any subject, as likewise venery, are hurtful." In turn, these psychological conditions exerted their own influence on human physiology. In such climatic circumstances, what he called the "passions of the mind" exerted "a much more quick and violent effect on the body, than in a purer and cooler air." As he further explained, "An excess of passion often brings on an instantaneous attack of a fever: a violent fit of anger, or grief, will immediately produce a jaundice, or the yellow fever."[24]

The mental lethargy and emotional exhaustion that tropical heat inflicted on white travelers likewise came to the fore in James Johnson's *Influence of Tropical Climates on European Constitutions*, another medical text we have earlier encountered, which first appeared in 1813. Here Johnson, a military surgeon in Calcutta, elaborated on the "sympathetic" association between the liver and the brain under a tropical sky. Johnson laid out what he had in mind by the idea of "sympathy" in a medical context in the first few pages of his treatise. Different, and often distant, parts of the body possessed some relation by which "when *one* is affected by particular impressions, the *other* sympathises, as it were, and takes on a kind of analogous action." Such strange alliances, he reported, had yet to be satisfactorily explained, though he was sure that the cause was only "locked up in the bosom of Nature" for the time being. Ignorance would soon yield to insight. Absence of a demonstrated explanation, of course, did not deter him from expanding on the mental and psychological state of Europeans in hot regions. Disorders of "the abdominal organs" had a great effect on the functions of the brain. Cognitive confusion, mental instability, a volatile temper, memory defects, and various psychological disturbances could all be traced to a "corporeal origin." Indeed, Johnson warned that Europeans in hot climates could well experience fits "of despondency, brooding melancholy, permanent irascibility, and still higher grades of intellectual disturbance, till, as sometimes happens, the point of temporary alienation is reached, and suicide terminates the scene."[25]

Many of the conditions that writers such as Lind and Johnson identified began to be gathered together under the label "neurasthenia" as a catch-all diagnosis for a range of symptoms loosely associated with nervous exhaustion. The term was beginning to circulate around the tail end of the 1820s but gained traction in North America during the late nineteenth and early twentieth centuries. In the late 1860s, it was being used by E. H. Van Deusen, a superintendent at the Michigan Asylum in Kalamazoo, and more particularly by the New York neurologist George Miller Beard, who popularized the term. For Beard, fatigue, depression, anxiety, and headache were all symptoms of neurasthenia—a condition of general enervation brought on by modern urbanization and industrialization.[26]

The construction of neurasthenia as a tropical psychopathology suffered by white settlers in hot climates owed much to the writings of Charles E. Woodruff (1860–1915), a lieutenant colonel in the U.S. Army Medical Corps. Indeed, it was from his experience as a military surgeon that he drew many of his more general conclusions about the climate's influence on mental health.[27] More concerned with excessive light than with temperature per se, Woodruff fastened on the influence of the sun's ultraviolet rays—what he called actinic rays—which he regarded as especially dangerous to fair-skinned blonds and

Aryans. Right from the start, Woodruff cast his analysis into the context of Darwinian evolution. Back in 1895, he had encountered some research by Jos Ritter von Schmaedel declaring that skin pigmentation had evolved for the purpose of "excluding the dangerous actinic or short rays of light which destroy living protoplasm." This explained the development over the generations of "nigrescence and blondness" and was the chief reason for the failure of Europeans to succeed in tropical environments. Evolution, race, and colonialism were thus seamlessly interwoven into his analyses. The thought that human intelligence had somehow allowed the human species "to escape the law of selection" was profoundly mistaken, for the human animal remained "just as much under the laws of evolution now" as it had ever been and could "never hope to escape natural law." This naturalization of colonial failure pushed him toward an extreme antiacclimatization stance. Simply put, "races always fail to colonize in a zone markedly different from the home land." Those claiming that acclimatization could be achieved by imperial settlers were simply flying in the face of "the anthropological and biological law that every living thing must remain in its zone to survive permanently." And these were not the only environments in which nature's cruel policy of extermination was on display. In modern cities, natural selection was callously eradicating those blue-eyed blonds "who are unfit for this environment as it weeded out the stupid and unfit when brains were increasing and anthropoid races were being changed to manlike races." What's more, natural selection was apparently a lot more color-blind than Woodruff himself: "Civilized negroes," he reported, "can only arise in millenniums, in the same way we arose, by the brutal method of killing off the stupid of each generation."[28]

A failure to comprehend the malign influence of direct sunlight meant that there was a deplorable lack of awareness of the degree to which nervous problems arose from insufficient skin pigmentation among the white races for life under hazardous actinic rays. Nowhere were these effects more dramatically on display than during the early stages of white settlement in the tropical world. During the first few months, before the light stimulation had brought on extreme nervous exhaustion, travelers enjoyed a great sense of well-being. But this experience proved to be treacherous for the simple reason that it lulled white visitors into a false sense of security and impelled them to work harder and thus expose themselves unnecessarily to the influence of the sun's scorching rays. This induced "tropical exhaustion," which Woodruff regarded as the very "picture of neurasthenia." Soon a host of pathological nouns were appended to the tropical adjective. Tropical apepsia, tropical insanity, tropical amnesia, tropical suicide, and many more were grouped together under the category of "tropical neurasthenia." Most of all it was the mental toll that neurasthenia exacted on those exposed to too much light that was diagnostic of the condition. Almost everyone, he remarked, suffered from tropical amnesia in the Philippines. Indeed, he went so far as to declare it "quite likely that every one who lives in the tropics over one year is more or less neurasthenic." In cataloguing these ailments, he drew on military medical statistics, and these encouraged him to comment on the degree to which far northern sunlight could be as challenging as its tropical counterpart. He reported that he had himself observed "the nervous irritability of soldiers, officers, and women, in our extreme northern army posts in summer, and cannot account for it except by the excessive light stimulation." He acknowledged the reality of "arctic mental depressions in the dark season" on account of its "profound psychic effect in the direction of soothing or depressing all functions" but in the last analysis insisted that it did not lead to nervous exhaustion, "for such soothing is preservative of energy in the hibernation of Eskimos."[29]

Taken in the round, Woodruff regarded tropical neurasthenia as paradigmatic of this nervous condition. But he could also see signs of it in high-light regions beyond the tropical zone. Among Americans, more generally, neurasthenia had become the main nervous disorder, as illustrated by the statistics he provided of seasonal suicides, nervous breakdowns, and childhood psychological disturbances. Blond women, he further lamented, were particularly prone to neurasthenic headaches, general debility, and neuralgias on account of overexposure to light rays. All of this was in keeping with his understanding of the role of evolution in the development of blond supremacy. It was in the cold and dark conditions of northwest Europe that the blond races had evolved the "large brain which has been an Aryan inheritance ever since" and had enabled them "to build up such high civilizations and now to control all the lower races who were so unfortunate as to escape too soon from that northern struggle for existence which we might call nature's brain factory." The mental largesse that tutelage in dull climatic regimes had gifted these peoples was now being squandered under the dire threats arising from their occupancy of sun-filled zones, not least in the United States itself. Because such regions were simply not "an Aryan climate," mental and emotional deterioration was inevitable. All of this fed into his declaration of what might be described as a prescriptive geography of global habitation—a kind of climatic apartheid. The "black man," he announced, should remain "within 25 or 30 degrees of the Equator," the "browns should be between 30 and 35 degrees," while the "olive (our

Semitic or Mediterranean type) flourishes best at 35 or 45 degrees." Blonds had arisen north of the 50-degree parallel, and that remained their natural habitat. Climate had dictated the racial boundaries of global occupancy, and the psychological problems that went under the label "neurasthenia" were the price to be paid for defying nature's will. "Every climate on earth is a splendid climate for the type of man physically adjusted to it by natural selection, and a bad climate for every other type of man," he explained. But at the same time, he concluded, "Nature is a brutal stock breeder and kills off all unsuited to any climate into which they may have wandered."[30] With these convictions, it is not surprising that Woodruff would shortly produce a text on the *Expansion of Races* (1909), in which standard environmental determinism was integrated with Darwinian vocabulary to produce the paradoxical conclusion that, as Kennedy puts it, "natural laws drove the white race to control, but prevented them from populating the tropics."[31]

Woodruff's diagnostics featured prominently in the presidential address to the Society of Tropical Medicine and Hygiene, delivered in 1913 by the Irish-born physician Sir Richard Havelock Charles, surgeon-general in the Indian Medical Service and surgeon to King George V. For the occasion, Charles took as his theme neurasthenia and its role in what he took to be the "decay of Northern Peoples in India." This was a subject he deemed "pregnant to many in this Imperial age" since it directly bore on the suitability of the tropics for white settlement. In a speech laced with stereotypical racial judgments, he turned to Woodruff's claims about the effects of actinic rays to confirm that white soldiers and settlers in the tropics "universally suffer from neurasthenia." This reinforced his own experience in the Punjab, where he encountered what he called "Punjab head"—a condition that included short-temperedness, forgetfulness, sleeplessness, and irritability. Missionaries, he added, had found the same thing. Intense light, heat, and humidity directly influenced the psyche and fostered parasitic diseases that further exacerbated the condition. Again following Woodruff, Charles insisted that European skin pigmentation was intrinsically unsuited to the rays of a tropical sun, and exposure to them resulted in wholesale emotional and mental breakdown—problems of "concentration, phobias, insomnia," as well as "lack of control of feelings, irritability and depression." So punishing were the atmospheric conditions that the only way a white race could "preserve its purity and predominance in a tropical climate" and retain "that vigour, intelligence, and physique which are its characteristics" was by ensuring new waves of immigration from the homeland.[32]

The neurasthenia of which writers such as these spoke was a remarkably flexible piece of medical nomenclature. It incorporated what Dane Kennedy calls "a bewildering range of symptoms." Everything from fatigue, depression, and lack of concentration to insomnia, memory loss, and sexual profligacy was gathered under the label. As Kennedy goes on, "What did this miscellany of afflictions share apart from an intangibility that placed it beyond the grasp of empirical investigation? Tropical neurasthenia, for all its aura of medical certitude, was exceedingly nebulous, a convenient repository for whatever bundle of obscure and often value-laden complaints [that] otherwise eluded classification and explanation. It was precisely this feature of the diagnosis that made it prone to climatic interpretation."[33] Under the influence of an unsuitable climate, it was widely believed, neurasthenia marked nothing less than the disintegration of a victim's entire psychic economy.

Mind, Metabolism, and Mass Psychology

Woodruff frequently called on the authority of Edwin Grant Dexter (1868–1938) in support of identifying connections between seasonal weather conditions and patterns of crime and drunkenness, incidences of insanity and suicide, and the prevalence of misdemeanors among schoolchildren. He also happily relied on Dexter's research to shore up his projection of neurasthenia as a recognizable medical condition. Yet there remained significant differences between the two men. While Woodruff fastened on the influence of actinic rays, Dexter was more taken with the whole question of the relationship between the weather and what he called "reserve energy." At the same time, Dexter exhibited much greater methodological sophistication than Woodruff and other contemporaries in the various analyses he undertook. He had been working on the relationship between weather conditions and human behavioral patterns since the late 1890s and completed a doctoral dissertation on the subject for Columbia University in 1899. During his career, he occupied positions at Brown University, Greeley Normal School (now the University of North Colorado), and the University of Illinois at Urbana-Champaign and in 1907 was appointed by President Theodore Roosevelt as chancellor of the University of Puerto Rico.

In his *Weather Influences*, published in 1904, Dexter sought to bring greater rigor to the biometeorological correlations he was assessing between weather conditions and a range of behavioral traits and psychological dispositions. Though his statistics remained descriptive, he used "goodness-of-fit" tests to interrogate potential connections linking atmospheric temperature, barometric pressure, and humidity with incidences of suicide, patterns of assault and battery, the performance and deportment of schoolchildren, the prevalence of insanity, and suchlike. In doing so, he drew on data from the records of the New York City Coroner, Chief of Police, and Superintendent of Schools. A good deal of Dexter's motivation in inquiring into the behavioral correlates of various weather conditions sprang from his lifelong interest in education and psychology. Indeed, he held the presidency of the National Society for the Scientific Study of Education in 1905–1906.[34]

The methodological aspirations that Dexter had for his undertaking may be gleaned from the fact that he recruited the distinguished Cleveland Abbe (1838–1916), chief meteorologist at the U.S. Weather Bureau, to provide an introduction to the volume. Abbe began by rooting the entire project of finding links between climate and human behavior in the deeper context of evolutionary history. But he was convinced that far too many authors attributed to meteorological causes, influences that in all likelihood sprang from entirely different sources. Ascribing "the intelligence, industry, frugality, and humanity of the New England people, the enterprise of New York and Pennsylvania, the haughty bearing of the southern people," as many did, to local meteorology, Abbe judged to be entirely wrong-headed. Political power and social forces were far more significant. At the same time, Abbe was convinced that climate did exert a profound influence on mental vitality. Migrants were a case in point. Moving from a "warm, moist, depressing climate" to a cool, dry regime, or "what is the same thing a gradual secular change of the climate in any locality in this same direction," always had the effect of increasing mental vigor. This meant that, as with other animals and plants, the human species had its own climatic optimum, and any departure from it brought about degeneration. As these remarks suggest, Abbe was sure that hot climates induced "lassitude, indolence, and mental inactivity" in those who were not accustomed to them. But this did not mean that those native to such environments were inferior. It was a simple fact, he declared, that "in all tropical countries as great a variety of men and talent" could be found "as in the temperate countries." For Abbe, it was just too easy to attribute to climate behavior that was far more likely to be due to inheritance or culture. But once the impact of other agents was taken into account, it would be possible to isolate those traits that were controlled by weather and climate. This, he concluded, "is what Mr. Dexter has done in the book before us now."[35]

In keeping with Abbe's strictures, Dexter made it clear that while the weather was certainly a causal agent, it was rarely "the immediate and exciting cause of any of the misdemeanors" he planned to address in his book. Instances of suicide were a case in point. "We cannot for a moment suppose that a low state of barometer ever drove a man to suicide," he commented, "though we shall see that suicide is three times as prevalent during conditions of low barometer as during high." Later he reiterated the point insisting that when saying that "high temperatures *cause* an excess" of some particular behavior, it was only to be taken in "a secondary sense."[36]

For all that, Dexter plainly felt that climate's causal powers were enormous. Even if the direct effect of the weather was restricted to human physiology, the way in which the metabolism of the body was altered under differing meteorological conditions brought about psychological responses. Given a physiological change, he affirmed, "mental change is sure to follow and the nexus is in every sense causal." Psychological states arose from biological operations. To Dexter, all this meant that seasonal patterns of behavior were clearly discernible. Suicide was most prevalent in the late spring and summer months and rose to a peak during extremely high and low temperatures. Schoolchildren's deportment was best in colder weather, and boys were more influenced by it than girls. Homicides and general assault increased during hot weather, because temperature more than any other condition affected "the emotional states which are conducive to fighting." And clerical errors, at least as reflected in data from New York City banks, escalated during the summer months when vital energies were depleted by excessive heat. More generally, despite Abbe's strictures, Dexter did not hesitate to extend influences of this kind to whole slices of the globe. "Inhabitants of hot climates are usually listless, uninventive, apathetic and improvident," he flatly stated. "No long-established lowland tropical people is a conquering race in the broadest sense of the word."[37]

What facilitated generalizations such as these were two abiding convictions that animated Dexter's entire project: his awareness of the power of aggregate data and his conviction that metabolic processes were affected directly by changing meteorological conditions. For the former, Dexter told his readers that however difficult it might be to provide a persuasive account of any individual's conduct, dealing with people en masse was decidedly less troublesome. As he explained, some laws of conduct could be formulated for a community, even if they lost their "validity when applied to the individual." Simply because some particular weather response was "not true for every individual" in a population did not for a minute invalidate the agency of temperature or humidity at the group level, "nor lessen their weight in the prediction of the conduct of the mass under given conditions." Mapping mood and meteorology was an exercise in tabulating the statistical frequencies of atmospheric and psychological variables. As for metabolic responses to meteorology, the power exerted by the weather on human behavior stemmed from the influence that temperature, barometric pressure, and the like exerted on the body's energy levels. The reserve energy that remained to human beings after all the vital operations were sustained and was utilized for mental processes and psychological functioning was "*effected most by meteorological changes.*" Herein lay the reason for the emotional turmoil experienced under certain atmospheric conditions. Indeed, he announced, as another primary finding that delinquency was the consequence of "*an excess of reserve energy, not directed to some useful purpose.*" In sum, it was the abundance or dearth of energy reserves in the human body that drove mental capacity, emotional response, and psychological behavior. The control that the weather exercised over the body's reserve energy supply was therefore of "the greatest importance."[38]

Reviews of Dexter's project were generally positive and his efforts to quantify the weather's effects on human behavior welcomed.[39] His work was picked up by Harvard's Robert DeCourcy Ward in his 1908 text *Climate: Considered Especially in Relation to Man.* We will have occasion to engage later with Ward's expansive climatic philosophy, but in the present context, it is a passage that addressed the "mental effects" of climate that is pertinent. Here he foregrounded frequent and sudden weather changes in the temperate zones, and more especially seasonal variation, as critically significant. Calling on the authority of Dexter to bolster his own impressions, Ward contrasted the vigorous effects of bright, crisp days, "when work is well and quickly done," with the sluggishness induced by a "dull, depressing, and enervating day." Generalizing to the peoples of the temperate zones, he urged that strong cyclonic winds blowing poleward from the lower latitudes were "proverbially disagreeable and irritating, in strong contrast with the cooler winds from higher latitudes." The sirocco from Italy was "deadly to human temper," as were the solano in Spain and the norte in Argentina. The warm foehn and chinook winds likewise stimulated nervous disorders, while the dry, dust-carrying zonda in the Argentine was said to induce temporary insanity, sometimes "leading to suicide."[40]

But perhaps not surprisingly, Dexter's greatest influence seems to have been on Ellsworth Huntington, whose biometeorological fixation with "climatic energy" has already been to the fore. To Huntington, Dexter's findings and his speculations on reserve nerve energy were gripping. And so, when cataloguing the effects of climate on the behavioral traits of the peoples of Asia in 1907, he paused to commend Dexter's efforts to link the disintegration of the nervous system to states of electric or magnetic tension generated by atmospheric events. "His results confirm the popular belief in the highly invigorating influence of clear, cool weather," Huntington reported. Damp, muggy weather, by contrast, so affected the "vital functions" that "no surplus energy" remained to engage in "anything very active." Much worse were extremely dry conditions. Now people's "nervous equilibrium" became unbalanced, and "the power to control emotional impulses of all kinds" was weakened. Characteristically, Huntington was not slow to press Dexter's conclusions into the service of racial stereotyping. The climatic regime of Persia and Chinese Turkestan had produced peoples who were "highly emotional, and very lacking in self-control." Likely that was the reason why "the Persians are so prone to lying."[41]

Dexter's name continued to crop up in Huntington's later writings. The 1915 *Civilization and Climate* and *The Pulse of Progress* of 1926 are illustrative. In these works, Huntington pushed the influence of the weather on mental life in many directions. Unvarying climatic conditions were exceptionally damaging. In California, it was because the climate was "too uniformly stimulating" that he deemed it "a factor in causing nervous disorders." Indeed, Huntington suggested that the "people of California may perhaps be likened to horses which are urged to the limit so that some of them become unduly tired and break down." In the tropical world, uniformity worked in a different way. In both tropical highlands and lowlands, the "same deadening monotony" prevailed. And that prevented white races from permanently settling there. For under a tropical sun, hard work—elsewhere a virtue—put colonists and travelers "in great danger of breaking down nervously."[42]

For decades, Huntington continued to plot a climatology of mental vigor. In later publications, he devoted attention to the geography of mental activity, seeking to ascertain the psychological status of individual nations and the world as a whole. Among the range of factors shaping mental acuity, Huntington urged that climate had played a key role in generating different psychological profiles in different periods of history. What he called the "Dark Ages" and the "Revival of Learning" occurred "at opposite phases of a long climatic cycle." "Storminess apparently reached a low ebb in the Dark Ages but was abundant and violent in the fourteenth century. These two periods were likewise times of psychological contrast. The Dark Ages were characterized by widespread depression of mental activity, whereas the Revival of Learning ushered in a period of alertness and hope."[43] In a similar vein, Huntington extracted several pages of Dexter's writing to support his own speculation that ancient Greece and Rome, and indeed "practically all the other great empires of antiquity," enjoyed a climate, at the height of their dominance, that possessed the very qualities that Dexter associated with an overabundance of human vitality.[44]

At a more detailed scale of inquiry, and seamlessly moving from historical epoch to seasonal variation, Huntington, like Dexter, was also preoccupied with the influence of seasonal weather changes on mental life. He thus inquired into seasonal comparisons in intellectual vitality by scrutinizing student performances, dates of patent filing, the outcome of civil service examinations, and patterns of library book circulation. By mapping monthly variation in hospital admissions for mental health problems, the onset of insanity, and incidences of sexual crimes, he also insisted that all of these peaked during the summer months in a variety of locations throughout Europe and the United States. To explain the results of his inquiries into weather–mind correlations, he focused on connections between mental activity and storminess, as well as on the influence of electrothermal forces on mass psychology. Here he found the electrical hypothesis of Bernhard and Traute Düll on the influence of solar activity on the human psyche to be strikingly suggestive.[45]

Huntington's influence on medical writers concerned with the health of tropical travelers and the question of acclimatization was considerable. Sir Andrew Balfour (1873–1931), for example, a Scottish medical officer, novelist, and first director of the London School of Hygiene and Tropical Medicine, drew liberally on Huntington's inquiries for his assessment of white tropical settlement in the pages of *The Lancet*. In three successive months—June, July, and August 1923—Balfour directed his readers' attention to "sojourners in the tropics" and the medical challenges of adjusting to atmospheric conditions in the "tropic zone."[46] He began this set of reflections by recommending that Huntington's *Civilization and Climate* should "be studied by all concerned with what is certainly an imperial aspect of the public health."[47] Of particular importance, Balfour noted, was Huntington's assessment of climate's influence on the nervous system, even though it departed from the views of the Tropical Diseases Board, which had assessed the psychological health of American soldiers in the Philippines. In surveying conditions in Kenya, Balfour called on the authority of Huntington to identify the optimum temperature for mental activity and to confirm that the ideal climate was characterized by moderate changes with air cooling at frequent intervals. Of all the effects the state of the atmosphere induced in human beings—physiological, psychological, and pathological—the greatest was undoubtedly the psychological. Indeed, it was the stress that a tropical climate inflicted on the nervous system of European settlers that presented "the greatest barrier to white colonisation."[48] While on several matters of colonial climate and mental health Balfour acknowledged the absence of medical consensus, he nonetheless presented Huntington as a key voice in the debate.

For all that, Huntington and Dexter, among others, came under critical scrutiny in the monumental analysis, *Contemporary Social Theories*, published in 1928, by the brilliant, if controversial, Harvard sociologist Pitirim Sorokin (1889–1968). In a lengthy exploration of what he called "the geographical school," Sorokin probed a range of proposals connecting weather conditions and other environmental factors with a range of human behaviors. He took a comprehensively dim view of their claims. Like the eighteenth-century Scottish thinker Henry Home—Lord Kames—he turned a jaundiced eye on "the endless number of writers who ascribe supreme efficacy to climate."[49] While the geographical school spent time looking for new theories and identifying new correlations between some aspect of climate and some dimension of human behavior, Sorokin was sure that what was really needed was a far more "rigorous analysis" and careful discrimination between "what is valid and what is childish." It soon became clear that the climatic theories of Huntington, symbolic of the whole environmental-causation camp, were on the childish side. Huntington's treatment of seasonal patterns of death, suicide in particular, was simply misinformed. What a contrast that was to Durkheim's classic work on suicide, which, Sorokin insisted, had conclusively shown the inadequacy of the climatic thesis even for cases in which "the parallelism between the fluctuations of climate and suicide" was striking. That "brilliant analysis" stood in dramatic contrast to Huntington's altogether "rough" methodology. The monocular vision that the likes of Huntington and Dexter took of the whole subject meant that correlation was routinely misconstrued as causation. Statistical crudities—not least Huntington's fallacious inferences, hasty generalizations, and attempts "to solve a problem of multiple correlation by the use of inadequate methods of gross correlation"—rendered his claims about the influence of climatic conditions on worker efficiency, criminal behavior, clerical errors, intellectual power, psychological disturbance, nervous energy, and much else besides little more than "fictitious correlations."[50] When the wheat was sifted from the chaff, there was precious little wheat to be had.

Notwithstanding such criticisms, causal connections between meteorology and mass psychology continued to be promoted by a range of writers, many of whom were closely associated with the development of the science of biometeorology. The British climatologist Austin Miller, whose comments on tropical denigration have already attracted our attention, told the students who worked through his textbook on *Climatology*, originally published in 1931 and in print for decades thereafter, that psychologically each climate tended to have "its own mentality, innate in its inhabitants and grafted on its immigrants." Miller's climatological dogmatics had a sharp political edge to it. To him, the direct causal relationship between a changeable climate and mental vigor meant that all the world's great civilizations were located in the mid-latitude cyclonic belt. For Miller, the outcome was a simple climatic fact: "the temperate zone governs the tropical zone by virtue of its infinitely greater energy and initiative."[51] In Germany, from at least 1911, Willy H. Helpach (1877–1955)—politician and academic—was also writing on climate and culture, seeking to identify how environmental and atmospheric conditions molded human psychology. Early in his career, he developed the idea of geopsychic phenomena when he published on the influence of weather, climate, and landscape on the psyche and later expanded this line of work into a more comprehensive account of Kultur und Klima. In doing so, he resisted the critique of climate determinism penned by the likes of Durkheim and Sorokin. To Hellpach, climate conditioned the personality traits of people in different locations. He was convinced, for example, that religious sentiments and spiritual meanings were affected by the climate, and thereby he explained the denominational divide between Protestants and Catholics in northern and southern Germany. All this fitted with the close association he urged between climate, psyche, and the *Volk*—a perspective resonating with the "Blut und Boden" philosophy of National Socialist Germany during the 1930s.[52]

In the United States, key figures involved with the promotion of biometeorology likewise continued to foreground the agency of weather and climate in the constitution of group psychologies. Clarence Mills, whose eugenic proclivities earlier captured our attention, remained oblivious to Sorokin-style critique and continued to hammer away at climate's psychic repercussions during the 1940s. In a chapter on "Bad Moods and Falling Barometers," he asserted that the widespread intuition that emotions and personalities reflected climatic conditions was well founded. Stormy weather induced in farm animals restlessness and irritability, horses behaved in unexpected ways "when the barometer is falling," and pet dogs became more easily agitated. People were no different. Domestic squabbles were more common during periods of falling atmospheric pressure, as were headaches, forgetfulness, traffic collisions, and industrial accidents. All kinds of mental activity declined on "falling-pressure days."[53] While the brand of weather wisdom that Mills dispensed in *Climate Makes the Man* was directed toward a popular audience, it built on his reflections the previous year on "mental function" in his 1939 *Medical Climatology*. Suicide, homicide, mental instability, and nervous breakdown were to the fore, and again he attributed these to "the intensity of climatic stimulation" markedly associated with frequent and sudden storm changes. A drop in mental efficiency and feelings of futility were associated with falling temperatures and rising atmospheric pressure. And when he mapped the patterns of suicide and murder across the United States over a five-year period, Mills was sure that they rather neatly followed the pathway of cyclonic storms. These and related data, he told his readers, confirmed the likelihood that the climate was "of major importance in the question of mental instability, and that, of all weather factors, falling barometric pressure is most disturbing." Living in the cyclonic storm belt had delivered high energy and bodily vigor to Americans in the mid-latitudes. But the price to be paid for those advantages was high. For climatic variability brought in its train "nervous exhaustion, neurasthenia, insanity and suicides."[54]

As we have seen, the influential German climatologist Helmut Landsberg, who moved to the United States in 1934, also played a pioneering role in the promotion of biometeorology until his death in 1985. Concerned with particulate matter in the atmosphere and its significance for pollution and public health, he directed much attention to the effects of climate on human behavior. In his 1969 *Weather and Health*, essentially an introduction to the whole science of biometeorology, he allocated a chapter to "Weather, Performance and Behavior." Here he spoke of "the far-reaching influences of weather on performance and mental attitude," not least during the passage between warm and cold fronts in cyclonic weather systems. Human reaction times were illustrative. During the worst weather phase—at the onset of a low-pressure system—human responses were at their slowest. When temperatures rose above 32°C, telegraphers made more mistakes, typists transcribed material more slowly, factory worker performance declined, and industrial accidents increased. Suicides seemed to be associated with the rapid pressure changes that characterized weather-front activity, and riots were linked to higher temperatures. In these reflections, Landsberg was sympathetic to the idea that ions in the atmosphere, which became attached to larger aerosol particles, could be a significant explanatory factor. The hypothesis that there was "an interaction between fluctuating electric fields and human brain waves," he believed, opened up a fertile line of inquiry.[55] Throughout, Landsberg remained rather more muted than some of the other figures we have examined in his tracing of climate's influence on mental activity, and this indeed reflects the circumspection for which he was known in the scientific circles in which he moved.[56]

However speculative many of the proposals linking meteorological conditions and mental life may have been during the earlier decades of the twentieth century, and however frail the methodological armory that encouraged commentators to move rather casually between correlation and cause, the belief that behavioral and mental states obey, in one way or another, the dictates of climate has resurfaced with renewed intensity since the dawn of the twenty-first century, spurred in no small measure by the threats arising from climate change. Some of the themes that wend their way through this body of thinking, and something of the dimensions of this most recent manifestation of psycho-climatology, now warrant attention.

Sunshine, Season, and Psyche

Recalling Huntington's claims about climatic optima and Dexter's studies of the influence of the weather on children's behavior, environmental health campaigner and award-winning journalist Pat Thomas drew the attention of the readers of her popular 2004 book, *Under the Weather*, to the mental health impact of climatic conditions. Her more general observations on the medical consequences of the state of the atmosphere have been rehearsed earlier. Not surprisingly, such figures as Dexter, Petersen, Sargent, Huntington, Landsberg, and Tromp, all of whom were closely involved with the development of biometeorology, featured prominently in her text. And like them, she also gathered within the sweep of biometeorology a range of mental health conditions that she considered directly or indirectly the consequence of prevailing weather. She reported, for example, that "hospitalisation due to nervous disorders increased with solar activity" because of its effects on the brain. Higher concentrations of positive ions in the atmosphere during the passage of weather fronts, she assured her readers, were "associated with depression, nausea, insomnia, irritability, lassitude." And what she referred to as the "thermic law of crime" also made an appearance.[57] No doubt this was an allusion to the "thermic law of delinquency" that Belgian astronomer and statistician Adolphe Quetelet (1796–1874) proposed in 1842 to describe the seasonality of criminal offences that seemed to follow the pattern that violent crimes were more common in hotter climates and seasons, while crimes against property more frequently occurred in colder conditions.[58]

Besides these, Thomas also sought to bring her readers up to date on seasonal affective disorder—both winter and summer varieties. While the symptoms associated with this syndrome had been noted by German psychiatrist

Emil Kraepelin (1856–1926), who found seasonal patterns in a range of emotional disturbances, seasonal affective disorder was first named in 1984 by South African psychiatrist Norman Rosenthal, who carried out research on the subject at the National Institute of Mental Health in the United States. While reaction to his diagnosis was initially skeptical, his 1993 book, *Winter Blues*, has become a standard introduction to the subject.[59] Seasonal affective disorder was named a major depressive condition characterized by a sense of worthlessness, lack of concentration, drop in energy levels, and potential risk of suicide. The bipolar mood swings characteristic of the syndrome displayed a seasonal pattern. At the time, these observations chimed with work on photoperiodism—the physiological reactions of plant and animals to changes in daylength—and on seasonal and circadian changes in melatonin secretion.

When Rosenthal first mooted his thinking on the disorder to the general public via a reporter at the *Washington Post*, the response was overwhelming. Thousands made contact, many offering themselves as subjects for his ongoing research.[60] Since then, his work has had a prominent place in public consciousness. As Thomas Dixon commented in 2013 at a Witness Seminar held at the Queen Mary University of London on the subject, "Seasonal Affective Disorder, or SAD, is one of those relatively rare conditions that's come into common parlance—people who are not experts in the field make reference to it, describe, and experience their own emotional lives through this medical label of SAD. It's something that's got into popular culture." Indeed, at that same gathering, Norman Rosenthal himself observed that recognition of the disorder "has been much more embraced by the general public who identify with the symptoms than by the medical profession."[61]

Since it was first put forward, the causal chain connecting season and psyche has been recast in a variety of ways in response to particular empirical investigations. Something of the range of these modulations may be glimpsed by inspecting a few, more or less randomly chosen, instances. In 1984, the very year in which Rosenthal constructed seasonal affective disorder as a distinctive mental health category, two psychologists, E. Howarth and M. S. Hoffman, embarked on a project to determine the effects of weather on human behavior. Synthesizing work by a range of writers mostly from the 1970s and early 1980s, they focused their research on the influence of meteorological conditions on such expressive states of mind as willingness to cooperate, the experience of anxiety, powers of concentration, the susceptibility to aggression, and the inclination toward skepticism.[62] Noting some of the tensions between the findings of previous researchers, they sensed the need to determine the influence of different dimensions of the weather on a disaggregated range of mood types. Matters of depression and anxiety, it turned out, were not tied to any specific weather variable, whereas a sense of optimism was heightened when the number of hours of sunshine increased. Lethargy and sleepiness were typically accompanied by higher levels of humidity, as was lack of concentration and poor academic performance. And skepticism, "a cynical, doubting outlook," was apparently "affected by the meteorological variables precipitation, hours of sunshine, and barometric pressure." Taken in the round, they argued that their data "pointed to the dominant influence of humidity on several mood dimensions" but insisted that a multidimensional approach to the whole subject remained a real desideratum.[63]

Other proposals have also been advanced to refine the ways in which the psychological effects of weather might be channeled. Contradictory findings from other investigations suggesting that high temperatures were sometimes associated with high mood, sometimes with low mood, promoted one group of researchers to revisit received explanations for seasonal affective disorder. Their suggestion was that the length of time a particular subject spent outdoors significantly modified results and that atmospheric influences operated differently at different points in the year. The more time people spent outside on warm, high-pressure days, the more their mood improved. What also emerged, they claimed, was that during springtime, higher temperatures increased a sense of well-being but had a negative effect during the summer season, particularly among those participants living in hotter southern climates.[64]

As yet another inflection on the mood–meteorology nexus, it has been suggested that the prevalence of seasonal affective disorder varies with distance from the equator. The association between "acute psychosis and climatic variation," some have urged, is particularly evident "in tropical countries." This means that the "severity" of mental health impacts is "in part determined by the adaptive capacity" of populations. Those in poverty, "those geographically vulnerable to extreme weather events," those dependent on agricultural production for their livelihood, and those with a prior disposition toward mental illness are at much higher risk from the psychological effects of climate variation.[65] Yet further modifications have also been proposed. Another team of researchers sought to divide people into a four-part taxonomy of what they call "weather reactivity types." Individuals whose mood improved with sunnier conditions were labeled "summer lovers." The "unaffected" named those who displayed only weak associations between weather and mood. "Summer haters" were identified as people whose sense of well-being declined during warmer seasons. And "rain haters" were those for whom wet days induced a particularly bad mood. In this way of thinking, the mental and psychological impact of the weather depended much more on individual disposition, rather than place of residence or economic power or the adaptive capacity of society more generally.[66]

Meanwhile, concerns over seasonality and the psyche have been leeching their way into other spaces. The turn, mostly by economists and politicians, to determining measures of well-being among workers and in society more generally, for example, has prompted some to urge that transitory weather conditions should be taken into account when analyzing subjective well-being data. Noting that women are "more responsive than men to the weather," that a sense of "life satisfaction decreases with the amount of rain," and that lower temperatures boost feelings of happiness, some researchers have pointed to the need to consider prevailing weather when inquiring into personal senses of well-being.[67] Of course, the perceived insights of earlier biometeorologists—Petersen, Huntington, Dexter—had already been seized upon by some students of economics in their appraisals of the changing fortunes of the market. Edgar Lawrence Smith (1862–1971), an investment manager with the brokerage firm Low, Dixon & Company, for example, devoted two chapters of his 1940 work on economic change, *Tides in the Affairs of Men*, to psychological topics. Convinced that "stock price movements are far more sensitive to changes in mass psychology than any other series," Smith sought to identify the causes of the mental state of populations. Relying on biometeorological sources, he declared that he found himself "following the path of Dr Ellsworth Huntington." Smith's sense was that "the *kind* of weather" that prevailed over certain seasons had a direct effect on "the mental outlook of the more nervously sensitive members of the business and financial community," as well as on the mass of people "whose purchases and actions are the substance of trade." Applying these impressions to the United States, he characterized the inhabitants of the "northern tier of states" as "nervous, energetic people" who had built a complex economic structure. To him, the wavelike movements of stock prices were "in reality" the "ups and downs of the business psychology of a nation." Indeed, he went so far as to describe the graph of stock price movements as "a fever chart of mass psychology."[68] In a relative key, and six decades later, Philip Parker called the attention of readers of his 2000 *Physioeconomics* to the economic significance of "psychological consumption." Because neurochemical imbalances were connected to levels of natural sunlight, Parker discerned a range of seasonal shifts in mood and behavior that were distributed by latitude. This suggested to Parker that the demand for what he calls "purely psychological products" such as electronic entertainment, phototherapy, and certain prescription drugs varies according to the prevailing climate. Appealing to him was the conjecture that "a person who watches a lot of television in colder countries is considered normal," whereas in warmer environments, the same individual would be judged to be friendless or even "psychologically unbalanced."[69] In their efforts to achieve mental and emotional equilibrium, it seems, those living in colder zones need to consume far more items deemed to deliver psychological comfort.[70]

In more popular outlets, the appetite for keeping up to date with ongoing research on seasonal affective disorder and kindred psychological responses to the weather is strong. Take, for example, PsychCentral, an independent mental health platform that provides news updates on its website. Overseen by mental health professionals and a component of Healthline Media, it reaches a wide audience and presents reviews of recent research to its readership. In bulletins issued by John M. Grohol, founder and editor-in-chief of PyschCentral, the findings of some of the projects mentioned above have been summarized for a broad audience in pieces with titles such as "Weather Can Change Your Mood" and "Can Weather Affect your Mood?"[71] Public investment in the whole subject is undoubtedly high and not least at a time when the future impacts of climate change are frequently dramatized in various mass media outlets. All of which confirms the judgment that links between mind, mood, and meteorology occupy a definite niche in twenty-first-century popular culture. One prominent theme that wends its ways through many such discussions is the widespread belief that higher temperatures arouse human aggression. Later we will scrutinize the manifestation of this idea in the context of the influence of climate and climate change on armed struggle, civil war, and international conflict. But now we turn to the ways in which weather conditions are thought to induce, in individuals and groups, hostility, anger, deadly assault, and violent crime of one sort or another.

Aggressive Atmospheres

In its landmark report on the race riots in the United States during the so-called long, hot summer of 1967, the National Advisory Commission on Civil Disorders, widely known as the Kerner Report after its chair, Otto Kerner Jr., governor of Illinois and a federal judge, paused in its analysis to make passing comment on the prevailing weather at the time. Bold and controversial in blaming white racism for the riots, the report became an instant bestseller, and over two million Americans bought copies of the 426-page document. "In most instances," the report noted, "the temperature during the day on which violence first erupted was quite high. This contributed to the size of the crowds on the street, particularly in areas of congested housing." An informational footnote appended to this comment reported that eighteen outbreaks of disorderly behavior were recorded during a day when the temperature reached a high of 79°F and that in nine cases the temperature soared to 90°F.[72] The bond between high temperatures and hot tempers had long been a prominent theme in public consciousness. But the Kerner Report's "support for the hypothesized relationship between temperature and aggression" stimulated Robert Baron to inquire into empirical evidence for this claim.[73] His work thus provides a convenient entry point to the recent rejuvenation of a theme that snakes its way through a good deal of writing on ambient temperature and outbursts of hostility and violence.

In 1972, Baron, then serving as a psychologist at Purdue University and later as a distinguished professor of entrepreneurship at Oklahoma State University, set out to subject this widely held intuition to empirical scrutiny. Under laboratory conditions, Baron set up an experiment to manipulate the anger arousal of his subjects through the administration of electric shocks and to determine the influence of cool and hot temperature ranges on their responses. Contrary to all expectations, Baron reported, his data "cast serious doubt upon the view that high temperatures per se often play an important role in the elicitation of instances of collective violence." Indeed, it appeared that extremely high temperatures might "actually tend to inhibit the performance of overt aggressive acts."[74] He did suggest, however, that the uncomfortably hot conditions produced in the laboratory could well have proven to be so unpleasant that participants simply sought to escape from the stifling atmosphere. In a follow-up investigation, with fellow psychologist Paul Bell, Baron suggested that "aggression would first increase but then actually decrease as subjects are exposed to increasingly aversive circumstances."[75] All this meant that the impact of environmental conditions on aggressive behavior was far more complex, relative, and context dependent than was usually assumed.

Craig Anderson, a psychologist who held professorships at several American universities, including Rice University and Iowa State University, took up the challenge a few years later.[76] In a 1979 investigation with J. Merrill Carlsmith, Anderson highlighted what he saw as the failings in a project on ambient temperature and violence by Baron and the Indianapolis lawyer Victoria M. Ransberger.[77]

Critical to Carlsmith and Anderson's riposte was the argument that they had failed to take into account the fact that the number of days in certain temperature ranges was greater than others and that therefore the results were simply an artifact of data that had not been properly normalized. Over the course of a year, riots were indeed most common when the temperature was between 81°F and 85°F and fell off at higher temperatures. But that was just because that particular temperature range was more frequent than any other interval. Put simply: more days, more riots. To Carlsmith and Anderson, the evidence strongly favored the view that the "probability of a riot increases monotonically with temperature."[78] Some years later, focusing on crime—murder, rape, assault, larceny—Anderson insisted that the data he had accumulated confirmed his prediction that "violent crimes would be more prevalent in the hotter quarters of the year and in hotter years." Uncomfortably hot temperatures, he urged, generated negative affect, which was "then transferred to a salient object in the person's immediate attention, an object that can be seen by the person as a reasonable source of the negative arousal." Although Anderson was well aware that treating correlation as cause was problematic and noted that the influence of temperature was frequently channeled through such things as vacation time and alcohol consumption, he still concluded that "much of our negative arousal is actually due to temperature."[79]

In subsequent inquiries, Anderson reinforced this claim. When he provided a synoptic taxonomy of the various ways—direct and indirect—in which temperature stimulated human aggression, it was with the intention of identifying the most compelling account of why hot tempers and hot temperatures were causally linked. Empirical evidence supported this claim, he insisted, at many levels of analysis. Hotter years, hotter seasons, and hotter days were all marked by "more aggressive behaviors such as murders, rapes, assaults, riots, and wife beatings." And the association held true spatially. "Hotter regions of the world yield more aggression," he concluded. Globally, heat and hostility were apparently welded together. That public policy recommendations could be extracted from these findings did not escape his notice. By controlling temperatures, unwarranted human aggression could be reduced. It "may be more cost-effective (as well as humanitarian)," he speculated, "to cool prison environments as a means of reducing inmate violence rather that to increase the supervision and segregation of prisoners."[80] That same emphasis continued to characterize his subsequent writings. In 1997, with two coinvestigators, he pursued the "heat hypothesis," as he termed it, concluding that it enjoyed strong empirical support. And even if the precise causal mechanics of the heat–hostility syndrome remained contested, the predictable association had immediate instrumental value for social policy. "Regardless of the cause of the relation," he suggested, "it may be helpful to policy makers at a variety of levels to know that hotter periods of time will produce increases in aggressive behaviour." The reason was simply because "law enforcement personnel as well as policy makers do not necessarily need the exact causal mechanisms in order to use the information about the existence of the temperature aggression relation."[81]

Looking to the future, Anderson also turned to the implications of global warming for crime rates. An increase of 2°F in average temperature, he noted, was likely to bring about "more than 24,000 additional murders per year in a population of 270 million."[82] Yet more recently, he further spelled out the ways in which global warming is likely to induce higher levels of violent aggression. In particular, the direct effects of uncomfortably hot temperatures on human irritability and hostility were highlighted. Now he suggested that the "temperature-aggression link may be 'hard-wired'" on the basis of biological studies claiming that high temperatures activated those parts of the brain governing thermoregulation and emotional response, as well as stimulating the production of extra adrenaline. But he also drew attention to indirect effects of climate change on adolescents and children, as well as on communities whose livelihoods are suddenly put at risk. Rapid climate change, he noted, increased the likelihood of growing up in poverty, having poor childhood nutrition, being in dysfunctional families, and experiencing exposure to war. Chief among these "indirect pathways to increased violence" were "food insecurity, economic deprivation, susceptibility to terrorism, and preferential ingroup treatment," all of which were sure to be exacerbated by climate change.[83]

Often building on the foundations of the earlier publications of Baron and Anderson, as well as their coauthors, a minor industry on the climatological psychology of heat and hostility has developed over several decades. A mere smattering of some of the directions in which this impulse has moved will give a flavor of the whole. A real-life experiment on the influence of ambient temperature on horn-honking, in response to a car stalled at a green light, for example, found that aggression increased in a linear manner with increases in the temperature–humidity discomfort index.[84] Using a formidable array of statistical wizardry, researchers have claimed that Major League Baseball players are more likely to strike the opposing team's batter on hotter days.[85] An inquiry into the relationship between weather and telephone calls to Brisbane police complaining of domestic violence claimed to find significant associations between police calls and maximum air temperature during all seasons, even though the causal mechanisms remained obscure.[86] By contrast, an investigation of temperature and aggression among Dutch police officers returned the verdict that high temperatures "did indeed produce negative affect, but this negative affect did not contribute significantly in explaining aggressive police behaviour."[87] In this case, emotion obviously did not translate into behavior. Other investigators have proposed amendments to the hot temper/hot temperature thesis. The need to moderate the influence of atmospheric temperature by taking into account diurnal and weekly patterns of criminal behavior, for example, has been urged. Although the afternoon tended to be the hottest time of day, criminal behavior was not at its peak because routine work activities and school attendance kept people indoors. This meant that, judging by the crime data from the Minneapolis and Dallas police departments, the relationship between temperature and assault was at its strongest during the evening hours. This challenged Anderson's claims about the ubiquitous effects of high temperatures but still retained the determining influence of temperature at particular temporal points.[88] The same combination of causes was also called upon to ascertain the role of temperature in explaining patterns of sexual violence in the United States.[89]

A few remarks on what I refer to as the developing thermic theory of aggression seem appropriate. First, the scenarios that psychologists and others concerned with climate and crime have plotted have become sufficiently significant to warrant inclusion in the *Encyclopedia of World Climatology*. In a brief review, the author of this piece tracked the association between hot weather and hostile behavior back to figures like the French lawyer and statistician André-Michel Guerry (1802–1866), whose moral statistics delivered a distinctive climatic geography of crime, and Adolphe Quetelet (1796–1874), who identified seasonal patterns of delinquency. The names of Dexter and Huntington also figured as precursors to more recent writings, though the author felt that the relationship between climate and crime was "more suggestive than definitive."[90]

Second, the identification of precisely what causal powers may be attributed to ambient temperature in inducing aggression has been treated in rather different ways by different commentators. Some remain cautious, others casual, yet others confident in the explanatory language they adopt. One team reports that they "have tried to avoid words that imply causation" and therefore remain content with identifying "a reliable correlation between temperature and assaults."[91] Another author simply declares that "weather has a strong causal effect on the incidence of criminal activity" and that "higher temperatures lead to higher crime rates."[92] Such irresolution suggests an ongoing ambivalence over the specter of climatic determinism. Perhaps because of their discipline's troublesome legacy of environmental determinism, two geographers addressed this question head on in the wake of their work on the connections between assault and heat stress in Dallas during the summer of 1980. Acknowledging that "there is some association between heat-related discomfort and aggressive human behaviour," they went on to argue that their analysis had the potential to "reconcile the hitherto polarised views represented by climatic determinism and folklore, on one hand, and the mass of work denying the validity of determinism, on the other." To them, a "complete rejection of deterministic ideas" was inappropriate because weather-induced discomfort played some role in accounting for aggressive behavior.[93] Whether the tensions exhibited in these remarks have been satisfactorily resolved remains to be seen.

Third, in light of widespread fears about current climate change, many students of climate and crime have felt the need to project their findings into the future. Despite commenting that his estimates were "highly uncertain," one author used findings from a fifty-year data set on monthly weather and crime statistics to predict, with remarkable precision, that "climate change will cause a substantial increase in crime in the United States," bringing an "additional 22,000 murders, 180,000 cases of rape, 1.2 million aggravated assaults, 2.3 million simple assaults, 260,000 robberies, 1.3 million burglaries, 2.2 million cases of larceny, and 580,000 cases of vehicle theft."[94] Another insists that climate change "will foster beliefs favorable to crime, contribute to traits conducive to crime, increase certain opportunities for crime." Among these, fear, anger, and frustration loomed large. While the whole subject continued to be "neglected by criminologists," this observer was sure that climate change "will become one of the major forces driving crime as the century progresses."[95] The list could certainly be extended. Another inquiry using two decades of monthly climatic and crime data from St. Louis, Missouri, came to the conclusion that changes in climate were having a greater impact on levels of violence in disadvantaged communities than in areas with more affluent residents. On the basis of these findings, the author predicted that in the future, 20 percent of the most disadvantaged neighborhoods would "absorb over 50% of climate change-related increases in violence."[96]

Of course, those interested in the influence of weather and climate on mind and emotion do not restrict their concerns to aggression, hostility, or related psychological disorders or one kind or another. Some have pursued the idea that personality traits more generally, positive as well as negative, come under the influence of the climate in different ways. Something of how this proposal has been developed now warrants a moment's attention.

Ambient Temperature, Personality Traits, and Climato-Economic Habitats

In 2017, an international team of researchers joined forces to investigate human personality traits across different geographical regions, drawing on data from China and the United States. "Because humans constantly experience and react to ambient temperature," they suggested, temperature was a promising candidate in accounting for the habitual behavioral patterns of individuals that underlay personality traits.[97] To get the project under way, they sought to collect the myriad of personality characteristics under five broad traits: agreeableness, conscientiousness, emotional stability, extraversion, and openness to experience.[98] This assemblage was further divided into "Alpha" qualities—the first three just noted—and "Beta" characteristics, representing the latter two. These were characterized as the socialization factor (Alpha) and the personal growth factor (Beta). The argument was that because human beings are a warm-blooded species and have "the existential need for thermal comfort," what they called "ambient temperature clemency"—they specified a "psycho-physiological comfort optimum of 22°C"—was a critical component in personality formation. By this they meant mild climatic regimes that encouraged inhabitants "to explore the outside environment where both social interactions and new experiences abound." Such conditions stood in marked contrast to environments characterized by extreme heat or cold. The causal chains through which these atmospheric conditions impressed themselves on individual and group personality were intimately associated with a number of proposals advanced to explain the "geographical variation in personality." The suggestion that different subsistence strategies produced "personality-related cultural constructs" could be deepened, the team argued, by taking into account the ways in which ambient temperature shaped agricultural practices. So too could the "selective migration theory," according to which "people selectively migrate to regions that fulfil and reinforce their physical and psychological needs." And it was the same with the "pathogen prevalence theory of personality," which suggested that people were less likely to migrate to regions with "a higher prevalence of disease-causing pathogens." For all of these, the investigators argued that "temperature clemency" provided a mechanism explaining how "macro-level environmental forces might shape individual-level personality." And they concluded that inhabitants of clement regions "scored higher on both the socialization factor . . . and the personal growth factor."[99]

As soon as this research was published, it was picked up by a science editor, Charles Choi, who drew its findings to the attention of the readers of LiveScience, a digital platform dedicated to bringing the fruits of scientific inquiry to online audiences. Choi focused on the thought that habits shape personality and argued that if temperature changes bring about behavioral adjustments, personality changes are sure to follow. This meant that climate change was likely to bring about shifts in personality traits around the globe. Choi reported that Wei and his collaborators had confirmed that "people who grew up in climates with milder temperatures were generally more agreeable, conscientious, emotionally stable, extroverted and open to new experiences." These findings held true for those surveyed in both the United States and China, despite differences in gender, age, and average income. As part of the report, Choi interviewed Evert Van de Vliert, a behavioral and social scientist at the University of Groningen, who works on cross-cultural psychology often in the context of climatic influence. Van de Vliert cautioned against the suggestion that "our ancestors, and we ourselves of course, are passive products of where we live." By economic means—"intelligently and actively using property and money"—he explained, human beings "can and do create their own identity and destiny in harsher climates." Due to different economic circumstances, harsher climatic conditions induced in some places a more "collectivist personality" and in others a more "individualist personality." Climate's influence was always filtered through the channels of local economic power.[100]

Indeed, Van de Vliert already had been long embarked on demonstrating that human psychology was rooted in what he called "climato-economic habitats." In 2013, he scrutinized the links between different climatic regimes and the distribution of what he called "fundamental freedoms" across the earth.[101] Crucial to his analysis was the claim that freedom from want, freedom from fear, freedom of expression, freedom from discrimination, and freedom to develop one's potential were all linked to the habitats that people occupied. Those habitats were characterized by their climatic conditions and by the available monetary resources they enjoyed to satisfy their needs, resolve their stresses, and fulfill their goals.[102] The habitat taxonomy he devised was thus a synthesis of seasonal climate and monetary wealth. "Threatening habitats" were those locations where the occupants were poor and the winters or summers were demanding. "Comforting habitats" were venues with undemanding climates in both poor and rich areas. And "challenging habitats" were depicted as rich areas with demanding winters or summers. Monetary resources plainly relativized the classification of climates. It was wealth that transformed "demanding" climates from threat to challenge. The psychocultural fallout of these experiential geographies was enormous. For poor regions, demanding climates "increase closed-mindedness and risk aversion" and promote behaviors that avoid "ambiguity by making relatively unfree choices that are necessary and routine rather than autonomous and adventurous." By contrast, demanding climates for the rich bring "open-mindedness and risk seeking" and deliver cultures celebrating ambiguity.[103]

Van de Vliert grounded his entire analysis in the thermal biology of metabolic processes, thereby giving priority to temperature as a crucial actant in human affairs. Above and below the thermoneutral zone, greater metabolic activity is required to maintain the body's core temperature. The cost of achieving existence needs is thus greater in the colder and hotter climates beyond the temperate zone. What drives the resulting psychological geography is the capacity to meet these climate-induced needs by the available wealth. Simply put: "Money has a crucial part to play in coping with bitter winters and scorching summers." Different climato-economic habitats, in consequence, induce distinctive "psychobehavioral adaptations" among their inhabitants. These traits, in turn, coalesce into distinctive cultural syndromes, enabling generalizations across different thermal and economic domains. For example, on the basis of a large cross-cultural research project, Van de Vliert claimed that while poor inhabitants of demanding climates focus on "existence needs," rich occupants of the same climatic habitat are able to dwell on "growth needs." As for goals, these two groups respectively aim for survival or self-expression. Politically, the first concedes to autocracy, while the latter encourages democracy. Van de Vliert worked through a sequence of such psychobehavioral adaptations to needs, stresses, goals, means, and outcomes and related these to the different climato-economic habitats he devised. One extract, emblematic of the whole, gives a sense of how each of the themes is treated:

> The analyses revealed that people residing in climates with more demanding winters or summers have no other choice but to endorse relatively unfree survival goals at the expense of moderately free easygoing goals and relatively free self-expression goals to the extent that their societies are economically deprived in terms of national income per head, household income, and economic growth.

Among other things, what this affirmation conveyed was a rejection of mono-causal climatic explanations and an insistence on conjoining atmospheric agency with economic forces. Climatic demands certainly influenced cultural groups, but the "effects can be observed only if we distinguish between poor and rich populations." In this way, Van de Vliert stood opposed to earlier forms of climatic determinism promulgated by the likes of Huntington. Fully aware that "psychobehavioral adaptions to climate are a sensitive subject given their history of single-factor determinism," he insisted that the demands of atmospheric temperature were always mediated through available economic wealth. Climate, capital, and cultural psychology were intimately intertwined.[104]

Not all of Van de Vliert's readers were convinced, though, that he had escaped the clutches of determinism. Two Estonian psychologists, Jüri Allik and Anu Realo, for instance, commented that Van de Vliert had worked hard "to convince readers that almost everything from charismatic leadership to driving performance is determined by climate" but complained that earlier efforts to causally connect climate with character remained unproven and that more recent proposals arguing that climate induced "aggregate personality traits" were no more successful. "If almost everything is explained by cold winters or hot summers," they quipped, "then nothing is explained."[105] Dominik Güss also discerned a number of limitations in Van de Vliert's analysis and concluded that the postulation of a direct relationship between climate, monetary resources, and psychological traits was an oversimplification that did not take sufficient account of intervening political and cultural processes.[106] By contrast, a group of psychologists from Florida State University read Van de Vliert's proposal as confirming that individual identity, forms of selfhood, and freedom of choice were predominantly "a creation of culture" given his emphasis on the conjunction of material wealth and climate stresses in explaining different psychological traits.[107] Yet again, tensions congregating around the explanatory nexus linking climate and the human psyche have resurrected themselves. Even those conscious of the burden placed on them by the history of climatic determinism have resorted to causal language, whether or not they claim to disavow its excesses.

The map I have constructed here of mind, mood, and meteorology is, of course, both partial and provisional and could be extended into other related realms. One arena in which climatic conditions abut on depression, anxiety, substance abuse, and related disorders centers on how extreme weather events and natural catastrophes, such as floods, hurricanes, drought, heatwaves, and wildfires, affect both individuals and communities. Such weather-related disasters, along with fears of climate change more generally, may incite anxieties arising from uncertainty about life in the future and the very survival of the human species.[108] Zones of prolonged drought, for instance, are reported to experience higher than usual levels of emotional distress.[109] What might be called therapeutic climatology is another such domain. By this I have in mind efforts to induce beneficial states of mind among mental health patients by exposure to certain climatic environments. One rehabilitation center in Thailand, for example, insists that "warmer weather means less stress." Cold conditions, this Wellness Centre contends, put stress on the human body and are less suitable for recovery from depression, bipolar disorder, general anxiety syndrome, and so on than warmer climates, where the human species originally evolved.[110] A related treatment is the use of light therapy to combat the seasonal affective disorder that is reported to afflict around two million people in the United Kingdom. So as to emulate brighter conditions, the Health Scotland website promotes the use of light boxes for a couple of hours a day for those suffering from seasonal depression.[111] In many ways, such practices continue a naturalistic tradition of mental health therapeutics that promoted natural history, appropriate climatic conditions, and out-of-doors fieldwork in rural settings as aids to the recovery of psychological well-being.[112]

Besides these, causal links have been identified between climatic conditions and intellectual creativity. In explaining what they refer to as "the geography of creativity," two psychologists, noting that "Nobel laureates, technological pioneers and innovative entrepreneurs are unequally distributed across the globe," looked to the influence of latitude-related cold and hot conditions as triggers for invention. Their argument was that climatic ecosystems incorporating "thermal necessities, monetary opportunities, and their parasitic repercussions" shape "interdependent pressures on creativity." This eco-theory of creativity proposes that further cooling of cold regions and the warming of hotter zones are likely to "hinder creativity in poor populations but promote creativity in rich populations."[113] To one degree or another, proposals like this connect with attempts, notably by Harold Dorn, to find in climate, hydrology, and topography the mainsprings of scientific ingenuity and technological innovation. In keeping with Karl Wittfogel's ecological reading of ancient history and his advocacy of the hydraulic hypothesis, Dorn tracked the emergence of early science to the need for water management.[114]

Efforts to determine the causal conduits through which weather and climate have reportedly conditioned the minds, moods, and mental health of individuals and populations have been manifold and persistent. Sometimes curious, frequently contradictory, often creative, these continue a long tradition of linking mental states with meteorology. Montesquieu's physiological theory of mind centered on the ways in which he believed the body's fibers expanded or contracted in response to climatic conditions. Falconer attributed intellectual advances to the effects of colder weather and likewise traced differences of religious conviction to climatic causes. Johnson's discourse on tropical medicine made much of the malign influence of a tropical sun on the alimentary system with its "sympathetic" connections to the human brain. For Woodruff, it was the role of the sun's actinic rays that induced nervous exhaustion and mental depression. An explanatory inventory of such proposals could readily be extended. Dexter turned to the influence of meteorological conditions on human reserve energy, which was vital to psychological functioning; Huntington made much of the influence of the deadening monotony of hot climates on the human psyche; Mills connected restlessness and irritability with falling barometric pressure and stormy conditions; Landsberg believed there were significant connections between electrical activity in the atmosphere and human brainwaves.

The genealogy of the idea that climates shape the minds, emotions, and beliefs of people turns out to be complex and multilayered. And yet the conviction that the empire of climate extends its influence into the deepest depths of the human psyche remains strong. Indeed, there have been continuing efforts to put to the test Montesquieu's eighteenth-century speculations about the climatic determination of human psychology.[115] In an era of unprecedented anthropogenic climate change, that fascination shows no sign of abating and may even be intensifying.

PART THREE

Wealth

7

Weather, Wealth, and Zonal Economics

"CAN THE climate of a country determine its wealth?" "The economic advantages of life in a cold country." "Hot climates may create sluggish economies."[1] These are just a sample of headlines calculated to catch the attention of readers of online blogs, academic journals, and news magazines. What underlies this preoccupation with welding weather, work, and wealth into a causal sequence is the thought that in some deep and important way, the global geography of poverty and affluence and the vicissitudes of national economies are shaped by climatic circumstances. Sometimes this takes the form of what has been christened the "sun and sloth" or "latitude and lassitude" theory—namely, the idea that high temperatures cause lethargy and listlessness among their populations and thus a disinclination toward hard work. At other times, the blame for poverty rests on the claim that hot climates encourage the proliferation of organisms hostile to the human race, which breed rapidly and foment a rampant transmission of infection and disease.[2] Sometimes it is the link between climatic regime and food production that is believed to massively govern the global geography of what has been dubbed History's haves and have-nots.[3] Because climate determines the incubation, development, and disease resistance of plants, this story goes, some locations enjoy insuperable advantages while others are condemned to sickness, scarcity, and poverty. On yet other occasions, the chain of causation runs along the train tracks of longitude inasmuch as invention and innovation are believed to be energized by more challenging, more stressful, yet more stimulating climatic regimes. As Don Pittis, business columnist for the Canadian Broadcasting Corporation, explains, "In northern climes, people need clothing and buildings and the fuel to heat them. We need systems for providing clothing, building materials and fuel. . . . The colder the climate, the more those things are needed. They are not luxuries; they are necessities."[4] In this narrative, it is the thermal environment that drives the economy. For necessity, after all, is the mother of invention.

Not all, of course, succumb to such raw reductionism. As Shreshtha Mishra, writing for *Qrius*, formerly *The Indian Economist*, tellingly observes, exceptions need to be considered. Countries like Malaysia, Thailand, and Singapore are in fact some of the rich nations in Southeast Asia. Plainly put, "correlation does not imply causation."[5] The New York University economist William Easterly concurs, noting that links between cooler climates and robust economies are merely accidents of time and place. To him, Europe and America benefited from the profits of the slave trade and, once ahead, stayed ahead.[6]

Criticisms notwithstanding, the idea that weather and wealth are causally conjoined has been both deep and lasting. In 2009, *Vox*, the policy portal for the Centre for Economic Policy Research, ran a piece entitled "Does Climate Change Affect Economic Growth?" by three economists—Benjamin Jones, Benjamin Olken, and Melissa Dell. Noting the long historical relationships between climate and national economic performance, they paused to comment that observations of this "phenomenon date at least to Ibn Khaldūn's 14th century *Muqaddimah*, appear in Montesquieu's 18th century *The Spirit of Laws* (which famously argued that an 'excess of heat' made men 'slothful and dispirited'), and have been confirmed in modern data."[7]

The confirmation to which these authors refer is to an article marking the election of the Yale economist William D. Nordhaus—later, in 2018, to receive the Nobel Prize for Economics—to membership of the National Academy of Sciences. In an account tracing the global patterns of macroeconomic activity to geographical factors, Nordhaus foregrounded the role that climate exerts in one way or another on economic performance. "In reflecting on the wealth of nations," he remarked, "early economists assumed that climate was one of the prime determinants of national differences." This was especially so in agricultural societies where the vast bulk of the population lived on farms and were thus subject to the seasonal and daily rhythms of the weather. He therefore agreed that earlier civilizations were far more reliant on climatic conditions and local resources than modern economies, and they were correspondingly less able to engage in specialization and complex forms of commercial exchange. Nonetheless, he was critical of recent accounts of international differences in contemporary productivity inasmuch as they had generally given "short shrift to climate as the basis for the differences in the wealth of nations." Through his own work and the writings of a number of other recent economists,

Nordhaus hoped to challenge this trend. Using new methods of handling macroeconomic data and deploying a range of sophisticated analytical tools, he came to the conclusion, put forward as a testable proposition, that "the relationship between temperature and output is negative when measured on a per capita basis and strongly positive on a per area basis." Two of his findings resonated particularly strongly with long-standing convictions held by champions of climatic readings of wealth: first, that "output per capita rises with distance from the equator" and, second, that "Africa's geography is indeed a major economic disadvantage relative to temperate countries."[8]

In these respects, there is rather compelling justification for directly connecting Nordhaus' analysis with the spirit, if not the letter, of the geographical philosophy of history elaborated in the fourteenth century by an Arabic scholar regarded by some "as the true father of historiography and sociology,"[9] Ibn Khaldūn, and codified in the eighteenth century by the celebrated Enlightenment jurist and man of letters, Montesquieu. The skeletal lineage that Jones, Olken, and Dell identify gestures toward a long tradition of thinking climatically about wealth, one whose intellectual provenance and political complexion are worth exploring in greater detail. For elucidating something of this history is crucial to plotting the genealogy of the more recent rejuvenation of the project to bring the wealth of nations within the compass of climate's expansive empire.

Zonal Wealth and Physiological Economics

In pursuing what he senses as an imperative to geographize macroeconomics, Nordhaus remarked that recent work by economists had confirmed "that geography (measured as distance from the equator) was among the most significant variables behind differences in per capita output by country" and that, with startling precision, "the 'optimal' temperature" for the maximization of individual economic output "is ≈12°C." This suggested to him that national economies wax or wane depending on their climates. Thinking in the categories of global latitudinal zones and climatic optima, he announced that his "major finding" was that "tropical geography has a substantial negative impact on output density and output per capita compared to temperate regions."[10] Economic performance, national wealth, and industrial productivity were simply expressions of how the earth's climatic zones have dictated the functioning of the global economy.

Zonal macroeconomics conducted in this key clearly echoes the philosophy of universal history put forward by the fourteenth-century Islamic student of civilization from Tunis, Abū Zayd 'Abd ar-Raḥmān ibn Muḥammad ibn Khaldūn (1332–1406), diplomat, professor, and judge, variously described as the father of history, economics, and sociology.[11] Ibn Khaldūn's reputation as a founder of the climate and civilization school of global history rests largely on his historiographical reflections in the second, third, and fourth prefatory discussions in his celebrated introduction to history, *Al Muqaddimah*, which he completed in 1377. As Gates observed, "It could well be said of him, as it has been remarked of Montesquieu, that he 'saw everything in climate.'"[12] Here, Ibn Khaldūn elaborated his sevenfold climatic division of the globe from the equator northward, with the Southern Hemisphere remaining beyond the reach of his zonal classification system. Focusing attention on those regions of the earth where advanced cultures flourished, he pondered why the northern quarter had a greater concentration of civilizations than its southern counterpart and why they had most fully developed between the third and the sixth zones, leaving the South "all emptiness." The reason lay in "the excessive heat and slightness of the sun's deviation from the zenith in the south."[13] This meant that moving from south to north from the equator, the first and second climatic zones were bereft of cultural refinement. By contrast, civilization burgeoned in the third and fourth zones and, beyond that, extended to one degree or another on into the fifth, sixth, and seventh.

In Ibn Khaldūn's mind, the conduit through which the forces of climate flowed was what I would call the physiology of culture. This was because the natural order and the operations of the human body—growth and decay, mortality and desire—were intimately intertwined and were, in turn, inseparable from his understanding of civilization. As Fromherz comments, "The clockwork of the body and the clockwork of nature were the clockwork of society."[14] To Ibn Khaldūn, the power that the climatic zones exerted on human society was on account of the "influence of the air upon the color of human beings and upon many (other) aspects of their condition." Because of the temperate climate in the fourth zone and the broadly similar conditions that inhabitants of the bordering third and fifth zones experienced, the culture of these regions displayed moderation rather than excess, sobriety rather than indulgence.

> The sciences, the crafts, the buildings, the clothing, the foodstuffs, the fruits, even the animals, and everything that comes into being in the three middle zones are distinguished by their temperate (well-proportioned) character. The human inhabitants of these zones are more temperate in

> their bodies, color, character qualities, and general conditions. They are found to be extremely moderate in their dwellings, clothing, foodstuffs, and crafts. They use houses that are well constructed of stone and embellished by craftsmanship. They rival each other in production of the very best tools and implements. Among them, one finds the natural minerals, such as gold, silver, iron, copper, lead, and tin. In their business dealings they use the two precious metals (gold and silver). They avoid intemperance quite generally in all their conditions.[15]

Welding together climate and character, Ibn Khaldūn now elaborated his global geography of civilization and thus his account of the wealth of nations. As Jean David Boulakia remarked, for Ibn Khaldūn, the wealth of nations did "not consist of the quantity of money that the nations have, but of their production of goods and services."[16] Because the denizens of the middle zones were temperate in both physique and character, as well as in their customs and behavior, they enjoyed all the natural conditions required for economic growth and the development of civilized society. Chief among these were "ways of making a living, dwellings, crafts, sciences, political leadership, and royal authority." Related cultural attributes attended these advances: prophetic leadership, dynastic succession, religious legislation, intellectual advancement, urban architecture, horticultural arts, "splendid crafts," and indeed "everything else that is temperate."[17]

The measured assets of life and culture in temperate regions stood in marked contrast to the inhabitants of the other zones.[18] There, everything was far removed from life in the moderate middle. In one sphere after another, Ibn Khaldūn itemized the devastating human consequences of life lived under the empire of climate at its cruelest edge. Native buildings were made from mere clay and reeds, and diet consisted of durra—a kind of millet—and herbs. Clothing came from the leaves of trees or animal skins patched together to serve as meager covering. Fruits and flavors, foodstuffs and seasonings, he presented as "strange and inclined to be intemperate." In their commercial transactions, hot climate peoples did not trade with "the two noble metals" but instead conducted transactions with "copper, iron, or skins, upon which they set a value for the purpose of business dealings." The black inhabitants of the first zone dwelt in caves and thickets, consumed herbs, lived in savage isolation, and practiced cannibalism. All in all, Ibn Khaldūn painted a picture of the world of the South as alien, distant, destitute; wholly other. Assigning blame for this degraded state of affairs was not a difficult task. It was their remoteness, both spatial and spiritual, from the temperate zone that had produced in these native peoples "a disposition and character similar to those of the dumb animals, and . . . correspondingly remote from humanity."[19] For those corruptions equally applied to their religious beliefs and practices. It was a universal truth. Occupants of what he described as the intemperate zones, both North and South, remained in a profound state of spiritual ignorance. Conditions indeed were so grim that Ibn Khaldūn came close to banishing them from the human family altogether. Dehumanized and animalized, they had no role to play in the household of nations. When guided by the edicts of climate, this was the world that grid-line ethnography delivered.

The chief channel through which climate condemned life on the zonal margins, and indeed beyond the zones altogether, was the physiology of the human body itself. Ibn Khaldūn's observations on the black races are illustrative. The operations of the climate on human biology had immediate implications for disposition and culture alike. Black "levity, excitability, and great emotionalism" were occasioned by the "expansion and diffusion of the animal spirit." In turn, the functioning of these psychoneurological mechanisms could be traced to the way in which "heat expands and rarefies air and vapors and increases their quantity."[20] Because they occupied the hot zone of the earth, their temperaments and ways of life were dominated by heat. In similar fashion, he disparaged white northern Slavs, and those further north still, declaring that civilization could never develop "in the area between the sixty-fourth and the ninetieth degrees, for no admixture of heat and cold occurs there because of the great time interval between them. Generation (of anything), therefore, does not take place."[21]

Even in those zones where the climate fostered civilization, atmospheric conditions continued to exert their influence at more local scales. In what he described as dense civilizations, Ibn Khaldūn was convinced that plagues arose on account of the corruption of the air—local atmospheric pollution—arising from "putrefaction" and "many evil moistures." Lessons could be learned from those epidemics that arose in the later stages of a dynasty. In order to facilitate the free circulation of air, it was vital that "empty spaces and waste regions" should be left between areas of densely populated urban settlement. In this way, the pestilences that occurred in heavily congested cities such as Cairo in the East and Fez in the Maghrib could be circumvented.[22] These sentiments confirm Fromherz's observation that although Ibn Khaldūn was reared in a rather elite urban setting, he "did not have the common prejudice of the urban intellectual who saw distance from the city as inversely proportional to intellect." Plainly, urban growth brought with it health hazards of a kind that rural communities escaped. This resonated with his belief that civilization could be tribal as well as urban, a claim that stood in opposition to those Arab philosophers who presumed that urban culture was intrinsically superior to Bedouin lifestyles.[23]

Besides the health risks of densely populated cities, Ibn Khaldūn was convinced that civilization was a precarious achievement in other ways too, however solid its climatic foundations. Surplus wealth and indulgent living could easily turn a people decadent and render them vulnerable to conquest. The ruling classes, all-too-soon mistaking luxuries for necessities, desires for needs, typically exerted more and more demands on laboring classes and imposed higher taxes, thereby fomenting rebellion. In the globe's middle zones, the rise and decline of civilization thus followed a cyclical pattern in which profligate urbanites were displaced by lean desert invaders who, in their turn, would find themselves supplanted as they succumbed to the enticements of a sedentary lifestyle.

Ibn Khaldūn's climatic philosophy of history, of course, was domiciled in a different explanatory universe from that of contemporary geographical macroeconomics. His version of scientific history was not divorced from, indeed was all of a piece with, his interest in divination, numerology, astrology, magic, talismanic art, and the like.[24] Not that he endorsed all of these practices. In point of fact, he was deeply critical of many aspects of the occult arts and grounded his refutations in theological, epistemological, and scientific reasoning. To some degree at least, his reflections on such subjects may well have been on account of the prophetic and messianic movements that swept through North Africa between the tenth and fourteenth centuries. As an orthodox Muslim, Ibn Khaldūn considered many such practices to be socially harmful. To him, astrology was not only false from an empirical standpoint but also harmful to human civilization, hurting "the faith of the common people when an astrological judgment occasionally happens to come true in some unexplainable and unverifiable manner. Ignorant people are taken in by that and suppose that all the other (astrological) judgments must be true, which is not the case," he warned.[25] And yet, as Musegh Asatrian has noted, Ibn Khaldūn's stance on astrology is entirely ambivalent. To be sure, he accepted the possibility that astral occurrences could influence earthly events. And when he turned his attention to prophecy and divination, he was sure that, when true, their pronouncements were consonant with astral conjunctions. Nevertheless, he remained chary of those seeking to bring "scientific scrutiny" to bear on such outcomes.[26] For

Ibn Khaldūn, if indeed astrological forces did influence human affairs, they remained beyond the scope of rational inquiry.

Ibn Khaldūn clearly believed that genuine knowledge could be attained through bodily senses or the inspiration of supernatural beings and that a distinction could be made between legitimate and illicit magic. With some justice, Hayden White remarked that the roots of Ibn Khaldūn's philosophy of history lie "not in a nascent secular science of society, but in the tenets of Asharite theology, Muslim jurisprudence and the Muslim adaptation of Greek philosophy."[27] Reading him as a lonely practitioner of positivist historicism therefore seems hopelessly presentist. Nevertheless, his foregrounding of climatic causes can be seen as something of a move toward naturalizing historical explanation. Explicitly concerned to depart from "blind trust in tradition" and to apprehend the deeper meaning behind events, Ibn Khaldūn sought to elucidate the universal laws of cause and effect in the life and times of human society.[28] What he was in pursuit of was a normative method that would enable the truth to be distinguished from falsehood when inquiring into the meaning of historical events. His rejection of popular accounts of the origin of the black races is a case in point. Referring to the Noahic narrative, he castigated those genealogists, who with "no knowledge of the true nature of things" imagined that "Negroes are the children of Ham, the son of Noah, and that they were singled out to be black as the result of Noah's curse, which produced Ham's color and the slavery God inflicted upon his descendants." To Ibn Khaldūn, that account betrayed ignorance not only of the detail of the Mosaic text but also "of the true nature of heat and cold" and of the means by which these elements had brought about differences in skin color. Black skin had nothing to do with execration and enslavement but was "the result of the composition of the air" and "the influence of the greatly increased heat in the south."[29]

The inclination to salute Ibn Khaldūn as a prime mover in the project of causally connecting climate, civilization, and culture owes a good deal to Franz Rosenthal's 1958 translation of *Al Muqaddimah* into English. So too, I suspect, does the recent propensity to venerate him as the inspiration for seeking in weather the wealth of nations. On its publication, Hayden White acclaimed the Rosenthal text "a masterpiece of the editor-translator's peculiar art" that had marvelously opened up the landscape of Ibn Khaldūn's mind to English-language readers. Noting that, in Ibn Khaldūn's vision, civilization only came into being in the central zones of earth where "a proper or harmonious mixture of heat and cold allow for the growth of nature and humanity," he announced that the *Muqaddimah* delivered "the most thorough examination of the relation between environment and society in historical literature between Herodotus and Montesquieu."[30] Shortly thereafter, in 1961, Ernest Gellner issued his judgment on the translation, declaring it "a magnificent production of a classic." Gellner fastened on the significance of the attention Ibn Khaldūn gave to the problem of social development and to his identification of an "antimony between civilization and social cohesion." In Gellner's rather Whiggish reading, what was particularly admirable about Ibn Khaldūn was his combination of "extreme tough-mindedness" in his espousal of "the proper method of social inquiry" with his "religious and anti-philosophical attitude to wider matters."[31] That same year, Gellner, reflecting on dynastic succession born of the tensions between preindustrial towns and dissident tribalism in Morocco, found Ibn Khaldūn's portrayal of this "circular destiny" so persuasive that he deemed his analysis to be of abiding relevance.[32] Subsequently, Gellner continued to rely on Ibn Khaldūn's diagnoses, not least in his *Muslim Society* of 1991, where he frequently resorted to the *Muqaddimah* in his account of tribal society, urban life, and the dynamics of political authority.[33] Meanwhile, Arnold Toynbee had encountered Ibn Khaldūn through the mid-nineteenth-century translation of the *Muqaddimah* into French by the Irish expatriate orientalist MacGuckin de Slane.[34] To Toynbee, Ibn Khaldūn had brought to birth a philosophy of history that was "undoubtedly the greatest work of its kind that has ever been created by any mind in any time or place," bearing comparison with Thucydides and Machiavelli for its "breadth and profundity of vision as well as for sheer intellectual power." Great praise went hand in hand with general disparagement of the Islamic world, however. To Toynbee, Ibn Khaldūn was "the sole point of light in his quarter of the firmament," the "one outstanding personality in the history of a civilization whose social life was 'solitary, poor, nasty, brutish, and short,'" a lone visionary "inspired by no predecessors."[35] Attracted by Ibn Khaldūn's cyclical—and frankly pessimistic—account of dynastic regime change, Toynbee admired the way in which the "panoramic vision" of "this mighty tree of thought" had made sense of "the rises and falls of empires and the geneses and growth and breakdowns and disintegrations of civilizations."[36] In Toynbee's mind, all of that dovetailed with the recent fashion, spearheaded by Ellsworth Huntington, for finding in climatic pulses the historical dynamic of nomadic irruptions.[37]

The retrieval of Ibn Khaldūn as the father of climate-shaped macroeconomics may well be challenged as an exercise in the retrospective construction of an intellectual lineage. But that he developed a comprehensive climatic philosophy of civilization and the wealth of nations certainly seems to be beyond doubt.

Climate, Character, and Commonwealth

According to Warren Gates, the spirit of Ibn Khaldūn's climatic history could well have been mediated to the West rather earlier through the writings of the traveler Sir John Chardin (1643–1712), who in turn is believed to have exerted considerable influence on Montesquieu. Gates infers this intellectual lineage from the close parallel between Chardin's claims and Khaldūn's convictions about the influence of high temperatures on the body, temperament, and craft industries of particular peoples. And indeed Chardin did maintain that "the hot Climates enervate the Mind as well as the Body" and "lay the quickness of the Fancy, necessary for the invention and improvement of the Arts." From the climate, he briskly deduced the entire intellectual condition and technological aptitude of those who populated these regions:

> In those Climates the Men are not capable of Night Watchings, and of a close Application, which brings forth the valuable Works of the Liberal, and of the Mechanick Arts. 'Tis by the same Reason likewise, that the Knowledge of the *Asiaticks* is so restrained that it consists only in learning and repeating what is contain'd in the Books of the Ancients; and that their Industry lies Fallow and Untill'd, if I may so express myself. 'Tis in the North only we must look for the highest improvement and the greatest perfection of the Arts and Sciences.[38]

In an era innocent of footnotes, Gates is surely correct to observe that it would not be surprising if Chardin delivered "a theory of climate which he had taken from the work of a XIVth-century Arab writer without mention of its source."[39] But whether or not that observation is well founded, there is no doubt that the passage above strongly resonates with Ibn Khaldūn's climatic philosophy even if their judgments on the so-called Asiatics obviously diverged. For both, if an explanation for the accrual of wealth were sought, climate was the place to go looking for it.

When the Parisian Chardin, traveling with fellow merchant-jeweler, the Lyonnais Antoine Raisin, spent some ten years in Persia, he determined to make himself an authority on the country before settling in England. He succeeded and subsequently received a knighthood from Charles II. His writings were extremely popular and widely cited, and he was elected to the Royal Society in 1682.[40] Convinced that the climatic conditions in which any people resided profoundly conditioned everything from preferences and practices to diet and disposition, he affirmed that the *Lebenswelten* of different societies were just as diverse as "that of the Temper of the Air is different from one Place to another." It was truly an expansive influence. From "a right Observation of the different Climates," he wrote, "one may form a better Judgement of the Food, Cloaths and Lodging of the several People of the World, as also of their Customs, Sciences and their Industry; and if one have a Mind to it, of the False Religions which they follow."[41] By way of summative encapsulation, Voltaire later readvertised Chardin's geopolitical verdict that high temperatures enfeebled mind and body alike and dissipated "that fire which the imagination requires for invention." "In such climates," Voltaire continued, "men are incapable of the long studies and intense application which are necessary to the production of first-rate works in the liberal and mechanic arts."[42]

Near contemporaries of Chardin also flirted with the idea that industrial and scientific progress was driven by the state of the atmosphere. The French philosophe Bernard le Bovier de Fontenelle (1657–1757), for example, was deeply suspicious of the spacious cultural territory surrendered to climate by Chardin and his like. He nonetheless observed that "it could be supposed that the torrid zone and the two polar regions are not well suited for the cultivation of the sciences."[43] By contrast, the Abbé Jean-Baptiste Du Bos (1670–1742) liberally quoted Chardin in support of the belief that it was only in the nations of the North that "the arts and trades" were to be found "in their highest perfection."[44] And in keeping with the Chardin mentalité, he was convinced that it was their climate that had given the Greeks "a greater sensibility of the pleasures of the mind, amongst which poetry and painting are the most bewitching allurements. Wherefore the greatest part of the Greeks were connoisseurs." By way of negative evidence, Du Bos declared that those in whose country the arts had never blossomed "are people who live absolutely in an improper climate." Otherwise, the high culture of civilization would have naturally and spontaneously arisen in those locations.[45] Climate, it was clear, explained both national attainment and cultural inertia. As for peasant societies, it was the "roughness" of the climate that rendered such people "gross and stupid, and the injuries of this same climate multiply their wants." In charting the progress of humanity's cultural estate, Du Bos plainly attributed much more to natural than to moral causes. Like Ibn Khaldūn, he embraced a latitudinal philosophy of civilizational accomplishment. The "arts and sciences have not flourished beyond the fifty second degree of North latitude, nor nearer than five and twenty degrees to the lines," he confirmed. To Du Bos, it was just a brute fact, a basic empirical truth, that aesthetic appreciation and mechanical achievement could only rise "of themselves in proper climates." To be sure, civilizations could migrate from one location to another. But, as he explained, they must "have their birth, their cradle, and their inventors, before they can be transplanted. Who is it that first brought the arts into Egypt? No body. But the Egyptians, favoured by the climate of their country, gave them birth themselves."[46]

In the interim, the influence of zonal climates on human society had been advanced in the writings of the French jurist Jean Bodin (1529/30–1596). Bodin directed his intellectual energies to the challenges of holding society together in the era of the European wars of religion, expounding the role of monarchy in governance, and finding ways of exercising sovereignty in a manner that took seriously differing local traditions, races, laws, and customs. Whereas other writers had elaborated a sevenfold zonal system, Bodin restricted his global vision to three fundamental climate types—frigid, temperate, torrid.[47] The purpose of this arrangement was to grasp the character of the diverse human societies spread across the inhabited world so as to distinguish what he believed to be the very great disparities between residents of the North and the South. For, as he put it, "there are as many types of men as there are distinct localities."[48] What animated Bodin's turn to climate in his inquiries into the nature of sovereignty was his sense that "no generalizations" could be made about legislative systems, religious practices, and social customs. In consequence, determining appropriate modes of governance should be drawn "not from the institutions of men, but from nature."[49] Nature's constancy stood in marked contrast to human caprice. For this reason, Bodin was convinced that wise rulers should fully acquaint themselves with the natural inclinations of their people before attempting to tamper with their constitutions or common laws. No single scheme of governance should be promoted. Instead, the "governments of commonwealths must be diversified according to the diversities of their situations."[50] To Bodin, knowledge of the climate and its occult ways was evidently as vital for the exercise of sound statecraft as a keen understanding of a kingdom's body politic.

This brand of legal naturalism and constitutional relativism brought with it a whiff of heresy. As Marian Tooley observes, "By seeming to call in question the moral responsibility of the individual," Bodin's proposals "struck at the root of the Christian ethic."[51] But if this was unorthodox in the eyes of some, in other respects, Bodin's zonal political philosophy rested on wholly conventional foundations. His entire anthropo-climatology, for example, was suffused with astrological dogmata. The locus of every spot on earth determined the celestial forces and intensity of radiation to which it was exposed and thus the astrological influences that were exerted on the body and the body politic alike. Mars and the moon dominated in the North, Saturn and Venus in the South, Jupiter and Mercury in the middle zone. At the same time, Bodin perpetuated the prevalent humoral theory, drawing directly on Galen. Those inhabiting polar extremities were phlegmatic; those in the extreme South, melancholic. Between these lay sanguine and choleric temperaments. The different proportions in the mixture of the humors in the various zones accounted for the notable contrasts between northern and southern peoples: the former were "modest and chaste," the latter "very libidinous as a result of their melancholy temperament."[52] Such variation could also be mapped vertically. As he explained in the 1566 *Method for the Easy Comprehension of History*, "There is the same difference between the natures of men from plain and from mountain as between those from the south and the north."[53]

Bodin elaborated the vast bulk of his climatic philosophizing in book 5 of his *Six Books of the Commonwealth*, which first saw the light of day in 1576. The purpose was clear. It was to determine the ways in which forms of governance should be adapted to the different dispositions of the various zones' human inhabitants. In contrast to other zonal philosophers, Bodin was convinced that equatorial regions were inhabited and that conditions there were healthy with abundant rainfall.[54] At the same time, the Franco-centric Bodin was in no doubt that those dwelling in the temperate zone were best fitted for the management of commonwealths. For here climate had conferred "the virtue of prudence, and prudence is the measure of human actions, a touchstone whereby men distinguish good from evil, justice from injury, honest proceedings from dishonest." In consequence, those inhabiting the "middle regions" of the hemisphere were ordained by nature and God alike "to bargain, trade, judge, persuade, command, establish commonwealths, and make laws and ordinances for the other races." By contrast, northerners were made for "labour and the mechanical arts" and those in the South disposed to metaphysical speculation and divine contemplation. The different zones housed different forms of governance, but it was clear that the true hallmarks of civilization—justice, trade, prudence—were the special provenance of temperate regimes. The arts of diplomacy were the preserve of the middle zone since rational argument, crucial to statecraft and negotiation, was "too mild for the crude northern races, and too prosaic for southerners."[55] It was for this reason, he insisted, that the world's great legislative systems—French, Roman, and Greek—had developed in this zone, as had the grand empires of the Assyrians, Medes, Persians, Greeks, Romans, and Celts.

The Montesquean Moment

In many ways, the zonal physiology and climatic philosophy of civilization that snaked its way through many of these writers, not least Chardin and Bodin, found its apogee in the celebrated 1748 *L'Esprit des Lois* of Montesquieu, whose thoughts on climate and national temperament earlier attracted our attention. Gates approvingly cites Muriel Dodds' view that "the elements which are considered to be the most original parts of Montesquieu's theory of climate are already present in the work of Chardin," who was a likely conduit through whom Ibn Khaldūn's climate theory found its way to the West.[56] Clarence Glacken, in turn, reflecting on the trajectory between Bodin and Montesquieu, quips that once we see through "the ponderous, scholastic style of the former," the "clear, fresh epigrammatic style . . . effectively concealed the lack of originality of the latter."[57] Indeed, Voltaire, already critical of the fashion for attributing the characteristics of government and religion to the influences of the climate, had announced that the "author of the 'Spirit of Laws,' without quoting any authority carried this idea farther than Chardin and Bodin." He continued, "A certain part of the nation believed him to have first suggested it."[58] Additional thinkers—such as Scottish medical practitioner and political observer John Arbuthnot;[59] Joseph Raulin, who was well read in contemporary medicine; and François-Ignace Espiard de la Borde, who wrote on the origin and character of nations—have likewise been identified as exerting dominant influences on Montesquieu.[60]

Physiologically, Montesquieu was convinced that the experience of pain mirrored the climatic conditions of the different zones. The coarse fibers of northern peoples meant that pain—caused by the tearing of some fiber or other in the body—was less distressing than injury to the fragile fibers of the peoples of hot regions. "A Muscovite," he announced, "has to be flayed before he feels anything"![61] And the characteristics of different climatic regimes marked nations in many ways. They manifested themselves in what could be termed their sexual ecology. In northern climates, Montesquieu considered that the sexual impulse was weak; in hot climates, by contrast, sex was vigorously pursued as "the sole cause of happiness." Climate-driven sexuality, moreover, profoundly shaped the reproductive economy of different realms. Female marriage age was a case in point. In hotter regimes, women were "marriageable" as young as eight, nine, or ten years of age; thus, they were "old at twenty." Accordingly, "reason in women is never found with beauty there." Reduced to an adage, all this meant that when "beauty demands empire, reason refuses it;

when reason could achieve it, beauty exists no longer." Under these conditions, polygamy flourished and women remained in dependent subjection. In temperate zones, marriageable age was higher, the charms of the female sex were "better preserved," and childbirth took place at a more advanced stage. Montesquieu thus insisted that in the temperate belt "a kind of natural equality between the two sexes has naturally been introduced, and consequently the law permitting only a single wife."[62] In the Far North, things were different on account of zonal birth patterns. In these realms, conditions were conducive to polyandry. "In the cold climates of Asia, as in Europe, more boys are born than girls," he remarked. "That is, say the Lamas, the reason for the law in cold countries that allows a woman here to have several husbands." Sexual politics, reproductive customs, and marriage practices were simply a reflection of climatic imperatives. On these assorted nuptial arrangements, Montesquieu passed no normative judgment, no attribution of what was better or worse. Whatever adjudications he may have issued on other matters, at this point, at any rate, he was disinclined to moralize the climates of marital union. As he insisted, "In all this I do not justify usages, but I give the reasons for them."[63] Montesquieu thereby presented himself as an inductive natural philosopher of procreative culture.

A prominent thread of what would now be called "orientalism" was woven into Montesquieu's sexual climatology. The servitude of women, he professed, was "very much in conformity with the genius of despotic government" in Asia, where domestic bondage and autocratic rule went hand in hand in every age. In the East, nature had induced that lassitude of spirit in the population that rendered inhabitants incapable of any action, effort, or application. The result was that their manners and mores had remained unchanged for a millennium and more. Oriental stagnation was the net result of a climate where "excessive heat enervates and overwhelms" and where "rest is so delicious and movement so painful that this system of metaphysics appears natural."[64] It is for reasons such as this that Celine Spector suggests that Montesquieu had a crucial role to play in the construction of the idea of Europe by juxtaposing European moderation over against Asiatic despotism, strength as opposed to weakness, progress in place of torpor, liberty instead of servitude.[65]

If national temperaments and sexual predilections alike fell within the arc of climate's compass, so too did morality.[66] "You will find in the northern climates peoples who have few vices, enough virtues, and much sincerity and frankness," he announced. Moving toward the South, he continued, "you will believe you have moved away from morality itself: the liveliest passions will increase crime; each will seek to take from others all the advantages that can favor these same passions. In temperate countries, you will see peoples whose manners, and even their vices and virtues are inconstant; the climate is not sufficiently settled to fix them." Tropical heat apparently had the effect of inducing a lethargy that found its way right to the heart of the human spirit. That meant that laziness prevailed, and so in hot climates, human nature was bereft of curiosity, "noble enterprise," and "generous sentiment."[67]

No doubt it was at least in part on account of the heretical tone of his moral climatology, with its attendant legal relativism and political naturalism, that Montesquieu's treatise was placed on the Vatican's Index of Prohibited Books in 1751. His climatic account of the global patterns of legislation meant that among Muslims, the prohibition against alcohol was simply a law arising from the nature of the climate of Arabia. But in cold countries, such a law would not be wise. "Drunkenness," he explained, was to be found "around the world in proportion to the cold and dampness of the climate."[68] Geographical variation and legislative relativism went hand in hand.

The temperamental sluggishness that Montesquieu perceived in torrid climates, coupled with the stimulus of more challenging climatic regimes, of course, bore significantly on the geography of global wealth. Particularly significant here was agriculture. Since the cultivation of land was, for Montesquieu, the grandest labor of the human race, it occupied a primary role in the evolution of civilization and the economics of wealth creation. Like the later physiocrats, Montesquieu considered the peasant farmer to be the fundamental producer on which the whole economy rested and thus that land was the primary source of wealth. Not that other forces were ignored: industry and commerce had their role to play. But agriculture supplied the principal impetus for economic growth since it was engaged in satisfying quotidian needs. Still, where a society failed to advance beyond the strictly agricultural, it remained at a primitive level of civilization with only rudimentary welfare at best. Agricultural surplus, then, was a necessary but not sufficient condition for the generation of wealth and the advancement of civilization, and where it was in jeopardy, that society remained economically retarded.[69]

By way of example, Montesquieu pointed to the natural indolence arising from the bad influences of the climate in the Indies. The inhabitants there shunned agricultural labor and inevitably reaped the sorry consequences. Their passivity, "born of idleness of the climate," fostered a legislative system that elevated rest and inactivity as virtues to a metaphysical level and thus delivered economic stagnation.[70] Harsher climates in the North, where landscapes were more barren, produced a population that was industrious and sober, prepared to acquire by diligence and drudgery what the earth refused to grant gratuitously. Climatic rigor and poor soil, as Alan Macfarlane puts it, "made the north into a kind of Spartan 'dura virum nutrix' (hard nurse of men) which seemed to Montesquieu to lead to hard work and liberty."[71]

Matters of labor, of course, abutted on the controverted question of slavery—an institution that Montesquieu deplored. In constitutional monarchies "where it is sovereignly important neither to beat down nor to debase human nature, there must be no slavery," he declared. Similarly, in democratic and aristocratic regimes, slavery was contrary to the spirit of human equality. Rationalizations for slaveholding, of course, were aplenty, and Montesquieu deployed his biting ironic wit in disposing of these. Yet for all that, Montesquieu conceded that there were circumstances in which slaveowners could find self-justification in the climate. For there were indeed regions where extreme heat enervated the body and sapped courage to the extent that workers only engaged in hard labor from fear of chastisement. In such conditions, he mused, "slavery . . . runs less counter to reason." To be sure, Montesquieu was convinced that "all men are born equal" and that "slavery is against nature," yet he did remark that "in certain countries it may be founded on a natural reason, and these countries must be distinguished from those in which even natural reasons reject it, as in the countries of Europe where it has fortunately been abolished." Evidently some kind of warrant for the institution could be marshaled in certain conditions. The thought that Europeans in the New World simply had to take on slaves to clear the land, and that sugar would have been too expensive if the agricultural labor required was not undertaken by slaves, was surely attractive to plantation owners and slave masters. Besides, given his convictions about the determining influence of climate on human constitutions, he was inclined to naturalize bondage, remarking that "the cowardice of the peoples of hot climates has almost always made them slaves and that the courage of the peoples of cold climates has kept them free. This is an effect that derives from its natural cause." In Asia, liberty never increased, and Africa, which had "a climate like that of southern Asia," displayed "the same servitude."[72] In sum, while he was sure that labor performed by freemen was motivated by hope of gain, he was no less reconciled to the idea that the climate and terrain in particular world regions well-nigh mandated slavery of one stripe or another—political, domestic, civil.

At the same time, whatever the moral, temperamental, and legislative contrasts between North and South, Montesquieu sensed a profound equipoise underlying the global economic system that found emblematic expression within Europe itself. As he explained in his inquiries into the nature and laws of commerce,

> There is a kind of balance in Europe between the nations of the South and those of the North. The first have all sorts of the comforts of life and few needs; the second have many needs and few of the comforts of life. To the former, nature had given much, and they ask but little of it; to the others nature gives little, and they ask much of it. Equilibrium is maintained by the laziness it has given to the southern nations and by the industry and activity it has given to those of the north. The latter are obliged to work much; if they did not they could lack everything and become barbarians.[73]

For Montesquieu, the distinctive natural endowments of North and South fitted snugly onto the temperamental contrasts between the inhabitants of high and low latitudes. Climate had ensured a beneficial reciprocity between natural geography and national psychology. As we shall later see, that very idea—that different scales of consumption are required to ensure comfort in different climatic regimes—has been resurrected and transplanted into twenty-first-century economic analysis.

In the Shadow of Montesquieu

In one form or another, the zonal geo-philosophy that Montesquieu did so much to perpetuate manifested itself in the writings of other Enlightenment *philosophes*.[74] His younger contemporary, Immanuel Kant (1724–1804), for instance, when not pondering on the nature of pure or practical reason, avidly devoured works of geographical exploration and anthropological inventory, not least the writings of Montesquieu himself.[75] In his 1775 reflections "On the Different Races of Man," he elaborated on what he called the "stem genus" from which the different race groups were derived, noting that the "portion of the earth between the 31st and 52nd parallels in the Old World," the home of this lineal root genus, was the zone "in which the most happy mixture of influences of the colder and hotter regions and also the greatest wealth of earthly creatures is encountered."[76] From this source, the four distinct human races—very blond, copper-red, black, and olive-yellow—were derived through migration, isolation, and exposure to differing climatic conditions. Latitudinal anthropology in Kant's racial portraiture carried with it both moral judgment and political prescription. Aesthetic and ethical sentiment—particularly as manifest in feelings of beauty and sublimity—for instance, were not evenly distributed across the climatic zones. The inhabitants of torrid regimes enjoyed "no feeling that rises above the trifling." Echoing David Hume, Kant went on to confirm that among freed African slaves, "still not a single one was ever found who presented anything great in art or science or any other praiseworthy quality." Given this verdict, it comes as no surprise to find Kant, in his posthumously published lectures on Physical Geography, elaborating a global cartographic in which the temperate world was exalted to the apogee of human excellence. Human life in hot climates did not attain the "perfection" of those dwelling in the temperate zones. In fact, humanity reached "its greatest perfection" in the white race while the peoples of the hottest zones were exceptionally lethargic. Kant was pleased, though, to report that tropical "laziness is somewhat mitigated by rule and force."[77]

For those in the middle zones, it was a different story. Italian genius had manifested itself in music, painting, sculpture, and architecture. French refinement was equally exquisite, though the sense of beauty was perhaps somewhat "less moving." Here "poetic or oratorical perfection" fell "more into the beautiful" whereas "in England more into the sublime."[78] Indeed, Kant seemed to think that the altogether admirable qualities of mind, body, and will that such nations exhibited provided good grounds for the predacious history of European expansionism. "The inhabitant of the temperate parts of the world, above all the central part, has a more beautiful body, works harder, is more jocular, more controlled in his passions, more intelligent than any other race of people in the world," Kant confirmed. "That is why at all points in time these peoples have educated the others and controlled them with weapons."[79] In just two sentences, he traversed the space between the character of the occupants of the mid-latitudes and their inevitable colonial dominance of the globe. All this confirmed that it was in Europe that civilization reached it headiest heights.

In some of these judgments, Kant was following in the footsteps of David Hume (1757–1838) as much as Montesquieu. The same year that Montesquieu's *L'Esprit des Lois* appeared, 1748, Hume brought out an essay entitled "Of National Characters." Here Hume, drawing on the support of Strabo's *Geography*, engaged in a detailed rebuttal of the view that national temperaments were influenced in any material way by physical causes such as climate. A whole raft of particularities, he insisted, simply undermined the idea that the character of a people could readily be reduced to climate. But his dissatisfaction with the more deterministic hue of some advocates of the climate thesis did not stand in the way of his conviction that there was reason to believe that all the nations inhabiting polar or tropical realms were "inferior to the rest of the species, and are incapable of all the higher attainments of the human mind." Moreover, he approvingly called upon the testimony of Francis Bacon to the effect that "where the native of a cold climate has genius, he rises to a higher pitch than can be reached by the southern wits." Nor did his professed aversion to climatic determinism prevent Hume from gesturing toward a global geography of mental and moral attainment. And nowhere was this suspicion more clearly intimated than in his comments on the distribution of certain "passions." These impulses were not simply the outcome of the different ways the climate directly evoked the affections in different latitudes. Nor did they arise from moral causes. Rather, they were the expression of the influence that atmospheric conditions exerted on the body's physical constitution. "The only observation, with regard to the difference of men in different climates, on which we can rest any weight, is the vulgar one, that people in the northern regions have a greater inclination to strong liquors, and those in the southern to love and women," he proclaimed. "One can assign a very probable *physical* cause of this difference. Wine and distilled waters warm the frozen blood in the colder climates and fortify men against the injuries of the weather. As the genial heat of the sun, in the countries exposed to his beams, inflames the blood, and exalts the passion between the sexes." What this admission disclosed was Hume's preparedness to countenance—as others had done before him—the influence of climate on those "passions" connected with what he called "the grosser and more bodily organs of our frame," even while exempting "the operations of the mind and understanding" from the self-same influences.[80]

Whatever the reservations of Hume, or the criticisms of the likes of Voltaire, Montesquieu attracted other supporters. Stephen Gaukroger points out that in his account of Jewish law, the Prussian biblical scholar Johann David Michaelis (1717–1791) provided a climatic explanation for different legal requirements in different regions in his *Mosaisches Recht* published during the early 1770s. So too did the eighteenth-century Lutheran church historian and biblical exegete Johann Lorenz von Mosheim (1693–1755), who connected liberty with colder climates, as did Isaak Iselin (1728–1782), a Swiss political philosopher and historian, who highlighted the contrast between strong, dogged northerners and weak, sensuous southerners.[81] And while they diverged in many ways, Adam Smith's (1723–1790) observations on the relationship between climate and wages resonated to at least some degree with Montesquieu's zonal outlook. In a supplemental note on "Wages of Labour" to his great work, *An Inquiry into the Nature and Causes of the Wealth of Nations*, which first appeared in 1776, Smith agreed that the "varieties in the climate of different countries occasion considerable variations in the necessary rate of wages." In colder climatic conditions, clothing was more expensive, housing more substantial, and the supply of fuel indispensable. Such necessities stimulated trade. In consequence, all other things being equal, he predicted that wages would be highest in locations where clothing and houses were at their most expensive and where the largest supplies of fuel were required. Comparing wages in Britain and Bengal, where food was "the simplest and cheapest imaginable," he found precisely the pattern that this principle projected. Certainly he acknowledged that countervailing forces, such as progress "in the arts of civilized life" or the location of productive coal mines, might be mitigating factors.[82] But taken in the round, Smith was sure that the association between climate and income made intuitive sense.

Decidedly more partisan, the French philosopher Victor Cousin (1792–1869), who championed a synthesis of German idealism and Scottish common sense realism, famously declared in his introduction to the philosophy of history, "Give me the map of any country, its configuration, its climate, its waters, its winds, and the whole of its physical geography; give me its natural productions, its flora, its zoology, &c, and I pledge myself to tell you, a priori, what will be the quality of man in that country, and what part its inhabitants will act in history;—not accidentally, but necessarily; not at any particular epoch, but in all." In thus pronouncing on the necessary influences of climate and geography over against "the common herd of thinkers" who see in human actions "nothing but the effects of glowing imagination," Cousin juxtaposed the genius of Montesquieu, "the man who, of all our countrymen, best understood history, and was the first who gave an example of the true historic method." To Cousin, Montesquieu was worthy of the highest admiration because he "hesitates not, to ascribe to climate an immense influence upon the human creation" and thereby reaffirmed the explanatory power of applying the laws of nature to humanity.[83]

Rather more ambiguous was Johann Gottfried Herder (1744–1803), whose monumental work on the philosophy of history appeared during the final decades of the eighteenth century. Herder was sure that there were clear ways in which climate—*klima*—had directly shaped the course of human history. Different regional geographies had distinctively molded human cultures. Yet at the same time, he resisted any crude necessitarianism by drawing attention to the freedom that the spirit of different peoples exhibited. After all, climate was "a chaos of causes," and its multifarious influences resisted simple specification.[84]

Besides, he discerned key ways in which human societies had transformed their environments and thereby modified the impact of nature's agency. Ancient Egyptians, for example, had been able to harness nature's forces and gain control of the Nile.

Herder's resistance to monocausal modes of thinking about history and society is scarcely surprising for one whose sympathies were undeniably particularist. As Isaiah Berlin remarked, "All regionalists, all defenders of the local against the universal, all champions of deeply rooted forms of life," owe something "to the doctrines which Herder . . . introduced into European thought."[85] Methodologically, Herder was always drawn toward what he called "an instructive examination of particular cases" and thus toward the history and geography of particular peoples.[86] For such reasons, he has often been hailed as the intellectual father of the historicist impulse to interpret local communities in their own cultural categories.[87] At the same time, his convictions about the unity of the human races and commonality of human nature were no less clearly advertised.[88] Between human groups, he discerned no essential differences and thus resisted Kant's division of humanity into several fundamentally different race types. Whatever varieties the human form assumed, Herder insisted "there is but one and the same species of Man throughout the Whole of our Earth." Complexions merged, forms followed a common structure, and "all are at last but shades of the same great picture."[89]

Yet, having laid out these universals, he soon turned to how climate and cognate natural agencies had shaped human beings, albeit within definite limits.[90] Crucial here was Herder's belief that while *klima* exerted its influence on the physiology and psyche of individuals, that did not mean that whole peoples responded in the same way to such external stimuli. Differences of scale and situation had to be taken into account. Besides, he recognized that human activity—the use of fire, clearing the forest, new plantings—could significantly modify the climatic experience of a people. "Once Europe was a dank forest," he wrote, "and other regions, at present well cultivated, were the same. They are now exposed to the rays of the Sun; and the inhabitants themselves have changed with the climate."[91] Plainly, there was reciprocity in the human–climate relationship.

Still, for all that, Herder wrote in the shadow of Montesquieu, albeit with equivocation. From Kamchatka to Tierra del Fuego, he reminisced, different "climatic pictures" could be exhibited of different nations. Wherever the sons and daughters of Adam "fixed their abode," they had "taken root as trees, and produced leaves and fruit adapted to the climate." Acquiescing to the "great

Montesquieu" in conceding that "we are ductile clay in the hand of Climate," Herder all the while insisted that "her fingers mould so variously, and the laws that counteract them, are so numerous, that perhaps the genius of man alone is capable of combining the relations of all these powers in one whole." Obviously, climate exerted significant influence, even if it did not compel. And so he resorted to the zonal philosophy that had long captured the attention of geohistorians. As he looked over the history of the earth from ancient times, Herder could see how the climates of the different zones had wielded "considerable influence on all the revolutions of the human mind and its operations." His evidence for this claim lay in his assurance that neither the torrid nor the frigid zone had produced any of those remarkable effects to which the temperate latitudes had given birth. This was simply an inevitability given the earth's ontology. For Herder could discern the "fine traits" by which "the finger of omnipotence has described and encircled all the changes and shades on the Globe. Had the earth's inclination to the Sun differed but a little from what it is, every thing on it would have been different." The latitudinal historiography to which Herder was sympathetic was overlain with a strong providentialism. It was the "father of the World" who had chosen "a favourable spot for our origin." Humanity had originated in "a milder climate, and with it a gentler nature, and a more variegated place of education: thence he let them wander by degrees, strengthened and well instructed, into hotter and colder regions."[92]

A comparable sense of vacillation is no less discernible in the writings of the Prussian geographer, scientific traveler, and Romantic philosopher Alexander von Humboldt (1769–1859), who, while eschewing the teleology that Herder espoused, shared his thinking about the mutual modifications of humanity and environment.[93] At the same time, Humboldt reaffirmed the view that challenging climatic regimes provided the foundations on which advanced civilizations were erected. This comes out, for example, in his 1805 *Essai sur la Géographie des Plantes* and its accompanying *Tableau Physique des Régions Équinoxiales,* where he proclaimed that the development of civilization was inversely related to the fertility of the soil in which it was embedded. The "culture du sol" in turn depended on climatic conditions. The more the nature of the climate presented difficulties to be surmounted, the more rapid was the development of the society's moral faculties. A harsh environment stimulated its population to engage in determined labor, and for this reason, civilization reached far greater heights in "les regions boréales" than in "la fertilité des tropiques."[94] And again in the narrative of his travels in Spanish America, which was first published in French between 1814 and 1825, his account of "the agriculture of the torrid zone involuntarily" reminded him of the intimate connection that pertained between the extent of land cleared and the progress of society. Thus, he paused to reflect on how vegetative abundance actually impeded cultural and economic advancement. In tropical zones, he remarked, the "richness of the soil, and . . . vigour of organic life, which multiply the means of subsistence, retard the progress of nations toward civilization." This was because in any climate characterized by mild and uniform conditions, the only urgent necessity was the supply of food. It was "the feeling of this want only" that excited the human race to engage in labor. For this reason, Humboldt considered, it was easy to conceive "why in the midst of abundance, beneath the shade of the plantain and breadfruit tree, the intellectual faculties unfold themselves less rapidly than under a rigorous sky, in the region of corn, where our race is in a perpetual struggle with the elements."[95]

Among the next generation of German geographers, something of the same zonal thinking found support in the writings of Friedrich Ratzel (1844–1904), who taught at the University of Leipzig and remains well known for his *Anthropogeographie*, which dwelt on the links between population, migration, and environment. Ratzel has typically been presented as a quintessentially determinist thinker, perhaps on account of how his geo-philosophy was translated into an American context by his disciple Ellen Churchill Semple (1863–1932).[96] Yet as Harriet Wanklyn pointed out more than half a century ago, the undoubted presence of determinist passages in his work should not be mistaken for the whole. In a similar vein, the celebrated American anthropologist Robert Lowie (1883–1957) reminded his readers that in fact, Ratzel repeatedly warned against naive determinism and the reduction of social life simply to climate.[97] Besides, Ratzel drew inspiration from the writings of Herder, and this manifested itself in his emphasis on the centrality of language and religion in the growth of civilization. At the same time, the influence of the explorer and naturalist Moritz Wagner (1813–1887) was crucial. For Wagner's migration theory provided him with a naturalizing impulse to root aspects of human history in their environmental circumstances and to evolutionize the study of humankind. In Ratzel's program, biogeography and political geography were all of a piece—not least in his conception of *Lebensraum*.[98]

Whatever the ultimate tensions in Ratzel's thinking on climatic agency, the zonal philosophy of history continued to shape his thinking. It came through, for example, in the introductory sections of his monumental anthropological work, *Völkerkunde*, the English translation of which was introduced by the distinguished English social evolutionist and cultural anthropologist Edward B. Tylor (1832–1917). Here, as he laid out his conception of the principles of ethnography, Ratzel issued commentary on the nature, rise, and spread of civilization. Drawing in part on the work of Henry Thomas Buckle, whose conception of the genesis and spread of civilization will later attract our attention, and perpetuating the Montesquieu vision, Ratzel learned that those natural conditions that allowed for the early "amassing of wealth from the fertility of the soil" and the labor associated with it were "undoubtedly of the greatest importance in the development of civilization." Humankind's origin could be traced to the warm, moist regions of the globe, which were "blessed with abundance of fruits," and thus he was inclined to the view that *Homo sapiens* was a tropical creature. The consolidation of civilization, however, was only achieved when the forces of human willpower could be harnessed and imposed on the natural order. Such capacities could only be acquired under much more favorable conditions where society had to more carefully attend to the necessities of everyday life than was the case "in the soft cradle of the tropics."[99] Progression to the advanced stages of social evolution could not be achieved without more challenging conditions for, in the awakening of the higher faculties, scarcity was infinitely more valuable than bounty. Austerity, it seemed, gave birth to ingenuity. For this reason, he pointed to the temperate zones as the cradle of civilization. The human race had certainly emerged in the tropics, but civilization was the preserve of the temperate world. All of this was, ultimately, controlled by climate, for as he put it,

> Beyond the historic operation of climatic peculiarities in favouring or checking civilization, differences of climate interfere most effectually by producing large regions where similar conditions prevail—regions of civilization which are disposed like a belt around the globe. These may be called civilized zones. The real zone of civilization . . . is the temperate. . . . Older semi-civilizations, whose relics we meet with in tropical countries, belong to a period when civilization did not make such mighty demands upon the labours of individuals, and when for that very reason its blossom sooner faded.[100]

Taken in the round, the narrative I have been recounting thus far marks a way of charting global wealth that gave pride of place to the natural environment, particularly to the forces of climate as the shaper of human destiny. With roots in the philosophy of history expounded by Ibn Khaldūn and its elaboration in the writings of figures such as Bodin and Montesquieu, this tradition burgeoned over the following centuries and has been revitalized with a good deal of enthusiasm in the past few decades. Courtesy of their contributions and others like them, climate was installed as a critical force, often *the* critical force, in generating the wealth of nations, in fostering the growth and spread of civilization, and in cultivating the infrastructure and institutions of the nations. Celebrating the temperate zone, while deprecating the frigid and torrid regions, they attributed the development of the arts and sciences, customs and resources, poetry and painting, to the influence of the mid-latitudes' moderate climate. At the same time, their message was not universally comfortable. Their inclinations toward legal relativism and thus the undermining of moral absolutism were troublesome to many. And yet they had little difficulty in justifying their own dispositional preferences by contrasting tropical torpor with temperate-zone vitality. As emblematic of this continuing fixation with zonal geo-history, I have focused attention on a number of leading thinkers—Kant, Herder, Humboldt, Ratzel—who espoused it in one form or another. No doubt that list could be extensively expanded. But exhaustive inventory has not been my intention. My aim has rather been to sketch in something of the range of ways in which the ties between climate and civilization, and thus between weather and wealth, have been charted as the deeper context for the more recent return to geographically reductionist readings of global economics.

At this juncture, two contrasting judgments on the most prominent anchor points of this whole way of thinking are worth recording. Both dating from 1959, they offer rather differing assessments of the kind of naturalistic histories of civilization and wealth that Ibn Khaldūn and Baron de Montesquieu had championed. First, the French Marxist philosopher Louis Althusser, in his celebration of Montesquieu's "revolution in method," foregrounded as seminal Montesquieu's incorporation of material forces, such as climate, into explanations of social dynamics and political formations. For such reasons, Althusser confirmed the judgments of Comte and Durkheim that Montesquieu should be regarded as the founder *par excellence* of political science. Inasmuch as Montesquieu dwelt on the material foundations and concrete behavior of human societies, rather than on normative metaphysics, Althusser regarded him as inaugurating the study of history as a scientific undertaking. Whatever other lacunae Althusser perceived in that project, and whatever modulations were required to render acceptable this resort to climate and other natural forces in his own day, Althusser remained enamored of the Montesquieu method. "Here undoubtedly stands revealed Montesquieu's radical novelty," he announced.[101] By contrast, the American historian and later author of *Metahistory*, Hayden White, felt compelled to write of Ibn Khaldūn that he "sought refuge . . . in the security of an abstract system constructed in such a way as to relieve man of his responsibilities of the tragedies . . . which he suffered."[102] Generalizing from that vantage point, White insisted that when "the historian has no genuine respect for men, he will took everywhere but to men for the cause of historical change, and, failing to center upon man as the agent responsible for human triumph and disaster, he will find it impossible to share, through his history, in the achievement of the former and the avoidance of the latter." In this telling, examining the role that climate plays in shaping the story of human civilization is not pernicious in itself, but as he remarked of Ibn Khaldūn, it "becomes so when it is used to justify escape from the burden of human freedom and its responsibilities."[103]

Throughout the nineteenth century and on into the twentieth, the climatic account of the origins and development of civilization continued to attract devoted followers. And the practical politics that flowed from the zonal philosophy of history were prominently displayed in discussions over the dilemmas posed by the problems of white labor in the tropical world, the institution of American slavery, and how the wealth of tropical resources could be managed from the temperate zones. Before siding with the judgment of either Althusser or White, we might want to ponder further on this posthistory of the Khaldūn–Montesquieu climatic credo.

8

Slavery, Sustenance, and Tropical Supervision

DURING THE nineteenth century, the zonal philosophy flourished, particularly in the context of European colonial adventures and the perennially troubling issue of how the tropical world was to be managed, not to say exploited, by European imperialists. At the same time, the problems of the suitability of hot climates for white settlers were a recurrent concern. Frequently this question, which as we have already seen abutted on the connections between medicine and meteorology over the possibility of human acclimatization, focused on the politics of colonial labor. Could the white race work in the tropical world? And if not, how could the wealth of the tropical and subtropical zones be extracted? In the American South, where whites had long settled, this dilemma manifested itself in the institution of slavery. And so here we will examine something of the ways in which the southern slavocracy was justified in the naturalistic language of weather and wealth. Meanwhile, on the other side of the Atlantic, a comprehensive climatic philosophy of food and civilization was developed by the English historian Henry Thomas Buckle, who championed a scientific history that obeyed the natural laws of human behavior. The perspective he brought to his history of civilization subsequently informed the writings of numerous commentators pondering on how the riches of the tropical and subtropical belt could be accessed and quarried by northern nations. During the final decades of the nineteenth century, and on into the early years of the twentieth, climatic accounts of colonial governance continued to merge with racial denigration among a circle of colonial administrators arguing for the moral and political necessity of European supervision of the tropics in the name of global trusteeship. This is terrain we will navigate in the pages that follow.

Sun, Soil, and Slavery

If echoes of the zonal philosophy of global economic wealth lingered in the writings of the likes of Herder, Humboldt, and Ratzel, they reverberated even more audibly across the Atlantic in a prominent series of lectures presented by the Swiss geologist and geographer Arnold Guyot (1807–1884) at the Lowell Institute in Boston in the early months of 1849. Known for his research on Alpine glaciers and meteorology, he had followed his colleague Louis Agassiz to the New World, where he took up a chair at the College of New Jersey, later Princeton University. Here he founded the Department of Geosciences and regularly delivered lectures to ministerial students at the Princeton Theological Seminary.[1] Neither as sophisticated as Herder nor as supple as Humboldt, Guyot's *The Earth and Man* was a work no less of theological apologetic than of geographical scholarship. For as he put it in the preface, "the chariot of human destinies," whose path he would presently chart, had been guided by the hand of Providence that orchestrated "the grand harmonies of nature and history."[2]

Like Humboldt and Ratzel, Guyot was convinced that where nature was "too rich, too prodigal of her gifts," there was no compulsion for "man to snatch from her his daily bread by his daily toil." The reason was that warm climates without a dormant season rendered forethought of little use. In such climates, nothing invited inhabitants to engage in the intellectual struggle against the forces of nature that raised human resolve and ingenuity to ever higher planes. By contrast, in temperate and, more particularly, maritime climates, activity, vitality, and dynamism were the warp and woof of all life. The alternations of warm and cool conditions, bracing air, and seasonal rhythms invigorated inhabitants of the temperate zone, prompting them to prudence, planning, and the vigorous development of all their faculties. On a global scale, this meant that climatic variability had generated radically different political economies in the two hemispheres and thereby had mapped out the ultimate destiny of the two worlds. In Guyot's cosmos, the physical, economic, and spiritual geographies of the globe were inextricably interwoven. The entire physical creation, he explained, corresponded to the moral and political order. In consequence, just as there were temperate and tropical zones, so too were there civilized and savage worlds. None of this was happenstance. For it was none other than the Creator who had "placed the cradle of mankind in the midst of the continents of the North, so well made, by their forms, by their structure, by their climate . . . to stimulate and hasten individual development and that of human societies; and not at the centre of the tropical regions, whose balmy, but enervating and treacherous atmosphere would perhaps have lulled him to sleep the sleep of death in his very cradle."[3] In Guyot's spiritual universe, "geo" and "theo" formed an indissoluble metaphysical unity.

However much Guyot sought to romanticize, even divinize, this hemispheric credo, the real-world implications were degrading and abusive. The northern continents were constituted to be the world's leaders, the southern made to be their eternal aids. This meant that people of the temperate continents would forever be "the men of intelligence, of activity, the brain of humanity." Correspondingly, the inhabitants of the tropical world were doomed to remain "the hands, the workmen, the sons of toil." The underlying reason was that the races inhabiting the three southern continents were held captive to an all-powerful nature, and their only hope of redemption lay in the promise that the "favored" races, the "privileged" races, would vouchsafe to them the glories and comforts of civilization. Tropical nature could not be conquered and subdued, or its riches released, save by the enlightened nations of the North, who were obligated to perform their duty "armed with all the might of discipline, intelligence, and of skillful industry."[4]

In conveying this vision, Guyot found it possible to bury callous sentiment in sermonic cadence. Brutal policies were couched in the language of a benevolent providentialism. For Guyot was sure that everything in nature had been orchestrated by a beneficent Providence to bring about the triumph of good through human instruments. As he portrayed it, the Southern Hemisphere awaited deliverance from the North. A divinely ordained global economy, based on the geographical asymmetries of North and South, had determined that it was only with the help of temperate-zone civilizations that the peoples of the tropics could participate in the universal progress that was the birthright of all humanity. Otherwise, it was forever impossible for "the inferior races" to escape from the "state of torpor and debasement into which they are plunged, and live the active life of the higher races." In fact, it was the solemn duty of the Southern Hemisphere to endure colonial subjection and economic tutelage for the global good. As he explained,

> Then shall commence, or rather shall rise to its just proportions, the elaboration of the material wealth of the tropical regions, for the benefit of the whole world. The nations of the lower races, associated like brothers with the civilized man of the ancient Christian societies, and directed by his intelligent activity, will be the chief instruments. The whole world, so

> turned to use by man, will fulfil its destiny. . . . Each northern continent has its southern continent near by which seems more especially commended to its guardianship and placed under its influence.[5]

What Guyot was effectively building was a scientific theodicy in which hemispheric hardships could be justified in the language of nature and grace, the climate and the creator.[6]

The intercontinental destiny of the Northern and Southern Hemispheres was not the only way Guyot conceived of what he called the geographical march of history. There was no less an East–West trajectory in humanity's story. The three great stages of humankind's trek across the ages marked a movement from Asia via Europe to North America. Asia was the cradle where humanity's infancy was spent, Europe the school where the human race was trained, and North America the sphere of adulthood where the potential of modern arts and industry enabled the overmastering of the natural world and the final unleashing of the human species from nature's long tyranny. Humanity's future now lay in the hands of the New World. Its yet-to-be-used wealth and its buoyant population meant that America was "made for the man of the Old World."[7] That was its manifest destiny.

Surging its way through Guyot's reading of global wealth and civilization was a racial climatology that traded in anthropometric stereotype and cultural denigration. The so-called finest races were domiciled in the continents of the temperate North and presented the most perfect types. In particular, the nations of Europe represented the acme of intellectual excellence never before achieved in any epoch by any race. They ruled over nearly every part of the globe and, mercifully, were pushing their conquests still further. Europe was evidently fated to follow an imperial imperative. Conversely, the Southern Hemisphere was populated by the most imperfect representatives of human nature. Guyot's denunciation was both anatomical and aesthetic. Moving south from civilization's temperate centers, the human types gradually lost "the beauty of their forms, in proportion to their distance, even to the extreme points of the southern continents," where "the most deformed and degenerate races, and the lowest in the scale of humanity" were to be found.[8]

Given these judgments, it was easy to move from ethnic portrayal to economic prescription. One climatic zone was destined for dominion, the other for drudgery; one to train, the other to toil. The implications for labor in the New World were obvious. And European settlers soon learned to replace native American labor with African manpower. The capacity of black Africans for hard work meant that they could easily be instrumentalized, brought into subjection, and harnessed to the ambitions of white society. With no sense of infelicity or mortification, Guyot could declare that the black laborer "endures, I had almost said, gaily, a degree of toil equal to that which destroyed the native of the country." None of this should be taken to imply, however, that Guyot gave his support to slavery. Far from it. Indeed, he went out of his way to repudiate the institution whether of ancient Roman or modern American provenance. Besides, he looked forward to the day when what he called "that fatal heritage of another age, which the Union still drags after, as the convict drags his chain and ball, shall have disappeared from this free soil, freed in the name of liberty and Christian brotherhood."[9] For all that, his climatic rationale for the institution made it intelligible and thus, for many, inevitable.

The links between climate, race, and labor were deep and lasting, and were indeed long mobilized in the cause of naturalizing slavery. We have seen how Montesquieu, however much he deplored it, hinted at a climatic rationalization for slave labor when he reflected on the challenges of northern peoples engaging in manual work in tropical regimes. In the New World, this argument could easily be marshaled by apologists for the southern plantocracy. Thomas Jefferson is a case in point. In Query 18 of his celebrated *Notes on the State of Virginia*, originally published privately in Paris in 1785, Jefferson took up the subject of the customs and manners of his own state. Here he remarked on what he described as the unhappy influence stemming from the existence of slavery in the new republic. As he explained, this was because "the whole commerce between master and slave is a perpetual exercise of the most boisterous passions, most unremitting despotism on the one part, and degrading submissions on the other." For all that, Jefferson noted that, in a warm climate, no one would labor for themselves if they could compel someone else to do it for them. This, it seemed to him, was just obviously true. Surely it was borne out by the simple observation that only the very smallest proportion of slave proprietors were ever seen to engage in manual labor. The fact that among Africans, a "greater degree of transpiration renders them more tolerant of heat, and less so of cold than the whites" provided him with an anatomical explanation for these prevailing political conditions.[10] The supposedly different structure of the black race's pulmonary apparatus and thus its capacity to more efficiently regulate body heat was a further indication of the naturalness of the arrangement. This climatically honed physiology, combined with less need for sleep, enhanced the ability of African Americans to labor long and hard under the heat of a tropical or semitropical sun. So too did their disposition toward late-night amusements, an indifference to forethought, and the transient nature of their sense of grief. These did not seem characteristic of a people who were crushed by slavery. Climatic conditions, Jefferson believed he had shown, had conspired with physiological constitution and psychological proclivity to render the southern slavocracy a virtual inevitability. Again, it seems, rendering the slave system comprehensible could make it seem inexorable.

Whatever the evils arising from a slave culture, Jefferson both naturalized and psychologized the institution. His declared commitment to emancipation, moreover, certainly did not entail any sense of racial egalitarianism.[11] For he was sure that nature had made real distinctions between the black and white races whether they were of monogenetic or polygenetic origin. A chief source of anxiety was Jefferson's phobia about racial hybridity. This was where a sharp contrast needed to be drawn between the liberation of slaves in ancient Rome and their emancipation in the American South. The freeing of white slaves in the Roman empire was not troublesome since they could mix without defiling the blood of their masters. It was different in the New World, and the only way to avoid the contamination of white blood was by removing free blacks beyond the reach of race mixture. For such reasons, he insisted that it was mutually essential for freed slaves to be repatriated; they simply could not occupy the land in which they toiled for their rulers. Whatever his political stance on manumission, he remained steadfastly committed to black inferiority, in both mind and body.

Ascribing slavery to the dictates of climate was an abiding theme, not least among the more venomous champions of black enslavement. In 1853, John H. Van Evrie (1814–1896), a New York physician and apologist for white supremacy, found in climate the major driving force of race history. The title of his tract for the times, *Negroes and Negro Slavery: The First, an Inferior Race—the Latter, Its Normal Condition*, made plain the tone and temper of his account.[12] This diagnosis was further supplemented in 1868, after the Civil War, in a much enlarged treatise entitled *White Supremacy and Negro Subordination or, Negroes a Subordinate Race, and (so-called) Slavery its Normal Condition*. Here he insisted that a northern climate was fatal to the black races since the tropics were "the natural center of existence of the negro."[13] Van Evrie, it seems, had made a career out of black denigration, turning racial scientizing into the ugliest propaganda. George M. Fredrickson described him as "perhaps the first professional racist in American history."[14] Indeed, he provided a preface to the infamous Dred Scott decision in 1857, which found that people of African ancestry could not claim citizenship in the United States, declaring that the

Supreme Court had now sustained the view that the Declaration of Independence only applied to whites.

To Van Evrie, emancipation was simply a popular delusion whose advocates were stubbornly pushing against the irresistible laws of nature and the purposes of God alike. Black Africans had been forced out of their native habitats and brought into unnatural relations with the white race. Abolition would simply mean that they would inevitably relapse into their old habits and ways of life. That perfectly fitted the rigid universe that Van Evrie inhabited. To him, the "eternal and immovable laws fixed forever in the heart and organism of things, can not be changed or modified by human folly, fraud, or power." The global patterns of climate, soils, and crops had all been ordained by the Almighty to contribute to human welfare. They were the preestablished harmonies that had been in place since time began. Divine decree, climatic control, racial supremacy, and agricultural necessity had combined to deliver a global economic system operating under the iron laws of God and nature. As he summarized this harshest of political economies: "The brain of the white man and the muscles of the negro, the mind of the superior and the body of the inferior race, in natural relation to each other, are the vital principles of tropical civilization without which it is as impossible that civilization should exist in the great centre of the continent, as that vegetation should spring from granite."[15]

What anchored this frightful arrangement were the climatic laws of labor and racial distribution. The Caucasian—or "master man" as he put it—certainly enjoyed an elaborate and near-perfect structure, permitting it to survive in all climates that could sustain animal life. But "Caucasian Man" was only fitted to till the soil or engage in manual labor in the temperate zones. The white races were forbidden by the laws of nature to toil under a tropical sun or to cultivate by their own physical efforts the crops indigenous to the tropics. Because black Africans were adapted to a tropical climate and thus perfectly suited to their physical wants and moral necessities, Van Evrie announced that the tropical zone was the natural center of black existence. Races obviously had their proper places. There was mutual reciprocity in this economic ecosystem . . . with all the benefits of wealth accruing to the whites. The natural laws of industry and climate had determined that "the negro element goes just where its own welfare as well as that of the white citizenship and the general interests of civilization demand its presence." That pattern had long satisfied the economic cravings of the New World. For American slavery was rooted in the need for labor that was of the "lowest kind" and for workers "whose imperfect innervation and low grade of sensibility could resist the malarious influences always more or less potent in new countries and virgin soils."[16]

In many ways, Van Evrie's ethnic climatology reflected the racialist creationism of naturalists like Louis Agassiz and other polygenists who insisted that organic life could only thrive in the climatic regimes for which they were designed.[17] All creatures had their own spaces of existence, their local habitats, their peculiar places in what he called the "mighty programme of creation." They were all specifically adapted to these great centers of life by their organic structures, their faculties, and "the purposes they were designed to fulfil, all harmonizing with their localities, the positions the Almighty has assigned to them." Harmonious though the language of this theological ecology seemed to be, in fact it delivered a brutal creationist economy in which mind and nature, anatomy and agriculture, soil and society were grimly entwined. The black race's "simple grade of intelligence, his physical organism, his specific, climatic, and industrial adaptations" all stood in "perfect harmony with the primitive soils, the simple products, and uniform atmosphere of the tropics, and in complete relation and perfect union with the circumstances that surround him in the center of existence where the Almighty had placed him."[18]

If Van Evrie might too glibly be dismissed as a rough ideologue lurking on the margins of science, however influential he may have been on some political activists, the same can hardly be said of Nathaniel Southgate Shaler (1841–1906), the prolific and polymathic Harvard geologist who wrote at length on the atmospheric determinants of African American slavery and black inferiority and emerged as a prominent public intellectual in Gilded Age America.[19] There is no doubt that Shaler considered slavery an evil, perhaps a necessary evil. But his account of its American incarnation was thoroughly naturalized in the regional distribution of sunshine and soil. As he told the readers of his popular text for schools, *The Story of Our Continent*, first published in 1894, sun, soil, and slavery were intricately entangled. In the development of the southern plantocracy, he believed he had detected some remarkable political effects arising from differing climatic and soil conditions. For slavery only flourished in those places where cotton and tobacco could be profitably cultivated. Both these crops demanded a long, dry summer and a ready supply of cheap labor. His 1891 volume, *Nature and Man in America*, reaffirmed this view. In the southern reaches of the Appalachian range, he remarked, the character of the soil, the local landscape topography, and the nature of the weather rendered the land entirely unfit for the widespread cultivation of either tobacco or cotton. In consequence, slavery had never gained a foothold as an economic institution in any part of this expansive terrain.

Shaler's environmental apologia for slavery was part of a more general isothermal theory of politics. Thus, on the western prairies, the climate ruled out any profitable cultivation of the staple crops on which slavery was founded. As a consequence, the economic development of that region had to depend on the export of grain. Had the isothermals fallen a hundred miles or so farther north, and the agriculture of the South extended into the grasslands of Middle America, "the evil of slavery might well have been fastened so firmly that it could not have been uprooted from our country." During the period of the Civil War, Shaler pointed out that where the soils were rich, a large slave population could be supported, the plantation system took hold, and in consequence the general populace made common cause with the South. Where the soils were poorer, slaveholding was rare. This meant that in the border state of Kentucky, a soil map could "serve as a chart of the politics of the people." Had Kentucky "possessed a soil altogether derived from limestone, there is no question but that it would be have cast in its lot with the South."[20] At the state level, politics mapped straight onto pedology. And the principle held good at an even more granular scale. As he had earlier reflected in his general account of his home state, Kentucky, a glance at its electoral map disclosed the remarkable influence of soil type on political affiliation. "The dwellers on the limestone formations, where the soil was rich gave heavy pro-slavery majorities," he commented, "while those living on the poorer sandstone soils were generally anti-slavery in their position."[21] What he dubbed the geological distribution of politics was not unique to Kentucky, of course; he believed it was common throughout the entire South. Indeed, in a real sense, the Civil War and its outcome could be reduced to matters of temperature and topsoil, rainfall and rock. For Shaler, then, the failure of the southern Confederacy to achieve its secessionist ambitions may well have depended less upon political leadership, available resources, collective willpower, or military expertise than "upon this matter of soils."[22]

However much Shaler declared slavery an evil, arguments of this ilk provided justification—both economic and moral—for the establishment and expansion of the southern slavocracy. For a start, climate had set strict limits on just where the white races could safely engage in manual labor. Atmospheric conditions and anatomical constitution were closely connected. For the whites, high temperatures and hard toil simply did not go together. Although it was certainly possible for them to labor in the fertile lowlands of the South, he was sure that the long, hot summers of the region set clear climatic limits on the extent to which the white race could work the land there. Shaler doubted that the region could have advanced economically at all if it had been forced to rely on white manpower and likewise suspected that in his own day, it could make little progress if it had to depend on that source for its workforce. "It is doubtful if this region could have been developed if it had depended on white labour," he mused, "or if it could now go forward with a people derived from the stocks of northern Europe."[23] Weather had simply determined that the wealth of the South could never have been generated without slave labor. And that was that. For the black race—"really a tropical people"—was constitutionally suited to grinding toil in tropical regimes.[24] Bluntly put, it was on the foundations of the slave trade that the "vast structure" of the cotton and tobacco empires had been erected. And the benefits were not confined to the Old South. King Cotton did nothing less than "promote the growth of our race on this continent in a very important way, for it provided the means for an extended trade with the Old World, and thus gave a degree of wealth to the New."[25] It was on the beaten backs of African Americans that the entire political economy of the United States rested.

Given the role that slavery evidently played in the runaway economic triumphs of the New World, it is hardly surprising that Shaler presented it not merely as a naturalistic inevitability courtesy of sun and soil but in ways that enabled him to modulate its ugliest tones. In an 1884 article on the so-called "Negro Problem," for instance, he announced that even though it shared the shortcomings intrinsic to all master–subject arrangements between people, American slavery was far and away the mildest and most decent system of slavery that ever existed. And with breathtaking insouciance, he added, "When the bonds of the slave were broken, master and servant stayed beside each other, without much sign of fear or any very wide sundering of the old relations of service and support. As soon as the old order of relations was at an end, the two races settled into a new accord, not differing in most regards from the old." In fact, in this same piece, Shaler went so far as to maintain that manumission was harmful for African Americans. For in "place of the old lash, his master had the crueler whip of wages and account books."[26] It pressed him to the conclusion that every experiment to free blacks on American soil had only delivered worse conditions than they had experienced in their state of bondage. Indeed, he even urged that slavery had actually presented African Americans with many advantages. Because, as a race, black people exhibited a disabling lack of willpower, he judged that the enforced consecutive labor to which they were subjected had accustomed them to a greater continuity of effort than they had ever known before. So, with cold indifference, he claimed that black Africans had gained, rather than lost, through the slave experience. Indeed, twenty years and more after making these observations, he told the readers of *The Citizen*, a posthumously published volume intended to instruct young men and women engaged in the management of public affairs, that "enslaved negroes were not debased by the change from their native countries to America," as many claimed. The truth was very much to the contrary. The slaves, he insisted, actually benefited from the opportunities granted to them through "contact with a superior race." By this means, they were "trained out of barbarism and into a Christian civilization" with the result that "they and their descendants had a chance of rising to a plane of life which has never come and most likely never will come to others of their kind."[27]

Slavery's tyrannies, to put it another way, were simply transmuted into training opportunities, its economic violence into educational provision. To the slaves, the planation system was, apparently, "exceedingly helpful." This was because providing tuition in human relationships, in domestic arts, in moral understanding, and in religious faith delivered virtues that the slaves could have acquired in no other way. Slavery, Shaler was assured, "lifted a savage race nearer and more rapidly towards civilization than had ever before been accomplished, and it gave to the people who did the work a peculiar moral training."[28] Moreover, the effects of climate's gifting of slavery to the New World brought distinct benefits for the social evolution of the whites. As he put it in Darwinian-sounding language, "In no other social condition was the survival of the fittest so evident as in slave-holding, and . . . as a result it developed a body of natural leaders, strong men who had won their way by quality of leadership handed down from generation to generation."[29] Darwinizing slavery could conveniently, if not obviously, upend morality. The ethics of human relations were submerged beneath the irresistible forces of natural selection. In the long run, of course, slavery had its downside. While it gave "the negro certain valuable elements of an education, in that it trained him in obedience to authority and in orderly consecutive labor," Shaler conceded that "it denied him nearly all chance of showing the peculiar capacities which he may have."[30] For the future, any further progress the black races might achieve depended on imitating the master race. That was because social and intellectual advancement could not spring from any innate drive within the race itself. Development under white tutelage was the only way ahead. But, happily, patronage and guardianship could do their work, for "this relatively simple species of our genus" was blessed with "a power denied to the higher race," namely, the "capacity for adjustment by a process of imitation," the very quality that had "enabled them to adopt the manners of their new-found masters."[31] Servile emulation, apparently, rather like adaptive mimicry in butterflies, was the African American's greatest natural endowment!

Shaler's tying slavery to the conditions of sun and soil, as well as his apologia for the Old South's plantation system, was not unique. Others followed much the same course and found in the climatic differences between North and South the primum mobile of the American Civil War. Not everyone was convinced, however. Needless to say, African American physicians certainly did not see things the same way. Two associates of Frederick Douglass, James McCune Smith (1813–1865), who studied medicine at the University of Glasgow, and the physician Martin Delany (1812–1885), who was also a major in the U.S. Army, developed their own rather different climatic philosophies of the body and racial politics. Smith, drawing on an impressive range of vital statistics, was convinced that all humans flourished best in temperate climes. He thus opposed moves to resettle freed slaves in the unhealthy climate of Liberia and urged his fellow African Americans to remain in the United States and to progressively colonize spaces where temperate conditions prevailed. Delany, on the other hand, insisted that blacks could flourish in any climate and were thus superior to whites, whose fragility was expressed in the ways in which they degenerated physically, mentally, and morally in tropical and subtropical conditions. That was why unhealthy whites had to resort to more vigorous Africans to build the southern economy. For him, as Colin Fisher has put it, "the institution of slavery was nothing less than a medical symptom that the white race had ventured outside its zoological province."[32] While there remained differences between both of these black public intellectuals, they were united in the conviction that African Americans could thrive in temperate regimes. In both cases, their political stances were erected on climatic foundations.

Despite these interventions, the wedding of slavery to the demands of medical and economic climatology persisted into the twentieth century. Writing in 1911, for example, the geologist-geographer Frederick V. Emerson (1871–1919), who taught at the Universities of Missouri and Louisiana State, attributed the development and spread of slavery to soils, topography, and climate but insisted that temperature and rainfall were the most critical determining causes.[33] No doubt other advocates of the climatic theory of servitude could be retrieved from the record.[34] In a subsequent chapter, we will examine how figures such as John William Draper, Ellen Churchill Semple, and Ellsworth Huntington turned to the determining role of climate in their accounts of America's internecine conflict. But for the moment, what is clear is just how deeply the sun–soil–slavery nexus had infiltrated American intellectual life and how widespread these "scientific" judgments had become. And this despite the cautionary observations of those like Cleveland Abbe, who, while acknowledging that different climate regimes had "contributed to the perpetuity of slavery," nevertheless hastened to add that "the origin of this social condition is traceable more directly to the wars and rivalries of domineering men, regardless of climatic conditions."[35]

Wealth, Weather, and the Philosophy of Food

The sun-and-slavery philosophy that dominated the thinking of these writers on the cultural politics of American race relations was not the only way in which the zonal mind-set manifested itself in nineteenth-century thinking about climate and commerce. And nowhere perhaps was this impulse more clearly advertised than in the writings of English historian Henry Thomas Buckle (1821–1862), deist and autodidact. In his magisterial *History of Civilization in England,* which first appeared in two volumes in 1857 and 1861, Buckle worked to further advance the naturalization of social life and cultural mores that Montesquieu had articulated.[36] "Are the actions of men, and therefore of societies, governed by fixed laws, or are they the result either of chance or of supernatural interference?" Buckle asked his readers in the first few pages of his magnum opus.[37] His answer was crystal clear: human behavior was regulated by natural law. The possibility of developing a genuine science of history was thus entirely achievable. From the Belgian astronomer and statistician Adolphe Quetelet, who applied statistical techniques to identifying the causal relations between crime and age, crime and poverty, crime and weather, and suchlike, Buckle had learned that there were intimate influences between the physical laws of nature and human actions. He was determined to put that insight to work in his magnum opus on the development of civilization.

The revelations of Quetelet and his like were far-reaching. On the conceptual level, they tended to favor some form of determinism over humanity's much-vaunted freedom. As Buckle saw it, over the course of time, an appreciation of the remarkable regularity of nature tended to destroy the doctrine of chance and to replace it with a sense of what he called "necessary connexion." And yet that was not the whole story. Given the task on which he had embarked, he urged that those who were persuaded about the viability of cultivating a science of history were not obliged to succumb to either a deterministic understanding of events or to an altogether radical human freedom. In so doing, he struck out on a path between what he described as "the metaphysical dogma of free will" and "the theological dogma of predestined events," all the while conceding that he himself had been driven to the conclusion that since human actions were universally determined solely by their antecedents, they displayed consistent regularity. Under precisely the same circumstances, he professed, the same results would always emerge. The positivist aura of this methodological revelation was not difficult to discern. The brute facts of statistical regularity and a penchant for uncovering natural laws of human behavior were the only materials on which a genuinely philosophical history could be erected. The proofs he believed he had adduced "of our actions being regulated by law, have been derived from statistics," he remarked, and these had "already thrown more light on the study of human nature than all the sciences put together."[38]

In sculpting this vision, Buckle turned to climate, food, and soil as the formative physical agents that had not only engineered the most significant elements of social organization but also determined the large and conspicuous disparities that existed between nations. In Buckle's telling, provender, pedology, and politics were tightly intertwined through their links with climate, and these in turn governed the global distribution of wealth. This was simply because of all the diverse components of social life that climate, food, and soil had originated, the accumulation of wealth was the first and indeed the most important. That was fundamental. With positivist predilections, he thus insisted with August Comte (1798–1857) that wealth was the foundation on which knowledge was erected, not the other way around.[39] "Although the progress of knowledge eventually accelerates the increase of wealth," he mused, "it is nevertheless certain that, in the first formation of society, the wealth must accumulate before the knowledge can begin."[40]

What is notable in Buckle's history is the dominant role that diet played in his scheme. His food philosophy of climate and culture was nothing if not comprehensive. To him, the laws of climate were directly connected with the laws of population through the medium of food provision and, therefore, in turn, governed the laws of the distribution of wealth. Food's effects were truly far-reaching. At the most basic level, the different foodstuffs available for consumption in different parts of the globe made all the world of difference to the supply of the "animal heat" necessary for the proper functioning of all life. Crucial here, Buckle believed, was the extent of nitrogenization or, as he put it, "azotized substances" present in certain foods, which repaired "the incessant decay" of the human body. In different climatic zones, the pattern was different:

> When men live in a hot country, their animal heat is more easily kept up than when they live in a cold one; therefore they require a smaller amount of that non-azotized food, the sole business of which is to maintain at a certain point the temperature of the body. In the same way, they, in the hot country, require a smaller amount of azotized food, because on the whole their bodily exertions are less frequent, and on that account the decay of their tissues is less rapid.[41]

To Buckle, the chemistry of food ingestion, when coupled with global eating patterns, had major implications for national temperaments. His own inventory of food consumption in Asia, Africa, and the Americas confirmed his conviction that "the colder a country is, the more its food will be carbonized; the warmer it is, the more its food will be oxidized." Fatty foods, to put it another way, were typical of colder zones; carbohydrates were more common in the diets of hot regions. And since carbonized food largely came from the consumption of meat, which was more difficult to obtain than oxidized food, those nations where a cold climate rendered a highly carbonized diet essential necessarily exhibited "a bolder and more adventurous character, than we find among those other nations whose ordinary nutriment, being highly oxidized, is easily obtained, and indeed is supplied to them, by the bounty of nature, gratuitously and without a struggle."[42] In turn, these different dietary habits had a direct effect on the degree to which wealth was distributed among the different classes. Food chemistry, atmospheric temperature, and national character conspired to generate the global patterns of wealth distribution.

Food's field of influence, plainly, extended well beyond the merely biological. For one thing, again like Comte, Buckle was sure that diminishing the precariousness of food provision had moral and metaphysical consequences. It gave impetus to the idea that there was regularity in nature and thus acted to curtail the inclination to resort to explanations mired in superstitious caprice. Besides, social institutions also mirrored the state of the supply chain. The price of corn, he argued, had a fixed and definite relation to the number of marriages annually contracted. Marriages, he was persuaded, were not determined by the wishes and desires of the individuals concerned but by elemental forces beyond their control. Over the course of a century in England, statistical analysis had revealed that "instead of having any connexion with personal feelings," marriages were "simply regulated by the average earnings of the great mass of the people." This solemn social and spiritual institution it seemed was swayed, if not completely controlled, "by the price of food and by the rate of wages."[43] Weddings were reduced to matters of eating and earning; sentiment seemed to matter scarcely a whit.

The connection Buckle identified between food price and wage rate in explaining marriage patterns was only one expression of a principle that operated universally throughout the global economy and indeed was the chief architect of the difference between the temperate regions and both the tropical and polar worlds. Where food was cheap and easily procured, populations rapidly increased and exerted greater pressure on wage funds.[44] The implications were immediate. Because food was less readily available and less easily acquired in cooler regions, Buckle believed that in hot countries, there was a powerful and consistent tendency for wages to be low, while in cold climes, they were routinely much higher.

The ties between climate, food, and wages manifested themselves in other ways that also reinforced latitudinal economics. Because of the influence of climate on labor patterns, Buckle was sure that high-latitude nations—like tropical peoples but for different reasons—had never acquired those staunch and steady work habits for which the inhabitants of the temperate zone were renowned. The reason was that severe weather and seasonal light deficiency meant that out-of-doors labor was largely impossible at certain times of the year. In these climatic regimes, working patterns were periodically disrupted with the consequence that the "chain of industry" was sporadically broken and economic impetus lost. As a result, far northern peoples, like the Swedes and Norwegians, possessed "a national character more fitful and capricious than that possessed by a people whose climate permits the regular exercise of their ordinary industry." In southern climes—Spain and Portugal were his examples—different weather patterns conspired to produce a similar state of affairs. Here, sustained labor was "interrupted by the heat, by the dryness of the weather, and by the consequent state of the soil." As a result, whatever their other differences, these four nations were all remarkable for their volatility and fickleness of character, which stood in marked contrast "to the more regular and settled habits which are established in countries whose climate subjects the working classes to fewer interruptions, and imposes on them the necessity of a more constant and unremitting employment."[45]

The operation of these self-same forces at a global scale meant that there was no tropical country where the inhabitants had ever managed to escape their climatic fate, even in those places where wealth had extensively accumulated.

Buckle just could find no instance in which a hot climate had failed to produce an abundance of food, which, in turn, generated an "unequal distribution, first of wealth, and then of political and social power." Tropical political economies were merely the outcome of the operation of the natural laws of climate and regional diet and demonstrated "the intimate connexion between the physical and moral world." Herein lay the explanation for why the earliest civilizations first developed in warmer zones, where surplus food, mostly in the form of cereal crops, was readily available. But these were rapidly overtaken when civilization spread to relatively less propitious climates, which generated much greater physical energy and mental agility. So determinative indeed were these forces that Buckle was convinced that knowledge of a few physical laws would be more than sufficient for students of human history "to anticipate what the national food of a country will be, and therefore to anticipate a long train of ulterior consequences."[46]

The laws of food that Buckle saw as foundational to economic and social life were supplemented by the influence exerted on human thought and culture by what he called the General Aspect of Nature, that is, the observed features of local landscapes in which human societies were domiciled. If climate, soil, and food controlled the forms of social institution and patterns of wealth, the combined constituents of the natural world acted more especially on the operations of the human mind. In particular, Buckle turned to the "General Aspect of Nature" to show how, by exciting the imagination, it conjured up in primitive minds "those innumerable superstitions which are the great obstacles to advancing knowledge." In consequence, just as "in the infancy of a people, the power of such superstitions is supreme," the way in which nature impressed itself on the human psyche in different locations had produced corresponding varieties in popular character. Even in advanced societies, nature was no less influential, imparting "to the national religion peculiarities which, under certain circumstances, it is impossible to efface."

Not surprisingly, the implications of this stirring of the imaginative faculty and the stimulus it gave to superstition displayed its own global geography. Tropical landscapes excited the imagination more powerfully than elsewhere and stood in marked contrast to European climes where rational understanding prevailed. Among civilizations beyond Europe, Buckle believed that "all nature conspired to increase the authority of the imaginative faculties, and weaken the authority of the reasoning ones." In Europe, by contrast, nature's operations tended to restrain the imagination and to animate cognition. This had the effect of inspiring confidence in the power of human reason by fostering the growth of knowledge and "by encouraging that bold, inquisitive, and scientific spirit, which is constantly advancing, and on which all future progress must depend."[47] In one way or the other, both superstition and science owed their existence not so much to spiritual powers or intellectual prowess but to the operations of Mother Nature.

All this stood in opposition to those downward-pulling forces that traded on theological dogma and religious sentiment. Calling on the authority of Charles Lyell (1797–1875), he noted how the rousing of the imagination among certain classes during earthquakes and other similar catastrophes induced a fearful state of mind and a sense of human powerlessness. Lyell indeed had recorded that "such awful visitations" fostered "a belief in the futility of all human exertions" and prepared "the minds of the vulgar for the influence of a demoralizing superstition."[48] Buckle further added that as the imagination was excited, "a belief in supernatural interference" was "actively encouraged." "Human power failing," he went on, "superhuman power is called in; the mysterious and the invisible are believed to be present; and there grow up among the people those feelings of awe and of helplessness, on which all superstition is based, and without which no superstition can exist." Even in Europe, for all its embrace of rationality and its cultivation of modern civilization, superstition still lingered in regions where volcanic and seismic eruptions were more prevalent, notably in Italy and the Iberian peninsula. These countries, he insisted, were precisely "where the clergy first established their authority, where the worst corruptions of Christianity took place, and where superstition has during the longest period retained the firmest hold."[49]

Unsurprisingly, not everyone was impressed with Buckle's endeavors. Christian writers, skeptical of rationalism, deism, and their attendant evils, were dismissive, as indeed were representatives of the new professional history. Lord Acton (1834–1902) thought Buckle the purveyor of absurdities, a mere positivist historian, and therefore, like all positivists, only half-educated. Mark Pattison (1813–1884), rector of Lincoln College, Oxford, complained that Buckle had removed all passion from human affairs. For his part, the historian and Whig politician Thomas Babington Macaulay (1800–1859) thought Buckle constantly overstepped his competence by declaring on subjects he did not know enough about.[50] On the other hand, Buckle's *History* was popular at the time. And since so much of his account rested on the natural laws of food, themselves regulated by climate and soil, it is understandable that materialist-orientated students of society would see much to admire in Buckle's scheme. George Bernard Shaw, for example, remarked in 1894 that few books made any "permanent mark" on human culture; among the great exceptions, he would certainly number "Marx and Buckle among the first."[51] Buckle's work, moreover, brought him introductions to such cognoscenti as Thackeray, Darwin, Spencer, and Huxley and earned him election to the Athenaeum and the Political Economy Club.

For all that, Buckle's reputation was already fast fading by the end of the nineteenth century. Nonetheless, some readers continued to find in his *History of Civilization* insights relevant to their own concerns. His analysis of the uneven distribution of national wealth in hot climates; the ways in which all-powerful rulers there denied their peoples any voice in the government of their own states, thereby stimulating long habits of servile submission; and the insurmountable obstacles that high temperatures placed on arduous labor—all these obviously raised questions over how, if at all, the colonial world should be governed and what was to be done about the untapped resources to be found there.

Climate, Tropical Resources, and Colonial Governance

Buckle's reflections on the problems of self-governance in hot regions and his insistence that their climates were conducive to despotism became a source for later writers concerned with how the temperate world could exercise control over the rich, but undeveloped, resources of the tropics. This was certainly the case with the Harvard Professor of Climatology Robert DeCourcy Ward, who equally welcomed Buckle's thoughts on the length of the farming season as a critical control on the returns gained from the soil and used the idea to insist on the inestimable value of the schooling in good habits that human society received from colder weather with its mental and physical disciplines. Ward, whose pronouncements on tropicality and acclimatization attracted our attention in chapter 3, will thus serve as a preliminary lens through which to inspect ideas about climate, civilization, and colonial governance that were in circulation in the United States around the turn of the nineteenth century. Recalling his role in the Boston Immigration Restriction League, Ward's views cannot easily be dismissed as the ramblings of an ivory tower irrelevance. He had long been alarmed about immigration and, along with Charles Warren and Prescott Fransworth Hall, friends from school days and students together at Harvard, he determined to move academic analysis of the problem into Realpolitik.[52] As a graduate student in the Harvard geology department, he had come under the influence of Shaler, who exerted a profound influence on his thinking,[53] and went on to assume the chair of that department and to produce polemical essays such as "The Fallacies of the Melting Pot Idea and America's Traditional Immigration Policy."[54]

Ward's *Climate*, which first appeared in 1908, was number 20 in the Science Series of books brought out by P. G. Putnam's Sons in New York, a publishing house that traced its history back to 1838.[55] The list of contributors to this collection boasted a range of prominent intellectuals, among them Alfred C. Haddon, St. George Mivart, James Geikie, Thomas G. Bonney, C. Lloyd Morgan, J. Arthur Thompson, and James Mark Baldwin. Like the other volumes in the series, *Climate* was aimed at a broad, literate audience. Indeed, Ward based the text on lecture notes he had accumulated for a decade and more for a course he taught on general climatology at Harvard and intended it as "supplementing the first volume of Dr. Julius Hann's *Klimatologie*," an English translation of which he had produced in 1903.[56] We can be sure that Ward's viewpoint was widely disseminated among several cohorts of early twentieth-century Harvard alumni. As a textbook, moreover, it represents a synthetic rendition of commonly accepted climatic dogma in the early twentieth century.

It is plain that Ward was no less captivated by the zonal philosophy of history that, in one way or another, Ibn Khaldūn and Montesquieu had espoused. For throughout his account, human life in all its forms was routinely connected with climatic conditions whether in the tropics, the temperate zones, or the polar regions. In Ward's opinion, there were certain common conditions of life that influenced people dwelling in the same climate regime in broadly similar ways. Calling on the testimony of the distinguished geographer-anthropologist Friedrich Ratzel, who, as we have seen, championed a form of geographical determinism, Ward urged that there was a universal climatic factor operating in the world to maintain differences between the people of the different latitudes, despite those population movements that constantly trended toward homogeneity. Those differences were profound. The French orientalist Ernest Renan, he recalled, had shown how desert landscapes fostered monotheism because of their monotonous yet sublime uniformity, which nudged witnesses toward a belief in the unity and singularity of an infinite God. The monotheistic religions were the desert's offspring.[57] Again, Ralph Abercromby, the Scottish meteorologist, had disclosed through his weather maps how the geography of Islam was remarkably correlated with mean annual rainfall. Presently, Ward himself declared that climate had played its part in determining the "broad, distinguishing characteristics of man in the temperate and tropical zones."[58] Here the links between zonal climate and zonal anthropology were prominently advertised.

The connections between zonal physiology and tropical denigration that have snaked their way through a good deal of writing on climate and wealth also manifested themselves with full force in Ward's analysis. We have already seen how his condemnation of "tropical monotony" was coupled with a stereotypical white anxiety over a litany of ailments afflicting colonials abroad: increased respiration rates, decreased pulse action, digestive troubles, psychological depression, and a general, ill-defined state of debility. Because the bracing effects of the cyclonic middle latitudes were lacking, as were the tonic effects of an icy winter, economic growth and the generation of wealth could not be expected from the native populations of the tropics. Writers of one sort or another had shown just how intimate the connections really were between weather and wealth. He told his readers that H. Helm Clayton of the Blue Hill Meteorological Observatory, for instance, had traced the relationship between patterns of rainfall and economic cycles in the United States, as had the pioneering econometrician Henry Ludwell Moore. In comparable vein, the Swedish chemist Otto Pettersson had demonstrated the influence of climatic variation on key economic events.[59] Lessons learned here were applicable a fortiori in the tropical world.

All of this, combined with the superabundance of tropical resources, explained why white governance of the tropical lands was not only an economic necessity but also a moral duty. In a climate that was both weakening and wearying, and where the profusion of resources relieved tropical inhabitants of any need for serious labor, the willpower to develop nature's riches was signally absent. Nature frankly did "too much" between the tropics, with the result that "little is left for man to do." The political imperatives were immediate, and Ward had no hesitation in spelling them out. "Voluntary progress toward a higher civilization is not reasonably to be expected" from native peoples, he announced. Consequently, "the tropics must be developed under other auspices than their own."[60]

In search of confirmation for this policy, he turned for support to the Wisconsin institutional economist, labor historian, and advocate for immigration restriction, John R. Commons, who, according to Ward, informed the readers of *The Chautauquan* in 1904 that where "nature lavishes food and winks at the neglect of clothing and shelter, there ignorance, superstition, physical prowess, and sexual passion have an equal chance with intelligence, foresight, thought, and self-control."[61] Ward pressed home the economic significance of such admissions. In the tropical world, there just simply was no superfluous energy for pursuing the higher things of life. Eagerly fastening on the arguments we have just examined by Arnold Guyot, he typified “tropical man” as a creature who “never dreams of resisting this all-powerful physical nature; he is conquered by her; he submits to the yoke, and becomes again the animal man,—forgetful of his high moral destination.”[62] The industrial fallout was as predictable as it was pitiable. Because the basic requirements of life were readily supplied without any need for manufacturing, and because black constitutions had long acquiesced to a life bereft of vitality and drive, native industrial activity had never flourished in the equatorial realm. Yet the future was far from bleak. Dynamic resource exploitation by the white race, benevolent supervision by the temperate nations, and growing native demand combined to convince Ward that tropical industries were beginning to develop. Crop harvesting and animal husbandry could also advance, but they too could only flourish under the direction of white overseers. Tropical self-governance was an impossibility. This, recall, was a message to which every Harvard student who sat in Ward’s classes was exposed.

Of course, achieving good governance was no easy task. What he dubbed the “Labour Problem in the Tropics” loomed large on Ward’s ideological landscape. It stemmed from his conviction that in the tropics lay lands that were highly valuable to the white races because of the wealth of their commodities. Indeed, these were the “spheres of influence”—colonies—that were numbered among the northern nations’ most coveted possessions. But the problem remained how to induce indigenous peoples to engage in continual steady work, especially when, as the British traveler and colonial commentator Alleyne Ireland had put it, “local conditions are such that from the mere bounty of nature all the ambitions of the people can be gratified without any considerable amount of labour.” With an ingrained disinclination to work of their own accord, there was simply no option but to turn to forced native labor or imported indentured labor to stimulate economic growth. There were problems with both, of course, but one way or another, local peoples had to shoulder the burden, for the obvious reason that the climate militated against white labor. Noting that the perpetuation of “negro slavery in the West Indies” was “aided by climatic controls,” he concluded that white “labourers are not likely to become dominant in the tropics for two reasons:—first, because the climate is against them; and second, because the native is already there, and his labour is cheaper.”[63] Climate, it seemed, had foreordained the class and color of the labor force in the colonies.

The climatic hurdles that tropical indigenes and colonial overseers alike encountered stood in marked contrast to temperate economies where mid-latitude civilization had flourished. Again following Guyot, Humboldt, and Ratzel, Ward reiterated his belief that it was in the seasonal pattern of the moderate latitudes that the secret could be found "of the energy, ambition, self-reliance, industry, thrift, of the inhabitant of the temperate zones." Seasonal variation was stimulating to mind and body alike and encouraged the development of higher cultural forms. A cold, stormy winter required a society to prepare during the summer months for winter clothing, food, and shelter. "Carefully planned, steady, hard labour," he explained, "is the price of living in these zones." But the outcome was steadfast and sure. Development was an inevitability. For, as he put it, behind "our civilization there lies what has been well called a 'climatic discipline,'—the discipline of a cool season which shall refresh and stimulate, both physically and mentally, and prevent the deadening effect of continued heat."[64]

In support of the belief that the tropical world could never be developed by indigenous peoples and that there was an unavoidable migration of the center of civilization from the tropics to the temperate world, Ward called on the testimony of the celebrated sociologist and evolutionist Benjamin Kidd (1858–1916). In 1898, Kidd had declared that while the human race may have "reached its earliest development where the conditions of life were easiest"—namely, in the tropics—slowly but surely "the seat of empire and authority" had moved "like the advancing tide northward." To Kidd, this migratory impulse was largely on account of the self-evident fact that "in dealing with the *natural* inhabitants of the tropics we are dealing with peoples who represent the same state in the history of the development of the race that the child does in the history of the development of the individual."[65] With such political and recapitulationist sentiments, Ward was altogether comfortable.

The Irish-born Benjamin Kidd was largely self-taught but, while first employed as a clerk at the Board of Inland Revenue in London, achieved fame for his *Social Evolution*, which appeared in 1894.[66] Conceiving of society and civilization in the language of Charles Darwin and more particularly August Weismann, he argued that religion was the driving force in the evolution of philanthropy and political enfranchisement. Human history was nothing but a saga of unadulterated struggle for survival, which could only be mitigated by religion, the enormous utilitarian value of which had been totally unappreciated by the likes of Spencer. To Kidd, it was Western civilization above all that was marked by its commitment to humanitarian sentiment, for it alone had given fulsome expression to a passion for justice, equality of opportunity, and the undermining of inherited privilege. At the time, the book was a roaring success and was translated into many languages, Chinese and Arabic among them, though its progressivist ethos meant it did not well survive World War I. Yet its sales were sufficient to enable Kidd to resign his civil service post and to travel as a public intellectual celebrity, even if there were many—such as Theodore Roosevelt—who severely attacked it. Something of its stresses and strains may be gleaned from Richard Hofstadter's judgment that it "was a peculiar mixture of obscurantism, reformism, Christianity, and social Darwinism"[67] and also from the fact that supporters and opponents alike were to be found among the ranks of his religious, evolutionary, secularist, and rationalist readers, as well as those on both the political left and right.[68]

In the present context, I want to dwell on the text to which Ward called particular attention—Kidd's 1898 *The Control of the Tropics*—the appendix to which had already appeared as chapter 10 of *Social Evolution*. The biological foundation on which human life rested was crucial to Kidd's outlook. "History and politics," he had claimed in 1894, "are merely the last chapters of biology."[69] And it is therefore no surprise that matters of geography and meteorology occupied a prominent place in his thinking about tropical governance. That the extended pamphlet had first seen the light of day as a sequence of articles published in *The Times* during July and August 1898 demonstrates Kidd's concern to guide the public mind on matters of tropical governance.[70] The statistics of international trade he had garnered illustrated just how vital tropical products were to the British economy. A huge proportion of Britain's trade was with the tropical world, and Kidd predicted that tropical regions would play an increasingly dominant role in the global economy. It was no surprise, he declared, that intense international rivalry for the control of the trade of the tropics would soon be widely seen. The problem was that over "a considerable portion of these regions at present we have existing a state either of anarchy, or of primitive savagery, pure and simple, in which no attempt is made or can be made to develop the natural resources lying ready to hand."[71]

For Kidd, the key question remained how the tropics should be governed, given the manifest failures of the past. "Plantation," or "possession," which often went hand-in-hand with the use of slave labor, had been disastrous. That had resulted in the more or less complete subordination of indigenous peoples to the interests of the imperial power and was, frankly, "incompatible both with the spirit and the forms of modern civilization."[72] Efforts by colonial settlers to work the tropical lands were just as wrongheaded. For that aspiration was based on the mistaken idea that white Europeans could acclimatize to the new meteorological regime in which they found themselves.[73] Climate forever dictated that this was an impossibility. France in particular had tried it . . . and failed. Indeed, those who had embarked on that route were soon possessed "of an overmastering conviction of the innate unnaturalness of the whole idea of acclimatization in the tropics, and of every attempt arising out of it to reverse by any effort within human range the long, slow process of evolution which has produced a profound dividing line between the inhabitants of the tropics and those of the temperate regions." The very attempt to acclimatize whites to the tropics was "a blunder of the first magnitude."[74] It was bad science and bad politics. Climatic conditions and natural selection, it seemed, had joined forces to create an unbridgeable chasm between the peoples of the temperate and tropical worlds.

If civic morality and climatic dominion ruled out the practices of plantation slavery on the one hand and colonial settlement on the other, the question remained how the tropics should be treated by the Euro-American world. It was a pressing concern given the tropics' vital role in world trade and their rich, but untapped, natural resources. To Kidd, it was plain as daylight that development could not be trusted to the hands of native peoples. Wherever the black races were left to their own devices, he insisted, they failed to develop the resource potential of the rich lands they occupied. Nor did they display any inclination to embark on any such scheme. Agricultural improvement and industrial growth could never emerge from that quarter. Neither, for that matter, could colonial settlers fulfill the need, even if they could acclimatize to the region. The reason was an evolutionary one. The people among whom expatriates lived and worked were often separated from them by thousands of years of biological history. In such circumstance, colonial officials should never be permitted "to administer government from any local and lower standard" that might emerge in imperial settings. The only realistic option for the tropical regions was to manage their resources, "not as in the past by a permanently resident population, but from the temperate regions, and under the direction of a relatively small European official population." European supervision of the tropical lands was the only viable alternative under the arrangements of a global trusteeship.

> If we have to meet the fact that by force of circumstances the tropics must be developed, and if the evidence is equally emphatic that such a development can only take place under the influence of the white man, we are confronted with a larger issue than any mere question of commercial policy or of national selfishness. The tropics in such circumstances can only be governed as a trust for civilization, and with a full sense of the responsibility which such a trust involves.[75]

Like many others, Kidd was troubled to ponder on the "the inexpediency of allowing a great extent of territory in the richest region of the globe—that comprised within the tropics—to remain undeveloped, with its resources running largely to waste under the management of races of low social efficiency." Happily, the laws of social evolution dictated the way ahead. Those laws had cultivated, over countless generations, "the energy, enterprise, and social efficiency" of the northern races, with the result that those "less richly endowed" lay abandoned in venues where the conditions of life were easiest. That was just "the cosmic order of things." And it could not be altered by wishful thinking, political will, or anything else. Cosmic forces and evolutionary science alike dictated that "the tropics must be administered from the temperate regions."[76]

Kidd's intervention speaks to a debate that raged during the later decades of the nineteenth century over how the tropical world should be treated and how its riches should be managed. The pro-empire *Spectator*, for example, found Kidd's contribution to be an "eloquent and convincing appeal for a new conception of the duties of civilized States towards tropical dependencies."[77] On the other hand, anti-imperialists were appalled at the arrogance of the thought that Europeans felt they could manage the affairs of others better than they could themselves. To them, talk of supervising the tropical world as a trust for civilization was nothing but a mask for continuing exploitation and naked European self-interest. At the same time, Kidd's views on acclimatization were challenged by the medical establishment in general and by Alfred Russel Wallace in particular.

Another figure to whom Robert DeCourcy Ward turned in support of his own views on tropical labor provides a further flavor of these ongoing turn-of-the-century fixations, as well as demonstrating the continuing legacy of Buckle's scientific history among those pushing for European supervision of tropical resources. In 1902, the British traveler, writer on tropical colonies, and sometime secretary to Joseph Pulitzer, Alleyne Ireland (1871–1951), at the time colonial commissioner for the University of Chicago, directed the attention of the American Academy of Political and Social Science to the question, "Is tropical colonization justifiable?" Here Ireland revealed the low esteem in which he held tropical peoples. His self-directed question on precisely what contribution they had made to human progress over the previous thousand years yielded to a stark suite of rhetorical interrogatives: "Have they produced a single poet of the first rank," he asked, "or a painter, or a musician, or an engineer, or a chemist, or a historian, or a statesman, or any man of the first eminence in any single art or science? Are we indebted to them for a single important invention, or for any new discovery in any branch of inquiry? Have they, during the past ten centuries, contributed a single great idea to the sum of human knowledge? The answer to all these questions is in the negative."[78] Clear as it was cruel, as arrogant as it was ignorant, Ireland's diagnosis left no doubt in his readers' minds where he stood on any indigenous development of the resources of the tropical world.

Just two or three years earlier in 1899, he had brought out *Tropical Colonization*—the text on which Ward called for support on the problems of tropical labor. The volume was essentially a survey of how a variety of different European states had approached the whole question of tropical governance, particularly dwelling on the problem of how a reliable workforce could be acquired for the successful development of any tropical colony. His inventory plainly exposed the viewpoint of its author. Ireland, cognizant of the huge value of trade between the United Kingdom and the British Colonies and Possessions—both imports and exports averaging around 45 percent over the previous twenty years—presented his readers with detailed international commercial statistics under the title "Trade and the Flag." All of this was to confirm just how vital the tropical world was to Britain's economic health.

Of course, European interests in the tropical world were forever running up against the problems of labor. Kipling might well be applauded for extolling "the capacity, the duty, of the men of the Anglo-Saxon race to do thoroughly the task laid on their shoulders, not for love of gain, not for hope of praise, but for the very joy of the accomplished thing," but the recurrent labor problems in the tropical zone, beginning in the sixteenth century and increasingly since then, routinely scuttled that seemingly laudable ambition.[79] No wonder Ireland affirmed, using the words of Sir Harry Johnston, that the "White and Yellow peoples have been the unconscious agents of the Power behind Nature in punishing the negro for his lazy backwardness." No moral qualms or humane restraint attended this verdict. It was simply the case that the laws of nature had ordained that humankind must engage in manual labor so as to wrest from nature "sustenance for body and mind." Neither tactful nor tenderhearted, he painted the bleakest of futures for African races if they refused to apply themselves under European tutelage to the development of their own continent's vast resources. Restaged for Ireland's audience, Johnston prophesied that the "force of circumstances, the pressure of eager, hungry, impatient outside humanity, the converging energies of Europe and Asia, will once more relegate the Negro to a servitude which will be the alternative—in the coming struggle for existence—to extinction."[80] Here Johnston's welding together of political hegemony, economic opportunism, and Darwinian vocabulary enjoyed the support of Ireland himself.

For the future, given the challenges of tropical climates and the engrained failing of indigenous peoples, the management of wealth generation in the equatorial lands lay in the hands of the northern continents. And for Ireland, that meant that serious consideration had to be given to indentured labor. To be sure, he conceded there had been abuses of that system in the past. But he was convinced that in his own day, when "conducted on humane and just principles," there was every reason to believe that contracted servitude could now be successfully established throughout the tropics in those regions where a steady labor supply could not be guaranteed.[81]

Appalled at the lack of progress among the millions of inhabitants of the tropics' vast domain, Ireland took refuge in the laws of history, laws of precisely the sort that Buckle had earlier elaborated. In condensed form, this meant that tropical civilization, such as it was, had always entirely depended "on the relation between nature and its own productiveness." Climate reigned supreme there. Under such conditions, the possibilities for advancement were inescapably narrow for human life was overwhelmed by nature's prodigality. To Ireland, Buckle's positivist history was entirely vindicated. The "enforced submission to physical environment" that "inferior peoples" had long experienced had indelibly impressed itself on their racial character. By contrast, the so-called superior peoples had benefited from the "slow and painful ascent of the ladder of progress." Ceaseless wrestling with a more unyielding environment was a blessing in disguise. It gifted them with a progressive enhancement of their "intellectual faculties and a corresponding decrease in the tyrannical power of external nature." For this reason, while it had become fashionable in some quarters to disparage Buckle, Ireland was convinced that "his theory in regard to the evolution of civilization remains unshaken."[82]

Further perusal of the ins and outs of the pro- and anti-imperialist altercation is beyond the scope of my concerns. Suffice to say that its literary archive could readily be excavated to reveal just how significant climate was thought to be for the growth of wealth and for the management of tropical assets. What is notable is what I describe as the bibliographic architecture of the belief that climate exercised inexorable power over the evolution of civilization and thus over the future relations between the temperate and tropical worlds. We have just witnessed the intertwining writings of Buckle, Ward, Kidd, and Ireland on such themes. The citationary network that has manifested itself here could readily be extended ad libitum. Let me illustrate.

In grounding his views about native labor in the scholarly literature, we have just noted how Ireland turned for support to Sir Harry Johnston (1858–1927), the British explorer, novelist, artist, and colonial administrator who published widely on a variety of African subjects. The lengthy excerpt that Ireland extracted from Johnston's *History of the Colonization of Africa by Alien Races* was replete with passages redolent with racial denigration. Plainly, he sided with Johnston's judgment that those races "that will not work persistently and doggedly are tramped on, and in time displaced, by those who do." "Let the Negro take this to heart," Johnston had warned; "let him devote his fine muscular development in the first place to the setting of his own rank, untidy continent in order." Lacking guile and grace in equal measure, he lashed out in dehumanizing language at the native inhabitants "of the vast resources of tropical Africa" who, if they failed to till and drain the soil under European tuition, would continue to lead "the wasteful unproductive life of a baboon." If that were to continue, Johnston gave notice in Darwinian diction that the irresistible forces of both Europe and Asia would inevitably reduce the black races to bondage, simply because serfdom was the only state in which they could survive in the impending brutal struggle for existence.[83] For the meantime, white Europeans were destined to be the custodians of black Africans and therefore entrusted with the task of ensuring that, with their newfound freedoms, equatorial peoples did not heedlessly squander their new opportunities. With these convictions, he provided precisely the kind of rationalization for European supervision of the tropics that Ireland sought.

For his part, Johnston affirmed his belief that the black races had been marked out by their "mental and physical characteristics as the servant of other races," and thus he could casually remark that "the negro in general is a born slave." Moreover, black Africans were constituted to toil under the punishing heat of an African sun and were fitted to labor in the unhealthy climates of the torrid zone. If properly supervised, they could serve as a local source of manpower. This was crucial, of course, since Johnston was on the lookout for locations in which whites could settle, as well as for those regions that could be managed from afar, that is, from the armchair comforts of temperate Europe. To that end, he produced a map of the "Colonizability of Africa" (see figure 8.1) on which he identified zones that Europeans could safely inhabit as well as those unhealthy stretches that lay beyond the range of potential white settlement but were blessed with "great commercial value and inhabited by fairly docile, governable races." These latter were the very sites of "the trader and the planter and of despotic European control."[84] To such venues Europeans

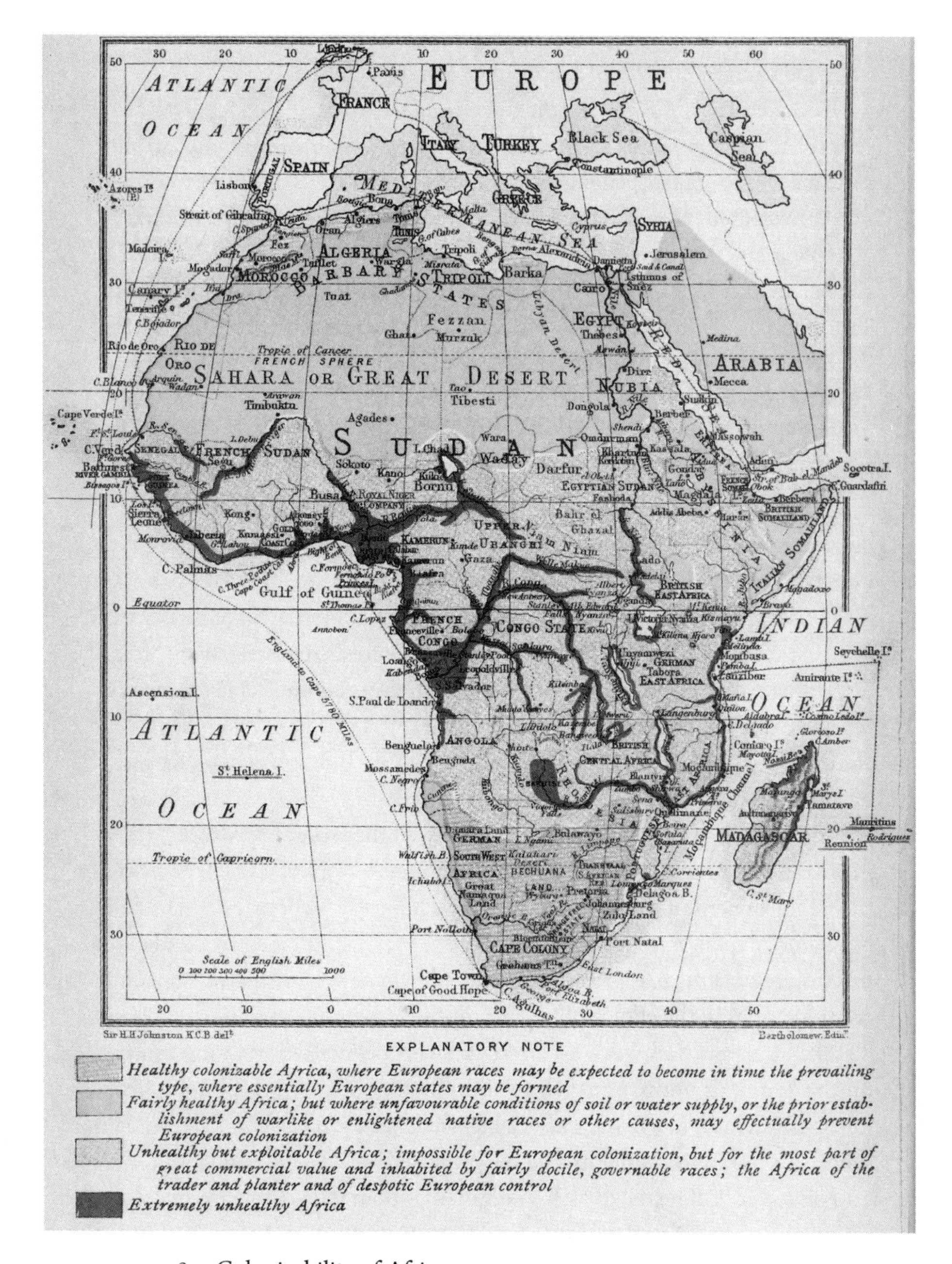

FIGURE 8.1. Colonizability of Africa

Source: Harry H. Johnston, *A History of the Colonization of Africa by Alien Races* (Cambridge: Cambridge University Press, 1899)

could come in small numbers for short periods of time to manage local economies and thereby cultivate lucrative commercial enterprises.

Throughout his *A History of the Colonization of Africa*, Johnston repeatedly referred to the contributions that Sir John Kirk (1832–1922)—physician, botanist, and British consul general in Zanzibar—had made to African affairs. Kirk had accompanied David Livingstone on his second Zambezi expedition from 1858–1864 and later was instrumental in persuading the sultan of Zanzibar to sign an antislavery treaty.[85] Kirk had provided Johnston with logistical advice and material support for his 1884 Kilimanjaro expedition, and later he reported in his obituary sketch of Kirk for the Royal Geographical Society that he "was more or less the promoter" of the entire operation.[86]

Kirk had taken up the subject of "The Extent to which Tropical Africa is Suited for Development by the White Race, or Under their Superintendence" at the Sixth International Geographical Congress held in London in 1895. His concern was to determine which areas of the continent could be safely inhabited by European colonists and which were suitable for periods of temporary residence without full colonization. Underlying both projects was the critical issue of determining just where it was worth the effort to embark on the project of guiding native peoples along the path of progress and of teaching them the meaning of hard work. Both were essential if the dormant resources of the country were to be extracted and utilized to the full. The pickings would indeed be anything but slim, for Kirk was convinced that the resource bounty of the African continent made it one of the richest in the world. No wonder he remained deeply distressed that in many locations, a tropical climate prevented the establishment of "permanent self-supporting white communities or colonies." For after all, European empires were themselves subject to the edicts of the empire of climate. As Kirk declared, "*Climate* is the most important of all considerations in the choice of a home for Europeans in Central Africa."[87]

Like Johnston, Kirk was convinced that where climate set strict boundaries on the limits of permanent white settlement, supervision by temporary European managers was the only viable way ahead. To be sure, Kirk thought Johnston's proposal about introducing Indian laborers into Central Africa was misconceived, but he relied on his judgments about the parameters of colonization in British Central Africa—Nyasaland. In sum, Kirk's piece concluded that it remained critically important for the future superintendence of the tropics that European government determined the most beneficial ways of chaperoning black Africans in their march toward modernity and instructing them on how best to develop the natural resources of their homelands. In this way, African prosperity could be achieved through reforestation, game protection, land reclamation, and proper agricultural management. "In all these, and in a thousand other ways, by the aid of steam and modern appliances," he concluded, "the Europeans will promote the prosperity of the land."[88]

The textual chain conjoining Ireland, Johnston, and Kirk stretched on into the early decades of the twentieth century. Another colonial administrator, Sir Frederick J. D. Lugard (1858–1945), later to be raised to the peerage and in retirement after a lengthy career occupying senior imperial positions in Nigeria and Hong Kong, brought out his influential *The Dual Mandate in British Tropical Africa* in 1922.[89] The book achieved fame for its championing of indirect rule in colonial Africa and providing what many took as a credible justification for tropical trusteeship.[90] To Lugard, this was the only way in which Britain—and other European powers—could benefit from the huge profits to be made from tropical resources. One biographer remarks that *The Dual Mandate* "won him instant recognition as the outstanding authority, in Europe as well as in Britain, on how colonial powers ought, morally and materially, to administer their possessions";[91] another commentator notes that it "guaranteed Lugard's position as one of England's most powerful elder statesmen."[92] In brief, Lugard had devised what many regarded as a broadly acceptable standard by which colonial stewardship could be exercised. Crucial to his proposal was the aspiration that the dual-mandate system would bring reciprocal benefits to colonial powers and colonized subjects alike.

Before remarking on its citationary apparatus, it is worth noting that Lugard, shortly before his first imperial appointment in Nigeria, had added the finishing touches to the published version of Kirk's 1895 speech to the International Geographical Congress. The printed version of the paper in the annals of the Congress notes Kirk's comment that the third part of the paper had been contributed by "Captain F.D. Lugard, CB., and embodies the result of his African and Indian experience."[93] Over a quarter of a century later, when *The Dual Mandate* appeared, Lugard, recently retired from his position as first governor-general of Nigeria, liberally referenced Kirk, Johnston, and Ireland in his treatise. He remarked on Kirk's "unrivalled experience" of East Africa, drew on his memorandum advising the British government on commercial relations with the sultan of Zanzibar, and described Kirk as "a man of very exceptional ability" and "the highest authority" on African slavery. Johnston was cited on the operations of colonial management, on land tenure in Uganda, and on settlement in Nyasaland. And Lugard called on the authority of Ireland when dealing with matters of representative government, different European systems of training for imperial administrators, and Javanese land legislation. Just as conspicuous, perhaps even more so, was his reliance on Kidd's *Control of the Tropics*. With Kidd, he was deeply conscious of just how vital the tropics were, not simply for its own peoples but to the future of the temperate world itself. More, they had actually become essential to world civilization. Tropical rubber, dye-stuffs, drugs, hardwood timbers, coffee, tea, rice, sugar, and jute were just a handful of the commodities on which the rest of the world so utterly depended. That was a subject that had been "admirably dealt with by Mr Kidd." He warmed too to Kidd's notion of social efficiency and its power to distinguish between inferior and superior race types and heartily approved of Kidd's belief that the "tropics will never be developed by the natives themselves."[94]

Two long-standing themes that have been prominent in our examination of tropical wealth and colonial governance—motifs echoing down the centuries from the likes of Ibn Khaldūn and Montesquieu—remained prominent features of the book's intellectual landscape. First, the determinative role of climate in shaping culture and politics continued to reverberate in Lugard's survey. Despite medical assurances that acclimatization was perfectly feasible, he remained skeptical, for example, about the possibility of rearing healthy children in tropical dependencies, even at higher altitudes. Again, he observed that in many locations, it was the climate that acted as a deterrent "to unrestrained white colonisation" though the same meteorological conditions were not inimical to "colonisation by Asiatics." In large measure, the nature and scope of imperial administrative arrangements were reduced to the demands of the climate. There just simply was "a fundamental difference between those countries in which there is a dense native population, and Europeans and others enter for a more or less temporary residence, and those in which the climate attracts European settlers, and the sparsity of the native population admits of the settlement of non-natives." Second, the zonal philosophy beloved of geo-historians of one stripe or another is readily detectable in Lugard's analysis. Like many before him, he casually confirmed that the tropical zone consisted mostly of regions "populated by backward races" and likewise urged that it was in the temperate zone along the northern fringe of the African continent that the most ancient civilizations had come to birth.[95]

Throughout the nineteenth century and well into the twentieth, the thought that the wealth of nations was in some profound way conditioned by the environmental particularities of the globe's great climatic zones continued to grip imaginations. Sometimes under different names—but still with the same import—the frigid, temperate, and torrid zones that had been enunciated in the ancient classical world, and whose political imperatives were championed by Ibn Khaldūn, Montesquieu, and their intellectual progeny, persisted with remarkable resilience in works traversing a range of distinct but related intellectual domains. One arena in which the Realpolitik of this zonal mythos manifested itself with particular urgency centered on the question of labor in challenging climatic regimes. For mid-latitude colonial powers, the most pressing issue was whether the white race could engage in the daily grind required to farm or mine in tropical or subtropical conditions. If, as indeed many believed, hot climates, whether by divine decree or physiological diktat, ruled out northern toil in the torrid zone, the exploitation of races accustomed to such conditions seemed inevitable. In these circumstances, many writers resorted to a kind of climatic theodicy to justify slavery. To them, climate provided a compelling apologia for the institution. Some even went so far as to suggest that the southern plantocracy in the United States was a gift that an urbane temperate culture had bestowed on African Americans, which propelled them along the path to civilization! Even among those who disapproved of the institution and supported abolition, the restraints that climate imposed on white labor and the favorable conditions it provided for the cultivation of cotton and tobacco delivered a post hoc rationalization of black servitude. Here climate's imperial rule over the generation of southern wealth found itself translated into the ugliest form of dehumanizing politics.

Meanwhile, in Britain, the determinative role that climate zones had reportedly played in the initial emergence and subsequent development of civilization also persisted. In the writings of Henry Thomas Buckle, this took the form of a philosophy of food provision that welded together weather and wealth into a scientific theory of historical change. Among those enamored of this positivist history, natural law, evolutionary theory, and the statistics of human behavior provided persistent frames of reference. To sustain this vision, Buckle turned to the differential impacts of cold and hot climates on human physiology and to the different dietary regimes required by the chemistry of food ingestion to sustain body heat and energy at different temperatures. A century and a quarter later, as we will presently see, precisely this species of argument has been enthusiastically rejuvenated to explain global economic development. In Buckle's case, the profound influences the climate exerted on human economy and physiology also found expression in national temperaments, moral sensibilities, and mental operations.

While to some degree, Buckle's positivist history fell out of fashion, his reflections on the problems of tropical self-governance became a significant reference point for many grappling with the politics of colonial management. A group of writers drawn from the ranks of academia, imperial administration, and public scholarship applied the same climatic thesis to the problems of foreign settlement in the tropical zone and to overcoming the obstacles that stood in the way of harnessing, for their own purposes, the rich natural resources of the intertropical belt. Among these commentators, the zonal philosophy prevailed and was resurrected in many different incarnations. Some argued that scientific progress would overcome the hazards of a tropical sun. Others mobilized the terrors of tropicality to curb imperial greed. Some insisted that hot climates had bred races lacking in both bodily vigor and mental agility and thus were in need of civilization. Some went so far as to portray the black races as destitute of all artistic, scientific, or intellectual attainment and thus as natural slaves fitted only for servitude. Still others suspected that the grueling climate of the torrid zone dictated governance by distant supervision as the only viable imperial strategy for overseeing natives incapable of profitably managing their own affairs. Many were sure that life in low latitudes was just too easy, and thus any impulse toward social progress was stifled at birth. But whatever stance the colonial powers adopted over their dealings with the tropical world, climate could be called upon to provide elemental support. For imperial nations believed and, indeed, frequently willingly believed that they were subject to the dictates of nature's first empire—the empire of the climate.

9

Climate, Capital, Civilization

MANY OF the strands of thought that have dominated the horizon of the previous two chapters persisted in a variety of forms into the twentieth century and well beyond. The whole question of the influence of weather on work, which figured prominently in concerns over colonial occupancy and the management of overseas empires by northern imperial powers, surfaced in other venues. Inquiries into the connections between climate and economy frequently converged on subjects revolving around working environments, productivity, and output and gave expression to a preoccupation with determining the most suitable atmospheric conditions for enhancing industrial efficiency. Such fixations were certainly dominant in the thinking of those captivated by the idea of identifying "ideal climates" and "comfort zones" that were suitable for white occupancy in colonial settings. Closely connected was a growing interest in how seasonal fluctuations, periodic droughts, intermittent flooding, higher than normal temperatures, and suchlike determined economic performance and consumer behavior. What might be dubbed the "meteorology of the marketplace" has been called upon to explain economic cycles and critical moments in the history of capital. Similarly, the long-standing resort to climate as the mainspring of civilization has continued to find devotees over the past century. Among these, the role that Buckle accorded to food production in the early accumulation of wealth and the genesis of civilization has continued to manifest itself, as indeed has the zonal economics that in one form or another still snakes its way through recent reformulations of the climatic philosophy of human history. Symptomatic of this legacy, and indeed of the perpetuation of physiocratic economics in one form or another, is the enduring resort to Montesquieu as the fountainhead of the climatic thesis. Even when the focus shifts to the human impact of climate *change*, genealogical continuity with these earlier plotlines remains conspicuous. Tracing something of these crisscrossing trajectories is my focus of concern here.

The Atmospherics of Employment

A good deal of the expansive territory that Ellsworth Huntington (1876–1947) ceded to the explanatory role of climate in human affairs has already attracted our attention. Patterns of global health, racial hierarchy, eugenic hopes and fears, the quality of human reproduction, and much else besides were all attributed to the causal powers of the climate. Here it is his pronouncements on work, weather, and wealth and the wider conceptual context within which he embarked on such inquiries that take center stage. Between 1910 and 1913, Huntington undertook an empirical study of the effects of climatic conditions on factory piece-workers over a four-year period.[1] He scrutinized the daily wage records for 550 Connecticut employees engaged in machine tending, sheet metal working, and the production of electrical fittings. Among the specific industrial tasks involved were rolling hot brass, the packing of hinges and screws, and the preparation of armatures and wire coils for electrical appliances. All employees were paid hourly wages according to their output and not at a fixed daily rate, and all were free from any union restrictions. By seeking to control these conditions and a number of other incidental variables, Huntington believed he could isolate the specific influence that daily weather had on industrial productivity, energy levels, and worker efficiency. In due course, he extended the reach of his inquiries by securing additional data from cigar makers and carpenters in Florida, as well as cotton mill workers in Georgia and South Carolina. Employees in an electrical apparatus factory in Pittsburgh soon joined their ranks.

Huntington's initial analysis centered on seasonal patterns of employment productivity. His aim was to determine worker efficiency at different seasons over a four-year period. The substance and style of his commentary on Bridgeport employees can be taken as emblematic of the whole. In "early January, 1910," he reported, "the efficiency of about 60 factory operatives in Bridgeport was 88 per cent as much as during the week of maximum efficiency that year. . . . Later it rose fairly steadily to 96 per cent at the end of April."[2] During the summer, it rapidly declined until, rising in the fall, it reached the highest point of the year in early November. On these slender foundations, Huntington generalized liberally. Physically, employee performance everywhere from New England to Florida mirrored seasonal changes in weather. Intellectually, it was similar as weekly test results of military cadets at West Point and Annapolis apparently showed. Whether dealing with factory operatives or mathematics students, the pattern was the same: seasonal variation mapped straight onto achievement. But it did not stop there. Ever keen to make explanatory mountains out of empirical molehills, he soon leapt to the global scale. From Connecticut to Denmark, from Maryland to Japan, he told his readers, physical and mental activity were at their height in spring and autumn, conspicuously dropping in winter and midsummer.[3]

To further buttress his conviction that temperature was the prime driver of economic efficiency, Huntington turned from seasonal patterns to daily changes in worker performance. Using the same data set, he found evidence to confirm his belief that uniform daily temperatures adversely affected workers. Provided the variations were not too great, a changing climate was emphatically more desirable for peak productivity. That meant, not too surprisingly, that industrial efficiency was at its best in locations that fell within the arc of the temperate cyclonic storm belt so beloved of Huntington. He was delighted. Not only did it fit snugly with his racial predilections, but it was consistent with the results of wider biological inquiries into plant life. Numerous experiments had apparently shown that unduly high or low temperatures retarded an organism's development. But, he happily reported, a moderate degree of change from the temperature optimum ensured that vegetation achieved maximum growth and functionality. Huntington had now invested his climatic philosophy of industrial efficiency with all the authority of natural law.[4]

Having connected up climatic conditions, plant physiology, and commercial productivity, Huntington pressed on to identify the ideal climatic conditions—what, as we have already seen, he called "climatic optima"—for human society. The best all-round climate was one where mean temperature never dipped below a mental optimum of 38°F nor rose beyond a physical optimum of around 65°F. Only four major world regions enjoyed such climatic excellence: England, the Pacific Coast of the northwestern United States and southern British Columbia, New Zealand, and southern Chile and parts of Patagonia. Not that these regions were all climatically perfect. The thermal environment was not the only relevant atmospheric element. Humidity and barometric pressure also had roles to play, even if they were less significant. Besides, the variability of conditions that prevailed in cyclonic zones was also of critical importance. When all these facets of the atmospheric environment were taken into account, Huntington stood prepared to hand down judgment on the world. So he took it upon himself to estimate the relative energizing—and, by implication, enervating—power of the globe's various climates. England came out on top, approaching nearer to the ideal than anywhere else. Its climate was uniformly stimulating on account of abundant storms and a moderate seasonal range. By contrast, the tropical world suffered from multiple atmospheric handicaps. Not only was the temperature mercilessly high, but there were practically no cyclonic storms. The zone's climatic uniformity—more "deadly than its heat"—bred "drunkenness, immorality, anger, and laziness."[5] The prospects for economic development there were forever bleak.

All of this had profound implications for worldwide industrialization. With the optimism of a would-be scientific soothsayer, he peered into the future. It was hopeful. A time would surely come when far greater numerical precision would be available for all the relevant climatic elements, and it would then be possible to construct a map of the economic efficiency to be expected in every part of the earth. That map would exert enormous power. It promised to be so accurate that manufacturers contemplating expanding their operations could use it to determine the efficiency of local labor, monetize the results, incorporate transportation costs, and compare potential locations with one another. Once the moral economy and the manufacturing economy were properly mathematized, industrial location could be put on a sound scientific footing.

The determining influence that the weather exerted on industrial employees remained a cornerstone of Huntington's economic philosophy. When he brought out with Frank E. Williams a textbook on *Business Geography* in 1922, for example, he turned again and again to climatic conditions as the explanans for an extensive range of economic matters. Atmospheric variability from day to day explained the different speeds at which workers completed their tasks. Different levels of "climatic energy" mapped straight onto the prevalence of initiative, the aptitude for intellectual creativity, and the talent for disseminating innovation. The nature of regional climate in association with local soils determined the basic agricultural wealth of nations. Here too he resorted to the effects of the seasons on human energy in his analysis of what he called "business capacity" and likewise reiterated the scientific underpinnings of his vision of economic geography by linking his convictions about climatic optima among factory workers to other forms of life. This time, he had crayfish in mind. "Man boasts that he is superior to nature," he announced, "but each advance in knowledge shows more conclusively than ever before that he is governed by the same laws which govern the rest of creation."[6] No matter how you looked at it, wealth generation was thoroughly naturalized in the raw facts of climate and biology.

To Huntington, the idea of climatic optima remained foundational. Like his physiocratic predecessors who contended that wealth was ultimately derived from agriculture, he introduced the whole subject of business geography by explaining the relationship between the world's agricultural products and their climatic optima, contended that foodstuffs were the most significant element in the global economy, and devoted a large portion of his text to worldwide inventories of domestic animals and plants in the various climatic zones. From the different types of society that different agricultural systems supported, Huntington believed he could reach conclusions about the character of their populations and did not hesitate to portray in rather stark language the shortcomings of regions that departed from what he declared was the climatic optimum.

The edifice that Huntington hoped to build on the foundations of the work habits of a few hundred factory operatives and on test results from a class of military students was—no doubt about it—enormous. His economic climatology was prosecuted in the broader context of a lifelong fascination with ranking societies on a hierarchical scale of excellence and identifying climatic sources of civilization. His early discussion of the effects of the weather on factory piece-workers came immediately after chapters entitled "Race or Place" and "The White Man in the Tropics." And his first enunciation of his doctrine of ideal climates immediately preceded his declarations on the distribution pattern of civilization worldwide. His inquiries into weather and work among his contemporaries were precisely intended to provide hard data to translate long-standing impressions about climate and civilization into firm scientific fact. In the place of mere intuition, Huntington promised solid science. When pondering on the season at which people worked fastest, it was with the explicit aim of pushing beyond "the generalizations" of Aristotle, Montesquieu, Humboldt, and Ratzel on the role of climate in the genesis of civilization. Because he had shown how atmospheric conditions influenced worker efficiency, Huntington convinced himself that he was well placed to declare what constituted an ideal climate *tout court*. Readers could surely take the next step. But to be certain they had got the point, he produced a map of global civilization that all too neatly correlated with the distribution of climatic energy. How, we might wonder, had he determined just what the distribution of civilization actually was? In fact, he had a register of the different grades of human achievement ready to hand. It was a hierarchical inventory based "on the opinion of well-informed persons" to whom he had written with the request that they assist him in drawing up a map ranking civilization on a scale from high to low.[7] If this was science, it was surely science of the softest sort.

Of the 209 historians, diplomats, colonial officials, geographers, missionaries, anthropologists, and travelers who had been approached, 137 replied. Some warmed to the task. Britain's ambassador to the United States, the Ulsterman James Bryce, found the whole idea ingenious and greatly looked forward to the results. He did pause, though, to point out just how "very complex" the idea of "efficiency" actually was.[8] J. Russell Smith of the Department of Geography and Industry at the University of Pennsylvania told Huntington that he relished taking "a half day off to sit in judgment upon the world" and confessed that it gave him "a sense of a sort of tyrannic despotism to hold a country" in his hand.[9] Others—especially anthropologists—were not at all persuaded. Franz Boas recoiled from the cultural imperialism of the whole operation. He felt "quite unable" to comply with Huntington's request. Huntington, he judged, had allowed subjective values to infiltrate their way far too imperiously into the entire scheme. As Boas put it, the "intrusion of the idea of our estimate" of the value of cultural forms undermined the intellectual integrity of the entire enterprise.[10] An unnamed Italian ethnologist objected that Huntington was operating with a European conception of civilization, which was limited and self-disabling. Indeed, some of the principles of categorization that Huntington had elaborated—the power of self-control, high standards of honesty and morality, respect for law, and suchlike—were, to this respondent, "characteristic of many savage and barbarian peoples, notably the North American Indians" who actually stood on a plane "very much higher than the average American citizen."[11] Alfred Kroeber frankly told Huntington that he would "obtain misleading results."[12] Several geographers were similarly perturbed. One deemed the whole thing "a very bad plan," another—a "Teutonic European"—adjudged the complete scheme "a failure," while the Edinburgh economic geographer George Chisholm confessed to his own "peculiar incapacity for forming judgments about peoples."[13] Yet others hesitated, less because of the demerits of the undertaking itself than because of their own sense of incompetence for the task. One was sorry to report that the whole effort only served to confirm "how little one knows" about the "big world this is" when "he attempts such a task as you have set. It is a most excellent means of taking the conceit out of one". Another found it "*the* most difficult and one of the most humiliating games I have ever tried to play. I always knew I was a fraud as a President of a Geographical Society, but I never knew before how great was my deception!"[14]

Whatever reservations these latter correspondents obviously harbored, Huntington relentlessly pursued his cartographic rhetoric. And the patterns of civilization that emerged from his impressionist survey fitted perfectly with his map of "climatic energy" (see figure 9.1). So precise indeed was the match that he confessed it "hard to think of any other" cartographic correlation that so closely mirrored the features of both distributions.[15] Indeed. First spelled out in *Civilization and Climate*, Huntington was determined to disseminate these maps far and wide, and so he reproduced them in his student textbook on *Business Geography*.[16]

Of course, Huntington stopped short of claiming climate as the only factor determining the condition of civilization. Yet he did insist that advanced cultures could only make progress in the presence of a stimulating climate. A high civilization, he reflected, could be transplanted from one location to another, but it makes "a vigorous growth and is fruitful in new ideas only where the climate gives men energy."[17] The shadow of Huntington's factory workers also extended back in time. Because the variable weather of the mid-latitude cyclonic zone was the greatest stimulus to American worker efficiency in his own day, Huntington was sure that wherever civilization had developed, the climatic conditions characteristic of the storm belt must have prevailed. And if that were indeed the case, climate change must have occurred; otherwise, the pattern of the shifting centers of civilization made no sense. We will have reason to explore other dimensions of Huntington's thinking on climate fluctuations and the fate of ancient cultures later in this chapter. But for now, it all pointed in one direction:

> Thus we are led to the final conclusion that, not only at present, but also in the past, no nation has risen to the highest grade of civilization except in regions where the climatic stimulus is great. This statement sums up our entire hypothesis. It seems to be the inevitable result of the facts that are before us. . . . Yet unless we have gone wholly astray, the surprising way in which independent lines of investigation dovetail into one another seems to indicate that a favorable climate is an essential condition of high civilization.[18]

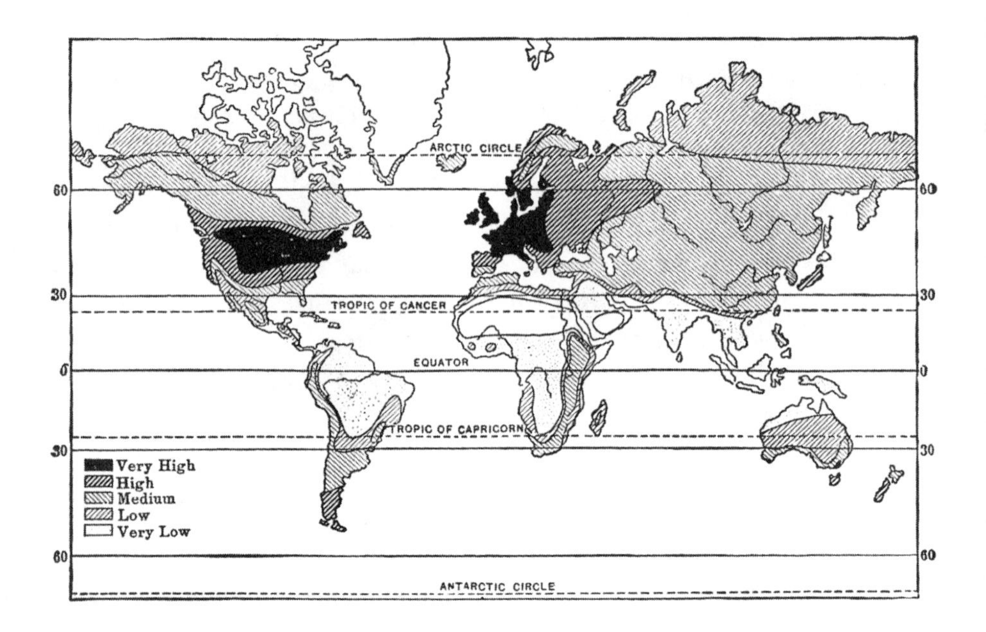

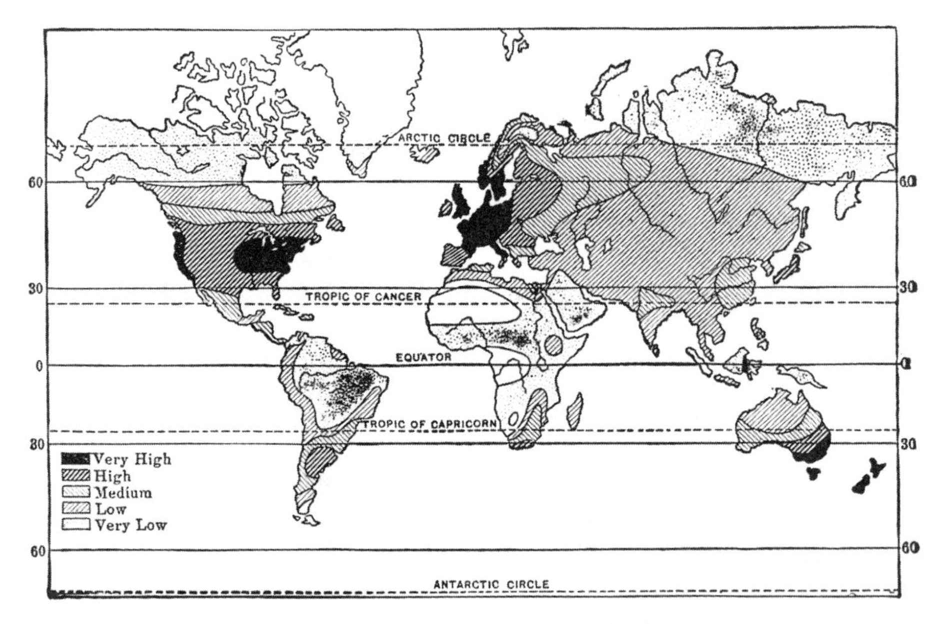

FIGURE 9.1. Huntington's Maps of Climatic Energy (top) and Civilization (bottom)
Source: Ellsworth Huntington and Frank E. Williams, *Business Geography* (New York: John Wiley, 1922)

Comfort Zones, Close Settlement, and Economic Development

If Huntington was drawn to the employment and civilizing significance of climatic optima, others—notably, the English-born Australian geographer Griffith Taylor (1880–1963)—were focusing on the importance of climatic "comfort zones" for economic development and the future of white settlement. Taylor's long-time efforts in this arena aroused public animosity during the 1920s as he sought to counter the prevailing idea that white settlers could readily and profitably occupy the Australian interior.[19] The White Australia policy, as it was known, was born of anxieties over Australia's sluggish postwar economy and fears that the continent's empty spaces invited a mass influx of Asian immigrants. Taylor's opposition to this white-only strategy aroused widespread condemnation, prompted accusations that he was unpatriotic, and acquired for him considerable notoriety at the time. Plastering "Useless" across vast stretches of western Australia on maps of the Commonwealth did little to endear him to the public (figure 9.2). Labels like that were not going to do much to attract white settlers and foster economic development, especially when he compared these interior regions to the Kalahari and the Sahara. Nor did the inflammatory newspaper articles he published left, right, and center. With a tin ear for diplomacy, Taylor seems to have had the knack of using scientific ideology, as Nancy Christie puts it, to "challenge every icon of the Australian national psyche."[20] The public gave his nation-planning venture short shrift, but it became a lifelong passion all the same, and he vigorously pursued it from North America, where, after 1928, he spent the bulk of his academic career, first in Chicago and then in Toronto.[21] His inquiries into the problems of climate and settlement in the Canadian North were simply a rerun of ideas birthed in the Australian outback.

Taylor's initial interest in what he called the "future close settlement" of Australia had first appeared in a volume entitled *Australia in Its Physiographic and Economic Aspects* that he published in 1911 in a book series edited for the Clarendon Press by A. J. Herbertson, Oxford's first professor of geography. Having charted something of the history of the exploration of Australia, its physical conditions and natural regions, he moved on to an analysis of its economic profile. In the final chapter, he turned his readers' attention to the Commonwealth's industrial future. Because no one could predict where profitable mining fields would show up, Taylor insisted that any policy dealing with future settlement had to be confined to locations suitable for agricultural and

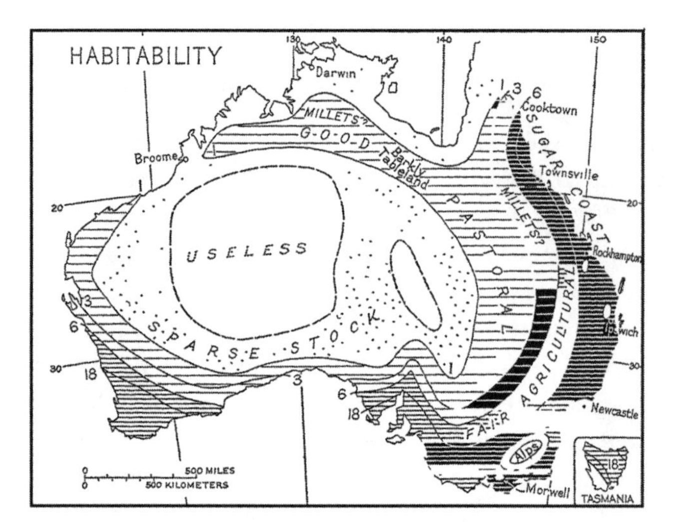

FIGURE 9.2. Taylor's Map of Australian Habitability
Source: Griffith Taylor, "Possibilities of Settlement in Australia." In *Limits of Land Settlement: A Report on Present-day Possibilities,* ed. Isaiah Bowman (New York: Council on Foreign Relations, 1937)

pastoral pursuits. That meant climate had ordained the limits of close settlement. Even a casual glimpse of a climatic map of Australia would disclose for those with eyes to see just where white settlement was feasible . . . and where it was not. Successful colonization boiled down to identifying Australian spaces that replicated conditions where Taylor's "white kindred" had flourished around the world. Climate had spoken, and any attempt to violate her edict was simply madness. It was a different matter for other racial types, of course, who might well be able to make something of certain tropical areas even if they could only deliver "inferior labour." The last sentence of the text summed up his message: "With regard to the tropics . . . may we not look forward to the time when—escaping the errors of less fortunate countries—we allow a confined but contented weaker race to develop our wasted northern areas."[22]

This early settlement ideology was to become a leitmotif of Taylor's determinist climatology. After serving as a geologist on Captain Robert Scott's fatal British Antarctic Expedition (1910–1913), he returned to his position as physiographer at the Bureau of Meteorology in 1914. And it was from that post that he more fully developed his theory of the climatic control of human settlement, hoping to move intuitive impressionism into scientific solidity. That was

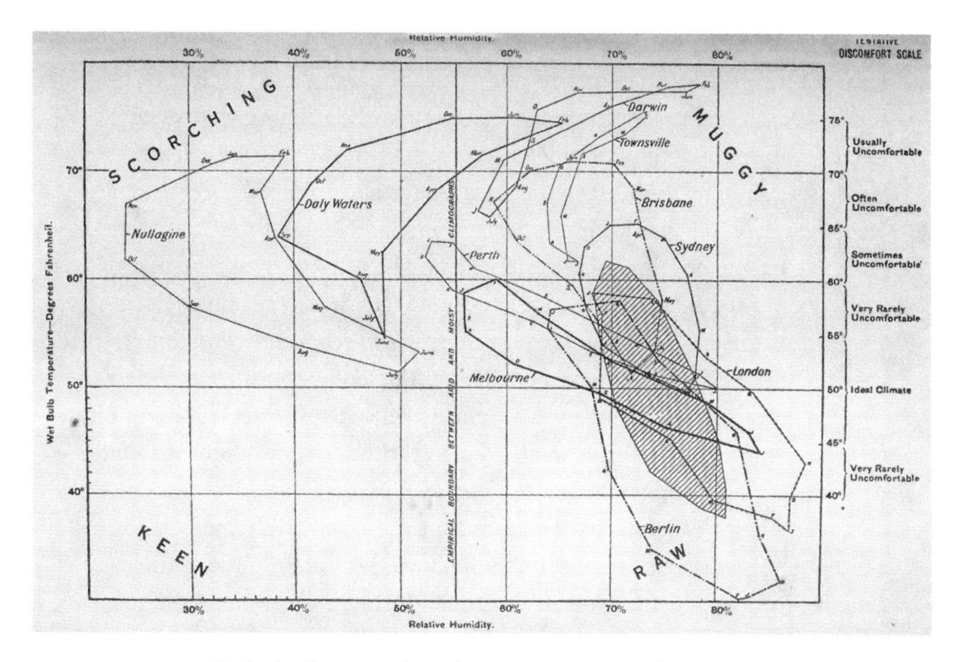

FIGURE 9.3. Taylor's Climograph with Tentative Discomfort Scale
Source: Griffith Taylor, "The Control of Settlement by Humidity and Temperature with Special Reference to Australia and the Empire." *Quarterly Journal of the Royal Meteorological Society* 43, no. 184 (1917)

entirely in keeping with the Department of Home Affairs' plan that the work of the Bureau should "inform a national strategy for settlement and economic development" and critically evaluate the nation's climate in relation to the government's racial immigration policy.[23] The central plank in Taylor's campaign of persuasion was an item of cartographic rhetoric to which he would perennially resort—the climograph (figure 9.3).[24] This was a chart on which he plotted relative humidity percentages and wet-bulb temperatures over the course of a year for a host of different venues. Superimposed on these graphs was a shaded zone formed from mapping conditions in a dozen cities around the world that Taylor deemed the most suitable for the white race. This composite white climograph constituted the gold standard against which the rest of the world was judged. Regional climatic regimes outside this white comfort zone were depicted as "scorching," "muggy," "keen," and "raw" and were allocated a place on what he called the "discomfort scale."

All these endeavors were intended to determine the most suitable climates for white settlement. For that purpose, he had found Huntington's maps of climate and energy particularly valuable and, while he demurred on some details, was delighted that his own findings conformed so closely to the Huntington creed.[25] It emboldened him to identify sites of optimum white energy—namely, zones where conditions were typical of the regions where white energy was best exemplified. And even if its perfect Platonic form was rarely instantiated, the idea of a climatic optimum acted as a kind of regulative ideal against which actual meteorological conditions could be judged. To add further rhetorical power to his analysis, Taylor developed the idea of the "homoclime"—an abstract collection of regions whose climatic and economic conditions were broadly similar. Identifying a host of different versions—the Siberian type, the Mediterranean type, the Sahara type, and so on—he portrayed the Western European type as "the ideal climatic region for the best development of the white race." The purpose of Taylor's entire climatic arsenal was manifestly clear. As the final sentence of his report put it, "It is with the hope that this attempt to give a scientific basis to the climatic aspect of Empire-building and Empire-welding will have some value in the forthcoming re-organization of all our resources that I have written this memoir."[26]

Taylor's investment in climographs and homoclimes, discomfort scales, and climatic optima was deep and lasting. Rendition after rendition appeared in the years that followed. And any sense of equivocation that had stolen into his earlier portrayals soon evaporated. A 1918 article for the *Queensland Geographical Journal,* soon to be reprinted in full in the American *Geographical Review,* is illustrative. Here he just flatly asserted that soil and climate controlled settlement and economic development. The old association between heat and inertia also reasserted itself. Calling on the authority of John Macmillan Brown (1845–1935), the prominent Scottish-born polymath and public intellectual at Canterbury College in New Zealand, he reaffirmed the tired trope that tropical moist heat fostered idleness and inertia. It was his unyielding faith that these were the fundamental facts of life that underscored his opposition to the White Australia policy of blocking settlement by "Mongols, Indians, or half-castes."[27] Not that this stance placed Taylor beyond the stain of racism. It was, rather, a pragmatic bid to find some use for land that climate had declared off-limits for white occupancy.

Never one to resist the siren call to universalize the particular, Taylor coveted the opportunity to bring the entire world within the reach of his scheme. So in 1922, he expanded his climate-and-colonization construct to the global scale. Under the rubric of "economic physiography" and explicitly confining his attention to "the so-called white race," he sought to assign quantitative values to the world's regions so as to determine once and for all their economic and settlement potential. Temperature was the most prominent of the four dominant controls—temperature, rainfall, location, and coal reserves. Even the most cursory "glance at an ethnographic map of the world" revealed it to be "the primary control in determining the distribution of the white race."[28] To put it gnomically, climatic optima governed colonial occupancy. In Huntington-like vocabulary, he declared that he had empirically established that a mean annual temperature of around 53°F was the optimum for close European settlement. White body comfort was of primary importance. But it was not the only means by which climate exerted its imperial rule over human settlement. There were definite temperature restraints on agricultural possibilities. For grain, it was the 35°F to 60°F belt that was most favorable. For stock, the optimum was 45°F for pigs; for cattle, around 50°F; and for goats, 56°F. Globally, these temperatures conspired to deliver distinct patterns of economic success. The optimum for wheat closely approximated the optimum for white settlement in northern lands. As for coal, the mother of industry and the other prong of economic progress, Taylor reminded his readers that its most striking feature was the clustering of all the major reserves close to the 50°F isotherm. This confirmed his hunch that past climates had not varied much at all and that the coal-forming vegetation had flourished under warm temperate conditions.

To hammer home his conclusions, Taylor developed what he called the "econograph." This graphic device, pyramidal in form, brought together location with the three paramount factors—temperature, rainfall, and coal—that regulated white settlement around the world. By this visual means, he believed he could show that charting climatic conditions predicted economic development. His identification of "the standard or ideal econograph" enabled him to speak of model regions and emboldened him to forecast the future of close white settlement across the globe by allocating a rank number to different world regions.[29] Grading venues by what he termed "Units of Habitability," Britain came out on top, sharing class I with places such as Franco-Prussia and the Northeast United States. By contrast, Uruguay and Natal along with eastern Canada and Oregon found themselves relegated to class IV. For all its seeming numerical precision and accompanying cartographic power, the results distilled down to how close conditions were to Europe, the "criterion" against which the rest of the world should be judged. With increasing population pressure and the demands placed on food resources, the need to find suitable venues for white colonial occupancy was acute. Taylor was sure that the implications of his future-gazing economics were enormous as he prophesied that "a vast world struggle between higher civilizations with a low birth rate and lower civilizations with a high birth rate seems to be foreshadowed. This would seem to be inevitable within the next two centuries if the white race is to maintain its dominant position."[30]

For the rest of his life, Taylor returned to these self-same preoccupations with at best only nugatory modifications. In the 1959 version of *Australia: A Study of Warm Environments and Their Effect on British Settlement*, which had first appeared in 1940, for example, he still resorted to his now forty-year-old homoclimes and climographs, his discomfort scale, his fixation with climatic optima, and his opposition to the White Australia policy. Now, reminiscing on why he had cultivated the science of economic climatology, he told his readers that his aim all along had been to tell Australian statesmen in no uncertain terms that encouraging white colonization of the arid and tropical parts of the continent was crazy. It was not difficult, he went on, "to make them agree that Calcutta or the mouth of the Congo were not ideal sites for white settlement; but they would not realize that Broome and Townsville are homoclimes of Banana and Calcutta respectively."[31] Homoclimes, climographs, econographs, and the like were powerful rhetorical devices of persuasion harnessed to subserve Taylor's political interests.

It had been a painful journey. Since 1906, he confessed, his work on tropical problems had been "one long period of continuous disillusionment." And he hit back at those who had made life difficult for him. Politicians irrationally wedded to the "naïve belief" that no part of their native land was unsuitable for white settlement were bound to disapprove of the sciences of climatology and anthropology. People like that wasted enormous sums of money on misdirected attempts to develop the tropics while at the same time withholding resources needed for the intensive settlement of lands in the South. Such policymakers typically justified their immigration programs by appeals for racial purity. They needed reminding that places like Athens, Rome, Paris, and London prospered precisely where different ethnicities met and mingled over the centuries. For twenty years, he had struggled to educate the Australian public on exactly these matters. But it was useless. The reason was simple: "all the applied science in the world cannot turn a sow's ear into a silk purse!"[32] Legislators who ignored his recommendations were not just opposing him. They were resisting the elemental forces of nature. For as he insisted in 1947, the "best economic program for a country to follow has in large part been decided by Nature."[33]

The scientistic cast of Taylor's politics had long been prominently on display. And it was merely a further expression of his decades-old credo that, as he told his mentor, the geologist and Arctic explorer Edgeworth David (1858–1934), problems long thought to be the preserve of ethics or philosophy must succumb to hard-headed empirical science. Citizenship was no longer under the control of the humanities; it was now the special possession of the natural sciences.[34] Critics like the anthropologist Ruth Benedict were disgusted. In the pages of the *New York Herald-Tribune*, she railed at Taylor's "Isothermic Anthropology," dismissing it as nothing less than a vulgar mission to flush human values out of the study of humanity.[35] Nevertheless, Taylor's influence was extensive not least among geographers during the early decades of the twentieth century. He maintained close links with Huntington, for instance, and his work on the climatic determinants of white settlement was registered by Isaiah Bowman, first director of the American Geographical Society and later president of Johns Hopkins University, who integrated it into his own research on the "pioneer fringe" even though he resisted its determinist ethos.[36]

Inasmuch as he brought a global cast of mind to the problems of climate, settlement, and economic development, Taylor can be thought of as internationalist in outlook. Yet the very transnationalism that he hoped would connect Australia to the rest of the world remained captive to what Carolyn Strange describes as "regressive determinism." Quoting Peter Fritzsche, she observes, "Moving from the nation to the world is not a guarantee of political virtue." For Taylor, transnationalism could never mean multiculturalism. For all its seeming inclusivity, his claim that the White Australia policy hampered the country's economic development remained "mired in centuries-old theories of climate and race."[37] Just so.

Meteorology and the Marketplace

Alongside the atmospherics of employment and the comfort zone philosophy of settlement that cast their spell over Huntington and Taylor, climatic explanations were often sought for other aspects of national and global economies. In 1916, Huntington himself reviewed something of the ways in which climatic variations influenced economic cycles at a range of different scales, calling on the contributions of several writers who found in climate the determining causes of economic performance.[38] Plainly, this intervention was intended to further extend climate's imperial rule in the economic realm and to keep his fellow environmental determinists abreast of relevant research in cognate fields of inquiry. Among those whose findings he canvassed were the meteorologist Henry Helm Clayton (1861–1946) of Harvard's

Blue Hill Observatory and the pioneering econometrician Henry Ludwell Moore (1869–1958).

In a 1901 assessment of "The Influence of Rainfall on Commerce and Politics," Clayton—later chief of the Argentine Weather Service—had made it clear that he wanted to challenge the popular assumption that political and economic changes were universally caused by human agency.[39] His aim was to uncover other forces that shaped the worlds of commerce and politics. Chief among these was the climate. This was simply because a sufficient food supply was the human race's prime necessity. Drawing on research that demonstrated the effects of drought on sugar export by the governor of Barbados, Sir Rawson William Rawson (1812–1899), sometime president of the Royal Statistical Society, and by Maxwell Hall (1845–1920), British resident magistrate in Jamaica and government meteorologist, Clayton foregrounded rainfall as the chief determinant of food production.[40] To support his case, he compiled tables of rainfall data for the Ohio and Mississippi Valleys taken from Julius Von Hann's *Handbook of Climatology* and correlated periods of drought with commercial crises. The severe drought that began in 1832 and lasted for the remainder of the decade, he urged, induced the severe financial panic of 1837. The self-same thing happened between 1853 and 1855, again in the early 1870s and yet again in 1893 to 1894. Such correspondences prompted him to formulate a lawlike generalization: "every severe financial panic has been closely associated with a protracted period of deficient rainfall." The resulting effects on civic life were dramatic. Precipitation, panics, and politics were bound together in a linear causal chain. No wonder that Clayton longed for a Darwin in the world of commercial and political explanation. For as he concluded, "If my assumption is correct that deficiency of rainfall is the paramount cause in this chain of events, then vast political and historical changes have been brought about, and the thoughts of men have been swayed by opinions which are akin to superstitions, because they attribute to human action what is largely due to natural causes beyond the control of man."[41]

We recall that Clayton's writings had also caught the eye of his Harvard colleague Robert De Courcy Ward. The following year, he too published an article in the *Popular Science Monthly* aiming to show how weather conditions from one week to the next influenced trade and industry in the United States. Poring over the available meteorological data for 1901, Ward noted that the first week of June had recorded unseasonably cold and wet conditions east of the Mississippi. This had directly "interfered" with both agricultural production and retail trade: crop growth was seriously retarded, and the manufacture of items like umbrellas, raincoats, and the like reported exceptional demand. The following week, when temperatures returned to nearly normal, the market for summer apparel quickly rebounded. Later in July, corn suffered from the effects of a prolonged drought, and while summer clothing boomed, the rest of the retail sector was badly hit. Week by week, the market mirrored meteorology. Whether it was the cost of coal or cotton, fur or footwear, potatoes or poultry, Ward could find in the weather the prime mover of shifting prices. Indeed, he was so taken with the way in which the tide of trade ebbed and flowed with weather trends that he declared it "possible to state in degrees, or in inches, the amount of excess or deficiency of temperature or of rainfall which were needed to stimulate or to depress trade."[42] Market fluctuation could be mathematized through meteorology.

Both Huntington and Ward also turned for inspiration to the weather-related research of the pioneering American economist Henry Ludwell Moore (1869–1958). In his efforts to uncover the natural laws of economic cycles, Moore brought his striking statistical expertise to bear on a project to connect aggregate trade patterns to cycles of rainfall. His enthusiasm for the physiocratic tradition was prominently advertised in his controversial 1914 volume on *Economic Cycles: Their Law and Cause*. Recalling the motto of the U.S. Department of Agriculture that "Agriculture is the Foundation of Manufacture and Commerce," Moore insisted that the wisdom of the old physiocrats had never been fully exploited. That was something he intended to remedy. How? "The rhythm in the activity of economic life," he explained, "the alternation of buoyant, purposeful expansion with aimless depression, is caused by the rhythm in the yield per acre of the crops; while the rhythm in the production of the crops is, in turn, caused by the rhythm of changing weather which is represented by the cyclical changes in the amount of rainfall. The law of the cycles of rainfall is the law of the cycles of the crops and the law of Economic Cycles."[43]

En route to this conclusion, Moore had deployed an impressive range of mathematical tools that far surpassed the analyses of his predecessors.[44] To equip himself for such work, he had taken courses with the biostatistician Karl Pearson at University College London and was steeped in the computational arithmetic of the French philosopher and mathematician Antoine-Augustin Cournot. He was equally well versed in Fourier's theorem and its refinement by the German-born physicist Sir Arthur Schuster to analyze periodic waveforms.[45] All this set him apart from the conventional resort to mathematical metaphors, which had been imported into economics from Victorian physics. He sought to circumvent what Philip Mirowski calls "the existing ersatz social physics" by grounding his theory in actual physics.[46] To Moore, the a priori ethos of conventional economic science simply left no room for the stochastic dimensions of real-world economic change. Instead, Moore found in the methods of harmonic analysis a means of giving precise mathematical expression to the cyclical patterns of rainfall in the Ohio Valley, using multiple regression, correlation coefficients, and the like. His empirical inquiries, drawing on data from reports published by the U.S. Weather Bureau, led him to the conclusion that over a seventy-two-year period, two interweaving cycles could be clearly detected. His purpose was not simply meteorological, of course. It was also to demonstrate the dependence of crops on weather cycles in the heart of the Ohio grain-producing region.

With the link between rainfall cycles and crop yield firmly in place, Moore pushed on to lengthen the causal chain by connecting the value of oats, corn, and potatoes to general economic performance using the production of pig iron as a particularly good barometer of the volume of trade. Now the ebb and flow of the market could be seen to roll on the vacillating forces of nature. In Moore's eyes, the general price index for the period from 1870 to 1911, derived from the Falkner Index and the Bureau of Labor series, provided ample confirmation of the natural laws of the economy: "the cycles of the yield per acre of the crops cause the cycles of general prices and . . . the law of the cycles of crops is the law of the cycles of general prices."[47]

Two things are conspicuous in Moore's analyses. First, just as Taylor was obsessed with turning settlement geography into a science, so Moore, though with markedly greater mathematical rigor, sought to inject a more exacting mathematical physics into economics. As the very title of his treatise made clear, his aim was to uncover the *law* of economic cycles. Because neoclassical dogma never appropriately addressed the ceteris paribus problem—the routine "all-other-things-being-equal" qualification—Moore was sure that previous theorists had mistaken the language and rhetoric of science for the real thing. Thus, as Mirowski put it, while "the neoclassicals were praising science to the skies," their deductive models were thwarting any serious empirical inquiry.[48] For all the talk of economics as a "calculus" of pleasure and pain, a "mechanics" of utility, and a "physique sociale," Moore was sure that the real, material laws of the natural order had not been taken with any real seriousness.

Second, Moore was every bit as concerned to naturalize as to scientize economic explanations. To him, and this confirmed his own diagnoses, nature and its laws were potent economic forces. But soon he began to think he had not gone far enough. Tracking trade patterns back to "the bounty or niggardliness of nature" was not the end of the causal chain.[49] Weather fluctuations themselves might be explained by the laws of physics and astronomy. Because he had freed himself from the constraints of neoclassical protocol, he felt free to look directly to the natural world. In *Economic Cycles*, he had flirted with the thought that annual rainfall cycles might be linked to the periodicity of sunspots, but it was in his 1923 work *Generating Economic Cycles* that he most fully developed his intuition that rainfall sequences could be traced to the effects of the transit of Venus. He also wondered if recent advances in solar physics connecting the condensation of water vapor with the production of sun-generated ionized gases might have explanatory potential in economics. It did not go down well. Ironically, for all Moore's scientific aspirations, critics thought that this move—a "final extravagance," as Stigler put it—smacked too much of astrology, even if the likes of Jevons had resorted to the influence of astronomical phenomena, in the form of sunspots, on commercial crises.[50] When the *Herald Tribune* ran an editorial entitled "Naughty Venus," and when astrologers gave Moore's work their benediction, critics rejoiced that the volume's doubtful reputation was sealed.[51]

No doubt with greater intellectual acuity, but nonetheless with comparable sentiment, Moore found himself recruited to the project that had captivated the likes of Huntington and Clayton—namely, binding the marketplace to meteorology. In the years that followed, their names cropped up in work that spiraled in many directions. Historians, ecologists, and economists found themselves, for one reason or another, attracted to their mission to climaticize commerce. Four instances from the 1920s and 1930s will serve as illustration.

In 1925, the American historian John D. Barnhart (1895–1967) picked up the climatic thesis in an article on the influence of rainfall on populist politics in Nebraska. Barnhart sought to explain the rise of populism in the state by foregrounding economic conditions among farming communities around the turn of the century. In doing so, he insisted on the crucial role of rainfall in shaping the political culture of Nebraska's agricultural region to the east of the Great Plains. The year 1890, he told his readers, recorded low rainfall levels in Nebraska and at the same time marked the demise of Republican supremacy in the state. The two events were surely related. The drought not only weakened the economic position of the farming community, Barnhart argued, but also made it receptive to the arguments of the independent populist party.[52]

A few years later, in 1931, Huntington and Moore also found their work attracting the attention of ecologists interested in analyzing fluctuations in animal populations. The sorry state of wildlife demography in Canada had prompted a group of ecologists—particularly the English zoologist and pioneer of animal ecology Charles Elton (1900–1991)—to convene a conference at a location near the mouth of the Matamek River on the northern side of the Gulf of St. Lawrence to examine the nature and causes of biological cycles. Elton, a key figure in the development of population ecology, had been turning to sunspots and their effects on climatic cycles as a possible explanation for often dramatic numerical fluctuations in wildlife. This intuition set the agenda for the Matamek conference.[53] At the meeting, the resonances between Eltonian ecology and the climate-driven econometrics of Moore were prominently displayed. Delegates also tuned in to a report by Huntington on the causal links between climatic conditions and commercial crises. That analysis reportedly attracted a great deal of interest as he drew parallels between the periodicity of business cycles and animal demographics.[54] Indeed, in collaboration with the editorial committee and with the approval of the Conference, it was Huntington who provided a report of the meeting for *Science* later that year. Here he announced, in a somewhat self-serving fashion, that the conference had been "impressed by the fact that droughts, panics and agricultural depression not only show greater regularity than the sunspot cycles, but seem to have a periodicity twice that of the very regular cycles found in sequoia trees, rabbits, grouse, foxes, salmon and many other animals."[55]

During the late 1930s, Moore's meteorological explanations for market vacillations frequently recurred in works dealing with business cycles. In 1938, for instance, Willford Isbell King (1880–1962), keeping a careful eye on forecasting potentialities, turned the attention of general readers to the causes of economic fluctuations. The early part of King's career had been spent as a statistician with the U.S. Public Health Service prior to becoming an economist with the National Bureau of Economic Research and then professor of political economy at New York University. Here, King devoted space to elucidating a variety of purported explanations for patterns of trade and concluded by calling for further research on the relationships identified by Moore, and indeed Jevons, since he suspected they had hit upon the underlying forces responsible for business fluctuations. It did not escape his notice, moreover, that Moore's analysis had the power to deliver pretty accurate economic forecasting. So when he turned specifically to an analysis of cycles—a chapter that his reviewer Jacoby highlighted as of special importance[56]—he drew attention to Moore's analysis of economic oscillations alongside the "most fascinating study" of Huntington.[57]

The driving force of Huntington's economic climatology was even more conspicuous in the thinking of the economist and investment manager Edgar

Lawrence Smith (1862–1971). His reputation largely rests on his 1924 *Common Stocks as Long Term Investments*, but here it is his later 1939 work, *Tides in the Affairs of Men*, that commands attention. Smith drew heavily on Huntington, Clayton, and King to deliver a comprehensive account of economic change. Smith had been very taken with the proceedings at the Matamek gathering on oscillating wildlife populations, no doubt because what he heard there chimed with the recurrent decennial patterns he had discerned in industrial stock price movements. This confirmed in his own mind that market fluctuations followed trends in rainfall, barometric pressure, and temperature. That naturalistic impulse came to drive Smith's analysis of periodic swings in share values. Indeed, it delighted him that economists were beginning to adopt the mathematics of harmonic analysis used by natural scientists.[58] He departed in one respect, though, from the likes of Clayton and Moore, who concentrated on the influence of the weather on crop yield and ultimately on pig iron production. Instead, Smith dwelt on the *psychological* impact of weather on the outlook of investors.[59] Deviations from the weather norm had an impact on mass psychology and thus on market optimism or pessimism, which in turn was reflected in the price level of industrial shares.[60]

Barnhart's political history, Eltonian ecology, and the economic analyses of King and Smith all bore the distinct marks of the Huntington–Clayton–Moore quest to find causal connections between meteorological phenomena and market performance. During the following decades, this belief became sufficiently commonplace in some circles that writers of climatological textbooks felt free to flag up the association for their student readers. When Thomas A. Blair (1879–1950), a meteorologist with the U.S. Weather Bureau and professor at the University of Nebraska, brought out his textbook on climatology, first in 1942 and in additional printings for a decade, he paused to comment on the climatic hypothesis of civilization. While he remained chary of crude determinisms, he was nonetheless sure that climatic differences were responsible for a large part of the world's commerce. "Business and economic cycles," he told his readers, were "closely related to rainfall and crop cycles."[61] This was because a plentiful food supply fostered the exchange of goods and thus stimulated the whole economy. On the other side of the Atlantic, it was the same with A. Austin Miller (1900–1968) of the University of Reading whose *Climatology* first appeared in 1931 and still remains in print. Miller inaugurated his text with some comments, no doubt as an apologia for the expansive significance of his subject, on "Climate and Trade." Climate, he assured his students, determined the choice of crops, sites for the cultivation of foodstuffs, and the supply of the raw materials of modern industry. These opening remarks fitted snugly into Miller's meteorological mind-set, an outlook that brought questions of race, colonization, health, and psychology within the ever-widening compass of climatology.[62]

Synoptic presentations of the climatology of trade for undergraduate students could all too easily mask the drudgery that its champions were prepared to undertake in the prosecution of their cause. Even as environmental determinism was attracting critics from many quarters by midcentury, there were those still prepared to put their shoulder to the wheel. It was as part of his grandiose program of Cycle Research, driven by his tenacious climatic construal of civilization, that Raymond H. Wheeler (1892–1961) undertook an investigation into U.S. business patterns from 1794 to 1950. What animated this particular project was Wheeler's determination to apply his more general scheme of "cold-wet"/"warm-dry" explanation of human culture to market fluctuations. For Wheeler, prosperity and precipitation went hand-in-hand. Wet phases, whether cold or warm, tended to be times of greater affluence, while dry phases were condemned to economic stagnation and depression. Warm-dry spells spawned dictatorships, absolutism, statism, and the like and induced cultural, moral, and economic decline; cold-dry phases were times of chaos, anarchy, and piracy. This was the explanatory landscape in which Wheeler's thoughts on the influence of weather trends on business cycles were domesticated.

Critical to Wheeler's analysis was the way in which he believed human behavior was conditioned by weather trends. It was not just—as Moore and others thought—that wealth increased in wetter times because of a more abundant crop yield. No. It was to do, rather, with the energy levels of human communities that were bound up with the effects of temperature and storminess on the body's metabolism. Climatic energy levels were the key to unlocking the mystery of the business cycle. In cyclonic regions, people were both invigorated and toughened and thereby energized during colder periods. In support of these intuitions, Wheeler laboriously scrutinized business trends over some fifteen decades using the percentage amplitude fluctuation around the normal derived from the Index of American Industrial Activity. He correlated the resulting undulating line with climatic records over the same period. Whatever other explanations might be advanced, Wheeler was convinced that the weather was always "an underlying, predisposing cause without which no explanation of a boom or depression is complete or adequate." The pattern relentlessly continued decade by decade with wet-year booms and dry-year busts. Plainly, something fundamental was missing from all purely economic business cycle theories. It was not the monetary factor, bank credit, production levels, or any other essentially fiscal force. The missing piece was the weather. It could be neatly summed up in a maxim: "The weather factors are sufficiently well synchronized with economic and business events to be the main primers and timers of business cycles."[63]

Civilization: Food, Fertility, and the Philosophy of Frost

Wheeler's determination to find in meteorological conditions the source of commercial cycles provides a particularly fitting segue to inspecting the continuing impulse to sustain the long-standing belief that, ultimately, climate made civilization. Indeed, Wheeler consciously positioned himself in that tradition by directly aligning his own climatic philosophy of culture with that of such figures as Hippocrates, Vitruvius, Ibn Khaldūn, Bodin, and Montesquieu.

However ambitious Wheeler's capacious vision was for research on climatic and cultural cycles, his undertaking synthesized many of the themes that had dominated the thinking of climate reductionists in different arenas for generations. Besides the generic influence of classical proponents of the climatic thesis, in all likelihood called upon to lend testimonial heft to his predilections, he culled the medical writings of figures like Mills and Petersen to fortify his claims for the influence of climate on physiology. Drawing on the testimony of these biometeorologists, he could inform his readers that disease, menstruation, conception, and productivity all bore the stamp of climate's influence. In fact, he supplemented Mills' research on hot- and cold-room rats with the findings of his own similar experiments, conducted with Dr. Leo Hillmer of the University of Kansas.[64] Apparently, rats reared at 55°F were larger, healthier, more intelligent, and more fertile than rats nurtured at higher temperatures.[65] On the foundation of rodent physiology, it seemed, a comprehensive philosophy of human civilization could be erected. Inhabitants of warm climates adhered to do-nothing philosophies. Their lethargy and laziness sharply contrasted with the hard work and accomplishments of peoples in colder climes. That meant that Americans living in the southern states lacked the "push" of their Yankee neighbors. It was all to do with the effects of climate on energy. Frost empowered; heat enfeebled. This was the same old story that advanced cultures exclusively developed in the temperate zone where the climate did "not drain the energies" of the population.[66] Hippocrates' and Ibn Khaldūn's zonal narrative of civilization had found its latest champion.[67]

Two additional lines of argument sustained Wheeler's climatic master narrative of civilization. First, he devoted his energies to mapping the location of major industrial centers across the globe. Having charted the distribution of 750 cities with the densest populations according to latitude, he plotted the results against their mean annual temperatures. What emerged was the "tremendous peak of cities" with temperatures ranging from 47° to 50°F. That impressed, though hardly surprised, him. It confirmed his judgment that high civilization could only develop in cyclonic storm zones with a "cool, humid, continental or maritime climate."[68]

Second, Wheeler was obsessed with climatic fluctuations and cyclical patterns over the long haul. This preoccupation, of course, stemmed from his determination to find climatic correlates for what he called "recurring outbursts of civilization."[69] He assiduously scoured paleoenvironmental climate records, not least the tree-ring data Huntington had accumulated from the California sequoia, to find correspondences between increasingly wet conditions and the blossoming of Mayan, Persian, Greek, and Etruscan civilizations. The complexities of the climatic and cultural record, combined with his "cold-warm, wet-dry" hermeneutic, led him to identify cyclical recurrences at a range of different intervals—1,000-year, 500-year, 100-year, and lesser intervals. At the end of the 1,000-year rhythms that terminated in 575 BC, 460 AD, and 1475 AD, he told his readers that "something very extraordinary has happened. *An old world has vanished each time and a new one has begun*." First the older civilizations of Egypt, Babylonia, Assyria, and China gave way to the world of the Greeks and Romans and to fresh empires in Persia, India, and China.[70] Roll on another millennium and the Middle Ages were ushered in. And so on. On the shorter end of the time scale, Wheeler, without any hint of self-doubt, connected the growth of democracy with cold climatic periods. A cold spurt in the early twelfth century witnessed the establishment of trial by jury during the reign of Henry II. It was the same for Magna Carta in the thirteenth century and the advent of the first written constitution during the English Civil War. The list could go on. And on. In the twentieth century, a warmer phase fomented such evils as "socialism, communism, decadence, and fanatacism."[71] History, it seemed, was written not so much in the stars as in sunshine and shower.

In pursuit of climate data for longer time scales, Wheeler called on the paleoenvironmental contributions of the Swedish chemist and hydrographer Otto Pettersson (1848–1941). His findings had already buttressed the climate philosophy of the likes of Huntington and Ward. In 1914, Pettersson, already known for his research on the chemistry of oceanic water and on the connections between salinity levels and the vicissitudes of the Hanseatic herring population, had turned to the Icelandic chronicles of the fourteenth and fifteenth centuries to demonstrate how various dimensions of Nordic culture were shaped by changing climatic conditions.[72] The old sagas, he told his readers, had routinely been dismissed by meteorologists as unreliable since the assumption had been that climate had remained unchanged in historic times. Pettersson thought that view entirely wrongheaded. Huntington's studies of ancient shorelines, the archaeological remnants of cities in Central Asia, and tree-ring data from Californian fir trees provided him with confirmation. For himself, Pettersson found in astronomical phenomena the cosmic cause of short-term climatic variations. The next step was short. Unprecedented icy conditions had precipitated the decline of European civilization in Greenland as outbursts of polar ice stirred up aboriginal insurrection. Besides, periodic ice winters had a profound impact on harvest yields and thus on the Scandinavian economy more generally. In complex ways, these cultural consequences of climatic variation brought about the decay of the Norwegian kingdom, less because of government mismanagement or political strife than frequent harvest failures.[73]

Wheeler's interest in Holocene climate cycles and their cultural out-turns drew inspiration from paleoenvironmental inquiries of this ilk. J. Russell Smith's endorsement of the old adage that cold conditions favored the growth of civilization also caught his eye.[74] In 1913, while he taught at the Wharton School of Business in the University of Pennsylvania, Smith (1874–1966) brought out his *Industrial and Commercial Geography*. His "food and frost" philosophy, as it might be styled, got Wheeler's attention. It was built on the conviction that colder conditions were a powerful stimulus to activity. That was as true of nations as of individuals. Both were at their most energetic when "frost forces them to activity and where the warm summer enables them to produce vast supplies of food."[75] With his physiocratic inclinations fully on display, Smith was certain that it was nature's niggardliness or bounty that determined the shape of human culture.[76] The former was much to be preferred, for civilization was the product of adversity. That simply meant that the torrid zone was entirely bereft of advanced cultures because human needs there were too easily met and, in any case, its enervating climate annihilated ambition. By contrast, in the temperate world, winter frost and summer fecundity had conspired to goad societies toward progress. In Smith, the zonal theory of global history remained alive and well. And the old justifications for colonial dispossession continued to burn bright. Tropical abundance all but invited imperial rule. Shiftless and slothful, tropical natives had fallen prey to the ambitious peoples of the temperate zone. That was an altogether handy set of hemispheric arrangements, at least for mid-latitude nations. And it was music to Wheeler's ears. Lamenting the recent warmer trend in the United States, Wheeler used his presidential address to the Kansas Academy of Science in 1943 to contrast cool-climate culture with the "passionate and 'sexy' behavior of warm-climate man."[77]

Within a couple of years of Pettersson's climatic reading of medieval Nordic life and Smith's frost philosophy of civilization, another critical source had made its appearance—Huntington's 1915 *Civilization and Climate*. In one way or another, large elements of this "magisterial" work, as Sörlin describes it, have already come within the compass of our concerns: the influence of weather on work and vitality, tropical deficiency, race and place, and climatic optima.[78] We have noted how Huntington built his theory of civilization on the productivity of factory workers under different atmospheric conditions and on how weather-determined levels of human energy were swept into his various pronunciamentos on cyclonic civilizations. One further dimension of his climatic credendum is pertinent here.

Like Wheeler, Huntington was attentive to the significance of climatic shifts and worked hard to square long-term climate *change* with his understanding of evolving civilization. Glacial periodicity had confirmed climatic variability, and this provided Huntington with a fertile line of inquiry centered on human responses to fluctuations in weather. While the German geologist-geographer Albrecht Penck (1858–1945) was formulating a theory of shifting climatic zones to explain ice ages, Huntington reported that he had been developing the same hypothesis for historic times. Earlier theories about progressive desiccation had given way to the idea of climatic pulsations of the kind detected by the American astronomer Andrew Ellicott Douglass (1867–1962) in his studies of annual tree rings and sunspots.[79] The shifting centers of world civilization could now be mapped onto migrating climate zones. In every case of the rise and fall of civilization, Huntington was sure that the migration of the storm belt in the Northern Hemisphere was of crucial significance. Whether it was with ancient Mesopotamian civilizations or Connecticut factory operatives, periodic storminess increased human energy and brought progress. That fitted well with the mantra that we have already noticed: "no nation has risen to the highest grade of civilization except in regions where the climatic stimulus is great."[80]

Even before *Civilization and Climate* appeared, Huntington was mesmerized by climatic and cultural pulsations. The final two sentences of his 1907 *Pulse of Asia* are emblematic: "With every throb of the climatic pulse which we have felt in Central Asia, the centre of civilization has moved this way or that. Each throb has sent pain and decay to the lands whose day was done, life and vigor to those whose day was yet to be."[81] This served as the capstone conclusion to a volume that labored long and hard to convince readers that human occupations, habits, and even character fell under the all-embracing sway of the climate.

It turned out to be a long-lasting love affair. In his 1911 volume on *Palestine and Its Transformation*, the outcome of the 1909 Yale Expedition to the Middle East, climatic pulsations, rainfall fluctuations, alternating epochs, climate waves, rhythmic weather swings, and similar cyclical synonyms all persistently reasserted themselves. Reminiscent of the scientific travelogue, descriptions of the landscape of Palestine dominated the report. Indeed, it is not surprising that he drew inspiration from George Adam Smith's *Historical Geography of the Holy Land*, and the text can plausibly be read as a contribution to the literary genre of scriptural geography.[82] But the underlying motif remained just the same. Palestine was simply a laboratory for testing his theory of climatic oscillations, for here the power of nature to mold human actions and thoughts was conspicuously on display.[83] His argument, prolix in detail and often inferential in reasoning, was that in order to make sense of the biblical narrative at many points the climate of the Holy Land must have been different from its present-day form. Abandoned ruins, remnants of former Dead Sea levels, population records in the Hebrew Book of Numbers, dynastic successions, and the like all supported that supposition. In some measure, the whole text can be seen as an apologia for what might be dubbed a biblical hermeneutics of climate change. The Exodus narrative, the Assyrian conquest, commerce under Solomon, Palestine fertility: these were only comprehensible if climate change had occurred. And from such particulars, Huntington readily extracted a law of historical change. Adverse climatic conditions caused war, migration, and dynastic collapse; favorable changes triggered expansion, economic growth, and the development of the arts and sciences. Every climate pulse brought poverty or prosperity.[84]

For Huntington, the pulse impulse was a lifelong fascination. It dominated a spat he engaged in with his ardent critic, the Assyriologist Albert T. Olmstead (1880–1945), a historian at the University of Missouri, in 1913.[85] Now he drew on recent work he had carried out with the leading American student of desert ecology and plant physiology, Daniel Trembly MacDougal (1865–1958), at Tucson's Desert Botanical Laboratory.[86] It was a case of new data, old argument. Civilizations, rising and falling, seesawed on the waves of climate change. As its title suggests, the same was true of the 1926 *The Pulse of Progress*. It had originally begun life as a course of lectures to Sunday school teachers on the geography of Palestine and revisited Jewish history as a telling illustration of Huntington's rhythmic rendering of civilization's sinuous story. "All history is a record of pulsations" was his first sentence—a law of life that governed the behavior of everything from lemmings to lawyers.[87] The desiccating conditions that explained the ancient Israelite exodus from Egypt and the Babylonian exile served as proxy for finding in the fluctuations of climate the global "pulse of progress" and the course of human civilization. The bulk of it was, by now, numbingly humdrum. By 1945, when Huntington's personal *Summa Geographica*—the *Mainsprings of Civilization*—appeared, climate pulsations remained center stage.[88] Palestine, Palmyra, and Peruvian depopulation were all here; so too were the Connecticut factory workers, the idea of climatic optima, and correlations between rain cycles and stock prices. Some new pieces of evidence were certainly added, but these did little more than give the impression that Huntington was simply combing the literature for every scrap of empirical fodder for his pet presumptions.

In the intervening years, though, he had added one final device to his stockpile of climatic explanations. He had long been convinced that the cyclonic storm belt was the birthplace of civilization and on this subject had corresponded with the Syracuse professor of German and independent student of weather systems, Charles J. Kullmer (1879–1942). Kullmer had written to Huntington on the connections he had perceived between barometric pressure and patterns of library book borrowing and had speculated that the electrical charges in the atmosphere during periods of barometric depression could well have psychological significance.[89] More recently, he had produced a couple of major publications on storm tracks, and Huntington incorporated these most recent findings into *Mainsprings* in his continuing hunt for a final meteorological metatheory of civilization.[90] And yet, as James Fleming detects, "something was still missing."[91] No doubt denunciations from critics about wild speculations, sloppy data collection, and overgeneralization from highly selective evidence had stung. He still hankered after a robust scientific explanation to hammer history down. Then in 1938, a hurricane passed through New England with devastating effects. In the midst of the mayhem, students at Massachusetts State College were taking a string of intelligence tests and—amazingly—when the results were announced, the grades were extraordinarily high. Huntington fastened on the thought that the storm had subjected them to an invigorating wave of ozone that enhanced their mental abilities. Perhaps the key to the civilization-and-climate pattern lay in a pale blue gas—ozone. Never slow to respond to any hint of support for his prized projects, Huntington was soon at work on the typewriter producing a piece for the International Congress of Biophysics, Biocosmics, and Biocracy the following year. His argument was that the passage of storms released trace amounts of ozone, which, through stimulating body and mind, could scientifically explain the triumphs of civilization. Immediately after the meeting, Huntington set to work on a much longer analysis, intended to become a book, under the title "Ozone, Weather and Health." It was never published. But now in *Mainsprings*, Huntington called his readers' attention to the importance of the ozone and incorporated it into his electrothermal cycle theory. The effect of atmospheric ozone now took its place in his fully orbed cyclonic theory of civilization.

In Huntington's analyses, what is noticeable is that both climate and climate change were so intertwined in his account of human life and culture that mapping any clear-cut boundary between them is impossible. To be sure, *anthropogenic* climate change was not on his agenda. But his interventions do serve as a transition to latter-day efforts to revitalize the climatic theory of civilization in an era of global warming. For frost, food, and fluctuating weather remain as fundamental explanatory forces among contemporary writers with big ambitions to sweep the story of civilization into a history narrated by climate.

Under Climate's Control: Fodder, Fortune, and the Fate of Civilizations

As the twentieth century drew to a close and a new millennium began to dawn, the passion to find in climate the fate of civilization resurrected itself with conspicuous vigor. In the interim, Huntington-style climaticism had fallen out of favor and found itself disavowed in large stretches of the academy. And then, perhaps because of the First Assessment of the Intergovernmental Panel on Climate Change in 1990, perhaps on account of new paleoenvironmental data on past climate change, perhaps for entirely other reasons, climatic readings of history and culture began to find new traction.[92] The weather-and-wealth program was back on the agenda. To get something of a sense of the dimensions of this latest retrieval, I want initially to turn the spotlight on the writings of three authors who have written extensively, and for a wide audience, on the subject. They are the economist and historian David Landes; Jared Diamond, an evolutionary ecologist and popular science writer; and the archaeologist and anthropologist Brian Fagan. This list is far from exhaustive, of course. But the acclaim that these authors have won make them appropriate vantage points from which to inspect the explanatory landscape they all inhabit. Ranging across economics, ecology, archaeology, and history, their fields of engagement are indicative of the scope of the territory over which climatic explanations are staking their claim.

When Landes brought out his monumental *The Wealth and Poverty of Nations* in 1998, a work widely applauded as much for its breathtaking scope as for its sophisticated apologia for Adam Smith's liberal market economics, he inaugurated his inquiries with some reflections on the sorry demise of geography as a university subject in the United States during the 1940s. He attributed its decline, not to say assassination, to the excesses of environmental determinism as championed by none other than Ellsworth Huntington. "In spite of much useful and revealing research," he reported, "Huntington gave geography a bad name." His folly was that he just "went too far." To Landes, geography had the misfortune to be "tarred with a racist brush" and therefore emitted "a sulfurous odor of heresy."[93]

In Landes' world, however, heresy is an imperative. To him, geography's discrediting was, in the last analysis, because of its irreducible honesty—it spoke "unpleasant truth." It told us that "life is unfair, unequal in its favors" and that "nature's unfairness is not easily remedied." It brought "bad tidings."[94] And realist that he considered himself to be, having just pronounced the death of geography in one incarnation, he immediately resurrected it in another. On a map of world income, he remarked, it was manifestly clear that the rich nations lay in northern temperate zone while the poor occupied the tropics and semitropics. It was precisely that species of cartography that Huntington delighted in. And so, not surprisingly, when laying out the global geography of economic failure, Landes' first port of call was climate. Its effects were both physiological and cultural, determining everything from the production of perspiration in what he called sweaty climates to the siesta, a cultural adaptation to keep people inactive during the heat of the day. That is apparently why tropical and semitropical zones induced slavery. But more. Hot climates also fostered the proliferation of hostile life forms that bred rapidly and quickly spread disease. Cold was better. Winter was humanity's great friend for it wiped out insects and parasites. A kind of snow ontology prevailed there. Once again, it seemed, good fortune and frosty weather went together. And so the globe was yet again divided, if not exactly into the pestilential and the paradisial, certainly into something fairly close. The consequences were far-reaching. As he summed it up in Hobbesian tones, "Life in poor climes . . . is precarious, depressed, brutish."[95] Not surprisingly, the adjectival modifiers that he appended to European climates stood in marked contrast. In the temperate world, the rain was "gentle," temperatures were "kind," and the climate was "privileged."[96] As Lucian Boia, noting the strong resemblance between Landes' venture and Montesquieu's project, comments, "Here we have geography and climate rehabilitated as the foundations of human history."[97]

If, in Landes' telling, the fortunes of the nations lay in the lap of the weather gods, for Diamond, it is the climatology of food production that has fashioned civilization. In his prize-winning panorama of global history, *Guns, Germs and Steel* (1997), he set out to demonstrate how geography, not race, shaped the destiny of—well—everybody for the past 13,000 years. Diamond persistently reiterated his conviction that the different histories of the world's peoples do not boil down to innate characteristics but to differences in their environments. The details of the case are, in one sense, overwhelming as he traveled with remarkable ease from the Fertile Crescent to Clovis hunters in the region of the Grand Canyon, from New Guinea to China. At the same time, the architecture of the argument is fairly straightforward.

From the book's title, it is obvious that Diamond attributes the broad patterns of global history to the forces of guns, germs, and steel. In the clash of different peoples across the world, outcomes have depended on the relative efficiency of weaponry, the uneven exchange of pathogens, and technological power. Yet from the outset, he makes it clear that lurking behind all these "proximate causes" of why some societies became disproportionately powerful and innovative lies a far more fundamental cause—the Primum Mobile of historical transformation. Taking the measure of that force is precisely what he is after—an ultimate explanation, a grand unifying theory that will unlock the mysteries of worldwide differences in human culture, social change, and everyday experience. His answer? Food production. In Diamond's vision, everything from the evolution of germs and technological innovation to the development of writing and complex political organization is connected by chains of causation to the mechanics of food provision. The narrative runs something like this. Food production required sedentary living and hence the accumulation of possessions. In turn, this facilitated more specialized forms of social organization and the emergence of groups not directly engaged in the food-generating process who could devote themselves to economic management, craft skills, political organization, and so on—practices that stimulated the development of writing and other technological advances. Hence, the patterns of global history are the story of food and fate. It was the means of food production that "determined which peoples became history's have-nots, and which became history's haves," where advanced civilizations would develop, and which languages would triumph in the worldwide clash of cultures.[98]

Underlying all this, of course, were elemental environmental realities. Again and again behind Diamond's comparative geography of domestic plant and animal cultivation lies the iron will of climate. That's because, in Darwinian patois, "each plant population is genetically programmed, through natural selection, to respond appropriately to signals of the seasonal regime under which it has evolved."[99] In the polity of plants, climate reigns supreme. Consequently, it rules no less imperiously in human affairs. Diamond's story of the fate of the Chatham Island Moriori is illustrative. Because tropical crops could not grow in the colder climate of the Chatham Islands, the Māori colonists from the northern parts of New Zealand simply had no option but to revert to hunter-gathering. With no surplus crops, neither a bureaucracy nor a military could be sustained. Cultural differences reduce to climatic circumstances.

For all his claims to eschew any form of cultural imperialism, Diamond does not exactly succeed in resisting the temptation to trade in the language of hierarchy. In the march of human history, he tells us, some stay "far behind," some wind up "technologically primitive," some remain in "a cultural backwater," and some demonstrate "significant failures to invent." Even with the self-conscious scare quotation marks, it is notable that within the space of a paragraph or two, Diamond rates New Guineans as more "culturally 'advanced'" than Native Australians even if "they had not yet 'progressed' as far as many Eurasians, Africans, and Native Americans" and wonders what might "account for the cultural 'backwardness' of Aboriginal Australian society."[100] Tricky language.

And critics have not been slow to issue challenges. It bothered the geographer-anthropologist James Blaut that Diamond—like Landes—simply assumed that it was the supposed superiority of Europe's environment that explained its dominance over other civilizations for thousands of years. Besides accusing Diamond of perpetrating numerous factual errors—about Chinese ploughs, draft animals in western Eurasia, tropical grains—Blaut insisted that his basic mistake was to treat an environmental condition as a cause of European progress when the truly important forces were cultural. More specifically, he was troubled by Diamond's opportunistic use of concepts like diffusion—mobilizing it when it suited, neglecting it when it did not, and all the while ignoring the role that culture itself played in shaping the transmission of farming practices. In the Cape of Good Hope, for example, the Khoi refused to adopt Xhosa agriculture simply because they "chose to remain pastoralists"; it "had nothing to do with the nondiffusion of Mediterranean crops, absence of domesticable crops, or nonadaptability of tropical crops. The decision to retain a pastoral way of life was an ecologically and culturally sound decision."[101] To Christopher Merrett, Diamond's view that the advent of civilization was "preordained" by environment was intellectually stultifying and politically crippling. With an eye to the future, he was bothered by the thought that the geography-is-destiny school of history provided little to draw on for future development. Those "unlucky enough to be in the wrong place are left with nothing but fatalism, or worse nihilism."[102] In reviewing Diamond's more recent offering, *The World Until Yesterday*, the cultural anthropologist Wade Davis expresses similar sentiments, arguing that Diamond failed "to grasp that cultures reside in the realm of ideas, and are not simply or exclusively the consequences of climatic and environmental imperatives."[103]

Whatever the strengths or weaknesses of these critical intrusions, a couple of further reflections on the epistemological structure of the Diamond project are worth making. His analysis does give off a strong aroma of teleological history, the sense of a linear directionalism that societies really ought to follow. So we hear that in some places, the development of the apparatus of food production "had to wait" for the advent of certain ecological or social conditions. One consequence of this way of proceeding is that Diamond not infrequently operates by counterfactual explanation. Why the social history of some particular group did *not* go in certain directions is what needs explaining. So, why did agriculture *not* independently arise in suitable areas such as California, Europe, temperate Australia, and subequatorial Africa? Why did tribal bands in some places *not* amalgamate into larger social units? This means that Diamond is not above judging local histories for failing to travel in the "right" direction. The fact that domesticated animals in the Andes did not reach Mesoamerica in pre-Columban times is considered an "astonishing failure" that "cries out for explanation."[104] This method is not surprising. For Diamond is convinced that often what did *not* happen in history is more important that what did. The end result is a kind of antihistory or, perhaps better, a history of absence.

Like Huntington, the anthropologist Brian Fagan has been drawn to climatic explanations for the fate of civilizations, ancient and modern. In a sequence of popular archaeological works, he has laid stress on the formative role of climate in human history. Subtitles such as *El Niño and the Fate of Civilizations* and *Climate Change and the Rise and Fall of Civilizations* advertise the priority he accords to weather in human destiny. What sets this corpus of work apart from figures like Diamond and Landes, however, is that it is not simply climate, but climate *change*, that is seen as the engine power of social transformation. In such scenarios, it is the instability of the climate, its rapidly shifting patterns and unpredictability, that has channeled the direction in which the human story has moved. Disavowing simple determinisms, he claims that the empire of climate does not govern an intrinsically impotent humanity by tyranny. When seeking an explanation for the shift from hunter-gathering to settled agriculture, he tells us that it would be naive to claim that sudden climatic shifts "'caused' people to become farmers." And yet in almost the same breath, he contends we became farmers "because we had to." We changed our ways of life "almost overnight because we were at the mercy of distant Atlantic currents that brought rain and mild winters to our homelands."[105] Comparably, we hear that the "seesaws of post-Ice Age global warming . . . turned foragers into farmers and villagers into city dwellers."[106] The sense of climate compulsion is certainly conspicuous in these potted scenarios.

In a number of popular works, Fagan has prosecuted this argument with both much conviction and a good deal of imagination. *Floods, Famines and Emperors* (1999), for example, took El Niño as its primary focus and traced its role in the fate of a range of early civilizations—Mesopotamian, Egyptian, and Southern and Central American. The argument is that fortunes varied with flooding, drought, food supply, and the ability of cultures to manage agricultural regimes in precarious times. *The Little Ice Age* dwelt on the impact of colder conditions on modern European history. Here Fagan traced everything from Norse exploration and the Industrial Revolution to "an endless zigzag of climate shifts.[107] Climate change conditioned the outcome of wars, human and animal migration, patterns of disease, and the development of urban culture. *The Long Summer: How Climate Changed Civilization* followed the same path. Typically taking in several thousand years at a bite and ranging widely across the surface of the earth, it rehearsed the now predictable story of the ways in which warmer conditions brought about the disappearance of Cro-Magnon hunting clans, the settling of the Americas, the emergence of cereal agriculture, the growth of the city state, and so on.

In his resort to climate as a primary agent of historical change, Fagan frequently prosecutes his case with what are basically inferential storylines. And so, what he thinks "may have," "would have," or even "must have" transpired in certain climatic circumstances frequently obtrudes. These modal verb devices serve to build plausible scenarios and compelling plot lines. Thus, we are told that early Europeans experiencing a warmer climate between 15,000 and 11,000 BC "may have cut back on hunting so that they could gather easily stored plant foods" and that they "might have" pulverized nuts, boiled them, and skimmed off the oil to overcome the harmful effects of high protein to pregnant women.[108] Similarly, we hear that when El-Niño "descended with primordial fury" on the Moche civilization, human sacrifices and elaborate rituals "would have served as devices to maintain social solidarity in the face of disaster and suffering." There is nothing wrong with deploying such speculative suggestions, of course. But conjecture can rapidly surrender to certitude and thus to an easy confidence in the causal powers of climate. "After A.D. 500," he writes, "prolonged droughts in the southern highlands far from the Moche homeland coincided with strong El-Niño episodes in the north, causing widespread political adjustments."[109] Here we move seamlessly from speculation to declaration, from insinuation to affirmation.

It seems too that Fagan plays rather fast and loose with the distinction between necessary and sufficient conditions in historical explanation. No doubt it is right to say that Gothic cathedrals and illuminated manuscripts depended on abundant food surpluses. Or that Norse hunting grounds expanded with the coming of warmer weather. These may well be *necessary* conditions for the activities they purport to explain. But they are far from *sufficient* to account for what he calls "a veritable orgy of cathedral building" or the production of gold-decorated texts.[110] Without exploring other relevant factors, and by resting content with correlations between historical episodes and changed climate conditions, Fagan's narratives tend to foreshorten explanation and to misconstrue preconditions as causal powers.

Lying beneath general accounts of the sort we have just been inspecting lie a host of specialist studies of the influence of climatic conditions on particular cultures. Some of these are certainly enamored of climate reductionism.[111] But others remain more cautious. Take the collapse of the Mayan civilization—a frequent port of call by those attributing its demise to drought.[112] Here claims for a straightforward climatic explanation are rather more muted. Gerald Haug and his colleagues, for example, suggest that a century-long decline in rainfall from around 810 AD multiplied the social stresses that resulted in the demise of the Maya. But they go on to insist that no single archaeological explanation is likely to "capture completely a phenomenon as complex as the Maya decline."[113] It is rather similar with David Hodell, who, with his coworkers, provided some of the earliest evidence of drought for the Mayan collapse from measurements of oxygen isotopes in shell and gypsum in the Yucatán peninsula.[114] Subsequently, however, he remarked that the interpretation of the data is far from straightforward, not least since the paleoclimate records disclose a good deal of regional and local variability in the temporal and spatial patterns of drought in the Maya lowlands.[115] Writing on the same subject in the same issue of *Nature*, the SUNY anthropologist James Aimers confirmed that support for the idea that drought explained a supposed pan-Mayan collapse is fraught with ambiguities. Rather than "marching synchronously towards oblivion"—what he calls a "tidy tale" of collapse—the Mayan demise resists any simple climatic-causal explanation.[116]

Similarly, but with wider empirical reach, the archaeologist-geographer Karl Butzer distanced himself from such accounts. To him, the collapse of civilizations is multicausal and rarely abrupt. Tracing the fascination with the rise and fall of dynasties back to the likes of Ibn Khaldūn and Ellsworth Huntington, Butzer carefully examined conditions contributing to the decline of the Egyptian Old and New Kingdoms and the Akkadian Empire in Mesopotamia. He concluded that climate change and environmental degradation must be set alongside such forces as institutional ineptitude, infighting, and infection. Butzer also remained sensitive to the fact that constructions of the nature of "collapse," akin to contemporary judgments about "failing states," are often embroiled in an unacknowledged Eurocentrism that imposes modern Western conceptions of stress and chaos on other times and places.[117] Comparable assessments remind us that while some human ecosystems do inexorably decline, others display remarkable resilience in the face of major challenges.[118] Such collective sentiments echo the earlier critique by the University of Pennsylvania anthropologist Clark Erickson, who in 1999 argued that the neoenvironmental determinism adopted by many archaeologists paints a picture of human society as much too passive and incapable of adapting to long-term climate change.[119]

For all that, reductionist headlines by journalists, whether in newspaper articles, magazines, or online blogposts, perpetuate a message of environmental determinism, frequently with an eye to current climatic challenges.[120] Titles such as "Ancient Civilizations Collapsed Due to Climate Change," "Climate Change Doomed the Ancients," and "Climate Change Could End Human Civilization by 2050" are only the tip of the iceberg, as it were.[121] At greater length, and with an altogether self-conscious quirkiness, the award-winning Russian journalist Alexander Nikonov claims to ventriloquize the climate research of Vladimir V. Klimenko, head of the Global Energy Problems Laboratory in Moscow, in a volume entitled *Civilization's Temperature: Effect of Climate on Humankind's History*. Drawing on Klimenko's multiarchive reconstruction of medieval climate change in northern Russian and northeast Europe more generally,[122] Nikonov's effort to bring the significance of paleoclimatology to a broader audience reduces to the declaration that worsening climate conditions have always been good for the human species. A representative sample of his rather grand proclamations include claims that deteriorating weather led to the "soaring of the human spirit, to new and unusual inventions, and to the birth of great empires"; that all "fifteen big migrations of people" between 3100 and 500 BC were caused by climate worsening; and that the Axial Age, so christened by Karl Jaspers to depict the "the era of unsurpassed flight of mind and human activities" from about the eighth to the third century BC, was triggered by a cooling climate.[123] Culling the findings of dendrochronology, palynology, glaciology, and limnology, Nikonov insists that human history must be viewed through the prism of climatology. Journalistic in tone and idiosyncratic in style, *Civilization's Temperature* is likely to survive only on the shelf of climate science ephemera. But it is emblematic of a persistent urge to find in temperature and tempest, flood and frost, the fate of civilizations.

Scarcity, Scapegoats, and Savage Weather

Even those professedly suspicious of finding in climate the mainsprings of civilization have felt the appeal of climatic determinism especially when working at a different scale of analysis. The German historian Wolfgang Behringer is a case in point. In a range of publications, he seems at once to resist and yet endorse a climatic reading of human culture. He is noticeably more circumspect in his assessment of climate's role in history than some of the writers we have just been considering. To be sure, Behringer is sympathetic to allocating considerable explanatory force to Holocene climate change in understanding the fate of civilizations. He speaks, for example, of a "climatic optimum" during the Middle Holocene that simulated the move from hunting to herding, thereby liberating humans from the insecurities of foraging, fishing, and fruit gathering. He is also sure that the other side of the story—the collapse of civilization—may no less fall within climate's domain. In one disaster scenario after another—whether in the Indus Valley around 2600 BC, Mycenae around 1200 BC, or the Chu dynasty in China—his eye is drawn to the triggering role of rapid climate change. But in making these claims, Behringer certainly knows he is flirting with old-style climatic determinism—something he avowedly has no desire to revive. And so he reminds his readers that climatic vacillations did not determine the direction in which events moved. They may have disrupted existing arrangements, but no particular outcome was foreordained. The rise and decline of the Roman empire, he insists, cannot be attributed to climate fluctuations. Its fall had much to do with mounting fiscal burdens and susceptibility to external attack. In the case of the Vikings, he is quick to point out that "climatic factors alone do not get us very far" for the simple reason that while cooling temperatures could devastate cereal-growing and pasture-farming economies, they did not have the same impact on hunting and fishing cultures.[124]

By the same token, Behringer has played a leading role in a subindustry that has been developing since the mid-1990s in the study of weather and witch hunting in sixteenth- and seventeenth-century Europe by synthesizing paleoclimatic data and archival records. In this work on weather and witchery, he drew on the writings of figures like Hubert H. Lamb on climate history, Jean M. Grove on the Little Ice Age, and Christian Pfister's reconstruction of climatic conditions between the sixteenth and nineteenth centuries in Switzerland.[125] Sources like these disclosed a marked "climatic deterioration between 1565 and 1629"—conditions that resulted in meager harvests, loss of wealth, and a range of other hardships.[126] Alongside these scientific revelations was a second source of data—documentary records of witch trials from the time. Witches were regarded as the cause of the harsh climatic and economic conditions and were frequently subject to scapegoating. One pamphleteer, writing in 1590 in southern Germany, for example, was chagrined to find that "so many kinds of magic and demonic apparitions are gaining the upper hand in our time that nearly every city, market, and village . . . is filled with vermin and servants of the devil who destroy the fruits of the field . . . with unusual thunder, lightening [*sic*], showers, hail, storm wind, frost, flooding."[127] Later, Behringer reported that broadsheets from the 1620s revealed that witches were held accountable for hailstorms, cattle diseases, poor harvests, and epidemics. All of this confirmed to Behringer that "witchcraft may be seen as the paradigmatic crime of the Little Ice Age."[128] Scarce supplies, scapegoating, and savage weather were tightly interwoven,

The idea that the witch-hunting craze was aroused by deteriorating weather has certainly caught on. Emily Oster, for example, used a variety of statistical correlations to conclude that "decreases in temperature led to more witchcraft trials."[129] Teresa Kwiatkowska perused a range of medieval religious documents to ascertain the attitude of learned divines to the belief that witches could bring about changes in weather. She stressed, though, that whatever the judgments of intellectual elites, a popular belief in magic and weather control persisted and that "at the various peaks of the Little Ice Age," many communities held "witches directly responsible for the high frequency of climatic anomalies and their impacts."[130] Lara Apps likewise sided with the argument that witchcraft persecutions increased "in response to the extreme weather events caused by the Little Ice Age," though she noted that this claim should not be taken to mean "that climate change is the only explanation for the witch-hunts."[131]

What this episode demonstrates is the length to which the climatic causal chain can be stretched to incorporate within its power other dimensions of cultural life. One related arena that climate was said to progressively colonize, for example, was the theater of theological disputation. This, it is claimed, was because during the late 1560s, at a time when natural occurrences in the heavens were interpreted as supernatural omens and frequently cast into eschatological anticipations of apocalypse, extraordinary weather conditions stimulated fresh theological reflection. With the advent of the Little Ice Age, trade in theological publication burgeoned. As Behringer noted, "Only the rigours of the Little Ice Age can explain why . . . disputatious theological literature was able to meet the spiritual needs of the age." "Sermons and other edifying literature," he continued, "often refer to natural phenomena and the disasters of everyday life; meteorological disturbances serve as a peg on which to hang theological considerations."[132] Since severe weather was often considered a mark of divine retribution, the need for personal regeneration and turning away from sin dominated religious discourse. In these circumstances, efforts were channeled into the passing of laws to avert God's wrath at sexual promiscuity, blasphemy, unscrupulous usury, and other offences.

What further encouraged such impulses, it has been suggested, was that harvest failures were frequently the backdrop to all manner of uprisings and insurgences. In these circumstances, streams of climatic influence percolated their way into the very depths of religious consciousness. The advent of extreme weather events that resulted in drought and deluge, decline in crop yield, and hunger crises prompted the cultivation among early modern theologians of what Behringer called "sin economics." This was a kind of transgression algebra by which penalties for misdeeds were meted out by the Creator through His providential management of the climate. This penitential mindset, itself a by-product of global cooling, in turn stimulated the composition of repentance music. Hymnody motivated by the struggle to come to terms with illness and death, and giving fulsome expression to worry, sorrow, and fear, assumed "a disproportionate role in the Lutheran body of song." In fictional literature too, the power of climate's rule was registered in "miraculous and spine-chilling" tales that dwelt on monstrous happenings as the signature of God portended in nature. Everyday life also came within the arc of the climate's reign. Changing fashions in clothing, the move from the gothic to baroque in architectural style, and new ways of managing fuel all registered the force of climate change. In the long run, according to this narrative, the challenges of the Little Ice Age encouraged a more rational perspective on understanding nature that eventually enabled society to find new ways of coping with "climatically determined emergencies." As reason relegated sin economics to the past and "pushed religious delusions into the background," as Behringer puts it, economic, political, and agricultural reforms fostered more efficient use of resources and the capacity to escape the tyranny of the climate.[133] Scientific method and technological expertise gradually brought a recalcitrant nature under control. Famines, for example, came to be seen as the consequence of human mismanagement rather than divine retribution. Regardless of the specifics of these particular causal pathways, the general idea that the climatic circumstances of the Little Ice Age provoked, to one degree or another, the scapegoating of witches and its knock-on effects for theological reflection and the turn to naturalistic explanations has recruited many historians to its cause.[134]

In presenting this picture, to repeat, Behringer claimed to resist the forces of climate determinism. "Whether it will ever be possible to elucidate the complex phenomenon of European witchcraft persecutions in monocausal terms appears increasingly doubtful," he insisted. The "synchronicity of subsistence crises and witch-hunts," he went on, "should not, indeed cannot, be interpreted as mechanical determinism. Times of crisis and disaster are historical constants, but external forces do not summon forth mechanical human responses, since they are constructed within modes of cultural perception."[135] In different locations, different mind-sets, cultural conditions, class relationships, and administrative structures all modulated just how bad weather was regarded and thus did much to either curb or encourage witch persecutions. Despite such declarations, however, Mike Hulme has detected rather contrary explanatory motions in Behringer's project. "The author is too sophisticated a scholar to be lured too far down the line of climatic determinism," he began, but immediately added that "at times he seems to get rather close to such reasoning." To Hulme, such epistemic swithering is rooted in the problems of scale and resolution. As he explained, "The more closely the history of an era, a period or a place is written, the more contingent and opaque becomes the role of any given weather phenomenon or climatic perturbation. The greater depth and resolution at which any historical event is examined, the more it emerges that human contingency dominates the direct physical effects of weather and climate. Only by staying at the macroscopic level of generalisation—at the synoptic scale—do cause-effect propositions about climate and society make any sense." And so Hulme suggested that, for all his carefulness, Behringer's *Cultural History of Climate* found itself suspended "between the twin dangers of determinism and epistemological slippage" and suspected that "without articulating a theoretical framework to keep all the pieces in place, there is a constant danger of allowing climate to intrude too far into his story."[136] Whatever conclusion may be reached about Hulme's diagnosis, there is no doubt that for many historians, bad weather, a decline in wealth, and witch hunting were tightly entwined during Europe's Little Ice Age.

Scions of Montesquieu: Zonal Economics Reborn

The strongly zonal outlook that has frequently pervaded accounts of the fortunes of civilizations has reincarnated itself of late in two further related but different guises. On the one hand, there has been a revitalization of Montesquieu's thermal physiology and, on the other, a return to what might be termed physiographical economics. In both cases, climate features as a primary explanatory principle.

Under the label "physioeconomics," the San Diego economist and international strategist Philip Parker contends that much of the economic variation between countries can be explained by the physiological effects of different climates. Parker's project is explicitly Montesquieuean in spirit, if not in letter. Hippocrates and Ibn Khaldūn also make guest appearances, as do David Landes, Jared Diamond, and Jeffrey Sachs. But it is Montesquieu who is hailed as the fountainhead of the impulse to ground economics in human physiology.[137] Because he rightly realized that the temperate world achieved higher levels of economic development than the tropical zone, Parker considers that Montesquieu was blessed with an intuition for thermodynamics and neurophysiology. What has prompted Parker's entire line of inquiry is what he refers to as the "equatorial paradox," that is, countries lying closer to the equator are more likely to have lower than average consumption per capita. That latitudinal inevitability sparked the thought that zonal differences in economic performance might be "a telltale sign of certain physics-based physiological mechanisms at work."[138]

With a passion to insert natural science into economics, Parker commends the aspiration of Nobel laureate Robert Fogel to find in biomedicine the key to economic growth. The term "physioeconomics" is intended to celebrate the marriage—physiology and economics joined in solemn union. It is an alliance that Parker is convinced can explain his primary postulate that countries closer to the human body's homeostatic steady state will grow more slowly than those farther away. How does that work? Enter the hypothalamus. Quoting the Cambridge neurophysiologist Roger Carpenter on the fundamental functions of the hypothalamus, Parker writes, "though scarcely larger than a peanut," that tiny portion of the underside of the brain "drives motivation, monitors homeostasis" and "determines all we do."[139] So, while Montesquieu turned to the different operations of fibers and juices in different climates to explain national differences, Parker looks to "the effects of *heat transfer,* or thermodynamics, and the workings of the hypothalamus on certain economic fundamentals."[140]

The economic fallout is direct. Thermal physics and physiology have dictated that the provision of goods and services—utility functions—will remain divergent across the different climate regimes. To Parker, this means that physiology is likely the part of economic growth that really matters. And so, speaking of comfort levels, he looks to the neurological mechanisms that regulate body temperature and thus consumption. What is critical here is that because humans first emerged in the tropical world, Parker insists that there are definite physiological limits to the degree to which we can adjust to conditions outside that zone. To survive in colder climes, robust *behavioral* responses are required. The autonomic system is not enough on its own. This means that the key to economic development lies deep in the evolutionary history of our species. The economic implications are plain: adaptive behaviors cost. Axiomatically put, achieving homeostatic comfort costs more in cold climates because we are naturally tropical animals. This is as true of housing and clothing as of food intake. Economic performance tracks latitudinal coordinates.

Hypothalamic economics, then, takes latitude to be every bit as physiological as geographical. Indeed more so. This is apparently what Montesquieu twigged when striving to explain the zonal wealth of nations. Speculative though his physiological reasoning was, Parker is sure that Montesquieu foresaw that higher temperatures lead to greater fluid, but lower caloric, consumption in a steady-state condition. "To Montesquieu," he writes, "even if bananas were easy to pick from the trees in all parts of the world, Germans would end up picking and eating more per capita in the steady state than Caribbeans." No matter what, variance in aggregate food consumption just *will* remain divergent in diverse thermal environments. Thermal regulation of food, clothing, and housing—the three greatest needs of humans—is thus the fundamental physioeconomic fact of life. The result? The equatorial paradox is here to stay. It will persist even when tropical disease is eliminated, even when tropical soils are improved, even when tropical institutions are reformed. Because economic behavior is always and everywhere responsive to thermal dictates, what Parker refers to as "the basket of homeostatic goods consumed will never converge."[141] That is just a brute fact of life, a climatic inevitability.

Physiological economics has continued to manifest itself in related ways. In projects reminiscent of Huntington's inquiries into the atmospherics of employment, some economists have directed their attention to the influence of temperature on individual and national productivity and thereby argued for a thermal philosophy of the wealth of nations. By way of example, consider a 2013 report for the American-based National Bureau of Economic Research, which presented findings on human task performance under thermal stress. It concluded that the links between temperature and income dramatically varied "with a country's position relative to an optimal temperature zone."[142] Not surprisingly, the names of Montesquieu, Huntington, and Nordhaus all featured. Drawing additionally on empirical investigations of the impact of heat on tasks such as mathematical reasoning and typewriting,[143] and reminding readers that humans had evolved to function more effectively in some climatic locations than others, the authors concluded that the physiology of temperature played a determining role in the wealth of nations. For the future, they urged that the further any society deviates from its thermoregulatory optimum will dictate the magnitude of the impact climate change will exert on its gross domestic product. For these researchers, as with Parker, thermoregulation is an elemental economic force.

A list of comparable projects could easily be elaborated.[144] But what is plain is that the thermal economics project that had its birth long before current misgivings about the effects of anthropogenic climate change has recently been revivified and continues to animate a significant corpus of academic output. When applied at the global scale, what often travels alongside the new physioeconomics is a related zonal mind-set—the "distance-from-the-equator" school of economic science—that turns to *physiographical* economics and likewise stretches from Hippocrates, Khaldūn, and Montesquieu to Huntington, Wheeler, and Nordhaus. This is particularly noticeable in the brand of geographical economics championed by the Harvard public policy analyst Jeffrey Sachs.

Sachs has long labored to insert geographical variables into the heart of thinking about income differentials and economic development. In 1999, along with John Gallup and Andrew Mellinger, he urged that while it has been almost completely neglected, geography actually mattered a great deal for emergence of advanced economies.[145] By this, Sachs does not mean the kind of economic geography that dwells on matters of scale, agglomeration economies, self-organizing patterns, and the like; what he has in mind is the kind of geography that concerns itself with different climates, transport costs, and physical landscapes. In pursuit of this brand of naturalistic restitution, he harks back to the geographical ruminations in Adam Smith's 1776 *Inquiry into the Wealth of Nations* and to the more recent writings of such thinkers as Fernand Braudel, William McNeill, and Jared Diamond, who connected Europe's global dominance to such matters as a temperate climate, soil fertility, and disease patterns.

To deliver his vision, Sachs and his colleagues adopted a taxonomic scheme for their global development program derived from the 1918 climate classification system of Wladmir Köppen (1846–1940). They divided the world into six major types of climate zone. What this cartographic exercise revealed, of course, was that the temperate zone contained "almost all of the world's economic powerhouse economies."[146] To explain this pattern, they deployed the category of the "ecozone" to characterize large-scale biogeographic realms of the earth's surface using such climatic descriptors as subtropical, tropical, dry temperate, wet temperate, and polar. Critical here were the climate-related differences that the temperate and tropical ecozones experienced in disease burden and food production. Feeding into these patterns were related geographical factors, such as coastal proximity, navigable rivers, and population density. Plainly, climate and physiography had a dominant role to play in regional wealth.

The strongly climatic impetus in Sachs' physiographical economics perhaps most conspicuously surfaced in an analysis he conducted with Gallup on why the tropics are "falling behind." Turning immediately to the inspiration of Diamond, they claim that what is central to the tropical poverty trap is the impact the climate exerts on productivity "through the channels of tropical disease ecology and agriculture." The entire analysis is couched in the language of climate zone comparisons. Drawing again on the Köppen scheme, they insisted that ecozone maps are the most compelling means of depicting patterns of agricultural crop yields. Tropical poverty ultimately boils down to the deficiencies of humid soils and thus "to the long term effects of the tropical climate."[147] While they certainly do allow for the possibility of technological improvement, they nonetheless insist that agricultural innovations such as new crop varieties do not easily colonize new ecological zones. Not surprisingly, then, when confronted with the choice between institutions and geography as the explanans for per capita income and economic growth, Sachs' sympathies lie solidly with the latter. Sure, he acknowledges that both institutions and geography have their parts to play in shaping economic development, but at bottom, he thinks that the choice is actually misconceived inasmuch as institutional decisions are themselves "based on the *direct* effects of geography on production systems, human health and environmental sustainability."[148] The title of their publication, "Institutions Don't Rule," certainly wears its meaning on its face.

Plainly, Sachs does not think that the history of colonial rule or matters of governance provide robust explanations for tropical underdevelopment. He emphatically rejects the contention that "the colonial interlude," as he calls it, can be the core reason for the zone's poor economic performance. This is simply because, for him, tropical poverty typically reduces to matters of food production and health. These are the elemental climate-driven realities that maintain the gap between temperate and tropical incomes. And they are exacerbated by the absence of agricultural and medical technologies to alleviate nature's tyranny. Immediate policy implications flow from this analysis. International agencies have tended to devote all their energies to market reform, he complains, "as if markets alone could address the special ecological and technological needs of the underdeveloped tropical world." To Sachs, that is an elementary mistake. Instead, the task for the global community is to "find new ways to harness global science to meet the challenges of tropical health, agriculture, and environmental management."[149] Tropical troubles are natural and therefore need natural science solutions. Until then, the peoples of the tropics are, and will remain, the captives of climate, the subjects of its deeply dark imperatives.

Perhaps not surprisingly, the label "geographical determinist" has often been applied to Sachs. He would likely demur. With a passion to end poverty and to "make it happen in our lifetime," it is clear that he believes in the power of human agency to effect transformation.[150] Still, critics remain convinced that environmental determinism of one stripe or another underlies his economic analyses. Armin Rosen, for example, writing for *The Atlantic* in 2013, casts aspersions on what he sees as Sachs' "environmentally and geographically deterministic view of African conflict"; David Correia likewise accuses him of espousing "crackpot but quintessential environmental determinist" arguments.[151] Critics such as these, especially those on the political left, no doubt irritated by his casting of socialism as a curse rather than a cure, remain unconvinced by his dismissal of colonialism as an explanatory cause of continuing poverty and rail against his audible silence on colonization's tropical legacy. They are likewise perturbed by what they consider his lack of sustained attention to the relationship between government and development, as well as to his seeming inclination to downplay the role of political leadership in economic affairs, in the interests of naturalizing poverty, disease, and income in the language of climate and physiography. Sachs' announcement in the pages of the *Economist* in 1999 that "if it were true that the poor were just like the rich but with less money, the global situation would be vastly easier than it is" was disquieting. His assertion that "the poor live in different ecological zones, face different health conditions and must overcome agronomic limitations that are very different from those of rich countries" seemed to replace real people with ecological descriptions, to discount the role the redistribution of wealth could play in reducing inequality, and to ignore the exploitative history of capitalist expansion.[152]

However these claims are to be adjudicated, Sachs' work, like Parker's, confirms that the zonal philosophy of wealth inherited in one way or another from the likes of Ibn Khaldūn and Montesquieu continues to flourish. Whether in the form of physiological or physiographical economics, the belief that climate has played the determining role *par excellence* in the genesis of the wealth of nations retains its grip. And in an era troubled by the threat of climate change, the language of climate reductionism continues to flourish.

In the early years of the twentieth century, research on the performance efficiency of factory workers at different temperatures formed the cornerstone for a comprehensive climatic theory of civilization. What was described as the climatic energy of the cyclonic belt was fastened upon as the engine power of social progress, wealth generation, and intellectual attainment. At the same time, the idea of human comfort zones was called upon to inform settlement programs and the needs of national planning in Australia. It is not difficult to see how this species of climate research could easily be harnessed in support of immigration restriction policies. What was also conspicuous about those championing these forms of climate determinism was their penchant for using cartographic rhetoric and the idea of climatic optima to advance their own cultural agendas.

Meanwhile, efforts to find causal links between meteorological conditions and market performance were beginning to attract the attention of historians and economists seeking to identify real *natural* laws of business cycles and thereby to place economics on a sure scientific footing. Correlations between rainfall patterns and market prices were examined with ever-greater statistical sophistication in hopes of identifying deep cyclical sequences in nature and culture alike. These efforts aroused interest among population ecologists investigating fluctuations in wildlife demography as well as among investment managers hoping to predict future market performance. They also fed the fascination of those striving to correlate major cultural events with weather cycles over both shorter and longer time spans.

Not surprisingly, the idea that weather and wealth were causally correlated found traction among researchers convinced that changing climatic conditions had been determinative in the rise and fall of older civilizations. During the early decades of the twentieth century, claims of this sort were advanced to explain cultural transformations in the ancient Near East, the decline of the Nordic economy in early modern Europe, and the fate of ancient central American civilizations. Even before the late twentieth century's realization of the challenges posed by anthropogenic climate change, the thought that climatic fluctuations had periodically occurred in prehistoric and historical times and that they had exerted a profound influence on human culture was already gripping imaginations. In the aftermath of the declarations of the Intergovernmental Panel on Climate Change, studies correlating climate events with the decline of civilizations mushroomed, as did inquiries into the economic, cultural, and political implications of both climate and climate change over several centuries. Among these, the climatology of economic failure, the geography of cultural misfortune, and the ecology of civilizational collapse loom large. Notable too is the revivification of a physiological form of economics, rooted in the thinking of Ibn Khaldūn and Montesquieu, that attributes economic growth to the influence the climate exerts on the human body's capacity to maintain homeostasis in different temperature regimes. Once again, the zonal philosophy of wealth continues to reassert itself.

We have now come full circle. Intensive research into contemporary links between thermal conditions and income levels resonates both with Huntington's project to determine the optimum atmospheric conditions for productive labor and with Taylor's fascination with comfort zones, settlement, and economic development. Similarly, early twentieth-century endeavors to connect the fortunes of ancient civilizations with climatic oscillations are echoed in recent accounts of climate change and cultural collapse. Equally, the research of figures like Clayton and Moore on the causal connections between rainfall cycles and stock exchange performance bears more than a passing resemblance to recent inquiries into the links between daily temperature fluctuations and market returns.

Again and again over the past few chapters, we have witnessed the multiple rebirthings of the climatic theory of the wealth of nations. Ideas once thought relegated to the museum of discredited curiosities have been rejuvenated, and writers, once popular but who fell into disrepute, have found their reputations resurrected, if not detoxified, in recent times. The authors of classic works of climate-driven history such as Ibn Khaldūn and Montesquieu are routinely reappearing in twenty-first-century bibliographies and serve as the architectural cornerstones of physiological and physiographic economics. The work of the once-despised Ellsworth Huntington regularly surfaces in assessments of the role of climate change in determining the fate of civilization. John Chappell's hopes, articulated in 1981, that Huntington's version of environmental causation would be rehabilitated are certainly being fulfilled.[153] Raymond Wheeler's preoccupations with climate and cultural cycles and his diagrammatic rhetoric are retooled for contemporary purposes as the climatologist Cliff Harris and meteorologist Randy Mann update his chart correlating climatic undulations with key moments in the long history of human culture.[154] Griffith Taylor, whose harsh determinism was widely repudiated in his own day, not only features as a character in Anne Michaels' *Fugitive Pieces* but is feted in Tim Flannery's *The Future Eaters* as "one of the greatest and most courageous scientists that Australia has ever produced."[155] No doubt the list could readily be extended. But enough has been said to demonstrate the continuing vitality of the weather-and-wealth school of global history. Born in ancient times, it seems that, in explaining the wealth of nations, climatic determinism is set to share the destiny spoken of in another ancient system of thought—the fate of the eternal return.

PART FOUR

War

10

Climate Wars

IN 2008, two books trading under the same title, *Climate Wars,* made their first appearance. As just the tip of a publishing iceberg, these volumes gave voice to a growing preoccupation among journalists, politicians, academics, and broadcasters with the intimate links between climate and conflict. In our day, this takes the form of a declaration that climate change will inexorably trigger armed struggle of various sorts—civil strife, international war, political unrest, insurrection. Here I want to suggest that this recent turn of events, while seemingly a response to the climate change crisis that faces the globe, actually perpetuates, in significant ways, a long tradition of causally connecting weather and warfare, albeit in a range of different keys.

The first book, by Gwynne Dyer, carried the subtitle, *The Fight for Survival as the World Overheats.* It was highly commended by a range of enthusiasts. James Lovelock, for example, thought it "truly important" and considered it a premonition of "how drought and heat may ignite wars, even nuclear wars," while Sir Crispin Tickell was sure that it confirmed his own conviction that "water and war have always been associated."[1] To give a sense of added urgency to his prophetic jeremiad, Dyer settled on a number of future scenarios—Russia 2019, northern India 2036, China 2042—to depict the "potentially apocalyptic" impact climate change will have on "human civilisation." The message was grim: "For every degree that the average global temperature rises, so do . . . the number of failed and failing states, and very probably the incidence of internal and international war." Along the way to a projected increase of 2.8°C by 2045, Dyer painted a picture of a collapsed European Union, its northern members struggling to close borders to refugees from Mediterranean countries stricken with famine; a dominant Russia now the acknowledged great power of Asia; and an ever-more fragile frontier between the United States and Mexico under intolerable strain. Border maintenance would be particularly difficult, with security patrols resorting to violence to fend off the hordes of hungry climate migrants. And so it went on. The tragic history of our future was reduced to the vicissitudes of climate's reign. In many places, Dyer noted, global warming will result in a falling food supply, and that means famine, death, and war. Here climate was writing a script that starred Malthus and Hobbes as the leading *dramatis personae*: exceed the world's climatically determined carrying capacity and countless millions live a life that is nasty, brutish, and short. The result will be a profoundly differentiated world—the fault line mostly running along a North–South divide—rife with violence and conflict, for as Dyer put it, "people always raid before they starve." Certainly some steps, requiring political willpower of Herculean proportions, might be taken to mitigate the very direst edge of despair. But the general direction of the tracks along which climate is propelling *Homo sapiens* was pretty clear. A radically new global political order is inevitable. Our current governmental systems, according to Leon Fuerth, were adapted to the current climatic era, but Dyer told his readers, quoting Fuerth, "If there are changes in the climatic background that are sufficiently vigorous, something is going to happen to the political systems," and this certainly does "not mean good news for representative democracy."[2]

The second *Climate Wars* author, Harald Welzer, a German sociologist and social psychologist with research interests in disaster memory, set out to explain *Why People Will Be Killed in the 21st Century*.[3] His tactics were rather different. His *Climate Wars* was more broadly conceived, and he was at pains to insist that, in the violent world of the new century, climate change is not the only game in town. But climate remained the specter overshadowing humanity's fate all the same by both direct and indirect means. For while the narrative strayed far and wide, climate wove its way like a high-voltage current through the account and served as the occasion for expanding on all kinds of issues and ills.

Rather than engaging in speculative forecasting, Welzer adopted an approach that historicized the future or, perhaps better, futurized the past. Because the future scenarios that are of greatest concern to the public were constructed out of historical data, Welzer informed his readers that he was not intending to engage in second-guessing the future but rather analyzing past and present episodes of violence in order to gauge just what the twenty-first century has in store for societies across the world. Killing yesterday and killing today tell us about killing tomorrow. War-torn Darfur in western Sudan, for example, showed our future arriving early. For here, Welzer insisted, "climate change is unquestionably one cause of violence and civil war." What was depicted on television news bulletins back in 2003 as a tribal conflict between African farmers and Arab horseback militias was really "a war by a government on its own population, in which climate change played a decisive role." Not that the lines of causal connection were direct. Rather, long-standing internecine clashes between settled agriculturalists and nomadic herdsmen were dramatically intensified by lack of rainfall, soil erosion, and other climate-related conditions inducing migration. Here was one case among many where "conflicts that have ecological causes are perceived as ethnic conflicts." Given this naturalistic inclination, it is understandable that Welzer spoke of the "ecology of war" to underscore his insistence that social breakdown is triggered by environmental degradation.[4] To him, what looked on the surface to be armed insurrection, internecine feuding, civic violence, and societal disintegration masked underlying climatic realities. Nor is it surprising that he resorted to Darwinian language to tell his audience that the result of rapid climate change is a grim struggle for existence in an increasingly fragile ecology where scarcity poses a monumental threat to survival. Things were particularly bleak for failing states where existing vulnerability is exacerbated by a diminishing space for survival in the face of greater and greater environmental frailty. Conflict inevitably ensues. To be sure, aggression may spring from many different sources, but Welzer was sure that "the consequences of climate change will reinforce and deepen survival problems and the potential for violence." And these effects would inevitably spiral beyond the likes of Africa and the Middle East: "The Western countries may remain islands of bliss for another few decades in comparison with less favoured parts of the world," he predicted, but "they too will inevitably be drawn into climate wars—or, to be more precise, into the waging of climate wars."[5]

For all the qualifications that Welzer interjected into his analysis, a sense of necessitarian inevitability wound its way through the narrative.[6] Whether it arose from conflicts over water or the consequences of Arctic ice melt, the potential for conflict was set not only to increase but also to "lead to new forms of warfare that the classical theories of war did not envisage." The lines of causation may be circuitous. Sometimes they find their center of gravity in the impact a changed climate has on soil degradation or water shortage limiting the means of making a livelihood; sometimes they spiral outward from increased flows of refugees and migrants and the resulting pressure on border controls; sometimes it is simply that climate change reinforces already existing inequalities and thus feeds terrorism. But whatever the causal channels, the effects of climatic change will ultimately issue in violent struggles of one sort or another. As he put it in his gloomy finale, "The sharpening of global asymmetries already offers examples of wars rooted in climate change and new forms of never-ending violence. Since climate effects strike hardest at societies least able to cope with them, migratory flows will dramatically increase in the course of the twenty-first century and push those societies into radical solutions." He continued, "The other side of this is a continual tightening of internal security . . . the keywords in this regard are extra-territorial detention centres, extra-judicial kidnapping, torture and execution, mercenary armies and privatized violence."[7] Taken in the round, the empire of climate, in these scenarios, continues to exercise its imperial rule over individuals, societies, and nation-states.

The avalanche of book-length pronouncements of this stripe gives every impression that attributing violence and war to the vagaries of climate is a novel preoccupation stimulated by current concerns over the human consequences of anthropogenic climate change. In fact, hunting for connections between climate and war is no new pastime at all. Rather, it has a long genealogy. And gleaning some familiarity with this deeper story will better enable us to set in context the present-day proclivity to sweep climate change into the theater of military operations and national security. At the same time, tracing something of this older tradition may also bring to light some of the pitfalls that critics of climate–conflict discourse continue to find troubling.

The inclination to find in climate a chief cause of combat can be traced back to the ancient world. Sometimes it has been prominently displayed; sometimes it lies buried in a fugitive literature that no longer sees the light of day. Here I want to provide a thumbnail sketch of some crucial moments in the chronicle of weather and war. Comprehensive survey is not my aim. Rather, it is to bring into view some key statements that illustrate how climate and conflict have been associated over many generations. The Hippocratic tradition of thinking on this theme, together with its early modern legacy, will serve as an introduction to a closer analysis of the way in which several nineteenth- and early twentieth-century commentators fastened upon climate as the driving force behind one major bloodbath: the American Civil War. Inspecting this body of interpretation will lead us to see how, in the minds of some, climate—and later climate change—came to occupy a dominant position in accounting for the genesis of armed conflict.

Climate and Conflict: The Hippocratic Tradition

The classic treatise on medical topography, which has come within our purview in earlier chapters, *On Airs, Waters and Places*, a work grounded in the thinking of the Greek philosopher-physician Hippocrates (ca. 460–370 BC), paused in its analysis of the impact environmental conditions exerted on human physiology and disposition to reflect on the fundamental differences between the European and Asiatic psyche. Typifying the Asiatic as displaying "pusillanimity and cowardice," the writer insisted that "the principal reason the Asiatics are more unwarlike and of gentler disposition than the Europeans is the nature of the seasons, which do not undergo any great changes either to heat or cold." The temperamental traits that the ancient Hippocratics attributed to Asiatic peoples were believed to be reinforced by their legislative arrangements and governmental systems, which gave no priority to the acquisition of "military discipline." In Europe, the Hippocratics believed, things were different. Here cowardice gave way to courage, timidity to pugnacity. Why? Because European climates were remarkably variable with hot summers, cold winters, frequent rains, and droughts. As *On Airs, Waters and Places* explained, "a climate which is always the same induces indolence, but a changeable climate, laborious exertions both of body and mind; and from rest and indolence cowardice is engendered, and from laborious exertions and pains, courage. On this account the inhabitants of Europe are [more warlike] than the Asiatics."[8]

It was not only through its influence on the humors and thus on human temperament, however, that climatic conditions exerted their influence on violence and warlike impulses. The impact of their climatic regime on a people's propensity for conflict could be mediated through the environmental conditions that it governed. The Hippocratic *On Airs, Waters and Places*, for example, explained that where "the country is bare . . . and rugged, blasted by the winter and scorched by the sun," there are found people "inclining rather to the fierce than to the mild" and thus "excelling in military affairs" (part 24). In similar vein, Herodotus (ca. 484–425 BC) put into the mouth of Cyrus the observation that "soft lands breed soft men; wondrous fruits of the earth and valiant warriors grow not from the same soil."[9] In both texts, rugged lands and bleak climates were connected with courageous people, while gentle terrain and balmy weather were home to weaker nations.[10] For Herodotus, climate worked less through the human constitution and more through the environment's subtle influence on behavior and customs.[11]

Echoes of the Hippocratic tradition are also detectable in the writings of the Greek geographer Strabo (64/63 BC–c. AD 24), who was sure that "where the climate is equable and mild, nature herself does much towards the production of these advantages. As in such favoured regions every thing inclines to peace, so those which are sterile generate bravery and a disposition to war." At the same time, Strabo resisted attributing bellicosity solely to atmospheric conditions. His disinclination to turn to climatic determinism could clearly be heard in his observation that, in Europe, the "number of those who cultivate the arts of peace" was largely attributable to "the influence of the government, first of the Greeks, and afterwards of the Macedonians and Romans."[12]

Either way, whether through immediate imprint or indirect influence, it seems that the links between climate and bellicosity passed into conventional wisdom, even though the specific connections remained anything but constant. The fourth-century Roman military strategist Flavius *Vegetius* Renatus, commonly known as Vegetius, for example, began his *Epitome of Military Science* with preliminary thoughts on the best regions from which army recruits should be enlisted. While it was common knowledge that people displaying cowardice and bravery were to be found in all manner of locations, he was sure that some nations were naturally more warlike than others on account of the "enormous influence" the climate exerted "on the strength of minds and bodies." This claim, he noted, was borne out by a variety of scholars. Because "peoples that are near the sun are more intelligent but have less blood" than their northern counterparts, they lacked the "steadiness and confidence to fight at close quarters." By contrast, the peoples of the North, being less intelligent and having "a superabundance of blood," were only too ready for battle. Accordingly, Vegetius insisted that military recruits should be "raised from the more temperate climes."[13]

The legacy of climatic humoralism as a means of explaining warfare and aggression was deep and lasting. Take, for example, Jean Bodin's (ca. 1530–1596) sixteenth-century synthesis of Hippocrates, Aristotle, Galen, Pliny, Vegetius, Vitruvius, and Leo Africanus, among others, in his celebrated account of political sovereignty.[14] Having scoured the writings of ancient historians and contemporary explorers alike, Bodin could proclaim that a close reading of the "histories of these various peoples" would readily reveal "that great and powerful armies have always been raised in the north." To Bodin, the reason was simple: cold climates bred valiant warriors. "Just as in winter, places underground, and the internal organs of animals, conserve the heat that is dissipated in summer, so people inhabiting the northern latitudes have a more vehement internal heat than those living in southern latitudes." Different climatic regimes evidently produced a worldwide constitutional geography that in turn delivered a distinctive global geopolitics: "Northerners succeed by means of force, southerners by means of finesse, people of the middle regions by a measure of both. They are therefore the most apt for war, in the opinion of Vegetius and Vitruvius. It is they who have founded all the great empires which have flourished in arms and in laws."[15]

As we have already noted, perhaps the most dramatic reworking of climatic Hippocratism crystallized in the writings of Montesquieu (1689–1755). Here his thoughts on climate, courage, and conflict take center stage. In *The Spirit of the Laws* (1748), he reported that inhabitants of cold climatic regimes exhibited a "bravery" signally lacking in the "inhabitants of warm countries" who were "timorous." This circumstance, of course, was directly relevant to the conduct of war. As Montesquieu explained, "If we reflect on the late wars . . . we shall find that the northern people, transplanted into southern regions, did not perform such exploits as their countrymen who, fighting in their own climate, possessed their full vigour and courage." The reason for this global geography of human temperament lay in how Montesquieu conceived the operations of the cardiovascular system. In keeping with his more general thinking on fibers, he explained that cold air "constringes the extremities of the external fibres of the body" and that "this increases their elasticity, and favours the return of the blood from the extreme parts to the heart." By contracting those very fibers, their force was increased. By contrast, warm air relaxed and elongated "the extremes of the fibres" and thereby diminished "their force and elasticity." The consequence? People were therefore "more vigorous in cold climates." This was because in these zones, "the action of the heart and the reaction of the extremities of the fibres are better performed, the temperature of the humours is greater, the blood moves more freely towards the heart, and reciprocally the heart has more power. This superiority of strength must produce various effects; for instance, a greater boldness, that is, more courage; a greater sense of superiority, that is, less desire of revenge."[16]

The Hippocratic inheritance likewise dramatically manifested itself in the 1781 treatise on climatic and related influences by the English physician William Falconer, whose more general climatic theory of national temperaments and latitudinal mind-sets has already emerged in our inquiries. In *Remarks on the Influence of Climate*, he paused to quote directly from the Hippocratic *On Airs, Waters and Places* in his examination of the "the Effect of Moderate Climates on the Temper and Disposition." Hippocrates' diagnoses confirmed in Falconer's mind that climatic circumstance had conspired with political arrangement to produce in Europe a people who were "ardently desirous of war, and freely and readily incur all its dangers and difficulties." They stood in marked contrast to the "Asiatics," who were "timid and dastardly." In making such declarations, Falconer also called on the authority of Vitruvius and the German Hippocratic physician Friedrich Hoffman (1660–1742), who ascribed cowardice to the effects of a hot climate, and of the likes of the Roman poet Marcus Annaeus Lucan (39–65 AD) on the bravery of cold climate peoples. But he most commonly resorted to the authority of Montesquieu, whose name featured on nearly every page of his exposition as he reprised the Baron's climatic philosophy. Already sure that war, considered an artform, was born of a hot climate, he learned from Montesquieu that pride, flattery, and overconfidence—all products of high-temperature environments—rendered the inhabitants of the torrid zone inferior in military affairs. For Falconer too, the medium through which the imperatives of climate insinuated their way into national character was the human glandular system. Heat had the effect of exciting "the action of the nervous system, in general, and of the cutaneous nerves especially" while cold conditions diminished the "secretion of the bile" and rendered its "quality . . . less acrimonious." The effects on national temper and mental disposition were marked. Inhabitants of moderate climates were more actively courageous, more enterprising, and far more capable of pressing home any advantages arising from success than those from regions characterized by extreme heat or cold. He thus commended Vegetius' insistence that soldiers from the temperate world were particularly well suited for military operations since they were endowed both with active courage and with the capacity to ascertain and capitalize on any favorable circumstance. By contrast, hot climates bred passion and malice, levity and timidity, alongside cowardice. Thus, he portrayed the Chinese as both violent and vindictive but happily pronounced that "the cowardly disposition of the inhabitants of the East" was so conspicuous that a "hundred Europeans . . . would without difficulty beat a thousand Indian soldiers." Correspondingly, inhabitants of cold climates were decidedly less volatile on account of the way in which cold tended "greatly to diminish the sensibility of the system." This meant that they displayed greater benevolence, which manifested itself in the "mild treatment of prisoners taken in war" and a relatively higher degree of bravery and stability.[17]

Just the year before the appearance of Falconer's treatise, James Dunbar (1742–1798), professor of moral philosophy at King's College, Aberdeen, had issued his *Essays on the History of Mankind in Rude and Cultivated Ages*. The sixth of these essays was devoted to what he called "the general influence of climate on national objects." Balancing Montesquieu's physical causes with David Hume's moral causes, he nonetheless insisted that climate acted both directly via its "irresistible impulse on the fabric of our being" and indirectly through its influence on local environments. Either way, it exerted its power on human affairs not least in the operations of war. "The nature of climates," Dunbar observed, and "the comparative fertility of countries, by determining the course of offensive war, and by affecting the measure of subordination in civil society, must be allowed no inconsiderable sway over the general fortune of the world." Conflict was thus a consequence of the ecologically differentiated world that climate had produced. "Nature, in some climates," Dunbar further explained, was "like an overindulgent parent" who "enervates the genius of her children, by gratifying at once their most extravagant demands" while in "other climates she dispenses her bounty with a more frugal hand, and, by imposing harder conditions, impels them to industry, trains them up to enterprise, and instructs them in the advantages of arts and regular government." The result was a polarized globe whose contours were traced by the imperatives of climate. The cultural differences that resulted from soil and climate and that were crucially dependent on their zonal location marked the "fundamental and fixed" distinctions between the human communities inhabiting the lower and higher latitudes. The consequences of this dispositional distribution, generated by different climatic regimes, were far-reaching. They explained, for example, "the difficulties encountered by the Romans in extending their conquests in Europe" compared with "their more easy triumphs on the theatre of Asia." To the compulsions of climate could be traced, "on the one hand, the astonishing career of the northern conquerors, who overturned all the governments of Europe, and on the other, the feeble resistance made to their progress by more opulent and luxurious nations."[18]

The interventions on climate and conflict in many of these works were typically interwoven with a more general physiological philosophy of history that resorted to a neo-Hippocratic vocabulary of humors, fibers, secretions, glands, and suchlike. Alongside these corporeal preoccupations, commentators from time to time identified what they believed to be the direct psychological impact of different climatic regimes on individuals and societies. In the years that followed, the attention of climatic determinists interested in the onset of violence and armed struggle turned more single-mindedly to the climate conditions that reportedly animated particular conflicts but now drew into the story further climate-derived causal agents such as agricultural productivity, labor power, regional energy, and economic conditions. As we will now see, climatic explanations of the American Civil War illustrate something of how such explanatory maneuvers played out during the nineteenth and early twentieth centuries.

Climate and the American Civil War

The spirit, if not the letter, of Montesquieu's climatic philosophy of warfare continued to snake its way through the nineteenth century and on into the early twentieth. And nowhere perhaps was this mode of explanation more on display than in two accounts of the American Civil War, one from the mid-nineteenth century the other from the early twentieth.

Between 1867 and 1870, the major three-volume *History of the American Civil War* by John William Draper (1811–1882) made its appearance. Right up-front, Draper announced "the great truth that societies advance in a preordained and inevitable course" on account of "uncontrollable causes." Chief among these was the climate. The sense of eschatological inevitability in humanity's following a script written in the main by climate, of course, brought with it a sense of moral relief. And Draper speedily embraced it, announcing its greatest gift—historical exculpation—on the very first page of the first volume. "Now when we appreciate how much the actions of men are controlled by the deeds of their predecessors, and are determined by climate and other natural circumstances," he began, "our animosities lose much of their asperity, and the return of kind feelings is hastened." An awareness of climate's influence on human affairs had the effect of relieving political history of the heavy burden of moral responsibility and thereby anesthetizing ethical recrimination. Naturalizing the causes of the Civil War through the agency of climate was thus a key means of fostering in the postbellum era "more philosophical, more enlarged, more enlightened, and, in truth, more benevolent views of each other's proceedings."[19] Perhaps it was because, as his biographer Donald Fleming put it, Draper had reduced the horrors of the Civil War "to the barest possible simplicity, a question of the weather," that his *History* was bereft of "the obsession with moral issues" that typified other contributions to the subject.[20]

The English-born Draper, president of New York University from 1850 to 1873, was a professor of chemistry, photographer, historian, and architect of the so-called conflict model of science and religion.[21] As perpetual president of the American Photographical Society of New York, moreover, he had offered the Society's services to the War Department in 1861 to cover the Civil War conflict. Given his scientific predilections, it is perhaps not surprising that in his historical writings, he would prioritize the influence of natural causes over the agency of "great men." Indeed, he had long been pondering on the effects that the physical environment, climate in particular, exerted on organic life more generally. Such reflections had pressed him to the conclusion "that the world is not governed by many systems of laws, or one part of it cut off and isolated or even at variance with the rest, but that there is a unity of plan obtaining throughout."[22] With these sentiments, it was only to be expected that the first volume of his Civil War narrative contained chapters devoted to the general effects of climate (chapter 3), its marked influence on the human race (chapter 4), the human production of artificial climates (chapter 5), and what he called the "Identification of National Character by Climatic Zones" (chapter 6). Besides, the whole third section of the treatise (chapters 13 to 16) was given over to the subject "Tendency to Antagonism Impressed on the American Population by Climate and Other Causes."

The argument flowed in predictable channels. Right from the start, Draper wanted his readers to be clear that the separation of the nation "into two sections, conveniently known as the North and the South, or the free and the slave powers," had been effected "chiefly through the agency of climate." This climate-driven disjunction had produced a geopolitical bipolarity ripe for internecine hostility. "A self-conscious democracy, animated by ideas of individualism, was the climate issue in the North" he declared; "an aristocracy, produced by sentiments of personal independence and based upon human slavery, was the climate issue in the South—an aristocracy sub-tropical in its attributes, . . . imperious to its friends, ferocious to its enemies, and rapidly losing the capacity of vividly comprehending European political ideas."[23] Like Montesquieu, Draper was convinced that climate exercised its global imperial power via human physiology, which, in turn, conditioned regional character. It delivered a literal *Weltschauung*—worldview—suffused with the lingering aftertaste of Renaissance geohumoralism:

> The nations of men are arranged by climate on the surface of the earth in bands that have a most important physiological relation. In the torrid zone, intellectual development does not advance beyond the stage of childhood; all the ideas correspond to those of early individual life. In the warmer portions of the temperate zone, the stage of youth and commencing manhood is reached. . . . Along the cooler portions of that zone, the character attained is that of individual maturity, staid sobriety of demeanor, reflective habits, tardy action. Fire, vivacity, brilliancy, enthusiasm, are here exchanged for coldness, calculation, perseverance. Present gratification, a life of ease, a putting aside of care, are the characteristics of the southern edge of this zone; contentment in the anticipation of a happier future, even though that happier future should imply a life of unremitting toil, is the characteristic of the northern.[24]

For Draper, then, zonal climate, bodily organization, mental attributes, and local temperament were tightly knitted together. Each climate, he announced, had its own "answering type of humanity." And this enduring geohumoral coalition, when combined with the direct impact of seasonal variation in different latitudes, translated into distinct economic regimes. In the northern states, the seasonal differences between winter and summer had allocated local populations distinct duties at periodic intervals throughout the year. Summer was the season of outdoor labor, while winter was mostly spent inside. By contrast, the patterns of work in the southern states were markedly different since more uniform weather conditions pertained all the year round. The psychosocial consequences of these natural arrangements were massive. "The Northern man must do to-day that which the Southern man may put off till to-morrow. For this reason, the Northern man must be industrious; the Southern may be indolent, having less foresight, and a less tendency to regulated habits." All of this was simply because Draper was sure that "uniformity of climate makes people homogeneous; they will necessarily think alike, and inevitably act alike."[25]

To Draper, climate's orchestrating of the temperamental disparity and economic divergence between North and South produced the conditions that fostered different labor systems. It was the southern climate, on account of the agricultural products it sustained, that "favored plantation life and the institution of slavery, and hence it promoted a sentiment of independence in the person and of state-rights in the community; that of the North intensified in the person a disposition to individualism, and in the community to unionism." It could all be captured in a formula approaching a climatic syllogism: "Climate tendencies facilitate the abolition of slavery in a cold country, but oppose it in one that is warm."[26] Now, to be sure, from time to time, Draper could rise above his inclination to reduce history to the weather and verge on the complex. But the lure of reductionism routinely proved too strong as he distilled the fate of the divided nation from the edicts of climate. It was climate that had "separated the American nation into two sections"; it was climate that "had made a North and a South"; it was climate that had cultivated "the distinctly marked" political instincts of each culture. As the second half of the nineteenth century dawned, the "labor-basis on which the two societies were resting had now become distinctly separate; in one it was machinery, in the other slaves."[27]

It is not difficult, of course, to discern the appeal that climatic destinism held for Draper. It delivered explanatory simplicity, political self-justification, and ethical absolution. For once he had demonstrated how climate had ghostwritten different histories in North and South and assured himself that there was "absolutely no hope of restoring equality between them," it merely remained for him to conclude that severing the Union was the only way for the South to avoid what else would be prolonged "humiliation." "Secession," he declared, "will release us from all farther vexatious entanglements with the North: it will leave our rivals free to pursue to its consequences their principle of human equality, and us to develop ours of subordination."[28] By thus grounding in climatic circumstances the South's right to make a nation of itself, Draper "drew the sting from any moral recrimination." "Indeed," as his biographer Fleming further observed, "one might suppose that the chief convenience—and possibly the chief defect—of the 'climatic' view of history was to by-pass ethical concerns altogether."[29]

More than forty years later, much the same stance was being advocated by the Kentucky geographer and student of Friedrich Ratzel, Ellen Churchill Semple (1863–1932), though without making any specific reference to Draper's endeavors.[30] In her systematic interrogation of geographical influences, she too turned to the irresistible power of climate, drawing on the work of "older geographers like Montesquieu." Semple remained certain that climate not only modified human physiology but also governed the temperament and "energy" of different peoples and thus their "efficiency" as "political agents." While the humoral underpinning of the neo-Hippocratic system was no longer prominently on view, Semple found it easy to rework her Montesquieu-inspired cosmos into Darwinian categories. What she called the "climatic control" of the ecology of human settlement only served to intensify "the struggle for existence" between human groups. However modern the evolutionary vocabulary, Semple's geography delivered the same kind of polarized world ruptured along a North–South divide reminiscent of the geohumoralists. For Semple was as certain as they were that the "greatest events of universal history and especially the greatest historical developments belong to the North Temperate Zone." The destiny of the world's nations was thus written in the naturalized language of zonal climate. "Nature has fixed the mutual destiny of tropical and temperate zones," she announced, and established them as "complementary trade zones." Economically, this meant that the "hot zone" acted as supplier to the "Temperate Zone," which—happily—enjoyed "greater industrial efficiency." The economic history of the United States bore witness to just such polarities. In New England, it quickened economic development, allowing access to "a nearby tropical trade in the West Indies" as well as to "sub-tropical products in the southern colonies."[31]

But the empire of climate did not simply stimulate a regionally diversified economy; it shaped population geography too. Through the actions of "natural and artificial selection," the warm, moist climate of the Gulf and South Atlantic states, she reported, was now "attracting back to the congenial habitat of the 'black belt' the negroes of the North, where . . . their numbers are being further depleted by a harsh climate, which finds in them a large proportion of the unfit." This Darwinian gesture notwithstanding, the Montesquieuean vision persistently reasserted itself. In the Old World, the "influence of climate on race temperament" had dramatically surfaced. For here "energetic, provident, serious, thoughtful" northerners stood in marked contrast to the "southerners of the sub-tropical Mediterranean basin," whom she portrayed as "easy-going" and "improvident," "emotional," and "imaginative"—all qualities, she feared, "which among the negroes of the equatorial belt degenerate into grave racial faults." It was not difficult to transfer such regional portraiture across the Atlantic. "The divergent development of Northerners and Southerners in America arose from contrasts in climate, soil and area," Semple announced. In the deterministic language of causal compulsion and historical inevitability, she continued, "It was not only the enervating heat and moisture of the Southern States, but also the large extent of their fertile area which necessitated slave labor, introduced the plantation system, and resulted in the whole aristocratic organization of society in the South."[32]

The climatic mind-set that Semple here elevated into lofty explanatory principle had long been installed in her geopolitical outlook. She had already applied its reductionist logic to explaining the American Civil War in her 1903 *American History and Its Geographic Conditions*. By causally coupling climatic conditions with agricultural production, on the one hand, and "an invigorating climate" with what she called "a population energetic" suitable for the "sustained labor of manufacture," on the other, Semple had a ready-made formula for explaining a war whose frontier zone ran along "a climatic line" dividing an urban North from a rural South. It was a simple enough equation: history reduced to geography. For to Semple, politics followed pedology, slavery was a matter of soil, and conflict boiled down to climate. As she explained in introducing the "geography of the Civil War,"

> The question of slavery in the United States was primarily a question of climate and soil, a question of the rich alluvial valley and fertile coast-land plain, with a warm, moist, enervating climate, versus rough mountain upland and glaciated prairies or coast, with a colder harsher, but more bracing climate. The morale of the institution, like the right of succession, was long a mooted question, until New England, having discovered the economic unfitness of slave industry for her boulder-strewn soil took the lead in the crusade against it. The South, by the same token of geographical conditions, but conditions favorable to the plantation system which alone made slave labor profitable, upheld the institution both on economic and moral grounds.[33]

Indeed, within the South itself, the same mandates prevailed. In Kentucky and Missouri, for example, Semple attributed diverse political allegiances to sun, soil, and stone. And to substantiate her claim, she called on the testimony of the Kentucky-born Harvard geologist-geographer, Nathaniel Southgate Shaler, who claimed that politics in the Old South was marked by a distinct "geological distribution."[34] By this he meant, as we have seen, that proslavery majorities were largely confined to rich limestone soils suitable for plantation agriculture. Semple readily concurred. "The whole region of the southern Appalachians," she observed, "had therefore no sympathy with the industrial system of the South; it shared, moreover, in contrast to the aristocratic social organization of the planter community, the democratic spirit characteristic of all mountain people, and likewise their conservatism, which holds to the established order."[35] By thus comfortably naturalizing both slavery and antislavery, and thereby sidestepping questions of moral accountability, Semple, like Shaler, could trace the source of the American Civil War to the dictates of climate's imperious rule.

Huntington, Conflict, and Climate Change

Given his belief in the ubiquitous power of environmental causation, it was only to be expected that in elaborating his climatic philosophy of history, the maverick Yale geographer Ellsworth Huntington would also cast his eye from time to time on slavery and the American Civil War. In his 1915 manifesto, *Civilization and Climate*, for example, he insisted that in accounting for the contrasts between the northern and southern states, "climatic effects" were the most "potent." Here again climatic imperative triumphed over moral sentiment. "Slavery failed to flourish in the North not because of any moral objection to it," he mused, "for the most godly Puritans held slaves, but because the climate made it unprofitable." What on the surface looked like a question of high moral principle turned out to be a case of low economic pragmatism. Take conditions in the northern states. "In a climate where the white man was tremendously energetic and where a living could be procured only by hard and unremitting work," Huntington judged, it just simply "did not pay to keep slaves, for the labor of such incompetent people scarcely sufficed to provide even themselves with a living, and left little profit for their masters." It was different in the South. For here, "even the work of an inefficient negro more than sufficed to produce enough to support him." Besides, in subtropical regions, the white race was far from energetic, and the manual work in which they engaged "was not of much more value than that of a negro." In these conditions, it was entirely understandable for whites to fall into the habit of using their "superior brain" power to secure control and to let "the black man perform the physical labor." Morality, apparently, was a mere matter of meteorology. "If the Puritans had settled in Georgia, it is probable that they would have become proud slave-holders, despising manual work."[36]

Thirty years later, in 1945, he was still pushing economic climatology as the fundamental reason for the South's underdevelopment. The "problems of slavery, the War between the States, and the present Negro question," he reaffirmed, "would scarcely have troubled the South if its climate had been like that of the North." And again the reduction of ethical sentiment to the vagaries of the weather surfaced with renewed vigor. "The suppression of slavery in the North was not due chiefly to moral conviction," he pronounced. "That arose after long experience had shown that slavery did not pay in a cool climate." Besides, good food, a stimulating climate, and a distinctive lifestyle conspired to make the white northerners "so energetic that it irked them to wait for slow-moving Africans." In the South, the weather was different, and so were the racial politics. It was the "warm, unstimulating weather" that had fostered entirely different "social ideals" south of the Mason–Dixon line and thus produced a culture in which the most successful antebellum families were those that looked down on manual labor even while enjoying all the benefits they derived from it. Naturally—or, better, naturalistically—that "system favored slavery and attached a social stigma to work with the hands." So, to Huntington, it was plain for all to see that "climatic contrasts paved the way for civil war."[37]

Huntington's climatic philosophy of conflict was not restricted to the American Civil War, of course. On numerous occasions, he attributed the outbreak of war to climatic energetics. Captivated by cartographic correlations, often at the most coarse-grained scale, he was forever finding causal powers in associations between climatic conditions of one sort or another and a range of human attributes—mental ability, labor power, morality, and the like. As part of this fixation, he fastened upon the idea of "climatic energy"—a kind of estimation of the stimulus human society supposedly received from atmospheric variability—and from it he derived distribution maps of what he called "human energy." Elaborating correlation upon correlation, Huntington soon connected his energetics charts with events recently unfolding in Europe. "As I write . . . at the end of October, 1914," he announced, "it is almost startling to see how the places of highest energy are the ones engaged in war."[38]

In 1919, just after the end of hostilities, Huntington turned to a climatic analysis of the Great War. Germany, he reflected, was endowed with a kind of nervous energy and an aggressive disposition—not unlike ancient Rome and contemporary America—which were largely attributable to "matters of climate." And this, combined with its geographical location, explained the German imperial spirit. Temperature, temperament, and territory, it seemed, had together ensured that as a state power, Germany was simply "bound to expand." In fact, it was a general principle that felt like a climatic rendering of Germanic *Lebensraum*: "the expansion of the great nations of the world is to a large extent determined by climatic conditions." Indeed, as he reflected on the European theater of operations, he was sure that the outcome was "in accord with the health and energy of the people" and therefore with the dictates of climate. Earlier, when Egypt extended its power southward into the territory of Arabia, it was at a point when the country enjoyed the health and energy that was fostered by the climatic variability that prevailed at the time. And it was the same with the ancient Babylonian and Assyrian conquests that occurred, he believed, when they benefited from the stimulus of a bracing climate. During the Middle Ages too, Italy enjoyed European dominance during those times "when the climate—or perhaps it would be better to say, the weather—was highly variable," but the center of political power moved to the northern nations "when they in turn were favoured with the atmospheric conditions that stimulate activity."[39]

In these pronouncements, Huntington's reading of warfare, and the American Civil War in particular, through the lens of climate was fully in keeping with the outlook of writers like Draper and Semple. But in other ways, Huntington moved beyond their rather static climatic historicism by headlining the role that climate *change* played in the history of state conflict and civil unrest.

As we saw in an earlier chapter, Huntington was long convinced by the evidence for historical climate change, not least from tree-ring analysis, and from at least as early as 1907, he urged that the climatic "pulsations" he believed he had detected played a profoundly formative role in human affairs.[40] And so, in *The Pulse of Asia*—a work part travelogue, part ethnographic depiction, part geographical description, and based on a Carnegie-sponsored expedition—he set out to show how "disorder, wars, and migrations" had arisen in concert with climatic oscillations and thus how in Central Asia, "widespread poverty, want, and depression have been substituted for comparative competence, prosperity, and contentment." Quite simply, it was the "climate which almost irresistibly tempts the Arab to be a plunderer as well as a nomad."[41] But his eye—or what James Rodger Fleming dubs his "overheated imagination"[42]—strayed into other spatial and temporal zones too. Europe's "relapse" during the Dark Ages was on account of "a rapid change of climate in Asia and probably all over the world" inducing violent "barbarian" migration. To Huntington, as he scanned the global horizon, this meant that there were contemporary lessons to be learned. Regions of China had experienced increasingly dry, and therefore less habitable, conditions over recent centuries, and Huntington prophesied that if progressive aridity continued, the West would find itself "in danger of being overrun by hungry Chinese in search of bread." It all reduced to one conclusion: "long-continuing changes of climate have been one of the controlling causes of the rise and fall of the great nations of the world."[43]

A few years later, Huntington presented his own apologia for climatic pulsations as historical agents, insisting that "any changes of climate which have taken place in the past or may take place in the future are of the highest importance." Migration was especially significant because great movements of population had "given rise to periods of invasion and anarchy."[44] Meteorology, migration, and militancy were tightly entangled. Here he readvertised what was becoming a dominant theme in his historical analyses—the violent political economy of progressive desiccation.[45] Climatic pulsations, economic downturn, population movement, and violent struggle were seamlessly bound together in a chain of interconnecting causes.

Huntington had already begun to voice his opinion on the subject in *The Pulse of Asia*, claiming that he had amassed data showing that "during the last two thousand years there has been a widespread and pronounced tendency toward aridity." With an all-too-eager proclivity for elevating the particular to the universal, he promptly outlined a general law: "In relatively dry regions increasing aridity is a dire calamity, giving rise to famine and distress. These, in turn, are fruitful causes of wars and migrations, which engender the fall of dynasties and empires, the rise of new nations, and the growth of new civilizations." Desiccation had determined the shape of human history time and time again by dictating the direction in which the whole narrative moved. "Insurrections, wars, and massacres" could all be tracked back to periods of "pronounced lack of rainfall." Obviously, the specifics of Chinese Turkestan on which Huntington was focused could stand as proxy for global history and so, with the enthusiasm of a new convert convinced that they had stumbled on an absolute truth, he announced, "Everywhere in arid regions we find evidence that desiccation has caused famines, depopulation, raids, wars, migrations, and the decay of civilization."[46]

Huntington returned to the subject in 1926 in *The Pulse of Progress*, a work that dealt centrally with Jewish history. Inferential though he admitted his theory to be, he was sure that population movements, invasions, and raids were all attributable to a drying climate. The Libyan and Edomite incursions into Egypt around the first millennium BC, for example, occurred during a time of pronounced aridity. In China too, the "half century of increasing aridity from 250 to 200 B.C. was a time of constant invasions on the part of the barbarian nomads of the north and west." In any case, he was convinced that anyone could see that "in our own day drought almost invariably drives nomads out of the desert and causes them to invade the better-watered borderlands." All this pointed to the obvious conclusion that periodic climatic changes needed to be accorded a far more prominent role in historical explanations of violence and war.[47] Later, in support of the thesis that desiccation, nomadism, and conflict were causally connected, he set about correlating Arnold Toynbee's catalogue of "historic migrations of nomads from the deserts and steppes of Asia and Africa" with cycles of tree-ring growth, lake-level changes, and related phenomena. This confirmed to him that there was a profound connection between "climatic cycles and outbreaks of nomads from the steppes" who surged out from their homelands as "destroyers and conquerors." Indeed, had it not been for the changing climate in Central Asia between 1000 AD and 1200 AD, Genghis Khan would likely have found it impossible "to arouse all the tribes of the steppes and deserts, and sweep over Asia with an almost unparalleled devastation." For it was "by nature" that the Mongols had been driven to a state of turmoil "favorable to his aspirations," even if it required his political genius to mobilize them.[48]

Desiccating climatic regimes of course did not just incite conflicts across territorial frontiers; they also provoked civil unrest. In Turkey, for instance, Huntington urged that the agricultural consequences of increasing aridity meant that local farmers often resisted tax officials and their "minions," who "would employ force and extortion" in their efforts to extract dues. Conditions like these could easily breed insurrection, and Huntington was sure that many civil upheavals were stimulated by the disgruntlement that prolonged periods of unfavorable weather and poor crops inevitably induced.[49]

Huntington's Long Shadow

While Huntington has often been disparaged for his tendency to overgeneralize ad libitum, to keep fact rather too subservient to theory, and to display a troubling methodological naivety,[50] his influence has continued to linger among those coupling climate and conflict in causal ways.[51] Arnold Toynbee, for example, confessed that he had been "enormously influenced"[52] by Huntington, whom he described as "one of our most distinguished and original-minded students of the physical environment of human life."[53] Certainly many had expressed doubts about aspects of his work, but Toynbee was convinced that even his critics had been deeply affected by his perspective.[54] It would be patently obvious to anyone reading his work, Toynbee confessed in his magisterial *The Study of History* in 1934, "that the writer . . . has the greatest admiration for Dr Ellsworth Huntington," even if he did query his interpretation of the rise and fall of the Mayan civilization. After all, he resorted on numerous occasions to Huntington's authority, not least in support of the thesis that the excessive monotony of Central Asian summers and winters with their severe heat and cold, and of tropical lowlands and highlands, with their extremes of humidity and aridity, all inflicted "on human spirits the same uniformly depressing and deadening effects."[55]

On the subject of violent conflict, at least for certain episodes, Toynbee likewise followed Huntington's lead. He was persuaded, for example, by Huntington's views on the stimulating effects of climatic pulsations and of the influence they exerted on nomadic movements. These climatic changes regularly ushered in conditions where, as Huntington had put it in *Palestine and Its Transformations*, "Conflict follows conflict." Toynbee thus found compelling the claim that weather drifting from desiccation toward humidity provided a convincing explanation of why "the Mongols erupted on all fronts with an unprecedented vehemence in the thirteenth century in order to recede with a bewildering tameness a century later." Just when they were embarking on their campaign of conquest that extended as far as Hungary in one direction and Burma on the other, "the demons of drought and starvation were spurring the Mongols on." Similarly, at the point when "they abandoned the Ukraine and the Transoxanian oases and China," they knew they could take refuge on their native steppes because the climate had rendered them "luxuriant once again with a heaven-sent vegetation."[56]

In his magnum opus, *A Study of War*, first published during World War II in 1942, American political scientist and international relations expert Quincy Wright (1890–1970), brother of the celebrated geneticist Sewall Wright, likewise found inspiration in Huntington's writings in general and his cartographic experiments in particular. Besides referring to his work on climatic oscillations, Wright turned to Huntington's *World Power and Evolution* in support, among other things, of the direct causal connection between climatic conditions and warlike impulses. Taken overall, Wright was rather more inclined to apply

Huntington's brand of naturalistic historicism to what he called "primitive peoples."[57] In more advanced societies, responses to climatic challenges, he believed, were less bound by necessity and more governed by rationality. Still, he considered that the tribal inhabitants of regions with temperature extremes—whether cold or hot—tended to be "unwarlike," though he did immediately concede that the "very warlike Bering Sea Eskimo lives in as cold a climate as the very unwarlike Greenland Eskimo." Such exceptions, however, only served to prove the rule. For in general, he continued, "a temperate or warm, somewhat variable, and stimulating climate favors warlikeness. . . . Among contemporary primitive people the largest proportion of the warlike live in hot regions of medium climatic energy." To substantiate the point, he produced a table on the "Relation between Climatic Energy and Warlikeness" correlating the different types of warfare—defensive, social, economic, and political—that "primitive peoples" engaged in with zones of high, medium, and low climatic energy.[58]

Wright's resort to the idea of climatic energy, of course, had a strongly Huntingtonian ring to it. And indeed, to support his thinking on the subject, he resorted to several maps that Huntington had long peddled. Of particular note was the global chart of the "Distribution of Climatic Energy," which Wright—like Huntington himself—positioned alongside a world map of the "Heights of Civilization in the Contemporary World."[59] The parallels were remarkable, with density shading identifying "very high" to "very low" closely corresponding on both sheets. As an exercise in cartographic rhetoric, it was designed to disclose the power of "climatic energy" as an explanatory tool in the analysis of war.

Huntington-style connections between climate and the "energy of nations" likewise captured the imagination of Sir Sydney Frank Markham (1897–1975), a British Labour politician, World War I veteran, and local historian. It is clear that Markham was in close contact with Huntington for he thanked him in the preface to his 1942 *Climate and the Energy of Nations* for "much excellent advice and for undertaking the arduous task of proofreading." Markham's declared aim in this volume was to "determine whether or not climate and climatic controls influence Civilization." In opposition to fashionable contemporary eugenic-type explanations that were rooted in genetic determinism, he was convinced that the mental and physical "energy" of different nations was indeed critically dependent on climate. For he was sure he had discerned a strong correlation between moderate temperatures and what he considered to be high energy levels across the globe. The reason was not hard to discern.

"Every second of the day," he mused, "the environmental factors of temperature, humidity, air movement, and radiation are having their effect upon our bodies and our energies, and there is not the slightest doubt that the ideal combination of these factors goes a long way towards enabling men to be healthier and more energetic."[60]

What especially marked out Markham's diagnosis was his bringing together of indoor and outdoor conditions into what Stan Cox calls "a single formula," namely a synthesis of natural and artificial climates, the former produced by the state of the atmosphere, the latter delivered by architectural innovations, heating techniques, ventilation systems, and the like.[61] And it was that tension between climatic influence and climate independence that Markham believed was crucial to explaining national success. "Civilization to a great degree depends upon climate control in a good natural climate," he declared.[62] In large part, his inquiry was therefore directed toward determining the degree to which different regions and states had been able to moderate the extremes of temperature by one means or another. As Jonathan Cowie elaborates, "His assessment focused on how nations through the ages had striven to become more independent from the environment and he argued (albeit obliquely) for the need to address what is known today as fuel poverty."[63]

Indeed, it was just that capacity to conquer cold winter temperatures and at the same time benefit from a stimulating natural climate that gave the northern United States crucial advantages over the South in the 1880s. Southerners, by contrast, had adapted rather too fully to the hot climate of the South by developing "a life-long habit of acting more slowly than the Northerners." That just stunted economic growth and retarded social progress. Back in the days of the American Civil War, the self-same forces had been at work. The North enjoyed massive climatic advantages, he noted, and therefore developed an "infinitely greater industrial capacity." This meant that "the whole economy of the South, from Virginia to rainy Louisiana," remained at a standstill while the vigorous and resourceful North continued to flourish. It was a universal rule: "all over the world areas similar to the South in climate were suffering or had suffered a like fate. The decay, decline, and defeat of Spain, Turkey, and Greece, and the rise of Germany and Japan showed that the sceptre was passing to those nations who had controllable climates and the means to control them." In global zones of conflict, military success would inevitably go to nations inhabiting regions with a sufficiently moderate climate where extremes could be controlled by human ingenuity and technological knowhow. "The British acquisition of India, the European conquest and partition of Africa, the wars between the United States and Mexico or Spain, all tell the same story," he announced: "victory was ever to those peoples whose home was in a climate that could be controlled."[64]

No doubt Huntington's name could be fished out of many climate-related publications in the decades that followed. An exhaustive trawl, while certainly illuminating, is not my mission here, however. Rather, I simply want to gesture toward the way in which his work remains an anchor point for discussions, both scholarly and popular, of climate and conflict. Recent writers, of course, routinely recoil from his judgmental historicist mind-set and eschew the racial biases that were woven into the fabric of his understanding of human culture. Nonetheless, his name continues to circulate in diagnostic statements on how climate is said to provoke warfare. A full century after its appearance, Zhang and his colleagues found the roots of their own neo-Malthusian project to ascertain the causal links between climate change and war frequency in eastern China in Huntington's *Pulse of Asia*, which claimed to demonstrate "the geographic basis of history."[65] It featured as the first item in the "long-standing scholarly tradition" showing how "organized armed conflicts and climate change are correlated."[66]

A couple of years later, in 2009, two other students of early Chinese history argued that climate change was a "dynamic forcing agent" that drove nomadic populations out of the steppes toward the South, and put enormous pressure on dry farming on account of both nomadic irruptions and worsening farming conditions. Their findings, they announced, provided "evidence in support of Huntington's hypothesis about migration out of central Asia" even if he had "focused on only one side of the issue."[67] And again, writing for *Science 2.0*, an online initiative using network technologies for information sharing and collaboration, Ed Chen resorted frequently, and favorably, to Huntington's work on climate and history in his assessment of "Climate Induced Migration and Conflict."[68] Jeffrey Mazo too, author of the 2010 book *Climate Conflict*, stages Huntington as the "first modern scholar to develop a coherent theory of environmental factors as a driver of history," not least over the role of desiccation on the Central Asian steppes and its influence on the periodic migration of tribes. While noting that the details of Huntington's thesis have been questioned, he nonetheless insisted that Huntington continued to inspire "more nuanced modern analyses of international relations on a global scale."[69] And in their 2013 quantitative assessment of the influence of climate on human conflict, Solomon Hsiang and his coauthors, Marshall Burke and Edward Miguel, likewise traced their project back to Huntington's more qualitative analyses, in this case his proposals about the role of climatic change and agricultural decline in the fall of Rome.[70]

Meanwhile, more than a half century earlier, the psychologist Raymond Wheeler, whose cyclical theories of climate, business, and civilization have already attracted our attention, was preoccupied with finding causal connections between *cold* climatic conditions and civil conflict. This is a theme that, albeit with much greater conceptual nuance, has much more recently captured the imaginations of a range of historians probing the influence of the Little Ice Age's freezing weather on the practice of warfare in the early modern period. While Wheeler's vision was both more expansive and correspondingly more imprecise, it nonetheless fastened on the association between icy conditions and armed conflict. In 1943, just the year following the appearance of Markham's *Climate and the Energy of Nations* and shortly before the death of Huntington, Wheeler delivered his Presidential Address to the Kansas Academy of Science. The shadow of Huntington was plainly on view as Wheeler reprised his thinking on climate and human ecology, seasonal change and human reproduction, the effects of ozone, and the links between "tree-ring curves" and "the culture curve."[71] Toward the end of his address, Wheeler turned to the relationship between climatic conditions and the advent of international and civil wars. Here he simply correlated the outbreak of hostilities with the different phases of global climate from cold to warm. Tree-ring data, of the sort that Huntington had assembled, from the United States, Canada, and Europe, alongside pollen analysis, sunspot occurrence, and pedological science, provided the raw data for his climate record. Together with military records, government sources, and monastic documentation, Wheeler identified what he called cold-dry, warm-wet, warm-dry, and cold-wet phases in world climate.

From these accumulated data, Wheeler constructed his "Drought Clock"—namely, a schematic diagram identifying the links over a millennium and a half between "Cold Droughts and Civil Wars" (see figure 10.1). As he explained, the key finding was that nation building took place during "shifts from cold periods to warm," that nations crumbled "on the shift from warm to cold," and that while international warfare occurred during warm phases of history, civil wars predominated during cold periods. Wheeler's warfare analysis was embedded in a broader climatic narrative of political history. What he termed fascistic regimes apparently only developed during warm phases and only faced the challenges of rebellion when temperatures began to drop. "So long as long as it remains warm," he continued, "people will not vigorously resist." By way of example, Wheeler made reference to the "unusually cold period" during the

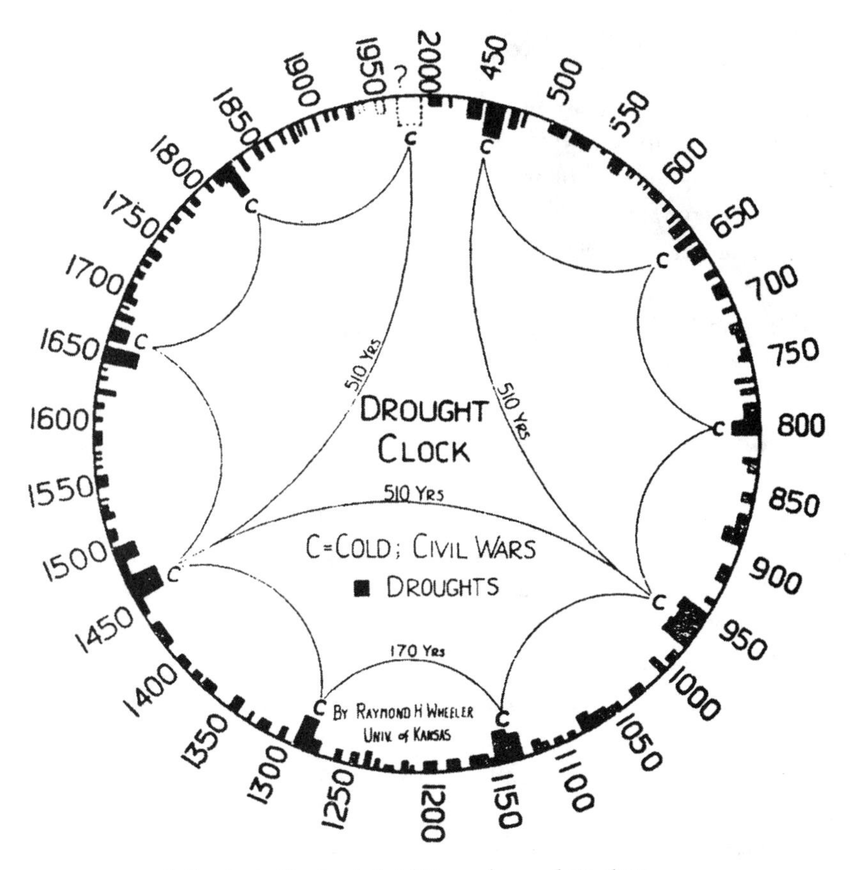

FIGURE 10.1. Wheeler's Clock of Cold Droughts and Civil War
Source: Raymond Wheeler, "The Effect of Climate on Human Behavior in History." *Transactions of the Kansas Academy of Science* 46 (1943): 33–51
Copyright © by the Kansas Academy of Science.

fifteenth century—a period that has continued to attract the attention of historians interested in connecting climate and political culture—with the Hussite Wars, the Wars of the Roses, rebellions in the Holy Roman Empire, and civil wars across "the Orient." The same conditions, he argued, later explained the "Wars of the Fronde in France, civil wars in Turkey, in the Sudan, Deccan, India, the Philippines, Poland, Russia, Hungary, Spain, Portugal, Naples, Scotland and the Crimea."[72] And it was the same for the centuries that followed as he brought the American Revolution, the American Civil War, and civil wars in Poland, Japan, Italy, China, and the Balkans all within the arc of cold phase history. As he summarized it in another piece entitled "Climatic Conditioning in History," cold phases of the climate cycle were "dominated by various kind of civil strife—civil wars, class struggles between rich and poor and between conservatives and liberals, religious wars, democratic reforms." These conditions arose, he claimed, because cold episodes were supposedly times of weak government and frequently, though not universally, of poor political leadership.[73]

The sweeping, not to say swashbuckling, style of the Wheeler warfare schema can readily be dismissed as idiosyncratic, driven by an obsession with natural and cultural cyclicity and lacking in empirical rigor. But it has continued to attract the fascinated eye of enthusiasts. Michael Zahorchak, for instance, who provided a postscript to the posthumously published collection of Wheeler's papers, pressed his mentor's formula into explanatory service in his own day by linking the civil unrest of 1960s America and increased resistance to the "nation's bureaucratic machinery" with colder environmental conditions.[74] Even more recently, Wheeler's "drought clock" and his scheme of global temperatures with corresponding cultural events has been reproduced on several internet websites.[75] Whatever the impressionism inherent in Wheeler's climatic ideology, the association between cold weather and warfare has, in much more recent times, invited the attention of a number of historians who have scrutinized in much greater empirical depth the ways in which the so-called Little Ice Age, which extended from the early fourteenth to the mid-nineteenth centuries, is thought to have shaped global history in the era of what has been denominated "the General Crisis." A brief inspection of this latest rendition of the climate–conflict thesis is worthy of attention.

Ferocious Weather, Frequent Warfare

Nowhere, perhaps, is the linking of ferocious weather and frequent warfare more evident than in the prodigious analysis—a "huge *tour de force*," according to Christopher Booker writing in the *Spectator*—of the "global crisis" of the seventeenth century by the distinguished historian Geoffrey Parker.[76] Subtitled "War, Climate Change and Catastrophe in the Seventeenth Century," Parker's *Global Crisis*—a work of truly epic propositions—was hailed as a monumental achievement when it first appeared in 2013. One reviewer described it as "essential reading for anyone interested in international relations in a warming world." Another, writing in the *Wall Street Journal*, mused that its emphasis on climate "instead of individual agency" was something that "future historians will inevitably have to take into account." Yet another insisted that

Parker's book "demands that historians of all stripes begin to reckon with the Little Ice Age when considering the foundations of the modern world."[77]

Parker's determination to get a handle on nothing less than the global history of the seventeenth century took the form of a two-pronged assault. First, he set out to further corroborate Hugh Trevor-Roper's depiction of the extraordinary political turmoil of the period between the Renaissance and the Enlightenment as the era of "General Crisis." Revolts, savage wars, insurrections, and other violent episodes take pride of place in this portrayal. Under this rubric, the Thirty Years War, the dissolution of the Spanish empire, the collapse of the Ming dynasty in China, the English Civil War, the revolt of Naples, the Cossack uprising, the "great shaking" of Russia, the Jewish genocide of 1648–1649, and much else besides all find a historiographical resting place. Taking 1648 as emblematic of these troubled times, Parker noted that in that year alone, Russia faced a surge of urban rebellions, the Fronde Revolt crippled Paris, Charles I in England was charged with heinous war crimes, and the sultan of the Ottoman Empire was executed in Istanbul. All in all, it was, as Parker remarks, quoting from Voltaire's 1756 *Essai sur les Moeurs et l'Esprit des Nations*, "a period of usurpations almost from one end of the world to the other."[78]

The second line of attack constituted the more innovative dimension of Parker's narrative. It centered on the part played by the challenging climatic conditions of what has come to be known as the Little Ice Age—an era when colder weather and increased glaciation occurred after the Medieval Warm Period with especially brutal conditions during the middle of the seventeenth century. Parker surveyed in impressive detail the global climate in the 1600s using both historical records and the scientific research of specialists in dendrochronology, sedimentology, and paleoecology more generally. The resulting weather inventory enabled him to juxtapose the uprisings in Russia and Ukraine with rapid weather changes from flooding to drought and fierce freezing during the 1640s, correlate the worst drought in Scotland with the beheading of Charles I in England, and link cold summers and icy winters with crop failure and popular uprisings across the world—in southern Portugal in 1637, Ireland in 1641, and in Sicily, central Italy, and Poland in 1647 and 1648. These instances constitute but a fragment of the extensive conflict catalogue that Parker put together. For the period from 1654 to 1667, he further reported that the winters were among the most severe of the past six centuries. All this undoubted industry duly impressed readers. One reviewer commented that when "global events are juxtaposed in this way, only a climate change denier seems capable of contending that nature did not play a foundational role in the generation of unrest during this dastardly decade."[79]

On a number of occasions throughout his narrative, Parker was at pains to explicitly disavow climatic determinism. He insisted, for example, in the opening pages that his intent was "to link the climatologists' Little Ice Age with the historians' General Crisis" but to do so "without arguing that global cooling 'must' have somehow caused recession and revolution." Again, celebrating the remarkable weather archive that climatologists had built, he urged historians to "integrate" the riches of these resources with political, economic, and social change while at the same time urging that "the new data, however abundant and however striking, must not turn us into climate determinists."[80] Indeed, when discussing how harsh winters and cool summers enter the explanatory terrain of historians concerned with state breakdown, he stressed that "we must not paint bull's-eyes around bullet holes and argue that since climatic aberrations seem to be the only factor capable of causing simultaneous upheavals around the globe, therefore those aberrations 'must' have caused the upheavals." What he is convinced of, though, is that there is conclusive evidence that "extreme weather anomalies triggered or fatally exacerbated major political upheavals."[81] And again, though perhaps a tad reluctantly, he paused to observe: "Admittedly historians cannot 'blame El Niño' for everything."[82] These abjurations have persuaded several interlocutors. Dagomar Degroot, for example, conceded that Parker's "detailed analysis of social resilience in the face of climate change" showed that "he largely avoids the environmental determinism that has undermined previous climate histories." Similarly, Christopher Booker considered that by "adding the vagaries of the weather to the overall equation, Parker is not suggesting that these were the cause of the general catastrophe." And Gregory Cushman claimed that Parker fully recognized that climate was "only one factor connecting together" the "underlying conditions, causes, and outcomes of this global conjuncture." Accordingly, he was of the opinion that the book, "when read carefully, actually serves as an antidote to climatic determinism."[83]

For all that, the lure of explanatory inflation proved hard to resist. Brendan Simms rather nervously noted that Parker's "focus on climatic factors, and the resulting social pressures" could "seem a bit deterministic as the examples pile up to the exclusion of other kinds of analysis." And others were yet more skeptical. Jan De Vries hinted at a likely flirtation with climate determinism by referring to the way in which—using scare quotation marks—Parker's "approach 'reduces' the economy to little more than a direct physical relationship between weather and harvest results." De Vries himself remained convinced that seventeenth-century economies in most of Eurasia did not behave in such a simple stimulus–response mode. "Technology, markets, and institutions," he remarked, "could, and did, buffer the effects of climate on food production and distribution."[84] Again, Anthony Pagden fully conceded that Parker had demonstrated that the seventeenth century was indeed an era of crisis and that it was, at the same time, a period of unusual and violent weather disruptions. What he had emphatically "not demonstrated beyond contention," Pagden went on, "was that the former was caused (or even precipitated) by the latter."[85]

The problem here centers on the nature of explanation. In part, the difficulty in identifying the precise role of ferocious weather in fomenting warfare lay in the multiform ways in which the Little Ice Age is omnipresent in Parker's text and performs different roles. Sometimes it features as mere backdrop, sometimes as an evolutionary challenge demanding adaptive responses, sometimes as a more directly active agent. Throughout, Parker remained careful about the language he used to "explain" events. He referred to the climate delaying events, exacerbating conditions, facilitating, triggering, permitting, intensifying, forcing, and the like. Still, the impression remained that time and time again among the candidates canvassed as critical in the explanatory nexus of frequent warfare, ferocious weather acted as the predominant force. There is also, as Pagden noted, the simple fact that the rebellions and uprisings that are projected as diagnostic of the General Crisis "were not always of the same kind." Some of these insurgences might well be "linked to, or exacerbated by" climate disasters, but others much less so. In De Vries' opinion, by transforming the Little Ice Age—a concept that he describes as an "abstraction unknown to contemporaries"—into an active historical agent, Parker sidestepped "the challenge of measurement and testing in favor of an evocative narration of catastrophe and survival."[86]

At rather greater length, Paul Warde weighed Parker's mode of historical explanation in the balances and found it wanting. Along the way, he quibbled over the problem of how to distinguish climate change from climate variability and whether extreme weather in other periods had the same outcomes. Warde also insisted that Parker's reliance on "the cumulative effects of exemplification" was questionable when what in fact was required was "not the accumulation of examples, but repeated clear-cut demonstrations of cause and effect." His deepest concern, however, congregated around reading the human impact of climate "as the history of 'various adaptive strategies taken to survive the worst climate-induced catastrophe of the last millennium.'" To Warde, casting the narrative in the language of evolution and survival risked what he referred to as the "adaptive fallacy"—namely, "falling into the idea that characteristics held by successful societies (or individuals, or species) but not less successful ones must be attributes that have been tested in the furnace of evolutionary selection, when in fact they might have survived or appeared for altogether different reasons (in other words, they were simply not actively maladaptive)." None of this, of course, was to suggest that any accusation of simplistic climate determinism could justly be leveled at Parker's account. But it did raise questions about just exactly how the role of the Little Ice Age as a formative influence was to be understood.[87]

Something similar might be said of the earlier *Little Ice Age* by Brian Fagan, who lauded Parker's intervention as a "masterly synthesis of history and paleoclimatology" that did nothing less than redefine the social and political impact of the Little Ice Age.[88] Like Parker, Fagan repeatedly eschewed climate determinism, allowing the changing weather conditions to occupy different but pervasive explanatory niches in this account of revolt, conquest, and various other crises in the period. Nonetheless, it is clear from its subtitle—*How Climate Made History, 1300–1850*—that explanatory pride of place goes to the changing weather. It is the same when we hear that, like Norse conquests, "cathedrals are a consequence of a global climatic phenomenon, an enduring legacy of the Medieval Warm Period."[89] The further juxtaposing of tables correlating what he calls historical events and climatic events similarly invites readers to find in climate the prime mover of history. None of this may be climatic determinism as postulated by earlier writers. But by resorting to such modes of expression as climate having "its sway in human events," of cycles of cold "affecting monarch, noble, and commoner," and of cold conditions bringing "immediate threats of social unrest," the narrative defaults to the pervasive influence of climate without delineating clear lines of causality.[90] What is needed in cases like these is a greater sense of the need to provide explanations for events under particular descriptions. To ascertain what role the weather played in Norse expansion, there is little reason to doubt that higher temperatures were involved. But to say that warmer conditions simply *caused* the expansion of Norse fishing grounds is to enthrone one dimension of the explanatory nexus above all others. It is the same with warfare. A changed climate may certainly be one element in a suite of conjunctive explanations proposed to account for hostilities. But its role needs to be evaluated alongside, say, the manufacture of arms, the quality of military leadership, the vicissitudes of government policy, the prevailing socioeconomic conditions, and much else besides.

In his monumental 2018 account of the Dutch Republic during the Little Ice Age, Dagomar Degroot made it clear that forces of precisely this sort shaped what he called "the frigid Golden Age," albeit in association with the effects of climate change. Weather and war dominated much of this volume, but before commenting on Degroot's perspective on that subject, something of his more general historiographical perspective needs to be registered. Right from the start, while admiring the power and industry of Parker's analysis, Degroot distanced himself from the way Parker's "landmark book" blamed the cooling climate of the seventeenth century for global catastrophes, including three Dutch coups d'état. In opposition to that portrayal, Degroot was sure that while "the weather that accompanied the chilliest phase of the Little Ice Age certainly did end lives and ruin livelihoods across the Dutch Republic," the changed climate "also offered important commercial, military, and even cultural benefits for Dutch citizens."[91] This immediately lifted the whole discussion about the impact of the Little Ice Age out of a narrative dominated by the language of a general crisis and opened up new interpretative possibilities. Certainly the weather of the Little Ice Age devastated some societies by contributing to harvest failure, famine, and "ultimately death on a vast scale," he conceded. But that was not a universal experience. Some prospered, and none more so than the Dutch Republic, which, between 1590 and 1750, he depicted as an extraordinary Golden Age. He thus routinely resorted to the language of "opportunity" as much as "catastrophe" in the chronicle he presented.

Ever concerned to distance himself from reductionism, he warned that historians "of all stripes can fall into simplistic determinism—the idea that a single force or set of forces predetermined the course of human history—but that trap may be especially dangerous for climate historians." Indeed, he remained convinced that "the big structures that give shape to human history—cultural, socioeconomic, and political—have always mediated, or channelled, how weather influenced by climate change affected individual people." Ultimately, Degroot's aim was to demonstrate that climate changes led to weather that "limited or expanded the choices open to people but did not determine their actions." And so the language he chose to capture the influence of the climate remained both muted and circumspect. A sampling. The Golden Age "coincided" with the coldest period of the Little Ice Age; the climate "presented challenges" for the Dutch citizenry but also "offered opportunities" that many happily exploited; climate challenges "exacerbated" unrest that had already been "provoked" by such forces as religious persecution and mismanagement of the political economy; the weather during the Grindelwald Fluctuation—a pronounced cooling phase from 1560 to 1630—"exacerbated" preexisting vulnerabilities of one sort or another.[92]

Several chapters of Degroot's analysis were devoted to conflict and climate change during the Dutch Golden Age. But from the outset, he made it plain that changing climatic conditions were a "catalyst for," but rarely a "cause of," military victories or defeats. At different points in time, moreover, the climate elicited different responses. Focusing on the Dutch Wars of Independence between 1564 and 1648, he observed that conditions only initially benefited the rebels. Climatic influence always rode in tandem with such forces as population growth, the prevailing fiscal circumstances, and the prevalence of firearms. Besides, Degroot remained sensitive to the ways in which soldiers could adapt to the changing weather, turning setbacks into successes. Climate's contrary motions meant that the weather "yielded a complex pattern of advantages and constraints" for the warring factions. Simply put, the "challenging environmental conditions did not determine military outcomes" as the opposing military forces "nimbly adapted to changing weather conditions."[93]

However imprecise the role played by the Little Ice Age in human affairs may actually be, the foregrounding of climate as the explanation for conflict, insurrection, and bloodshed may well be a consequence of latter-day fears about anthropogenic climate change, the increasing availability of historical data from a burgeoning scientific industry in paleoclimatology and climate change modeling, and the urgent need in our own time to develop effective mitigation strategies. Parker's analysis, for example, was widely welcomed on its publication, not only because of its major contribution to the global study of the seventeenth century but on account of its lessons for the looming climate crisis of the twenty-first century. Lisa Jardine, for instance, began her reflections with the comment that Parker's "magisterial book argues that 17th-century climate change led to social unrest—and offers a timely 21st-century warning." And in similar vein, she concluded her assessment with the plea that "we should now turn our attention to anticipating and developing remedies for what, on the evidence of the earlier, comparable climate change in the 17th century, are likely to be recurring calamities of this kind with grave long-term consequences." With a corresponding eye on contemporary concerns, Degroot too welcomed Parker's ushering climate into the arena of historical explanations. It was not "a moment too soon," he added, "for past relationships between humans and climate change can inform how we adapt to global warming today. *Global Crisis* is therefore essential reading for anyone interested in international relations in a warming world."[94]

These observations were entirely in keeping with the aspirations that Parker and Fagan, and indeed Degroot, had for their own books. Parker, for example, hoped that his emphasis on adaptation to Little Ice Age conditions could serve to instruct the twenty-first century on climate strategy. And he provided an epilogue to his study dwelling on the recent challenges of extreme climatic events. While no human intervention could have prevented past climate disasters, he insisted that "a better warning system, better popular education about evasive strategies, and faster and more effective emergency responses would surely have mitigated the consequences." At a fundamental level, whatever the differences between the seventeenth and twenty-first centuries, "governments during the Little Ice Age faced the same dilemma."[95] Fagan too dwelt on lessons to be learned from European history during the Little Ice Age. His take-home message was clear. History had shown that climate change "is almost always abrupt. Shifting rapidly within decades, even years, and entirely capricious."[96] Complacency is not an option. Neither is optimism. For his part, Degroot inaugurated *The Frigid Golden Age* by presenting his readers with an alarming litany of the climate crisis of the early twenty-first century with its intensified storms, protracted heatwaves, and flooding. Yet he remained more cautious, reiterating that his way of thinking about climate change was "at odds with the kind of determinism that can enter assessments of the present-day and projected impacts of global warming." Those who talked of climate change as "a straightforward cause of human events," Degroot went on, brought society no closer to understanding its true impact. "We only learn how to adapt to climate change," he concluded, "by understanding how its influence filters through the socioeconomic, cultural, and political systems in which humans structure their lives."[97] Either way, this body of historical research and writing is being conducted in the shadow of the urgent challenges that climate change is presenting to twenty-first-century society.

———

With roots in the ancient Hippocratic anchoring of violent impulses in atmospheric conditions, mediated through the influences of the four humors on national temperament, the idea that climate and conflict are causally connected has gripped the imagination of generations of philosophers, historians, geographers, and students of war. The legacy of this neo-Hippocratic vision has been remarkably persistent flourishing, throughout Renaissance and Enlightenment Europe, in the writings of figures like Jean Bodin, Montesquieu, William Falconer, and James Dunbar. More recently, in one modified form or another, it was mobilized by such writers as Ellen Semple, Nathaniel Shaler,

John William Draper, and Ellsworth Huntington as the key explanatory cause of the American Civil War. Here the web of causation was entangled with the conviction that the mainsprings of African American slavery lay in the differing climatic conditions north and south of the Mason–Dixon line and in the inability of the white race to engage in manual labor under the extreme temperatures of the Old South. By thus naturalizing the conflict and rationalizing the southern plantocracy, not infrequently in the language of Darwinian natural selection and adaptation to environment, questions of moral probity could easily be subverted. Later, Huntington's climatic worldview, consciously grounded in the emerging science of paleoenvironmental reconstruction, attracted the attention of Arnold Toynbee, Quincy Wright, Sir Sydney Markham, and Raymond Wheeler, all of whom were captivated by the idea that, in one way or the other, climate and climate change were potent causes of violence, invasion, and armed conflict.

Throughout this history, violence and warfare have been moored to climate by somewhat different anchor lines. One series of links laid emphasis on human physiological responses to different climate regimes. Here the ways in which weather and war were connected was bound up with a medical discourse that dwelt on how bodily fluids, fibers, glands, and the like performed their functions under different weather conditions. Another related strand of thinking concentrated on how global climates supposedly fashioned the psychological dispositions of whole nations, predisposing them to pugnacity or timidity, courage or cowardliness. A further line of connection laid emphasis on how climate could stimulate war or peace through the ways it dictated the nature and scope of agricultural production, patterns of resource plenitude or poverty, and the advent of times of feast or famine. All of these explanations, while accented differently, remain connected through genealogical succession and thus display a discernible family resemblance.

Comprehensive cataloguing, as I have said, has not been my quarry here. And this is especially so for twenty-first-century interventions on the subject, where my treatment is intended as diagnostic rather than encyclopedic. My aim has simply been to redraw attention to the lengthy shadow that the neo-Hippocratic social philosophy and, latterly, Huntington-shaped environmental causation continues to cast over writing on climate and war in many different registers. For in this intellectual genealogy, the force lines connecting classical climate determinism with contemporary climate change reductionism are exposed with particular clarity. Even among those chary of the cruder reductionist impulses in the Huntington school of climate history, there remains an inclination toward attributing causal powers of an overwhelming sort to the climate in accounts of violence, unrest, and civil war. In part reflecting the abundance of historical climate data, together with the understandable anxieties that attend contemporary anthropogenic climate change, several writers have found in the weather of the Little Ice Age the prime mover in the genesis of war in that era. And while there are those who remain suspicious of a too ready resort to climatic explanations independent of the social structure and the political culture of the times, the impression remains that climatic determinism continues to lurk in the recesses of the historical imagination. Something of how these intellectual forces are continuing to play out in our own day, especially among those concerned with national security agendas, will be our next port of call.

11

Securitizing Climate Change

"CLIMATE CHANGE will lead to an increased threat of war, violence and military action against the UK and risks reversing the progress of civilisation."[1] So readers of the *Guardian* newspaper were told on July 6, 2011, on the authority of Chris Huhne, then Britain's secretary of state for energy and climate change. The article advertised a speech the member of parliament would deliver the following day on "The Geopolitics of Climate Change" to the Future Maritime Operations Conference at the Royal United Services Institute. The cabinet minister's message was crystal clear, if profoundly disturbing. The devastating impact that climate change would have on global food, water, and health meant that "unstable states" would become "more unstable. Poor nations poorer. Inequality more pronounced, and conflict more likely."[2] In thus casting climate change as a major player on the world's national security stage, Huhne was giving voice to a refrain echoing its way through the corridors of government, academia, and journalism.

Even the most cursory glance at the internet will show how indeed matters of climate and warfare are making headline news. A mere sampling of news report titles is indicative. "Did Climate Change Spark the War in Syria?" asks Kristina Chew on the *Care 2 News Network*.[3] *Newswise* tells visitors to its website that researchers at the University of Iowa have just shown "How Global Climate Change Affects Violence."[4] In November 2009, environment correspondent Richard Black announced on the BBC News channel that "Climate 'Is a Major Cause' of Conflict in Africa."[5] Online readers of *Nature* learn from Quirin Schliermeier's headline that "Climate Cycles Drive Civil War" and that "tropical conflicts double during El Niño years."[6] That same day, August 24, 2011, Wynne Parry spread the message to visitors to *Live Science*'s webpage under the title "Climate Fluctuations May Increase Civil Violence."[7] The following day, Rosanne Skirble conveyed the same story to the Voice of

America, declaring that "Climate Is Major Violence Trigger."[8] In July 2017, under the title "The Uninhabitable Earth," in the *Intelligencer*, an offshoot of *New York* magazine, David Wallace-Wells told his readers that famine, economic collapse, and "a sun that cooks us" were just a sampling of the havoc that climate change would "wreak." And the whole story, he promised, was "worse than you think." "Doomsday" best described it, not least because the future held "perpetual war" for humanity with the "violence baked into the heat."[9] With headline stories like these, it is hardly surprising that climate has been catapulted to the forefront of national security consciousness.

This is the terrain over which we will pass in this chapter, examining something of how academic research, climate-change journalism, and the politics of national security have interacted in sometimes mutually reinforcing, sometimes contrasting, ways. Critics of the climate–security coalition will also come within our purview as we examine the arguments of those resistant to the deterministic ethos of a flourishing research industry invested in predicting conflict as the inexorable, and devastating, consequence of climate change.

Climate Change and Contemporary Conflict: Configuring Connections

Underlying graphic revelations like those just mentioned is a suite of research projects devoted to exposing how conflict of one stripe or another is an inevitable response to some dimension of climate change. Consider first the findings of one project that appeared in the August 2011 issue of *Nature* and, in turn, formed the basis of several of the news reports noted above. Here two postdoctoral fellows in public policy and a climate scientist from Princeton and Columbia Universities joined forces to produce an article entitled "Civil Conflicts Are Associated with Global Climate." The central thrust of the undertaking was to detect causal connections between the interannual mode of the El Niño Southern Oscillation and patterns of global civil conflict. Deploying a formidable array of statistical techniques and using war and climate data from 1950 to 2004, they came to the conclusion that organized political violence "arising throughout the tropics" in situations where at least twenty-five deaths occurred in clashes between governments and oppositional organizations "doubles during El Niño years." They expressed their conclusions with remarkable, if not forensic, mathematical precision: El Niño "may have had a role in 21% of all civil conflicts since 1950."[10] Arresting though the correlations they derive may be, the precise chains of causal influence linking weather conditions to violent conflict remain black-boxed, however. One journalist reported that these researchers themselves admitted they had yet to "untangle the mechanisms that link a change in sea surface temperature with, for example, a guerilla war."[11] And in their article, they did indeed note that adverse economic conditions, such as loss of income or increasing food prices, arising from changes in the weather, might play a pivotal role, as indeed might disease outbreaks triggered by climatic circumstances. They also claimed that "altered environmental conditions stress the human psyche, sometimes leading to aggressive behaviour."[12] The manifold imprecisions breaking surface here, however, did little to mute the confidence that the lead author, Solomon Hsiang, displayed when he announced in an interview, "Not only does the climate affect conflict, it's a major factor in determining global patterns of violence."[13] It is precisely that style of pronunciamento that propels research into the limelight.

It was much the same with a subsequent analysis that Hsiang, working with Marshall Burke and Edward Miguel, put out in 2013. In this synthetic account, the authors scrutinized a large body of relevant literature and claimed that they had found "strong causal evidence linking climatic events to human conflict." "The magnitude of climate's influence is substantial," they went on, and again with extraordinary statistical exactitude declared that "for each 1 standard deviation change in climate toward warmer temperatures or more extreme rainfall . . . the frequency of interpersonal violence rises 4% and the frequency of intergroup conflict rises 14%." Even though they inserted at the tail end of their study a statement that they did "not conclude that climate is the sole—or even primary—driving force in conflict," the deterministic tone their report adopted, alongside its computational ammunition, meant that it was widely picked up in the media.[14] And in that arena, even this nuance rapidly evaporated. A headline in the *Los Angeles Times*, for example, informed its readers that "violence will rise as climate changes."[15] As the *Washington Post*'s correspondent, Brad Plummer, observed, the original article bore "little resemblance to the media stories written about it."[16]

Several years earlier, in 2007, the National Academy of Sciences featured in its *Proceedings* an article charting "Global Climate Change War, and Population Decline in Recent Human History" by a team of researchers from a range of disciplines—environmental science, geography, international affairs, and anthropology. The message here was similar but focused more exclusively on the influence of temperature—using paleoclimatic data—on war and demographic decline in the preindustrial era. Once more, a daunting battery of statistical devices featured as the lingua franca of their claim that "long term fluctuations of war frequency . . . followed cycles of temperature change."[17] In this analysis, David Zhang and his colleagues drew particular attention to the significance of falling temperatures. On the basis of their data graphs of the half millennium between 1400 and 1900, they contended, in language reminiscent of earlier writers we encountered in the previous chapter, "It can be seen that all war peaks occurred in a cold climate." Again, this conspicuously broad-brush declaration achieved notable algebraic precision: "the worldwide war ratio during cold centuries was 1.93 times greater than that of the mild 18th century, and 1.77, 1.91, 1.50 and 2.24 times greater for the NH [Northern Hemisphere], Asia, the arid areas of the NH, and Europe, respectively."[18] Again too, the threads of causal connection were complex, with various intermediary agents exerting their power. The story went something like this. Rapidly cooling conditions during the Little Ice Age adversely affected agricultural production with the result that food prices dramatically increased. These conditions of ecological and economic stress brought about a dramatic drop in population size, increased migration, and the adoption of new diets. And the outcome of this cocktail of climatically induced economic woes was violent conflict. Colder temperatures during the preindustrial era were thus the source of population change, shifts in agricultural production, and price adjustment cycles, which, in turn, were causal conduits to the outbreak of war.

Zhang and his coauthors saw all this as a variant on the Malthus–Darwin model of social history in which carrying capacity and adaptation played dominant roles, but this time powered by the imperatives of "a fundamental driver, climate change." In this scenario, warfare emerged as "an adaptive mechanism" that preindustrial societies put in place to cope with rapid environmentally driven social change. Of course, they did pause to observe that with greater technological sophistication, other responses to the same stimuli might mean that things could pan out differently. But they remained convinced that positive adaptive choices had not enabled the human race to escape the social and economic calamities that were "caused by severe cooling at both global and continental scales." For the future, they anticipated that both the direct and indirect impact of climate change could mean that the economic burden of global warming would "cause conflict over resources and intensify social contradictions and unrest as we have seen in the past." Still, the burden of their investigation was to confirm from historical analysis that "cooling . . . would cause wars, forced migration, or even famine."[19]

That same year, 2007, Zhang and his colleagues published another report focusing specifically on eastern China, arguing again that warfare over the past millennium was causally connected with falling temperatures. Even though there could be time lags of between ten and fifty years between cooling conditions and the outbreak of war, they remained certain that climatic "oscillations drove historic war-peace cycles."[20] Here too they reiterated their Darwinian-sounding conviction that "warfare is an 'adaptive ecological choice' under circumstances of population growth and resource limitation." What is also notable here is that despite the sophistication of their statistical maneuvers—the data are averaged, smoothed, and otherwise normalized—the authors continued to root their project in a tradition of climate determinism running back to the early twentieth century and indeed much earlier.

According to Richard Tol and Sebastian Wagner, the broad contours of Zhang's Chinese narrative linking conflict to cooling were also applicable to Europe. These authors, however, were rather more tentative in their claims and more carefully circumscribed their declarations. They acknowledged, for example, the critical role played by forces other than climate as determinants of violent struggle, they clearly discriminated between correlation and cause, and they itemized additional components in the "causal chain" of historical explanation. Still, while insisting that their "results do not show a clear-cut picture" and that their analysis "implies that future global warming is not likely to lead to (civil) war between (within) European countries," they remained persuaded that their data revealed close and sometimes ultimate causal connections between outbreaks of violence and lowering temperatures at least in Europe's preindustrial era.[21]

While Zhang, Tol, and their coworkers dwelt on the causal connections between *falling* temperature and rising hostilities, others traced the outbreak of war to *warmer* conditions. In 2009, for instance, another team of researchers, led by the agricultural economist Marshall Burke, published the results of an inquiry into the impact of global climate change on armed conflict in the sub-Saharan world under the title "Warming Increases the Rise of Civil War in Africa." Mobilizing an assortment of intricate model-based projections, they claimed that by 2030, rising temperatures are likely to bring about a 54 percent increase in armed conflict in the region and an estimated 393,000 deaths. Every one degree of temperate increase leads to a 4.5 percent rise in civil war within the year. To be sure, Burke and his team allow that climate's formative influence is mediated through temperature-related crop decline; agricultural performance is the channel through which a warming climate generates war. But these authors were also quick to declare that hot conditions act directly too—inducing violent crime, for example. Of course, mitigating circumstances may blunt the sharpest end of heat's imperious regime, not least through the forces of increasing productive capacity and political will. But in the last analysis, they remained sure that in sub-Saharan Africa, "the adverse impact of warming on conflict by 2030 appears likely to outweigh any potentially offsetting effects of strong economic growth and continued democratization."[22]

Even this mere snapshot of recent research discloses the multitudinous ways in which climate—and climate change in particular—is believed to incite violent conflict. Others could doubtless be added to the portfolio. Changes in rainfall, for example, have attracted the eye of researchers. Some focus on long-term trends in precipitation, others on short-term triggers. For some, outbreaks of violence are a consequence of scarcity resulting from the impact climate change has on natural resources and the ensuing decline in carrying capacity.[23] For others, what is critical in triggering civil conflict is the impact of interannual rainfall variability on agriculturalists.[24]

The cumulative effect of these inquiries is to underscore the ways in which conflict may be sparked by climatic conditions. What is also notable, however, is the markedly different rhetorical stances that researchers adopt toward the causal networks that are adduced to explain various episodes. Some deny the salience of mediating circumstances and attribute to the climate direct causal agency that overrides issues of governance, mitigation measures, human cooperation, and so on. Others carefully circumscribe climate's influence, emphasizing its indirect character and declaring that climatic conditions, as one study puts it, are "not associated with the onset of conflict in the absence of economic, political and demographic control variables."[25] When headlined for public consumption or marshalled in the interests of security agencies, however, such precautionary caveats rapidly evaporate as the following case illustrates.

History Lessons: Rain, Raiding, and the Rise of Genghis Khan

National security investment in the science of climate change is large. For if indeed climatic conditions are a major trigger for armed conflict at a range of scales, both national security agencies and the public more generally need to be kept abreast of the real and present danger that uncontrolled climatic change will inevitably bring. Given the epic consequences of climate change that are habitually envisioned, it is perhaps understandable that the public appetite for stories finding in some climatic condition or weather pattern the cause of violence, warfare, and military conquest seems to be limitless, even when they are about events that took place seven or eight centuries ago. The case of Genghis Khan (Chinggis Khaan) in the early thirteenth century is illustrative. A mere sampling of a spate of journalistic headlines indicates something of the craving for climatic explanations of violent episodes in human history. "Genghis Khan's secret weapon was rain," announced Roff Smith in the *National Geographic* in March 2014.[26] "Warm, wet times spurred medieval Mongol Rise" was Sarah Zielinksi's headline in the *Smithsonian Magazine*.[27] Doyle Rice of *USA Today* declared that "Genghis Khan rode climate change to take over Asia" and that "his sprawling empire was propelled by a temporary run of nice weather."[28] And Steve Connor of *The Independent*, writing under the title "How climate change helped Genghis Khan," asserted that "a long period of warm, wet weather spanning several decades helped one of history's most famous tyrants to conquer most of Asia . . . Genghis Khan owes his place in history to a sudden shift in the Asiatic climate."[29] The list could easily be extended: Nathanael Massey used the strapline "Changing precipitation patterns helped Genghis Khan rise to power—and rise to victory across Eurasia" to inform the *Scientific American*'s audience;[30] in *The Australian*, Emma Reynolds reported her story under the headline-grabbing title "'Climate change' drove rise of Genghis Khan";[31] and the Asia section of the BBC News website informed visitors to the Genghis Khan story that "Good weather 'helped him to conquer.'"[32]

These news bulletins, press releases, and general interest commentaries were all reports based on a 2014 study of tree-ring records of moisture during the rise of the thirteenth-century Mongol empire. The original research article appeared in the *Proceedings of the National Academy of Sciences*. The research team related the findings of their tree-ring reconstruction of changing weather patterns from Siberian pines in central Mongolia using a standard drought severity index. But the article's novelty lay in its claim that whereas "the demise of complex societies" had conventionally been linked to "deteriorating climate conditions," it had solid evidence to show that an "ameliorating environment" fostered "the rise of empires." Mongolia was a case in point. Here, fifteen consecutive years of higher-than-average wet weather "promoted high grassland productivity and favored the formation of Mongol political and military power." The explanatory chain is a short one. These unusual climatic conditions resulted in a period of high grassland productivity that facilitated a significant increase in domesticated livestock, not least horses, which were fundamental to the military might on which Genghis Khan's empire was built. "Warm, wet conditions enabled the Mongol leadership to concentrate political and military power in designated localities," the authors observed. "It is this condition that was a necessary and important factor in the successful mobilization of nomadic power in Chinggis Khan's military expeditions."[33]

The intellectual lineage in which this recent paleoecological investigation stands is itself interesting. It is part of a long tradition of finding in changing climatic conditions the root cause of violent episodes. The earliest piece to which they resorted dates from the mid-1970s—an account by Gareth Jenkins of the connection between climatic cycles and the rise of Genghis Khan. While the empirical detail of the latest report differed from Jenkins' emphasis, which attributed political agency to fluctuations in mean temperature rather than to precipitation or desiccation cycles, the structure of the argument about climate's historical agency remained center stage. What is noticeable is that while Jenkins paused to note that "objections against wholly mechanistic or deterministic arguments in which impersonal forces, such as weather changes, dictate the fate of nations are, of course, appropriate," he nonetheless located his analysis in a body of work snaking its way back to figures like Ellsworth Huntington and Owen Lattimore. In particular, he engaged with their desiccation thesis while acknowledging that their work had in part anticipated his own hypothesis.[34] His own conclusion was that "Chinggis' success in welding his quarrelsome, infighting countrymen together and in his subsequent exploits may have been as much the product of climatic deterioration at home as of Chinggis' alleged charismatic personality, capacity for leadership, or sense of world mission."[35]

Several things are conspicuous about this whole episode, both about the original academic publications and the way their conclusions have been picked up in popular outlets. First, they all displayed, to one degree or another, their commitment to what might be called the "may-have, might-have, must-have" school of historical explanation. This is a mode of writing history that prominently displays the inclination to drift from the tentative to the definitive, from the speculative to the declarative. And it is certainly discernible in the original research publications. Pederson, Hessl, and their coworkers, for example, move seamlessly from telling their readers that "climatic variability, hypothetically, could have provided ample resources for strengthening the new leadership" under Genghis Khan to declaring that changed climatic conditions were "necessary" for the triumph of his military exploits.[36] But the approach is yet more dramatically exhibited in the journalistic reports of their work. Again and again, the conjectural transmutes into the actual. Sarah Sheffer of PBS Newshour, for example, drifted effortlessly from the observation that "nice weather might have helped him [Genghis Khan] succeed" to the expansive proclamation that "nice weather is great for empire building."[37] The title of Matthew Stinson's report in *Liberty Voice*, "Genghis Khan Aided by Climate Change to Conquer Asia," ran unselfconsciously alongside the comment that "a warm, wet climate . . . possibly aided the Great Khan's domination of massive swaths of territory."[38] Bryan Walsh, writing for *Time* magazine, slid from noting that "the Mongol Empire wasn't solely the product of Genghis's will . . . [but] may have owed just as much to beneficial changes in the climate" to the proclamation that "he couldn't have done it without a change in climate."[39] *The Japan Times* began with the comment that "mild weather may have propelled Genghis Khan to power" but presently quoted the original study's remark that it "wasn't the only thing, but it must have created the ideal conditions for a charismatic leader to emerge out of the chaos."[40] The indicative mood soon conceded to the subjunctive when Sarah Zielinski's confident pronouncement that this "great empire was made possible not by brilliant leadership alone, but by a 15-year period of abnormal moisture and warmth" eventually retreated to the rather more conjectural inference that these "conditions would have provided a surplus of grass" that, without "the warm, wet years," Genghis Khan "wouldn't have had the resources for building a strong government" and that the "world may have been a very different place."[41]

Working scientists, of course, cannot always be held accountable for the way in which journalists of one stripe or another represent their findings through mass communication media. The slippage gap between scientific publication and web journalism can be wide indeed, as provisional findings transmute into solid facts. But increasingly, it seems, scientists happily involve themselves in the marketplace of popular commentary. In this case, two members of the original research team worked hard at getting their story into popular outlets. In a press-release statement picked up in several magazines, Hessl presented the project's conclusion in customary academic-speak: "it seems likely that he [Genghis Khan] benefited from that unusual bout of climate change."[42] Yet writing for *The Conversation*, an online pilot information channel branding itself as blending "Academic Rigor" with "Journalistic Flair," Pederson and Hessl heralded their take-home message with declarative certainty under the title "Wet Climate Helped Genghis Khan Conquer Asia."[43]

A second noticeable feature of the way this breaking news about Genghis Khan has spread across the internet, and one that also fuels public interest in such stories, is the persistent way in which the narrative is used to inject fears for the future into the conversation. The original research report, of course, concluded with projections into the twenty-first century. Temperatures are anticipated to rise markedly in Central Asia, the authors briefly noted, with social, economic, and political consequences to follow. The reporting journalists fastened on this forecast to further dramatize their narratives. The National Science Foundation, for example, ran its press release under the title "Climate of Genghis Khan's Ancient Time Extends Long Shadow Over Asia of Today." The whole report was animated by the thought that while "the ramifications for past history are significant, so, too, are they for today's."[44] *Time*'s Bryan Walsh likewise wrapped up his piece with the menacing portent: "Climate change is still putting Mongolians on the move—but this time, there's no end in sight."[45] Further examples could easily be itemized. Two will suffice. Nathaniel Massey explained that the "tree ring study puts" recent climate "changes in troubling perspective": the most recent drought identified by the researchers, the one that persisted from 2002 to 2009, turns out to be "the hottest on record."[46] And Sarah Zielinksi, having noted what she calls the "new invasion" of rural migrants "flooding into Mongolia's capital" on account of harsher climatic conditions, rounds off with the ominous generalization: "Weather, it seems, can make invaders of us in one way or another."[47]

Third, it is worth recalling that climate-propelled explanations for the rise of the Mongol empire have been far from consistent. The Huntington–Toynbee scenario, for example, which caught our attention in the previous chapter and doubtless was symptomatic of what Diana Davis has called the "desert blame game," placed emphasis on the critical role of aridity in accounting for the surge of nomadic "destroyers and conquerors" from the steppes.[48] As Huntington had already insisted, it was "when a comparatively rapid increase of aridity was forcing the Mongols out of the mountains, the wandering nomads raided their neighbors in the villages of the plain so mercilessly as to drive away practically the whole population."[49] All of this runs directly counter to the current claim that increasing rainfall was the crucial cause of Genghis Khan's success. Evidently, different advocates of the climatic thesis attribute the same outcome to very different climatic conditions.

While the avalanche of publicity broadcasting the most recent climate-driven account of Genghis Khan's empire leaves little room for dissenting voices, Deborah Netburn, writing for the *Los Angeles Times,* did pause to wonder whether unusually warm wet weather caused, or merely coincided with, the expansion of the Mongol empire. To at least register the thought with her readers, she turned to the anthropologist Jack Weatherford, now living in Mongolia and well known for his book *Genghis Khan and the Making of the Modern World*. He told her that while the weather findings were undoubtedly intriguing, he was not convinced "that they had much to do with Genghis Khan's ascent to power." "I am skeptical," Weatherford observed, "of any single variable as the explanation for the Mongol rise."[50] Besides, he wondered why additional rainfall would have benefited Genghis Khan but not his Chinese enemies or other tribal groups in Mongolia at the time.

Netburn's reserve over the claim that rainfall was the major weapon in Genghis Khan's armory is conspicuous by its rarity. And the fact that so few journalistic reports have expressed qualms about reducing warfare to climatic circumstances demonstrates the degree to which many are content with naturalizing violence and conflict in the language of global warming, desertification, land degradation, sea-level rise, and the like.[51]

The Securitization of Climate

Given the plethora of studies linking climate and warfare in causal ways, it is little surprise that the U.S. military came to see its next major operation, according to Gwynne Dyer writing in 2010, as "the long struggle to maintain stability as climate change continually undermines it." Quoting from such high-ranking military figures as General Anthony Zinni of the U.S. Marine Corps and General Gordon Sullivan of the U.S. Army, Dyer writes, "What they are selling is a mission. The 'war on terror' has more or less had its day and besides, climate change is a real, full-spectrum challenge that may require everything from special forces to aircraft carriers. So it's time to jolt the rank and file of the officer corps out of their complacency, re-orient them towards the new threat, and get them moving."[52] Obviously impressed with the power of Dyer's analysis, CTV News Channel excerpted this diagnosis, and indeed the entire first chapter of his *Climate Wars*, for its readers.[53]

The securitization of climate has been developing for at least thirty years. A major world conference on "The Changing Atmosphere," which convened in Toronto in June 1988, placed climate firmly on the global security agenda. At the time, the Canadian political scientist and now president of the World Refugee and Migration Council, Fen Osler Hampson, commented on the event under the title "The Climate for War." Here he noted that the "conference underscored the need for governments to refine their national security and military spending priorities, and to address the geopolitical dimensions of climate change in resource allocation decisions."[54] Diminishing food security, environmental refugees, political instability, and resource conflicts were only some of the prognosticated effects of climate change compelling government strategists to turn militaristic.

A few years later, a now-famous article in the *Atlantic* by Robert Kaplan, an American journalist and later distinguished professor in national security at the U.S. Naval Academy, staged the environment as "a hostile power" and therefore as the "national-security issue of the early twenty-first century." While his focus was more on the environment in general than on climate in particular, his dismal prognoses further opened the space for ushering climate concerns into the security arena. Drawing on the neo-Malthusian analyses of Thomas Homer-Dixon,[55] whom he credited with officiating at the marriage of "military-conflict studies and the study of the physical environment," Kaplan insisted that "future wars and civil violence will often arise from scarcities of resources." Because to him "nature is coming back with a vengeance," it would necessarily have "incredible security implications."[56] With these predilections and his enthusiasm for the "geography-is-destiny" school of history, it comes as no surprise that Kaplan's later *Revenge of Geography* advocated a Halford Mackinder–inspired geographical fatalism intended to illuminate, as he put it in the subtitle, "What the Map Tells Us about the Coming Conflicts and the Battle against Fate."[57] Here his liking for environmental determinism, albeit hedged about with such adjectival modifiers as "hesitant," "partial," "probabilistic," and "quasi," dramatically materialized, as he pushed for what one reader called "the pivotal role of geography in global strategy and how to understand American interest in it."[58] The heart of the argument, in fact, had already seen the light of day in an article of the same name published in the *Foreign Policy* magazine a year or two earlier. Here, reprising Fernand Braudel's 1949 *The Mediterranean and the Mediterranean World in the Age of Philip II*, Alfred Thayer Mahan's *The Influence of Sea Power Upon History* (1890), Nicholas Spykman's ideas about geostrategic containment developed during the 1930s and '40s, and Halford Mackinder's famous article "The Geographical Pivot of History" published in 1904, he lauded "the wisdom of geographical determinism." "Such determinism," he avowed, "is easy to hate but hard to dismiss." And, given that in driving human history, human agents turned out to be "trivia compared with the deeper tectonic forces of geography," Kaplan plainly wanted this most materialist of geographies installed at the heart of America's security agenda.

As for the future, he concluded, "We all must learn to think like Victorians. . . . Geographical determinists must be seated at the same honored table as liberal humanists." Why? Because "denying the facts of geography only invites disasters that, in turn, make us victims of geography."[59] In that scenario, whatever its qualifications, it is clear that Kaplan's geography, to use his own vocabulary, decrees, destines, and dictates.

Since then, and with increasing intensity since the dawn of the twenty-first century, as political analysts and policymakers have plied the tools of their trade, the militarization of climate discourse has proceeded apace.[60] Not invariably, though. A 2002 report for the German Federal Ministry for the Environment, Nature Conservation and Nuclear Safety (BMU), authored very largely by Hans Günther Brauch, stressed that "climate change impacts will not be the single or main causes of future conflicts" and laid emphasis on cooperative strategies. Moreover, the report insisted that "climate change impacts do not pose a military threat nor can they be solved with the traditional mindsets nor by the means of military services."[61]

The prominence given in this document to conflict avoidance, partnership building, and mitigation is hardly typical, however. In October 2003, for example, Peter Schwarz and Doug Randall, two futurists, presented the U.S. Department of Defense with a bleak report on humanity's future. To them, climate change would usher in a grim world of internecine war, economic disaster, and social dislocation. As Maria Trombetta observes, this report for the Pentagon "increased the sense of urgency that contributed to the transformation of climate change into an existential, urgent threat, both because it focused on abrupt changes and because it was supposed to be confidential."[62] Declaring its purpose to be "imagining the unthinkable," the report intentionally set out "to dramatize the impact climate change could have on society" using as its point of departure the transformations climate change reportedly induced around eight millennia ago when global temperatures suddenly decreased. The "plausible scenario" they laid out included the prediction that by 2011, the United States would be engaged in serious disagreements with Canada and Mexico over water, that the world would witness a significant movement of Scandinavian populations southward by 2012 on account of severe drought and cold, that floods of refugees would be streaming into the southeastern United States from the Caribbean the same year, and so on. Skirmishes between France and Germany over the Rhine were forecast by 2022, civil wars in China by 2025, and increasing migration to the European Mediterranean zone from Algeria, Morocco, Egypt, and Israel by 2027. All of these and many more were laid out as "potential military implications of climate change" in a world where climate-caused disruption and conflict were confidently predicted to be "endemic features of life."[63]

Climate change was thus firmly located at the heart of national security.[64] Indeed, just a couple of years later in 2007, British Foreign Secretary Margaret Beckett declared that climate change went "to the very heart of the security agenda" in her presentation to the United Nations Security Council first-ever debate on the impact of climate change.[65] That same year, the Center for Naval Analyses (CNA) produced a report by a Military Advisory Board convened in 2006 to assess the impact global climate change would have on national security. It was composed of eleven retired three- and four-star admirals and generals. The report staged climate as a "threat multiplier" fueling instability in some of the world's most volatile regions. The entire project was premised on the assumption that changing climatic conditions "have the potential to disrupt our way of life and to force changes in the way we keep ourselves safe and secure." Given this concern with impacts on American lifestyles, it is perhaps not surprising that concerns about environmental conditions in Africa initially focused on the continent's "strategic value to the U.S. as a supplier of energy, oil, and strategic minerals such as chrome, platinum, and manganese." One general added, "Well, when we go looking for oil, we're really looking for trouble." Disruption arising from migration and the need for far tougher border security also featured strongly. And given the militarist thrust of the whole communiqué, its concerns over the impact of coming climate change on the American war machine were predictable. "Climate change will stress the U.S military by affecting weapon systems and platforms, bases, and military operations," it explained. Rising sea levels would challenge the readiness of military bases around the world, while energy supplies and the transport of fuel would be compromised by extreme weather events. As a consequence, political leaders and military chiefs could well "face hard choices about where to engage." As for tactics in the teeth of such anxieties, one thing is clear—as the report repeated a couple of times: "military leaders . . . don't see the range of possibilities as justification for inaction"—a policy ominously reminiscent of the "Draw-Fire-Aim" school of political strategy.[66]

The Age of Consequences, a Pentagon-orientated report prepared for the Centre for Strategic and International Studies and the Center for a New American Security in 2007, further dramatized the issue. Here the cultural fallout of climate change featured prominently. The aim was to lay out the security implications of three plausible climate change scenarios for American foreign policy, respectively dubbed "expected," "severe," and "catastrophic." Acknowledging that the way in which climate might shape future history remained contested, the report nonetheless announced a range of consequences in robust language. An example or two will illustrate. Expected changes, arising from a global average temperature increase of 1.3°C, were depicted as "not alarmist" but rather "to a large degree inescapable." For Europe, the focus was on immigration from Africa and South Asia. That meant Muslims. And their presence would "exacerbate existing tensions and increase the likelihood of radicalization among members of Europe's growing (and often poorly assimilated) Islamic communities." In consequence too, readers were told, "the viability of the EU's loose border controls will be called into question, and the lack of a common immigration policy will invariably lead to internal political tension." Among the severe outcomes of a 2.6°C rise in temperature was moral revolution. "Massive social upheaval will be accompanied by intense religious and ideological turmoil," the report declared. "Among traditional religious beliefs, the 'losers' are likely to be those faiths that have formed the closest association with the secular world and with scientific rationalism. . . . This intensified search for spiritual meaning will be all the more poignant under conditions of severe climate change." At 5.6°C increase, a range of ills would converge "in one conflagration: rage at government's inability to deal with the abrupt and unpredictable crises; religious fervour, perhaps even a dramatic rise in millennial end-of-days cults; hostility and violence toward migrants and minority groups; . . . and intra- and interstate conflict over resources." In the light of these eschatological prognostications, the overall conclusion reiterated the conviction that the security repercussions of climate change were a "greater foreign policy and national security challenge" than "reversing the decline in America's global standing," "rebuilding the nation's armed forces," and "finding a responsible way out from Iraq while maintaining American influence in the wider region."[67]

Since then, and especially in the wake of the Fourth Assessment Report of the Intergovernmental Panel on Climate Change (IPCC) in 2007, pronouncements on the security implications of climate change have mushroomed.[68] And indeed, the IPCC's later 2014 report devoted a whole chapter to the subject, concluding that "climate change and climate variability pose risks to various dimensions of human security."[69] In the meantime, the Center for Climate and Security, which provides weekly updates on "the security risks of climate change," was founded in 2010. With an advisory board largely composed of senior military personnel, its declared mission is to recognize "that climate threats to security are significant and unprecedented" and thus to act "to address those threats in a manner that is commensurate to their scale, consequence and probability."[70] And the following year, when Chris Huhne took up the subject in April 2011 at the Royal United Services Institute, the *Telegraph*'s political correspondent, James Kirkup, summarized his message for the public with the alarming strapline: "Climate change will threaten Britain's national security by causing wars, mass migration and food shortages."[71]

The stakes could scarcely be higher. Not unexpectedly, one inquiry into media reporting on the theme found that over a fifteen-year period, advisory bodies, think tanks, and nongovernmental organizations at both national and international levels had significantly intensified commentary on climate as a national security issue. This study, which covered over 100,000 newspaper articles, however, also revealed diverging international trends. While Western industrialized nations like the United States, the United Kingdom, and Australia had increasingly focused on the securitization of climate change, in the emerging economies of India and South Africa, discussion of the national security implications of climate change had markedly declined and were far less prominently advertised than concerns over food and water. One implication of this project's findings, the authors commented, was that the dramatic increase in the securitization language used in Western media reportage "could be used to justify security experts preparing for the coming threats and engaging in long-term military planning in security-related disasters and violent conflict possibly turning talk of 'climate wars' into a self-fulfilling prophecy."[72]

Dissenting Voices

The fashion for reducing war to climate has plainly reemerged with renewed vigor in recent years, stimulated in part by the proclivities of funding agencies and the priorities of national governments.[73] Once the preserve of classical thinkers, Enlightenment philosophers, and geo-historians, "it is now seducing"—in Mike Hulme's words—"those hard-nosed and most unsentimental of people . . . the military and their advisors."[74] But this reductionist impulse has certainly not met with universal approval. Over the past decade or so, critical commentary has intensified. A team of research ecologists based mostly at Colorado State University, for example, has challenged the suggestion that warming increased the risk of civil war in Africa, arguing that attributing such causal powers to climate "oversimplifies systems affected by many geopolitical and social factors." They pointed out that a range of unrelated geopolitical trends—most notably decolonization and the vicissitudes of the Cold War—which "perturbed the political and social landscape of the African continent," tended to be ignored in climate reductionist agendas. What was even more disturbing to them were the political consequences. In climate determinist scenarios, they urged, "Africa is predestined to additional strife related to global warming," and the result could well be to "discourage the kind of meaningful engagement that is so important for political and economic stability, economic development, and peace in Africa."[75] To be sure, authors finding climatic conditions at the headwaters of violent conflict do, from time to time, issue cautionary asides intended to mute absolutist readings of their work. Burke and his colleagues, at whom this particular critique was aimed, for example, claimed that they did "not argue that temperature is the only . . . determinant of civil war."[76] For all that, the eye-catching titles and energetic headlines under which such research is often presented give every impression of one-dimensional historical causation.

Halvard Buhaug, a political scientist at the Peace Research Institute Oslo, shares this same skepticism about climate's supremacy. Reworking a range of models used by advocates of climate's determining role in civil war, he contended, in an article for the National Academy of Sciences, that "climate variability is a poor predictor of armed conflict" and that civil wars in Africa are far better explained by such conditions as "prevalent ethno-political exclusion, poor national economy, and the collapse of the Cold War system." The prehistory of a particular violent episode is relevant too for, he argued, a recent history of violence in a region could well "affect the likelihood of a new conflict breaking out." In fact, he went so far as to claim that his data showed that since 2002, "civil war incidence and severity in Africa have decreased further while the warming and drying of the continent have persisted." What feeds into such critiques are such basic issues as just what passes as "civil war," what variables are considered to fall within the explanatory frame, which parameters and controls are appropriate, the spatial scope of aggregate data and the significance of local anomalies, and the different ways in which independent variables can be modeled. Buhaug's conclusion was clear: "The simple fact is this: climate characteristics and variability are unrelated to short-term variations in civil war risk in Sub-Saharan Africa."[77]

Several other critics based, like Buhaug, at the Peace Research Institute Oslo have added their voices in support. In introducing a special issue of the *Journal of Peace Research*, Nils Petter Gleditsch raised a range of queries that cast doubts on any tidy reduction of warfare to climatic conditions. For a start, he pointed to the distinction between climate and weather. In many such narratives, he observed, the findings are based on time scales that are much too short to satisfy the IPCC's definition of "climate as 'average weather,' usually over a 30-year period." He was concerned too to "disentangle the causal chains" that attribute war to climatic forces. On closer inspection, tensions quickly emerge. Consider rainfall. According to some, incidences of conflict and violence rise during wetter seasons, while others insist that war is more likely to break out during anomalously dry periods. What is often overlooked too in climate determinist scenarios about war is that "the largest killer in the 20th century . . . was one-sided violence," notably genocide. The geography of knowledge doubtless plays a role in these portrayals.[78] A large proportion of empirical work linking climate to war in recent decades has had Africa as its focus of concern, whereas the "bloodiest wars in the second half of the 20th century occurred in East and Southeast Asia" where "by the turn of the century there were fewer conflicts in these areas and those that remained were at much lower levels of severity."[79]

Insights from disaster sociology have added force to these reservations as Gleditsch's colleague, Rune T. Slettebak, explained in his contribution to the same journal. For having inspected research on postdisaster behavior, he argued that, if anything, the relationship is "opposite to common perceptions: Countries that are affected by climate-related natural disasters face a lower risk of civil war." The reason was that sociological investigations of human responses to climatic catastrophes had reported evidence showing that such natural disasters could induce a greater sense of communal solidarity and a growing concern about the welfare of others. Besides, how climate-driven calamities were handled by leaders turned out to be critical: competent management could increase support for a government rather than igniting a struggle to overthrow it. Far more significant predictors of civil wars than the brute occurrence of a weather-related tragedy were such mediating factors as population size, semiauthoritarian state governance, and poverty. Findings such as these were confirmed by Slettebak's own modeling exercises using variables from the Armed Conflict Dataset managed by the Uppsala Conflict Data Program and the Peace Research Institute Oslo. But he also added that ethnic factionalism and whether or not the country in question happened to be an oil exporter were factors of major significance in explaining battle-related deaths. The complications such things added to the climate reductionist story of civil war were far from trivial. As Slettebak observed, "There is a serious risk of misguided policy to prevent civil conflict if the assumption that disasters have a significant effect on war is allowed to overshadow more important causes."[80]

Investigations of particular regions, emanating from the same research institute, were intended to foster further skepticism about simply reducing conflict to climate. Consider the case of Kenya. Ole Magnus Theisen set out to test the neo-Malthusian thesis that resource scarcity, stimulated by the vicissitudes of the weather, explained the advent of civil war and the outbreak of intergroup violence. That thesis, he noted, fed a popular "crisis narrative" about sub-Saharan Africa that depicted the continent as desperately poor, miserably resourced, and prone to violence provoked by periods of drought. Kenya, he urged, provided a useful test case, since it had been home to some of the bloodiest civil conflicts in the recent history of the entire continent. Focusing on the period from 1989 to 2004, and using geographically disaggregated data on armed conflicts, he came to the arresting conclusion that "climatic factors do influence the risk of conflicts and violent events" but that "the effect is opposite to what should be expected from much of the international relations literature." Dry conditions, it became clear, far from inciting violence, actually had "a peaceful effect"—"a temporary cooling effect on tensions." In certain areas, for example, raiding was much less likely to occur during periods of drought; at such times of stress and hardship, cooperation and intergroup reconciliation were far more common. What was markedly more significant in fueling hostilities were ethnic antipathies, the advent of a recent election, and the hope of "political gain," not "desperate scrambles for scarce land, pasture and water resources."[81]

In a separate inquiry in which he was joined by Helge Holtermann and Halvard Buhaug, Theisen further shored up his argument. The claim that "climate change is a 'threat multiplier' for instability and conflict" was widely presumed. But, they asked, was it based "on solid scientific evidence"? Their answer was crisp: "Not so." In their view, pronouncements on the "security implications of climate change" had "run far ahead of the scientific evidence." Using a high-resolution data set for Africa between 1960 and 2004, they concluded that there was "little evidence of a drought-violence connection." Far more significant drivers of civil violence, they claimed, were political exclusion, asymmetric social structures, and economic vulnerability. In sum, they were sure that their findings countered the prevailing notion that critical water shortages increased the local risk of violent conflict. "Instead," they continued, "the location of civil war in Africa can be explained by generic political, socioeconomic, and geographic factors: a politically marginalized local population, high local population density, the country capital, proximity to an international border, and high infant mortality."[82]

Much of this line of reasoning had already appeared in earlier work by Theisen. Assessing the neo-Malthusian argument that climate-induced eco-scarcity is a major determinant of civil conflict and revisiting the models used by advocates, he urged that "proximity to regime change, the level of development, population size and oil exports" were far more tightly correlated with civil unrest. Such predisposing factors sat alongside a range of other influences increasing the risk of violent conflict, including rural population density and youth bulges. In concluding his assessment, Theisen declared his agreement with Salehyan that in accounting for such violence, "exclusive emphasis on the role of resource scarcity tends to let governments off the hook."[83]

Empirical inquiries like these,[84] which challenge the assumption that climate and climate change are prime causes of violence, raise troubling concerns about the ease with which an ideology of climate reductionism has infiltrated its way into national security consciousness. Critics of this determinist turn, particularly of the presumption that increased environmental scarcity and migration "weaken states" and "cause conflicts and violence," continue to express grave concerns about the lack of attention devoted to ascertaining "the ways that environmental violence reflects or masks other forms of social struggle." For one thing, such scenarios take outbreaks of violence as merely the natural consequence of social-evolutionary adaptation. Climate reductionism thus facilitates the sense that war can be readily "naturalized and depoliticized," with local culture and social relations relegated to the sidelines.[85] In the case of Darfur, it has been said, "blaming the violence . . . on climate change also means overlooking the destructive role of the Sudanese government."[86] Similarly challenging the 2007 claims of the United Nations Secretary General Ban Ki-Moon that the Darfur conflict began as an ecological crisis, Erin Owain and Mark Maslin have cast doubts on the climate–conflict thesis, drawing on climatic, economic, and political data they had amassed for ten East African countries. They concluded that climate variations, notably local drought and global temperature, did not exert a significant impact on the level of regional conflict or on the number of displaced people. Instead, they argued, the major forces fostering conflict were population increase, negative economic growth, and the relative instability of political regimes. None of this was to deny that changed climatic conditions, particularly in fragile ecosystems, could well exacerbate violent unrest, but the authors were emphatic in their insistence that the failure of political systems, rather than climate change per se, was the primary cause of conflict.[87]

The credibility of the prevailing neo-Malthusian mind-set has also come under critical scrutiny. Working through the ways in which armed conflict is purported to result from changes in precipitation, temperature, and sea level, some have concluded that the evidence for a climate–conflict link are very far from established. This, it has been suggested, is not simply a scientific defect. It is more serious than that. Neo-Malthusian explanations, it is claimed, could well "lead peacemaking astray" not least because the "climate-conflict discourse can . . . be exploited by actors who want to escape responsibility."[88] Cultivating fear, anxiety, and insecurity may help to grab the headlines, but, as Luis Alves writes, such tactics can have "negative consequences, undermining the role of civilian institutions in the search for democratic and sustainable solutions."[89] In comparable vein, the online title of a 2018 editorial in *Nature* charged its readers, "Don't Jump to Conclusions about Climate Change and Civil Conflict." Climate change, the article reported, "is never the sole cause of war, violence, unrest or migration. Syria and Jordan have been stricken by drought this decade. But it's clear that different social, political and economic factors in the two nations explain why people are desperate to flee from Syria and not from Jordan."[90] Resorting to simplistic climatic causation, the author further added, can have the effect of undermining peacekeeping efforts.

Acutely aware of the perils of a resurgent climatic fatalism, John O'Loughlin, Andrew Linke, and their colleagues, already known for their claims that, unlike cooler temperatures, warmer than normal temperatures increase the risk of violence notably in East Africa, have also joined the chorus of dissent.[91] Political disquiet dominated their horizon of concern. Some students of environmental security, they reported, were "in danger of promulgating a modern form of environmental determinism by suggesting that climate conditions directly and dominantly influence the propensity for violence among individuals, communities and states." Such "depoliticized analyses" had the effect of removing violence from its "local, social and political contexts," thereby reducing violence to "an immediate and unmediated function of physical, biological and physical-geographical signals." When analysts ignore the role of politics in the creation of environments that magnify risks for poorer, marginalized communities, they fail to take into account "the complex political calculus of governance" and the remarkable ways in which human societies actually do cope with challenging environments. In so doing, they reached conclusions that were "little different from those ascribing poverty to latitudinal location or lessened individual productivity to hot climates, as was common in European and American scholarship about a century ago." Such proclivities were "most troubling" when they found expression among environmental security agencies who "presume that individuals and communities cannot engage in positive coping behaviour to attenuate climate risks."[92]

Doubts have also been cast on the adequacy of the databases that have been employed to provide evidence for the climate–conflict thesis. And at the same time, there have been calls for much greater precision in defining what passes as violent conflict in a good deal of that literature. Having issued warnings about such matters, a group of researchers from the University of Hamburg, for instance, has also spoken of the need to explore the "multiple pathways and feedbacks" among climate, resources, security, and stability and conclude their diagnosis with the following observation: "If the debate on the securitization of climate change provokes military responses and other extraordinary measures, this could reinforce the likelihood of violent conflict."[93] Interventions in fragile states, tougher border security, and a scramble for access to resources could themselves become sources of violent conflict at various scales.

Given these perceived drawbacks, some have suspected that the motivation toward tracing conflict to climate change may lie well beyond the empirical state of nature's affairs. In Jon Barnett's judgment, for instance, much recent writing on environmental conflict "is more theoretically driven, and motivated by Northern theoretical and strategic interests rather than informed by solid empirical research." And the consequences spill far beyond academic discussion. To him, what is particularly troubling is the degree to which casting the question of climate change into the arena of national security risks "making it a military rather than a foreign policy problem and a sovereignty rather than global commons problem"—a particular irony too because, as he did not hesitate to point out, the world's militaries "are major emitters of greenhouse gases."[94] Moreover, as he points out elsewhere, much of this literature trades on "the ethnographic assumption that people in the South will resort to violence in times of resource scarcity," which perpetuates a long tradition of "scripting of people from the South as barbaric."[95] What is also obvious is the inclination to cast the environment as a hostile power and thereby to naturalize resource scarcity and obscure the role played by social institutions.[96]

These wider concerns characterize other commentaries also claiming a too facile resort to climate as the explanation for war. Take the set of reflections introducing a special issue of the journal *Climatic Change* dedicated to "Climate and Security." Across the intellectual and political landscape as they sketched it, several troubling features were brought into focus. Among these was the claim that the whole question of asymmetrical power relations in the study of the security dimensions of climate change remained conspicuous by their absence. What they called "deterministic mechanisms" routinely assumed to operate between climate change and national security, as well as the "apocalyptic visions" that typically accompanied them, were just too one-dimensional to do serious explanatory work. Philosophically, correlation had been too easily misconstrued as causation and explanatory pluralism too readily sacrificed to a monistic reductionism. Empirically, to these authors, the seemingly obvious links between climate, famine, and migration had been undermined once and for all by Amartya Sen, who had shown that "famines have political roots and are overwhelmingly caused by failures of entitlement to food and resources" rather than by absolute scarcity.[97] Missing too were the possibilities for peacemaking and social cooperation that climate change could sometimes afford. To put it another way, the rhetoric of threats and risks had triumphed far too fully over human capabilities and new institutional possibilities.

Besides these particular critiques, there have been a number of interventions challenging the entire methodological apparatus deployed by students of climate change and conflict in their attempt to provide robust statistical evidence for the causal links between weather and war. First, the 2018 editorial in *Nature*, to which I have already referred, alluded to problems of sampling bias, "including a statistically and politically dubious focus on mainly African countries formerly under British colonial rule." As the writer saw it, the eyes of researchers had been too willingly drawn to easily accessible regions with a violent history rather than to locations with the most severe experiences of climate change. One consequence of these sampling problems has been to further "stigmatize troubled countries as being prone to even more instability in the future."[98] The resulting traducement, which depicted certain locations as naturally violent and chronically disordered, it was suggested, could well "contribute to the re-production of colonial stereotypes."[99]

At much greater length and with more analytical precision, Jan Selby, an international relations expert, subjected what he called "positivist climate conflict research" to critical scrutiny, arguing that the entire research program is flawed in three fundamental ways. First, correlations between the variables are never unproblematic for the simple reason that they are always subject to coding conventions and causal assumptions ranging from the arbitrary to the untenable. Second, even if the correlations were persuasive, they remained intrinsically incapable of supporting sound predictions. Third, the research program *in toto* "reflects and reproduces an ensemble of Northern stereotypes, ideologies and policy agendas." Accordingly, Selby found that the climate–

conflict research tradition routinely perpetuated a neo-Malthusian outlook that he deemed fundamentally flawed. Along the way, he paused to highlight a range of incoherences and inconsistencies in the published output. Some find that higher temperatures are linked to conflict; others that lower temperatures stimulated civil unrest. It was similar with rainfall data. Conflict has been attributed both to high and to low measurements. Indeed, some researchers had noted that higher rainfall was associated with reduced conflict. With the effects of El Niño and climate-related disasters, similar tensions are notable. What was also conspicuous, moreover, was that even when the data were interpreted as pointing in different directions, the researchers operated, even when they affect to distain it, with a "deterministic understanding" of the connections between "climate and conflict." That mind-set Selby judged to be wrongheaded, for the simple reason that environment, weather, and climate are themselves objects of "purposive social action—willed activity that is oriented variously to transforming the environment, to adapting to it, or to exploiting opportunities by it." All this he found particularly troubling because the "more alarming and incautious findings" of champions of the climate–conflict thesis frequently enjoyed extensive media coverage.[100]

No doubt other observers could be called upon to give voice to similar dissenting commentary. But multiplying testimonials is not my aim. Suffice it to note the conclusions from an analysis, reported in the pages of *Nature,* of the views of some fourteen experts on the subject of the connections between climate change and armed conflict within states. The reporting article summarized the collective judgments of these experts, noting that while they agreed that climate had indeed "affected organized armed conflict within countries," they were of the opinion that "other drivers, such as low socioeconomic development and low capabilities of the state," were "substantially more influential."[101]

Despite its perhaps unfortunate title, unfortunate because it might seem to further pathologize a geographical zone already burdened by a surfeit of European denigration, Christian Parenti's *Tropic of Chaos,* published in 2011, provides a challenging, if chilling, exploration of what he calls "Climate Change and the New Geography of Violence." Here, while fully acknowledging the role that climate change may play in exacerbating or triggering conflict, he strenuously resisted what he saw as an "exclusive tribalism" that is emerging as "the only solidarity forthcoming in response to climate change." "This is not 'natural' and inevitable," he went on, "but rather the result of a history." The inclination to militarize climate change has induced a sense that the "politics of the armed lifeboat" is the only viable form of response. To adapt one of

Parenti's telling observations: planning too diligently for war can preclude planning sufficiently for peace.[102]

Connecting climate and conflict is no new enthusiasm. With roots stretching back to classical times, the association between weather and warfare has routinely attracted the gaze of devotees. In recent decades, in large measure in response to the all too real challenges of anthropogenic climate change, the idea has reincarnated itself with renewed vigor. In these circumstances, climate change has found itself firmly lodged on the agendas of national security agencies. Supporting this turn of events is a burgeoning research industry intent on disclosing the ways in which violent conflict is determined, in one way or another, by climate change. Sometimes observers are content to rest with demonstrating a correlation between changed atmospheric conditions and the incidence of strife. On other occasions, some particular causal chain is identified as the pathway from climate change to warfare and violence. It might be climate-induced resource scarcity, the effects of El Niño oscillations, or the consequences of ecologically forced migration. Sometimes the limelight falls on inadequate water supply or climate-related health problems as the triggering cause of hostilities. Among researchers stressing the direct effects of temperature change, differing stances continue to be adopted. For some, it is a lowering of the temperature that is thought to be the catalyst for the outbreak of hostilities, while for others, the aggravating cause of aggression is identified as warmer conditions. Either way, contradictory claims are bolstered by elaborate sets of statistics. In the latter case, some still resort to the malign influence of higher temperatures on the human psyche inducing stresses of one sort or another.

That claims of this sort should capture media headlines is understandable, and all the more so as researchers, often considered ivory tower academics, actively seek such public exposure. The boundary between scholarly caution and caption drama is frequently breached as the causal connections between climate and conflict are short-circuited. At least in part, the pressure on universities to display social relevance in their research work has doubtless a role to play in these most recent maneuvers. Attention-grabbing news banners routinely give their readership every impression that human history has been determined by the vagaries of the climate and that, for the future, our species remains in the grip of climate's iron will.

This powerful coalition of academic, political, and security interests is not without its critics, of course. They have been much less persuaded by what they see as the hasty habit of resorting to climate reductionism in accounting for war and indeed by the seemingly insatiable appetite of the media for the ominous op-ed and menacing headline. These detractors have expressed their disquiet along several different lines of dissent. Some remain resolutely empirical, pointing to situations such as Jordan, where climatic conditions comparable to those experienced in Syria have not induced civil conflict. To them, members of the climate–conflict school of thought have simply not paid careful enough attention to their counterfactuals. Others have dwelt more centrally on methodological issues such as systemic sampling problems or the inclination to mistake correlation for causation. Yet others have been troubled by the political implications of climate-change determinism, notably, disabling the development of effective mitigation strategies and stereotyping people and places as inescapably mired in conflict and violence. Still others are disturbed by the lack of attention to places where climate change has been an occasion for peacebuilding and thus to lessons that might be learned from situations where potential conflict has been transformed into peaceful cooperation.

In one way or another, reducing war to the state of the weather turns out to be a moral enterprise. For if indeed conflict can simply be attributed to the state of the atmosphere, then—as John William Draper realized a century and a half ago—humanity is well-nigh absolved of the responsibility of seeking political solutions to climatic challenges. That outcome, of course, flies in the face of a recent comment by a leader writer for the *Economist* intent on instructing readers on "How to Think about Global Warming and War": "No conflict occurs without leaders to give orders and soldiers to pull triggers. No atrocities are committed unless human beings choose to commit them."[103] In that context, it is not inappropriate to recall the observation of Thomas Hobbes in the mid-seventeenth century who made it clear that the impetus for warfare mostly lay in the "dispositions" of the human will:

> WARRE, consisteth not in Battell onely, or the act of fighting; but in a tract of time, wherein the Will to contend by Battell is sufficiently known: and therefore the notion of Time, is to be considered in the nature of Warre; as it is in the nature of Weather. For as the nature of Foule weather, lyeth not in a showre or two of rain; but in an inclination thereto of many dayes together: So the nature of War, consisteth not in actuall fighting; but in the known disposition thereto, during all the time there is no assurance to the contrary. All other time is PEACE.[104]

Conclusion

12

Immortal Bird

REFLECTING ON the troublesome state of his own discipline in 1960, Oskar Spate, writing from his position as foundation professor of geography at the Australian National University, paused to comment that his academic tribe had long been "assured that determinism is as dead as the dodo." If so, he mused, it seemed "singularly tough for an extinct bird." Instead, he suspected that environmental determinism was more accurately described as "an Immortal Bird, not born for death." "How often," he went on, "has its Positively Final Appearance on Any Stage been billed, how many self-appointed exorcists have laid the ghost to their own satisfaction! And yet it still walks abroad, a lively spirit."[1] The journey we have been on through what I have called "the empire of climate" bears witness to the force of Spate's shrewd prognosis. For the ancient idea that climate controls or conditions, determines or directs, human life in a myriad ways has resurrected itself in new incarnations in our own Anthropocene-conscious era.

The Anthropocene Alter Ego

There are, to be sure, discontinuities between then and now in the histories I have sought to relate. Dipesh Chakrabarty, for instance, is doubtless right to dwell on the rupture introduced into human self-understanding by the revelation that our species is a potent geological agent. The very idea of the Anthropocene has rendered fragile, if it has not entirely fractured, the traditional distinction between human history and natural history. As he puts it, revisioning *Homo sapiens* as an aggressive environmental force "severely qualifies humanist histories of modernity/globalization." There is certainly much to these claims.[2] But I have been emphasizing the other side of the equation—namely, the continuing role that climate, even if changed through collective human activity, is believed to exert on a renegade humanity and its determining agency in shaping history and society. Here it is the affinities, rather than the breaches, with the past that I find conspicuous. The view that humanity is "a prisoner of climate," as Alfred Crosby put it quoting Braudel nearly thirty years ago, seems to me to still retain significant cachet in our own time.[3] To put it another way, climate determinism is often the flip side of human-induced climate change, the *alter ego* of the Anthropocene.

The real and present danger that now confronts humanity has prompted many to warn of how climate change will transform, if not devastate, human society. Indeed, global warming is now regularly presented as the single greatest challenge facing the human race. Sir David King, formerly the U.K. government's chief scientific adviser, for example, considered climate change to be "the most severe problem that we are facing today—more serious even than the threat of terrorism."[4] The assumption behind this warning, and a myriad comparable *pronunciamentos* in newspaper headlines, scholarly articles, and political commentary alike, is that in some profound and menacing way, climate change will, at the very least, radically redraw the map of civilization or, at its worst, bring about the extinction of the human race altogether. Sometimes, it has to be said, these forecasts are presented with astonishing prophetic precision.

The devastating, albeit unintended, effects that our species is having on climate thus runs in tandem with a deep fear about the influence a changed climate will exert on human society. Indeed, some have resorted to the language of "revenge" to underscore the impact future climate will have on the human race or to a meteorological theology of "climate sins" in need of "confession," "repentance," and "atonement."[5] Whether or not the choice of such language is taken as metaphorical overkill, the sentiment that changing weather conditions are exacting their toll on human health and culture is entirely in keeping with the rule that the empire of climate has long been believed to exercise over human life. From Noah's flood in the Hebrew Bible to James Lovelock's *The Revenge of Gaia*, extreme weather has been regarded as an instrument of retribution. There are other continuities too. Mike Hulme, for instance, describes the preoccupation with climate security in some quarters as an expression of what he calls "the new determinism." For he detects echoes of "the old deterministic" way of thinking among those who portray climate change as "a driver for conflict" through exacerbating border disputes, fueling mass migration, and intensifying social trauma and humanitarian calamities. To him, it is a case of "old idea, new idiom."[6] Another climate scientist, Simon Donner, writing for the *Scientific American*, similarly senses that the "ugly history of climate determinism is still evident today." To him, "Rhetoric about climate refugees robs people of agency in their own future" and perpetuates the long and altogether nasty history of climatic determinism.[7] In yet another key, Timothy Sweet has observed that the "openness of the human body to the environment posited by climate determinisms resonate with recent eco-materialist accounts of matter and agency that understand the body as porous and agentially dispersed." At the same time, the hierarchical political economy of the zones imagined by earlier advocates of latitudinal determinism "persists in the conceptualization of history in terms of a Global North and Global South, a conceptualization that is inextricably bound up with questions of race and demographics."[8] Meanwhile, Edward Gibbon's late eighteenth-century reading of the decline and fall of the Roman Empire, haunted as it is with the spirit of Montesquieu, continues to be invoked for moral resources to face the threat of climate-driven decline in our own time.[9]

When the London *Times* reported in 2001 that "months of record rainfall are to blame for rats, divorce, political plots, overweight women and dead bees, according to the French," the underlying sentiment was entirely in keeping with the ways in which scholars have attributed to climate forcing, everything from the practice of witch hunting to patterns of economic growth, from senses of national identity to the quality of human resources, from dramatic increases in the number of nomadic refugees to catastrophic cultural collapse.[10] The list could be extended ad libitum: pestilence and poverty, anxiety and aggression all continue to be traced to the vicissitudes of climatic circumstance. All of these are in harmony with Montesquieu's "empire of climate" ruling over the fabric of human flesh, the geography of legislative governance, and the dynamics of social and political life alike.

In this book, I have sought to take the measure of this empire by inquiring into the intellectual architecture of the persistent conviction that climate exerts an ineluctable power over the human species, shaping everything from our deep evolutionary past to recent cultural history, from the quotidian affairs of everyday life to our mental and physical well-being. We have noted how the health consequences of exposure to particular climatic conditions, increasing now on account of climate change, have long attracted the attention of medical practitioners. We have explored how climate has been critically implicated in the politics of imperial control, labor power, and race relations, as well as inspecting something of the ways it has been used to explain market performance, national character, cultural collapse, and economic breakdown. We have remarked on how it has been denominated a critical factor in psychological disorders, patterns of suicide, and the prevalence of acute psychosis, as well as the ways it has been called upon to explain human evolutionary pathways and various adaptive features of human physiology. And we have examined some of the various ways in which climate and, more recently, climate change have been drafted in as explanations for civil war and internecine aggression.

In mapping something of this territory, I make no claim to comprehensive cover. There are doubtless distortions and silences on my map, as there are on every map.[11] Some vicinities remain *terrae incognitae* and in need of careful inspection; others are likely to require surveying at a different scale with a different legend. Nor have I presented a straightforward historical narrative that proceeds in a remorselessly chronological fashion. Rather, I have sought to plot a set of genealogies of the inclination to attribute causal agency to climate and to work between different time periods and different spatial arenas along a sequence of diverse, but related, axes. Because the complexities, consonances, and contradictions that have surfaced here defy simple explanation, I have resisted the temptation to explain the attractions of climate determinism by reference to any single cause. Critics will likely find this unsatisfactory. But I am reluctant to commit the mistake, all too ironic as it is, of providing a deterministic account of determinism by reducing climate reductionism to some particular set of social, philosophical, or scientific conditions.[12]

Cross-Cuts

The compass points I have employed to orientate my survey of the empire of climate were derived from four fixations that routinely absorb the attention of those concerned to catalogue the impact of rampant climate change on human well-being: health, mind, wealth, and war. My hope is that by elucidating the intellectual history of these preoccupations in the past, the lineage of their present-day incarnations may be illuminated. Viewing the empire of climate from these vantage points, moreover, has revealed a multitude of ways in which different strands of thought intertwine and interconnect, merge and mingle, at different times and in different settings.

A number of these cross-cutting trails continue to thread their way through the empire of climate. One theme that has persistently reasserted itself is the zonal or latitudinal philosophy of history and culture. This impulse has encouraged many to take up the project of finding a climatic explanation for the rise and fall of civilizations. Stretching back to the writings of Herodotus and Hippocrates, and later Ibn Khaldūn and Montesquieu, the resort to distance from the equator as an explanatory mechanism for economic growth, the flourishing of urban culture, and technological sophistication has persisted albeit in different registers. In pursuing this goal, some have found persuasive the sociology of Wittfogel-style hydraulic management as the mainspring of the scientific enterprise, while others have drawn scientific research on climate pulsations, desertification, soil salinization, and the effects of El Niño into the frame for determining the fate of civilizations. Linking past and future, it is noticeable how frequently investigations into the dramatic collapse of ancient advanced civilizations—notably the Akkadian, Egyptian, Mayan, and Moche declines—have provided resources for warnings about the future impact of climate change on contemporary society. At the same time, some distinguished economists have revivified a zonal theory of wealth in which latitudinal location performs the chief explanatory role.

Another train of thought that tracks its way through many of these different discourses is the influence of weather and climate on the human body. With roots in the tradition of the Hippocratic humors and further developed in the writings of Bodin and Montesquieu, the anatomical registration of climate's influence has continued to resurface in a wide range of arenas. The question of white acclimatization to different climatic regimes, together with the evolution of the tradition of medical topography and the emergence of distinct disciplines like tropical medicine, for example, kept questions about climate's influence on human flesh to the fore. Such preoccupations had a major influence too on a wide range of spheres of practical engagement, including medical therapeutics, the manufacture of dedicated types of clothing for particular climatic zones, the justification of slave labor, and the evolution of architectural design in what were regarded as sinister climates. Indeed, some of those pushing for the establishment of a new science of biometeorology in the early twentieth century reconceived of the human body as a suite of meteorotropic zones or centers that registered climatic stimuli in different ways. Some found in this way of thinking resources that could be marshaled in the cause of eugenics.

Others have sought to blend the findings of medical climatology with economics. Identifying the most suitable atmospheric conditions for worker efficiency and industrial productivity preoccupied some early twentieth-century researchers. More recently, the relationship between ambient temperature and the body's thermoregulatory system has encouraged efforts to reconstruct economics on a physiological foundation—physioeconomics, as it has come to be called. The anatomy of climate has also come to the fore in a rather different tradition of scientific inquiry—paleoanthropology. This field of research has recently witnessed a turn to periodic ice ages as the explanation for increased human brain size, bipedalism, encephalization, prolonged infancy, and much else besides. Among members of this academic community, climatic constraints on the success or failure of hominid lineages have attracted attention in the endeavor to establish how climate directed evolutionary pathways. Conspicuous too has been the revivification of climatic explanations for the disappearance of the Neanderthals. Perhaps not surprisingly, these projects have been self-consciously promoted as providing resources for facing up to the hazardous implications of the current global warming crisis and its future consequences.

A closely connected refrain that has echoed down the centuries is a fascination with identifying the influence of temperature on temperament and the ways in which moods, minds, and emotions may be affected by meteorology. In the aftermath of the Hippocratic insistence that different climates could induce wildness or serenity, pugnacity or passivity, and Montesquieu's development of a physiological theory of mind and mentality, the idea that individual and national dispositions mirrored atmospheric conditions flowed in many different directions. In some registers, it facilitated the typecasting of whole swathes of the globe as indolent or industrious, energetic or lethargic. Such bipolar judgments could and did encourage the naturalization of slavery, the politics of colonial trusteeship, and the mapping of what was often called "climatic energy."

In a more specifically medical lexicon, the influence that meteorological conditions were long believed to exert on mental life has reemerged in the identification of a range of weather-related psychological disorders. That quintessentially tropical malaise—neurasthenia—was seized upon to confirm the intimate connection between mood and meteorology and to reinforce the insurmountable difficulties that colonial settlers faced in hot climes. This syndrome was undergirded by a range of different scientific assertions. Some attributed it to the monotony arising from extreme heat, others to the influence of actinic rays, and yet others to the effects of atmospheric conditions on human reserve energy. In a comparable vein, the identification of conditions such as seasonal affective disorder perpetuate the earlier fascination of biometeorologists stressing the links between sunshine and psyche. More generally, seasonal variation, flooding, drought, high winds, and other extreme climatic events have been seen as the cause of mental distress and emotional turmoil expressing itself in bipolar disorders, mood swings, incidences of suicide, and various other psychoses. As for the benefits that more salubrious environments might instill, the planning of institutional venues for the treatment of mental illness has sometimes taken account of prevailing climatic conditions in the architecture of hospital provision, in identifying spaces for indoor and outdoor therapies, and in finding locations for psychiatric wards.

The specification of climatically induced nervous exhaustion and mental illness has fed into related concerns in a different sphere—criminology. Seasonal and diurnal patterns of homicide, assault, riot, and rape, as well as such predisposing behaviors as aggression and anger, have all been scrutinized with the aim of uncovering a kind of thermal theory of crime. When mapped at the global scale, some have urged that a distinctive climatic geography of criminality comes clearly into view. Perhaps not surprisingly, the future implications of the heat-and-hostility school of criminal behavior in a rapidly warming world have not escaped the notice of champions of this theory. What has been called psychoeconomics has also recorded the influence that weather patterns are reported to have on mass psychology. Fluctuations in the stock market have been inspected to see if they correlate in any way with changing weather. The assumption here has been that market sentiment expresses the mass psychological state of investors at any given point in time. Deviation from normal weather, for instance, was proposed as a key factor in stockholder optimism or pessimism and consequently on share price. The psychological effects of changes in barometric pressure and variations in radiation have likewise been put forward as determining periodic fluctuations in commodity prices.

Since classical times, the perceived association of weather and war has likewise captivated many writers in one way or another. For some, the disposition toward warlike behavior reflected how climatic conditions were thought to shape human constitutions. In the ancient world, it was widely believed that colder climates bred bravery and bellicosity. Later it was the determining influence of high temperatures that captured the imagination of those seeking a climatic explanation for human aggression and violent behavior. For others, the chain of causes linking climate and conflict was channeled through the ways in which forms of agricultural production generated different regional cultures. In the United States, differences between the plantation economy of the Old South and an industrializing North were attributed to the different climatic conditions and geographical landscapes on either side of the Mason–Dixon line. This provided many with an inclination toward environmental determinism to find in these conditions the prime cause of the American Civil War. More recently, a neo-Malthusian struggle for scarce resources has been widely espoused as the cause of civil violence and full-scale war. Not surprisingly, in the light of dire warnings about future conflict and climate change, there have been calls for global warming to be incorporated within national security agendas. These claims have not gone unchallenged, however, and a considerable number of dissenting researchers have subjected the available statistical records of weather and war to scrutiny and found reason to cast doubts on the weather-and-war thesis. Others have baulked at the seeming fatalism of naturalizing warfare and contend that focusing on climate as the culprit masks underlying social and political forces. It is noteworthy, however, that the Sixth Assessment Report of the Intergovernmental Panel on Climate Change, which came out in 2022, has insisted that "non-climatic factors are the dominant drivers of existing intrastate violent conflicts." In some regions, the report continued, "extreme weather and climate events have had a small, adverse impact on their length, severity or frequency, but the statistical association is weak."[13]

Down through the generations, commentaries on climate have routinely been freighted with moral cargo of different kinds. The manifestation of moral sentiment in diverse meteorological settings has been expressed in rather different guises. Frequently, climates have been judged to be deleterious because of the way in which they supposedly induced undesirable behavior and immoral conduct among their populations. Such vices as laziness, cowardice, drunkenness, lasciviousness, promiscuity, indolence, and suchlike have routinely been attributed to denizens of the so-called torrid zone. And not surprisingly, during the colonial era, many expressed fears that the moral fiber of imperial officialdom would soon decay in hot climates. Indeed, moral probity was, from time to time, integrated into the very definition of successful acclimatization to a tropical regime. Accordingly, authors of medical texts for overseas travelers frequently included sermonic advice on moral hygiene, sexual discipline, and the inculcation of virtue among the children of the empire. In these ways, climate was long believed to have crafted a global geography of virtue and vice. In another register, blaming the climate for war, famine, violence, crime, and other social ills has had the politically pleasing effect of transferring moral responsibility from human agents to atmospheric conditions. As Mike Hulme has put it, "It may be politically 'tidy' and convenient to use climate in naturalistic explanations for such social disasters, but it has an eviscerating effect on political accountability"—especially in the face of compelling counternarratives.[14] All this means that climate or climate-change determinism can satisfy the desire to shift culpability from humanity to nature. Of course, anthropogenic climate change itself constitutes one of the greatest moral challenges in the world today as humanity stands guilty of bringing about climate change on an unprecedented scale. Just how the human race should take responsibility for its destructive actions and pursue more ecologically accountable behavior remains crucial to the ethics of twenty-first-century living. Meteorology, it seems, has always been a moral enterprise of one sort or another.

What our inquiry has also disclosed is something of the way in which climate all too frequently has been implicated in questions of racial difference. Explanations of human monogenesis—whether based on the Genesis narrative of Adam and Eve or advocating a more secular scientific account of race divergence—typically attributed racial variability to the influence of the sun. Among those attracted to the racialization of the climate were such prominent philosophers as Kant and Hume who, like many others in the eighteenth and nineteenth centuries, resorted to the thought forms of climatic determinism to sustain their views about racial hierarchy and senses of national identity. Among such figures, the moralization of global space was readily mobilized in the interests of political ideology. During the twentieth century, a scientized version of this same creed was called upon to explain the character of races and to inform eugenic practices and immigration policies. And notwithstanding protestations to the contrary, late twentieth- and early twenty-first-century writers have not always escaped the shadow of racial politics in the language to which they have resorted. The sense of hierarchy implicit in their commentaries on climates, cultures, and civilizations certainly gives every impression that in judging climates, they are no less engaged in judging people.

It surely goes without saying that the cross-cutting themes that I have just identified do not exhaust the intellectual stream networks that wend their way through the empire of climate. But they do demonstrate something of the continuities that have persisted over the centuries among those who find in climate the principal driver of human life and culture. Is it possible that attending to the stresses and strains inherent in these enduring strands of thought may throw a little light on the current ethical and emotional challenges of unbridled climate change and even provide a critical resource for cultivating responsible strategies for the future? Some have thought so. Dagomar Degroot, for instance, used his analysis of the Dutch republic during the Little Ice Age to uncover the complex and sometimes counterintuitive relationships between climate change and human activity in the early modern period. His findings, he insisted, were "at odds with the kind of determinism that can enter assessments of the present-day and projected impacts of global warming." And yet, he continued, "We can only learn how we might adapt to climate change by understanding how its influence filters through the socioeconomic, cultural, and political systems in which humans structure their lives."[15] Whether that outcome will prove to be the case remains to be seen. One thing is certain. For all the differences between the recent tempo of global warming during what has come to be called the Anthropocene and earlier rates of climate change, there remain remarkable continuities in the history of ideas about climate's influence on human populations.

Apocalypse, Now and Then

One appropriately final resonance between contemporary writing on climate change and earlier pronouncements on extreme weather lies in the resort to apocalyptic language to convey a sense of fear and alarm about the devastating consequences for humanity of runaway environmental disaster.[16] The intention of such rhetoric, of course, is to use dread, fright, perhaps panic, as motivational sentiments to promote changed behavior as humankind faces one of the most serious challenges to its very survival. In many ways, this form of declamation reprises a dominant theme in the political theology of climate of earlier centuries. The title of Slavoj Žižek's 2011 *Living in the End Times* identified worldwide ecological catastrophe as one of the four horsemen of the coming apocalypse. In an introduction styled "The Spiritual Wickedness in the Heavens," the Slovenian Marxist philosopher and cultural critic elaborated a whole series of contemporary woes that convinced him that "the global capitalist system is approaching an apocalyptic zero-point." Among the signs of the times heralding the troubling news of "Apocalypse at the Gates," Žižek identified the coming of the Anthropocene, which has presented humanity with such an "unthinkable catastrophe" that it has crippled human agency in confronting the ecological results of humanity's own actions.[17]

Apocalyptic-sounding announcements about the devastating effects of climate change have been commonplace for quite some time. On April 3, 2006, for example, *Time* magazine ran a special report on global warming under the front-cover headline "Be Worried. Be Very Worried" and exactly a year later produced a similarly menacing issue under the title "The Global Warming Survival Guide." This alarmist genre of writing has been growing in intensity since the 1980s, so much so, it has been suggested, that the Kyoto environmental summit in 1997 was anticipated in apocalyptic terms reminiscent of the biblical Book of Revelation. According to a couple of commentators, "John of Patmos intoning his visions of a Day of Judgment" could almost be heard in the run-up to the convention. In due course, moving into the twenty-first century, "apocalyptic readings of threat have become more strongly institutionalized, particularly among governmental policy elites," the same authors claim, and not least in the widely publicized Stern Review on the *Economics of Climate Change* that appeared in October 2006.[18] Since then, the leitmotif of imminent civilizational collapse has continued to attract the attention of distinguished authors producing volumes with titles such as *The Collapse of Western Civilization: A View from the Future* and *Learning to Die in the Anthropocene: Reflections on the End of a Civilization*.[19] Of course, apocalyptic environmentalism is not just a twenty-first-century phenomenon. Environmental classics such as Rachel Carson's *Silent Spring* (1962) and Paul Ehrlich's *The Population Bomb* (1968), in Greg Garrard's opinion, were already speaking in apocalyptic patois. The "horrifying apocalyptic projections" such works promoted, he suggested, offered little hope in the face of nature's tyrannical powers.[20] Even if the aim was to avert disaster through behavioral transformation, the deterministic rhetoric seemed to leave little space for human agency. The price of persuading the public of imminent disaster may be to engender hopelessness, and thus apathy, in the face of Fate.

By contrast, Amitav Ghosh judges official climate-change policy documents, notably the Paris Agreement negotiated at the 2015 United Nations Climate Change Conference, to be too lukewarm, too lackluster, to convey to the public the ruinous situation in which humanity finds itself. Much to be preferred, he believes, is the tone of Pope Francis' encyclical letter *Laudato Si'*, which does not hesitate to describe the coming predicament in decidedly more robust language. While the Paris Agreement, he writes, is "tepid in its naming of the conditions that it is intended to remedy," the encyclical letter unhesitatingly resorts to the vocabulary of catastrophe and disaster. In doing so, Ghosh remarks, it strove "to make sense of humanity's present predicament by mining the wisdom of a tradition that far predates the carbon economy." Ghosh obviously finds that *Laudato Si'* better evokes what has been dubbed our "catastrophozoic" era than the staid scientific recycling of "our dominant paradigms."[21]

What has further reinforced the sense of apocalyptic doom that is already greeting humanity in a climate-transformed world is a whole tradition of ecofiction that works hard to imagine a dystopian future.[22] Certainly, some contributors to the climate fiction oeuvre display a degree of optimism about the future, but many are more drawn to cataclysmic spectacles. The merest sampling must suffice. As early as the 1960s, and long before climate scientists had identified the threats of global warming, J. G. Ballard portrayed a postapocalyptic world where climate change had rendered most of the planet uninhabitable.[23] More recently, the deluge theme dramatically unfolds in Maggie Gee's *The Flood*, which presents readers with a vision of a world precariously balanced on the verge of environmental disaster and of human hubris exposed for the shallowness it is when faced with the unrestrained power of nature.[24] Ilija Trojanow's *The Lamentations of Zeno,* "a modern fable" that dramatizes "the high stakes of our current climate gamble," as Colum McCann puts it, has been described as "an apocalypse with elegy" and a "requiem for the future."[25]

The resonances here between contemporary climatic apocalypticism and classical Judeo-Christian eschatology have been noted by many commentators. Indeed, remarking that biblical eschatology was bound up with the idea that the Day of Judgment would be presaged by thunder, lightning, earthquake, famine, and pestilence, Stefan Skrimshire suggests that the experience of episodes such as the Little Ice Age from the end of the fourteenth century "became central to the success of the popular imagination of end-term belief in early modern Europe."[26] Structurally, Greg Garrard tracks environmental apocalypticism to early Christian millennialism and to the end-of-time theologies developed by what he calls "Mathematicians of the End Times." The spirit of apocalyptic eschatology, he suggests, continues to manifest itself in "the popular journalistic propensity to interpret every drought, every hurricane or ice storm as a crypto-Biblical 'sign' of catastrophic global warming."[27] So it is perhaps not surprising that, as Adam Trexler sees it, Gee's *The Flood* "reworks Judeo-Christian, individual judgment for the Anthropocene. Where Genesis imagines divine rage at moral sins, in the Anthropocene humanity condemns itself by miniscule trespasses on the atmosphere, accumulated over decades and centuries. *The Flood*'s final tsunami is both a punishment and an opportunity for humans to resume animality."[28] Another set of commentators has noticed how "movies such as *The Day After Tomorrow* draw on powerful visual metaphors" deeply rooted in "Christian apocalyptic symbols to narrate a story of apocalyptic climate change."[29]

In depicting the character of what they call "apocalyptic tragedy in global warming discourse," Foust and Murphy make much of the distinction between two strands of apocalypticism—the tragic and the comic.[30] While the former, the tragic variety, sees "global warming as a matter of cosmic Fate," the comic version—comic in the Aristotelian sense—acknowledges that evils arise from foible and folly and, accordingly, that "humans have a capacity to influence (within limits) the global warming narrative's end." To put it another way, the comic envisages time as open-ended and grants that change is possible, while the tragic closes down human agency in the face of inevitable doom. In both scenarios, scientists "are cast as prophets who interpret ambiguous signs and complex data which foretell a grim future."[31] Both fatalist eschatology and what I would call provisionally deferred apocalypticism are deeply rooted in

Christian theology. One draws on a premillennial vision to warn of the earth's inevitable ruin by cataclysm. The other finds hope in passages where the healing of the land is promised if people humble themselves and turn from their wicked ways. In recognition of such diversity, Stefan Skrimshire has noted that in the history of Christian apocalypticism, there "has never been a consensus or a monopoly on *how* this source of imagination was to direct attitudes and behaviors." To him, one of the attractions of apocalypticism as applied to the crisis of climate change "is precisely its ability to attract both pessimism *and* optimism with regard to human action"—a synthesis that at once curbs yet calls for human agency.[32] This is the "properly apocalyptic stance," the "enlightened catastrophism," that Žižek advocates.[33] And it chimes too with the attempts by Erik Swyngedouw, who argues that "the ecological Armageddon is already a reality" in many parts of the world, to recondition apocalyptic imaginaries for a radically anticapitalist politics.[34]

In her analysis of the Enlightenment origins of the idea of Nature's revenge, and of the contributions of the Italian physician and naturalist Antonio Vallisneri (1661–1730) in particular, Lydia Barnett observes that early modern theologies of climate change could profitably attract the scrutiny of environmental historians. In these early scenarios, she finds that, while the devastating effects of a changed climate were registered, human beings were granted "a relatively high degree of agency" to meet the challenge. Her analysis takes off from the narrative of the Noachian flood and its transformation of the globe, noting in particular the widespread belief that it induced health problems and shortened life spans. In Vallisneri's view, the reproductive damages brought on by flood conditions "transformed humankind from a race of nearly invincible giants to a group of stunted weaklings, newly vulnerable to death and disease." All of this reveals that, to these climatic theologians, human beings "played a dual role as both agents and victims of climate change."[35] To be sure, Enlightenment philosophical theology of climate is not what dominates today's horizon. But whatever the differences, Barnett senses that climate-change discussions in the twenty-first century likewise cast human beings in the Janus-faced roles of villain and victim, culprit and casualty, in their encounter with global warming.

Dialogue between contemporary and historical horizons of the apocalyptic imaginary has opened up illuminating affiliations and disjunctions in the long history of the political theology of climate. At the same time, it has provided rhetorical resources to reconceptualize how we might profitably think and speak and behave in a time of climatic cataclysm.[36] My hope is that the narrative I have provided may serve a similar purpose by affording some conversation partners from the past for thinking over, talking about, and acting in our own climatically distressing times.[37] These dialogues with the past may sometimes be encouraging, sometimes intriguing, more frequently disquieting. Some will reveal how the human spirit has been diminished; from time to time, hopeful hints of a better way may be glimpsed. Taken in the round, though, the history of climatic determinism with its aura of fatalism and fanaticism reveals just how crippling a force it can be on human agency, equality, and empowerment. So it could be said, no doubt with some justice, that a good deal of the intellectual landscape I have painted is cast in somber tones. But the dark spaces I have tried to mark on the map of "the empire of climate" may nonetheless provide signposts toward some of the toxic topographies that most need to be avoided. By shedding light on how climate has been implicated in the reproduction of racial ideology, the justification of slavery, the rationalizing of conflict, the nurturing of eugenics, and the psychological and cultural stereotyping of major zones of the globe, to name but a few, I dare to hope that *The Empire of Climate* may make it a little more difficult for a guilty humanity to stray into the habit of blaming the weather instead of undertaking to weather the blame.

NOTES

Preface

1. Lorraine Daston, "Nature's Revenge: A History of Risk, Responsibility, and Reasonableness," Inaugural Lecture, Humanitas Visiting Professorships at the Universities of Oxford and Cambridge, May 2013, https://podcasts.ox.ac.uk/inaugural-lecture-natures-revenge-history-risk-responsibility-and-reasonableness.

Chapter 1: A Matter of Degree

1. Montesquieu, *The Spirit of Laws* (1750), Book XIX, §14.

2. Montesquieu, *The Spirit of Laws* (1750), Book XXIV, §26. In 1851, the work was placed on the Index of Prohibited Books compiled by the Vatican's Congregation for the Doctrine of the Faith.

3. Ostrow, "'Six Degrees' Charts Climate Apocalypse."

4. *National Geographic*, "Six Degrees Could Change the World."

5. Lynas, *Six Degrees*, front matter.

6. Sir John Houghton, House of Lords, Economic Affairs, Minutes of Evidence, available online at https://publications.parliament.uk/pa/ld200506/ldselect/ldeconaf/12/5011804.htm.

7. Diamond, *Guns, Germs and Steel*.

8. Landes, *The Wealth and Poverty of Nations*, 470.

9. Fagan, *The Little Ice Age*, xv; Fagan, *The Great Warming*, 56; Fagan, *Floods, Famines, and Emperors*, 109.

10. Le Roy Ladurie, *Times of Feast, Times of Famine*, 7. This work was originally published in French in 1967. The foreword to the English translation was provided by environmental scientist Gordon Manley. Manley is the subject of a study by Endfield, "Reculturing and Particularizing Climate Discourses."

11. Degroot, *The Frigid Golden Age*, 308.

12. Something of these developments may be gleaned from Wigley et al., *Climate and History;* Walker and Lowe, "Quaternary Science 2007"; White et al., *The Palgrave Handbook of Climate History*.

13. For a critique of what he calls "dataism," see Han, *Psychopolitics*. See also the reflections of Brooks, "The Philosophy of Data."

14. Guldi and Armitage, *The History Manifesto*, 109.

15. Hulme observes, "Although climate determinism is easy to caricature and easy to dismiss as hopelessly naïve, there are more sophisticated variants circulating today," which he refers to as "climatic reductionism." Hulme, *Weathered*, 69.

16. Some of the complexities were identified in Lewthwaite, "Environmentalism and Determinism."

17. Walsh, *Scientists as Prophets*.

18. Fagan, *Floods, Famines*, 308.

19. Fagan, *The Long Summer*, 10.

20. Fagan, *The Great Warming*, dustjacket.

21. Behringer, *A Cultural History of Climate*, vii, viii.

22. Diamond, *Guns, Germs and Steel*, 17.

23. Blaut, "Environmentalism and Eurocentrism."

24. Sluyter, "Neo-Environmental Determinism," 815, 813.

25. Judkins et al., "Determinism within Human-Environmental Research," 23. See also the critique by Meyer and Guss, *Neo-Environmental Determinism*.

26. Hulme, "Reducing the Future to Climate," 249, 247, 248, 264. Elsewhere, along with Martin Mahony, Hulme has called for a greater awareness of the geographies of the knowledge politics of climate change. See Mahony and Hulme, "Epistemic Geographies of Climate Change."

27. Midgley, *The Myths We Live By*, 44, 47.

28. Smith and Howe, *Climate Change as Social Drama*, 53.

29. Kaplan, *The Revenge of Geography*, 45, 59.

30. Hulme, *Weathered*, xiii.

31. Something of this ongoing legacy may be gleaned from the special issue of the *Bulletin of the History of Medicine* on "Modern Airs, Waters and Places," edited by Alison Bashford and Sarah W. Tracy. See in particular Bashford and Tracy, "Introduction."

32. It has been argued that perceptions about salubrious climates were as much cultural as medical. See Carey, "Climate and History"; Janković, "The Last Resort."

33. For overviews of climate and colonialism, see Endfield and Randalls, "Climate and Empire"; Mahony and Endfield, "Climate and Colonialism."

34. Ghosh, *The Great Derangement*, 87.

35. For wider reflections on the zonal or latitudinal theory of climate and culture, as well as on climate and race, see LaFauci, "Climate Theories," and Parrish, "Climate and Race." An innovative critical history of the zones, mediated through the career of the pineapple as a tropical commodity and a "trophy of empire," is provided in Okihiro, *Pineapple Culture*.

36. For a sustained critique of naturalistic inevitability in the context of nineteenth-century famines, see Davis, *Late Victorian Holocausts*.

37. De Waal, "Is Climate Change the Culprit for Darfur?"

Chapter 2: Heirs of Hippocrates

1. Webster, "Curb Climate Global Warming or Face a Health Catastrophe."

2. Centers for Disease Control and Prevention, "Health Effects."

3. Met Office, "Climate Change and Healthcare."

4. Liggins, "Climate Change and Health."

5. World Health Organization, "Climate and Health."

6. U.S. Environmental Protection Agency (EPA), "Human Health Impacts and Adaptation."

7. Natural Resources Defense Council, "Extreme Weather."

8. The *Scientific American* reported on this in Morello, "Impacts of Climate Change Extend to Human Health."

9. Portier et al., *A Human Health Perspective On Climate Change*, 3.

10. Portier et al., *A Human Health Perspective On Climate Change*, 63.

11. For the merest sampling since 1999, see Epstein, "Climate and Health"; Patz et al., "The Potential Health Impacts of Climate Variability"; Patz et al., "Impact of Regional Climate Change on Human Health"; Haines et al., "Climate Change and Human Health"; Frumkin et al., "Climate Change."

12. McMichael et al., eds., *Climate Change and Human Health.*

13. Corvalán et al., "Conclusions and Recommendations," 271.

14. Haines et al., "Climate Change and Human Health," 585.

15. Frumkin et al., "Climate Change," 435.

16. McMichael, "Global Climate Change and Health," 8. The same lineage also appears in Diaz et al., "Climate and Human Health Linkages." See also McMichael and Woodruff, "Climate Change and Human Health.'

17. Rosenberg, "Epilogue: *Airs, Waters, Places,*" 661, 666.

18. Githeko and Woodward, "International Consensus on the Science of Climate and Health," 43.

19. Martens, "How Will Climate Change Affect Human Health?" 534.

20. Epstein and Ferber, *Changing Planet, Changing Health*, 3.

21. Epstein and Ferber, *Changing Planet, Changing Health*, 197, 273.

22. Castree, "A Climate of Ill Health," 418.

23. Thomas, *Under the Weather*, xiii, xiv, 190. See also Editorial, "Climate Change—The New Bioterrorism."

24. Sargent, *Hippocratic Heritage*, 51.

25. There are numerous surveys of Hippocratic environmentalism. See, for example, Hannaway, "Environment and Miasmata"; Boia, *The Weather in the Imagination*, 24–30; Marrum, "Hippocrates and the Hippocratic Tradition"; Porter, *The Greatest Benefit to Mankind*, 55–62; Glacken, *Traces on the Rhodian Shore*, chap. 2.

26. Hippocrates, *On Airs, Waters and Places*, part 23.

27. Porter, *The Greatest Benefit to Mankind*, 53.

28. Hippocrates, *On Airs, Waters and Places*, parts 3 and 4.

29. Marks, *The Aphorisms of Hippocrates*. See also Sargent, *Hippocratic Heritage*, 56.

30. Hippocrates, *On Airs, Waters and Places*, part 11.

31. The following brief account draws on Porter, *The Greatest Benefit to Mankind*, 60.

32. Aristotle, *Problems*, book 14. See also Martin, "Experience of the New World and Aristotelian Revisions."

33. Glacken, *Traces on the Rhodian Shore*, 99–100.

34. Nova, "The Role of the Winds in Architectural Theory."

35. Vitruvius, *The Architecture*, 14, 15. For a discussion of resonances between Vitruvius and Victor Olgyay, the twentieth century architect and champion of bioclimatic design, see Gelder, "Design with Climate in Ancient Rome."

36. Deming, *Science and Technology in World History*, 92–94.

37. Porter, *The Greatest Benefit to Mankind*, 118.

38. Something of the fortunes of Hippocratic medicine in the early modern period is traced in Wear, "Place, Health, and Disease."

39. Riley, *The Eighteenth-Century Campaign to Avoid Disease*, 1.

40. See Dewhurst, "Thomas Sydenham"; also Dewhurst, *Dr Thomas Sydenham*.

41. See Cook, "Sydenham."

42. Sydenham, *The Works of Thomas Sydenham*, vol. 1, 16, 11.

43. Porter, *The Greatest Benefit to Mankind*, 230.

44. Sydenham, *The Works of Thomas Sydenham*, vol. 1, 15, 75.

45. Sydenham, *The Works of Thomas Sydenham*, vol. 1, 34.

46. Sydenham, *The Works of Thomas Sydenham*, vol. 2, 138.

47. Sydenham, *The Works of Thomas Sydenham*, vol. 1, 29.

48. See DeLacy, *The Germ of an Idea*; Keele, "The Sydenham-Boyle Theory."

49. Dewhurst, "Thomas Sydenham," 113.

50. Ross, "Arbuthnot, John." See also the discussion of Arbuthnot in Glacken, *Traces on the Rhodian Shore*, 562–565.

51. Janković, *Confronting the Climate*, 19. See also Shuttleton, "'A Modest Examination': John Arbuthnot and the Scottish Newtonians."

52. Golinski, *British Weather and the Climate of Enlightenment*, 143.

53. Arbuthnot, *An Essay Concerning the Effects of Air*, vii.

54. For a discussion of the role of fiber theory in the period, see Ishizuka, "'Fibre Body': The Concept of Fibre in Eighteenth-Century Medicine."

55. Arbuthnot, *An Essay Concerning the Effects of Air*, 63, 134, 164.

56. Arbuthnot, *An Essay Concerning the Effects of Air*, 146.

57. Arbuthnot, *An Essay Concerning the Effects of Air*, 148, 149.

58. Arbuthnot, *An Essay Concerning the Effects of Air*, 151, 152.

59. See Golinski, *British Weather and the Climate of Enlightenment*, 175.

60. See the discussion in Fleming, *Historical Perspectives on Climate Change*, 12–16.

61. Golinski, *British Weather and the Climate of Enlightenment*, 80.

62. Lane, "Wintringham, Clifton."

63. Moore, "Huxham, John."

64. Huxham, "Observations on the Air and Epidemic Diseases," vol. 1, 1.

65. Satchell, "Rutty, John."

66. Rutty, *A Chronological History of the Weather and Seasons*, xiii.

67. Rutty, *A Chronological History of the Weather and Seasons*, xv.

68. Golinski, *British Weather and the Climate of Enlightenment*, 56.

69. See the discussion in Miller, "'Airs, Waters, Places' in History."

70. General reflections on the history of medical geography as an enterprise include Ackerknecht, "Changing Attitudes toward History and Geography of Disease"; Paul, "Approaches to Medical Geography"; Barkhuus, "Medical Geographies"; Rupke, ed., *Medical Geography in Historical Perspective*.

71. Chalmers, *An Account of the Weather and Diseases of South-Carolina*, vol. 1, 6.

72. Chalmers, *An Account of the Weather and Diseases of South-Carolina*, vol. 2, 1–2.

73. Currie, *Memoirs of the Yellow Fever Which Prevailed in Philadelphia*.

74. "Biographical Notice of the Late Dr. William Currie," 205; Currie, *An Historical Account of the Climates and Diseases of the United States of America*, 2.

75. Currie, *An Historical Account of the Climates and Diseases of the United States of America*, 2, 3.

76. Here Currie seems to be referring to chlorosis, an iron-deficiency anemia primarily affecting adolescent girls.

77. Currie, *An Historical Account of the Climates and Diseases of the United States of America*, 58, 212, 328.

78. See Gerbi, *The Dispute of the New World*.

79. Currie, *An Historical Account of the Climates and Diseases of the United States of America*, 407–8.

80. Valenčius, "Histories of Medical Geography," 14. On 'big data" and historical explanation, see Guldi and Armitage, *The History Manifesto*.

81. Numbers, "Medical Science before Scientific Medicine," 217.

82. Horine, *Daniel Drake*. See also the discussion in Sargent, *Hippocratic Heritage*, 261–68. Drake's work is also discussed in Barrett, "Daniel Drake's Medical Geography"; Barrett, *Disease and Geography*; Dorn, "Climate, Alcohol, and the American Body Politic."

83. Drake, "Some Account of the Epidemic Diseases."

84. Drake, *Notices Concerning Cincinnati*, part 2, 33, 37–38.

85. Drake, *A Systematic Treatise*, 447.

86. Drake, *A Systematic Treatise*, 716, 516.

87. Drake was already the author of *A Discourse on Intemperance*.

88. Drake, *Systematic Treatise*, 670, 645.

89. See Dorn, "Climate, Alcohol, and the American Body Politic," 219, 241.

90. Dorn, "Climate, Alcohol, and the American Body Politic," 87.

91. For an overview of Haviland's contributions, see Barrett, *Disease and Geography*, 323–31.

92. Haviland, *Climate, Weather and Disease*, 14.

93. Haviland, *Climate, Weather and Disease*, 90–144.

94. See Barrett, "Alfred Haviland's Nineteenth-Century Map Analysis."

95. Haviland, *The Geographical Distribution of Heart Disease and Dropsy*, 35.

96. Haviland, *The Geographical Distribution of Disease in Great Britain*, vii.

97. Haviland, *Geographical Distribution of Disease in Great Britain*, ix. Here Haviland was referring to Latham's *The Varieties of the Human Species*.

98. Haviland, *The Geographical Distribution of Disease in Great Britain*, 40, 331.

99. Smith published on this subject in Appleton's famous International Scientific Series. Smith, *Foods*. On this series of volumes, see Howsam, "An Experiment with Science for the Nineteenth-Century Book Trade."

100. Smith's contributions are discussed in Chapman, "Edward Smith"; Lavie, "Two 19th-Century Chronobiologists."

101. Smith, *The Aortic System*.

102. Smith, *Health and Disease*, 131, 132, 135, 208.

103. Smith, *Health and Disease*, 232.

104. Bordier, *La Géographie Médicale*; Poincaré, *Prophylaxie et Géographie*; Davidson, *Geographical Pathology*. The work of Bordier and Poincaré is discussed in Osborne, "The Geographical Imperative in Nineteenth-Century French Medicine," 31–50.

105. Rupke observes that Hirsch was "one of the greatest representatives of medical geography." Rupke, "Humboldtian Medicine," 293.

106. On various dimensions of Hirsch's work and career, see Sargent, *Hippocratic Heritage*, 269–79; Barrett, *Disease and Geography*, 331–344; Barrett, "August Hirsch." Barrett spends considerable time focusing on matters of definition and the distinctions between medical geography and geographical medicine.

107. Hirsch, August, *Handbook of Historical and Geographical Pathology*, vol. 1, 2.

108. Finke, *Versuch einer Allgemeinen Medicinisch-Praktischen Geographie*; Mühry, *Die Geographischen Verhältnisse*; Boudin, *Essai de Géographie Médicale*; Boudin, *Traité de Géographie et de Statistique Médicales*; Lombard, *Traité de Climatologie Médicale Comprenant la Météorologie Médicale*.

109. Hirsch, *Handbook*, vol. 1, 31.

110. Hirsch, *Handbook*, vol. 1, 247.

111. Rupke, "Adolf Mühry."

112. Rupke, "Humboldtian Medicine," 297.

113. Rupke, "Humboldtian Medicine," 298, 301.

114. Vaj, "Medical Geography and Phthisic Immunity."

115. Eylers, "Planning the Nation: The Sanitorium Movement in Germany."

116. Baikie, *Observations on the Neilgherries.*

117. See the discussion in Vaj, "Medical Geography and Phthistic Immunity."

118. Tschudi, *Travels in Peru*; Lossio, "British Medicine in the Peruvian Andes."

119. See Vaj, "Medical Geography and Phthistic Immunity"; Lombard, *Les Climats De Montagnes Considerérés au Point de Vue Médical.* Lombard is discussed in Osborne, "Geographical Imperative in Nineteenth-Century French Medicine."

120. On British thinking about health travel and "the choice of air," see Janković, *Confronting the Climate,* chap. 5.

121. A biographical sketch is available in Baigent, "Jackson, Robert Edmund Scoresby-."

122. Scoresby-Jackson, *Medical Climatology,* 2–3.

123. Scoresby-Jackson, *Medical Climatology,* ix.

124. Scoresby-Jackson, *Medical Climatology,* 25, 30.

125. Scoresby-Jackson, *Medical Climatology,* 137.

126. Scoresby-Jackson, *Medical Climatology,* 148, 144. Here Scoresby-Jackson seems to have been quoting loosely from Martin's discussion of the "salubrity of New South Wales" in Martin, *History of the British Colonies Vol IV,* 281.

127. Scoresby-Jackson, *Medical Climatology,* 194, 212.

128. Osborne, "Resurrecting Hippocrates," 80.

129. Willinsky, *Learning to Divide the World.*

130. Sargent, *Hippocratic Heritage,* 170.

Chapter 3: Tropical Terrain

1. On the shifting ideas about health and imperialism, see Curtin, *Death by Migration.*

2. Smith, *Health and Disease,* 167.

3. Scoresby-Jackson, *Medical Climatology,* 6.

4. Sullivan, *The Endemic Diseases of Tropical Climates.*

5. Hirsch, *Handbook of Historical and Geographical Pathology,* vol. 2, 502, 107.

6. Valenčius, "Histories of Medical Geography," 15.

7. See Porter, *Greatest Benefit to Mankind,* 295; Roddis, *James Lind.*

8. Bartholomew, "Lind, James."

9. Lind, *An Essay on Diseases Incidental to Europeans in Hot Climates,* 2, 3, 9.

10. Lind, *An Essay on Diseases Incidental to Europeans in Hot Climates,* 36, 57.

11. On the search for healthy niches where Europeans could live in the tropics, see Wear, "The Prospective Colonist and Strange Environments."

12. Lind, *An Essay on Diseases Incidental to Europeans in Hot Climates,* 192, 198, 195, 196, 204.

13. See Booth, "Hillary, William."

14. Hillary, *Observations on the Changes of the Air,* ii. See also Hillary, *Treatise on Such Diseases as Are the Most Frequent.*

15. Hillary, *Observations on the Changes of the Air,* v.

16. Trapham, *A Discourse of the State of Health in the Island of Jamaica.* See Kiple and Ornelas, "Race, War and Tropical Medicine in the Eighteenth-Century Caribbean"; Sheridan, *Doctors and Slaves.*

17. Kiple and Ornelas, "Race, War and Tropical Medicine in the Eighteenth-Century Caribbean," 65.

18. Huxham, *An Essay on Fevers, and Their Various Kinds*, 18; Bisset, *Medical Essays and Observations*, 11.

19. Harrison, *Climates and Constitutions*, 115. See also Golinski, *British Weather and the Climate of Enlightenment*, 189.

20. Lind, *An Essay on Diseases Incidental to Europeans in Hot Climates*, 146, 188.

21. Bisset, *Medical Essays and Observations*, 17.

22. What follows draws in part on Livingstone, "Human Acclimatization." See also Kennedy, "The Perils of the Midday Sun"; Harrison, "Tropical Medicine in 19th-Century India"; Harrison, "'The Tender Frame of Man'"; Harrison, *Climate and Constitutions*; Duncan, *In the Shadows of the Tropics*; Duncan, "The Struggle to be Temperate."

23. Johnson and Martin, *The Influence of Tropical Climates on European Constitutions*, 108.

24. Johnson and Martin, *The Influence of Tropical Climates on European Constitutions*, 1, x, 2.

25. Thomson, "On the Doctrine of Acclimatization," 69.

26. Thomson, "Could the Natives of a Temperate Climate Colonize." Thomson, who also served as a military surgeon in New Zealand, was the author of *The Story of New Zealand: Past and Present, Savage and Civilized* in 1859, the first major history of the country. For biographical details, see Belgrave, "Thomson, Arthur Saunders."

27. See Livingstone, "Human Acclimatization"; Richards, "The 'Moral Anatomy' of Robert Knox"; Biddiss, "The Politics of Anatomy."

28. Knox, *The Races of Men*, 106, 107, 108.

29. On acclimatization in the period, see Livingstone, "Human Acclimatization"; Anderson, "Climates of Opinion"; Arnold, ed., *Warm Climates and Western Medicine*; Osborne, "Acclimatizing the World."

30. Mair, *Medical Guide for Anglo-Indians*, 4. Mair's *Medical Guide* was often appended to Hull, *The European in India or Anglo-Indian's Vade-Mecum*.

31. Cullimore, "On Tropical and Sub-Tropical Climates," 437.

32. Cullimore, *The Book of Climates*, 3, 5.

33. Cullimore, *The Book of Climates*, 13, 14. In the South African context, Georgina Endfield has explored the way in which South African climate was constructed around 1900 as heroic, healthful, and a site for "climatic therapy" and was marketed as such to attract British women as settlers. See Endfield, "The British Women's Emigration Association and the Climate(s) of South Africa." On climate and women's health in colonial settings, see Endfield and Nash, "'Happy Is the Bride the Rain Falls On.'" See also Endfield and Nash, "'A Good Site for Health.'"

34. Kenny, "Claiming the High Ground," 656. See also Kenny, "Climate, Race and Imperial Authority."

35. See Webb, "Parke, Thomas Heazle."

36. Cullimore, *The Book of Climates*, 262, 263, 265.

37. See the discussion of Moore in Winther, *Anglo-European Science and the Rhetoric of Empire*, 117–28.

38. Moore, "Is the Colonisation of Tropical Africa by Europeans Possible?" 36, 38, 40.

39. Ravenstein, "Lands of the Globe Still Available for European Settlement."

40. White, "On the Comparative Value of African Lands," 192.

41. I have discussed this inclination in Livingstone, "The Moral Discourse of Climate," and in "Race, Space and Moral Climatology." Some of the political dimensions of what I call moral medicine are discussed in Tavárez, "'The Moral Miasma of the Tropics.'"

42. Johnson, *The Influence of Tropical Climates on European Constitutions*, 531.

43. Martin, *The Influence of Tropical Climates on European Constitutions*, 139.

44. Martin, *Notes on the Medical Topography of Calcutta*, 44. In the main, the sentiments expressed here were repeated and extended in James Ranald Martin, *Official Report on the Medical Topography and Climate of Calcutta*.

45. Martin, *Notes on the Medical Topography of Calcutta*, 46, 51, 96, 52.

46. Felkin, "Tropical Highlands," 162. See also Felkin's other contributions, including "Can Europeans Become Acclimatised in Tropical Africa?"; "On Acclimatisation"; and "The Distribution of Disease in Africa."

47. Crawfurd, "On the Connexion between Ethnology and Physical Geography," 5, 6. Crawfurd had been peddling such judgments for over forty years. In 1820, for instance, he told his readers that a "luxurious climate" and soil fertility explained the peace-loving character of the Javanese. See Crawfurd, *History of the Indian Archipelago*. Opinions such as these prompted Savage to observe that Crawfurd "was a prisoner of Classical environmental thought, cultural smugness, and western stereotyped conceptions of the East." Savage, *Western Impressions of Nature and Landscape in Southeast Asia*, 182.

48. Hunt, "On Ethno-Climatology," 53. Hunt also presented the same material at the British Association. See *Report of the British Association* 31 (1862): 129–50.

49. Thomson, "Niger and Central Sudan Sketches," 582, 584.

50. Hutchinson, *Impressions of Western Africa*, 32, 52, 234, 235, 241.

51. Cullimore, *The Book of Climates*, 275.

52. See Chowdhury, "Climatic Determinism and the Conceptualization of the Tropics in British India."

53. For developments in France, see Osborne, *The Emergence of Tropical Medicine in France*. See also see Osborne and Fogarty, "Medical Climatology in France."

54. Naraindas, "Poisons, Putresence and the Weather," 3. For general background to Mansonian tropical medicine, see Worboys, "Manson, Ross and Colonial Medical Policy"; Worboys, "Germs, Malaria and the Invention of Mansonian Tropical Medicine"; and Anderson, "Immunities of Empire." On Manson, see Manson-Bahr, *Patrick Manson*. See also Li, "Natural History of Parasitic Disease."

55. See Livingstone, "Tropical Climate and Moral Hygiene." Sambon's parasitology is discussed in Meulendijks, "The Descent of Malady."

56. For example, the following pieces by Sambon contain very substantially the same material: "Acclimatization of Europeans in Tropical Lands"; "The Possibility of the Acclimatization of Europeans in Tropical Regions"; "L'acclimatation des Européens dans les Régions Tropicales." See also Sambon, "Climate and Colonisation."

57. Sambon, "The Possibility of the Acclimatization of Europeans," 66.

58. The combination of geographical analysis, epidemiology, and parasitology that Sambon brought to the study of acclimatization also characterized his research on pellagra, a vitamin deficiency disease. See Gentilcore, "Louis Sambon and the Clash of Pellagra Etiologies."

59. On Sambon's "solario," see Kennedy, "The Perils of the Midday Sun."

60. Arnold, "Introduction: Tropical Medicine before Manson." See also Arnold, *The Problem of Nature*, chap. 8 on "Inventing Tropicality."

61. Arnold, "India's Place in the Tropical World," 2.

62. Adamson, "'The Languor of the Hot Weather.'"

63. Kennedy, *The Magic Mountains*. See also Carey, "Climate, Medicine, and Peruvian Health Resorts"; Kenny, "Climate, Race, and Imperial Authority"; Carey, "Inventing Caribbean Climates"; Jennings, *Curing the Colonizers*.

64. Njoh, "Colonial Philosophies, Urban Space, and Racial Segregation."

65. Chang and King, "Towards a Genealogy of Tropical Architecture," 284. See also Chang, *A Genealogy of Tropical Architecture*. For an architectural critique of the assumption that temperate climates are normative, and an argument for the value of tropical architecture in the age of the Anthropocene, see Ferng et al., "Climatic Design and Its Others."

66. Waterhouse, "Smith, Thomas Roger."

67. See Metcalf, "Architecture and the Representation of Empire."

68. Smith, "On Buildings for European Occupation in Tropical Climates," 197, 198–99.

69. Smith, "On Buildings for European Occupation in Tropical Climates," 208. Smith repeated this sentiment word for word in Smith, "Architectural Art in India," 286.

70. Smith, "On Buildings for European Occupation in Tropical Climates," 202.

71. I have relied here on Johnson, "Commodity Culture." Here Johnson cites the Burroughs Wellcome & Co. comment.

72. See Johnson, "European Cloth and 'Tropical' Skin."

73. Chang and King, "Towards a Genealogy of Tropical Architecture," 291.

74. Jeffreys, *The British Army in India*, 59–93. See also Zuck, "Jeffreys, Julius"; Zuck, "Julius Jeffreys, Pioneer of Humidification."

75. Jeffreys, *The British Army in India*, 213, 203.

76. See Callaway, "Dressing for Dinner in the Bush."

77. Jeffreys, *The British Army in India*, 14.

78. See Collingham, *Imperial Bodies*.

79. See Kneale and Randalls, "Invisible Atmospheric Knowledges in British Insurance Companies." What follows in this paragraph relies in large measure on this contribution. See also Kneale and Randalls, "Imagined Geographies of Climate and Race in Anglophone Life Assurance."

80. Allen, *Medical Examinations for Life Insurance*, 13.

81. Brinton, *On the Medical Selection of Lives for Assurance*, 29.

82. Lister, *Medical Examination for Life Insurance*, 151.

83. See Howell, *Exploring Victorian Travel Literature*, 83–108.

84. Horton, *Diseases of Tropical Climates and their Treatment*, viii, 633, 634.

85. Horton, *Physical and Medical Climate and Meteorology of the West Coast of Africa*, v, 234, 206.

86. I have discussed monogenism and polygenism in Livingstone, *Adam's Ancestors*. See also Stocking, "What's in a Name?"; Stocking, *Victorian Anthropology*; Stocking, "The Persistence of Polygenist Thought."

87. Horton, *West African Countries and Peoples*, v, i, vi, 51.

88. Horton, *West African Countries and Peoples*, 73, 74.

89. Darwin, *The Descent of Man*, vol. 1, 238.

90. Abercromby, *Seas and Skies in Many Latitudes*, 366.

91. Seacole, *Wonderful Adventures of Mrs Seacole in Many Lands*, 1.

92. Basic biographical information is available in Palmer, "Seacole [*née* Grant], Mary Jane."

93. See Seaton, "Another Florence Nightingale?"

94. For example, Robinson, "Authority and the Public Display of Identity"; Paravisini-Gebert, "*Mrs. Seacole's Wonderful Adventures in Many Lands* and the Consciousness of Transit"; Robinson, *Mary Seacole*.

95. Seacole, *Wonderful Adventures of Mrs Seacole in Many Lands*, 69, 171.

96. Seaton, "Another Florence Nightingale?"

97. Seacole, *Wonderful Adventures of Mrs Seacole in Many Lands*, 27, 5, 26, 77.

98. Howell, *Exploring Victorian Travel Literature*, 29.

99. Seacole, *Wonderful Adventures of Mrs Seacole in Many Lands*, 59–60.

100. Seacole, *Wonderful Adventures of Mrs Seacole in Many Lands*, 14, 41, 42–43, 79.

101. The use of the discourse of climate, race, and tropical medicine for anticolonial purposes in early twentieth-century India is discussed in Bashford, "Anticolonial Climates."

102. Biographical treatments include Farwell, *Burton: A Biography*, and Kennedy, *The Highly Civilized Man*.

103. See Grant, *Postcolonialism, Psychoanalysis and Burton*; Howell, *Exploring Victorian Travel Literature*, chap. 2.

104. See the discussion in Kennedy, *The Highly Civilized Man*, chap. 5, "The Racist."

105. A F.R.G.S. [Burton], *Wanderings in West Africa*, vol. 1, 5.

106. Pratt, *Imperial Eyes*, 201, 204.

107. See Lovell, *A Rage to Live*.

108. On this general subject, see Curtin, *Death by Migration*.

109. A F.R.G.S. [Burton], *Wanderings in West Africa*, vol. 1, 262, 227, 296.

110. A F.R.G.S. [Burton], *Wanderings in West Africa*, vol. 1, 36, 66.

111. A F.R.G.S. [Burton], *Wanderings in West Africa*, vol. 2, 166, 167; vol. 1, 245.

112. Burton, *The Lake Regions of Central Africa*, vol. 1, 92; vol. 2, 130; vol. 1, preface.

113. A F.R.G.S. [Burton], *Wanderings in West Africa*, vol. 1, 200, 242, 243.

114. Burton, *The Lake Regions of Central Africa*, vol. 1, 161, 158.

115. Burton, *Wanderings in West Africa*, vol. 2, 77.

116. On the Boston Immigration Restriction League, in which Ward was heavily involved, see Solomon, *Ancestors and Immigrants*; Higham, *Strangers in the Land*.

117. The scope of Ward's climate endeavors is surveyed in Rohli and Bierly, "The Lost Legacy of Robert DeCourcy Ward."

118. For example, Ward, "Physiological Effects of Diminished Air Pressure"; Ward, "The Classification of Climates"; Ward, "Weather Controls over the Fighting in Mesopotamia, in Palestine, and Near the Suez Canal"; Ward, "Land and Sea Breezes"; Ward "Fog in the United States"; Ward, "Weather Controls over the Fighting During the Autumn of 1918"; Ward, "Meteorology and War-Flying."

119. Ward, "The Crisis in Our Immigration Policy"; Ward, "Immigration and the Three Percent Restrictive Law"; Ward, "Immigration and Eugenics"; Ward, "The Second Year of the New Immigration Law"; Ward, "The Immigration of Orientals"; Ward, "Fallacies of the Melting Pot Idea." This whole subject is treated in Lavery, "Situating Eugenics."

120. Ward, *Climate Considered Especially in Relation to Man*, 76, 81, 106, 183.

121. Ward, *Climate Considered Especially in Relation to Man*, 203, 204–5. See also Ward, "The Acclimatization of the White Race in the Tropics."

122. Fleming, *Historical Perspectives on Climate Change*, 99. On Huntington's determinism, see also Livingstone, "Changing Climate"; Davis, "The Coming Desert." For biographical details more generally, see Martin, *Ellsworth Huntington*.

123. Huntington, "The Adaptability of the White Man," 185, 187, 196. Huntington's philosophy of climate and civilization is treated in chapter 7.

124. Huntington, "Adaptability of the White Man," 198, 199, 202.

125. Tower, Review of *Civilization and Climate*.

126. Whitbeck, Review of *Civilization and Climate*.

127. Tenney, Review of *Civilization and Climate*, 634.

128. Huntington, *Civilization and Climate*, 38, 45, 47, 42.

129. Miller, *Climatology*, 2.

130. I have discussed this in "Race, Space and Moral Climatology," 173.

131. The continuing debate over the suitability of tropical environments for white settlement in the context of early twentieth-century Australia is discussed in Bashford, "'Is White Australia Possible?'"

Chapter 4: Climate, Eugenics, and the Biometeorological Body

1. Huntington, *Pulse of Asia*, xv (from 1919 preface).

2. Huntington, *Pulse of Asia*, 15, 323.

3. Huntington, *Civilization and Climate*, 2, 121, 109.

4. Huntington, *Civilization and Climate*, 68, 77.

5. Huntington, *Civilization and Climate*, 7, 8. Huntington's climatic theory of economic productivity and civilization is discussed in chapter 7.

6. Huntington, Review of *Health and Social Progress* by R. M. Binder.

7. In discussing climate and upper respiratory infections, Winslow and Herrington insisted that the subject had been "illuminated by the fruitful contributions of the late Ellsworth Huntington" but "confused and misrepresented by other and less scientific writers." Winslow and Herrington, *Temperature and Human Life*, ix. On Winslow, see Viseltear, "C.-E.A. Winslow and the Early Years of Public Health at Yale."

8. Huntington, "The Control of Pneumonia and Influenza by the Weather."

9. Huntington, "Influenza and the Weather in the United States," 471.

10. Huntington, *Weather and Health*, 9, 120.

11. Huntington et al., *Economic and Social Geography*, 118.

12. Huntington et al., *Economic and Social Geography*, 120.

13. These are identified in Lavery, "Geography and Eugenics"; Lavery, "The Power of Racial Mapping."

14. Huntington, "The Effect of Overpopulation on Chinese Character."

15. This work was reviewed by George S. Schulyer, who took the opportunity to attack its racial and class bias and to argue, as Ewa Barbara Luczak puts it, that "eugenics served as a front for a group of privileged Americans to protect their racial and class interests." Luczak, *Breeding and Eugenics in the American Literary Imagination*, 174.

16. Huntington and Whitney, *Builders of America*, 1, 283.

17. Huntington, *Tomorrow's Children*, 4, 13, 29.

18. Huntington, *Season of Birth*, v.

19. At times, Huntington displayed neo-Lamarckian tendencies, and at others, he resorted to the more selectionist vocabulary of the Darwinians. There is reason to think that he moved between these perspectives to suit the topic at hand. See the discussion of his Lamarckism in Campbell and Livingstone, "Neo-Lamarckism and the Development of Geography."

20. Huntington, *Season of Birth*, 290, 260, 420.

21. Huntington, *Season of Birth*, v, 3.

22. Huntington, *Civilization and Climate*, 131.

23. Huntington, *World-Power and Evolution*, especially chaps. 5 and 8.

24. Huntington, "The Control of Pneumonia and Influenza by the Weather," 6, 23.

25. Petersen, *Man–Weather–Sun*, 363. In his work *Lincoln–Douglas*, Petersen also drew on Huntington.

26. Sargent, *Hippocratic Heritage*, 394ff.

27. See Dalén, *Season of Birth*; Nonaka, "Effect of Delivery Season on Subsequent Birth"; Miura, "The Influence of Seasonal Atmospheric Factors on Human Reproduction"; Norris and Chowning, "Season of Birth and Mental Illness"; Janerich et al., "Season of Birth and Neonatal Mortality."

28. The whole question of the geography of excellence also attracted the attention of Huntington's colleague and coauthor Stephen Visher, who drew attention to what he called the "ecology of American notables." See Visher, "Ecology of American Notables"; Visher, "Geography of American Notables." Visher's eugenics is discussed in Lavery, "Stephen S. Visher."

29. Huntington, *Season of Birth*, 15, 16.

30. Huntington, *Season of Birth*, 18.

31. Huntington, *Season of Birth*, 327, 348, 408, 445.

32. Huntington et al., *Economic and Social Geography*, 126.

33. Ellsworth Huntington to Madison Grant, August 17, 1925. Series III, box 50, no. 1630 General Correspondence. Ellsworth Huntington Papers, Sterling Memorial Library, Yale University, New Haven. Quoted in Lavery, "Geography and Eugenics," 86.

34. Huntington, *Tomorrow's Children*, 94, 95.

35. See, for example, Bouma, "A Short History of Human Biometeorology"; Sargent, *Hippocratic Heritage*; Tout, "Biometeorology."

36. "Obituary: William F. Petersen." See also Sargent, *Hippocratic Heritage*, 524–558. There is a brief sketch of Petersen in Bischof, "Introduction to Integrative Biophysics," on 39–40.

37. Petersen, *Lincoln–Douglas*, 131–150; Petersen, *Man–Weather–Sun*, 237.

38. Petersen, *Man–Weather–Sun*, 284, 286.

39. For example, Petersen, *Protein Therapy and Non-Specific Reactions*; Petersen and Levinson, *The Skin Reactions, Blood Chemistry and Physical Status*.

40. Sargent, "The William F. Petersen Foundation."

41. Petersen, *The Patient and the Weather*, vol. 2, 11.

42. Sargent, *Hippocratic Heritage*, 353.

43. Sargent, *Hippocratic Heritage*, 533.

44. Petersen, letter to Ellsworth Huntington, March 30, 1933, cited in Sargent, *Hippocratic Heritage*, 534.

45. Petersen, *The Patient and the Weather*, vol. 3, 327; Petersen, *The Patient and the Weather*, vol. 1, part 2, 159.

46. Sargent, *Hippocratic Heritage*, 375.

47. Petersen, *Man–Weather–Sun*, 79, 84, 265, 266.

48. Petersen, *Man–Weather–Sun*, 362, 365, 364.

49. Petersen, *Man–Weather–Sun*, xii.

50. Huntington, *Season of Birth*; Lombroso, *The Man of Genius*.

51. Petrie, *The Revolutions of Civilisation*.

52. Petersen, *Man–Weather–Sun*, 421.

53. Petersen, *Lincoln–Douglas*, cited on dust jacket.

54. Petersen, *Lincoln–Douglas*, viii.

55. Petersen, *Lincoln–Douglas*, 156.

56. Petersen, *Lincoln–Douglas*, 12, 13, 14.

57. Petersen, *Lincoln–Douglas*, 166.

58. Petersen, *Man—Weather—Sun*, ix.

59. Sargent, *Hippocratic Heritage*, 386–390.

60. Quoted in Sargent, *Hippocratic Heritage*, 546.

61. William F. Petersen, *Hippocratic Wisdom*, 11, xiv.

62. Petersen, *Hippocratic Wisdom*, xv–xviii. On Henderson more generally see, Cannon, "Biographical Memoir of Lawrence Joseph Henderson."

63. Petersen, *Hippocratic Wisdom*, 15, 19, 24, 127.

64. Sargent, *Hippocratic Heritage*, 389–390.

65. Meyer claims that Mills was "second only to Huntington as a promoter of determinism." Meyer, *Americans and Their Weather*, 171.

66. See the discussion in Sargent, *Hippocratic Heritage*, 408–21.

67. See, for example, Mills, *Air Pollution and Community Health*; Mills, *This Air We Breathe*.

68. Mills, *Climate Makes the Man*, 29.

69. Mills, *Climate Makes the Man*, 50, 65ff; Mills, *Living with the Weather*, 8; Mills, *Medical Climatology*, 124ff.

70. Mills, *Climate Makes the Man*, 36ff, Mills, *Medical Climatology*, 34ff.

71. Mills, *Climate Makes the Man*, 59, 62; Mills, *Living with the Weather*, 75, 82.

72. Mills, *Climate Makes the Man*, 96.

73. Mills, *Medical Climatology*, 89ff, 124ff, 164ff.

74. Huntington, Review of *Medical Climatology*.

75. Spengler, Review of *Medical Climatology*.

76. Review of *Medical Climatology*, 2263.

77. "Links Fascism with Weather."

78. Mills, *Climate Makes the Man*, 8, 10.

79. Mills, *Living with the Weather*, chap. XI "Weather Stimulation, Human Energy, and Business Cycles." This theme is treated in more detail in chapter 7.

80. Mills, *Living with the Weather*, 120, 133.

81. Mills, *Living with the Weather*, 120, 133, 119–20.

82. Mills, *Climate Makes the Man*, 3, 27.

83. Mills, *Medical Climatology*, 12.

84. Mills, *Living with the Weather*, 188; Mills, *Medical Climatology*, 14.

85. Mills, *Medical Climatology*, 14.

86. Mills, *Living with the Weather*, 13, 14. Later he confirmed that "from the standpoint of human achievement and basic progress, present day African climates offer little encouragement." Mills, *World Power and Shifting Climates*, 64.

87. Mills, *Living with the Weather*, 5.

88. Mills, *World Power and Shifting Climates*, 128.

89. Mills, *Climate Makes the Man*, 8.

90. Mills, *Climate Makes the Man*, 141.

91. Mills, *World Power and Shifting Climates*, 13.

92. Mills, *Climate Makes the Man*, 136, 140; Mills, *Living with the Weather*, 182, 183, 184; Mills, *Medical Climatology*, 53; Mills, *World Power and Shifting Climates*, 99–101.

93. Mills, *Living with the Weather*, 7–8.

94. Mills, *Medical Climatology*, 53, 69–70, 72–74.

95. Mills, *Living with the Weather*, 47. He repeated this in *World Power and Shifting Climates*, 38.

96. Mills, *World Power and Shifting Climates*, 10. See also Mills, "Temperature Dominance over Human Life."

97. Mills, *Climate Makes the Man*, 111–12. See also Mills, *Living with the Weather*, 47–48.

98. Mills, *Medical Climatology*, 139.

99. Mills, *Living with the Weather*, 149–50.

100. Mills, *Living with the Weather*, 171–72.

101. Mills, *Climate Makes the Man*, 114. In the United States, these examinations routinely included a blood test to detect syphilis and became increasingly common during the late 1930s. Some of this is discussed in Brandt, *No Magic Bullet*.

102. Mills, *World Power and Shifting Climates*, 39.

103. Mills, *Medical Climatology*, 35.

104. Mills, *World Power and Shifting Climates*, 37.

105. Mills, *Living with the Weather*, 22, 38.

106. Mills, *Medical Climatology*, 271–72.

107. Weihe, "Review on the History of the International Society of Biometeorology," 12.

108. Tout, "Biometeorology," 474. See also personal historical reflections on the Society by Folk, "The International Society of Biometeorology."

109. Tromp, *Biometeorology*, 1.

110. Tromp, *Biometeorology*, 159, 217.

111. This is a theme that has resurfaced in the twenty-first century as an explanation for global patterns of economic growth. See chapter 7. For an attempt to find in the adaptive significance of thermoregulation a mediating position between climatic determinism and human free will, see Auliciems, "Human Adaptation within a Paradigm of Climatic Determinism."

112. See Folk, "International Society of Biometeorology"; Tout, "Biometeorology"; Weihe, "Review on the History of the International Society of Biometeorology."

113. Sargent, *Hippocratic Heritage*, xxxviii.

114. Sargent, "Nature and Scope of Human Ecology," 1, 5.

115. Audy and Dunn, "Health and Disease," 337.

116. Sargent, "Nature and Scope of Human Ecology," 18.

117. There are numerous biographical sketches, among them, Droessler, "Helmut Landsberg"; Baer, "Helmut E. Landsberg"; Liebowitz, "Landsberg, Helmut Erich"; Baer et al., eds., *Climate in Human Perspective: A Tribute to Helmut E. Landsberg*. See also Henderson, "The Dilemma of Reticence."

118. Landsberg, *Weather and Health*, 102, 39, 64.

119. Landsberg, *Weather and Health*, 28, 114.

120. Landsberg, "Atmospheric Variability and Climatic Determinism," 22.

121. Tisdale, "Confessions of a Cosmic Resonator."

122. Sargent, "The William F. Petersen Foundation," 105.

123. Rosen, *Youth, Middle-Age*.

124. Rosen, "Guestwords: Spring Fever."

125. Rosen, *Weathering*, xi, 113, xii, xiii.

126. Rosen, *Weathering*, 94.

127. Rosen, *Weathering*, 98, 103.

128. Kennedy, "The Perils of the Midday Sun," 123. See also Crozier, "What Was Tropical about Tropical Neurasthenia?"; Anderson, *Colonial Pathologies*; Kennedy, "Diagnosing the Colonial Dilemma."

129. Rosen, *Weathering*, 91. I discuss Draper's climatic interpretation of the American Civil War in chapter 10.

130. Rosen, *Weathering*, 91, 92. Here Rosen drew on an article by Wheeler, "The Effect of Climate on Human Behavior in History," that originally appeared in 1943 in *Transactions of the Kansas Academy of Science*.

131. Baum, *When Nature Strikes*, 135, 137.

132. Moss, *Weather Shamanism*.

133. DeBlieu, *Wind: How the Flow of Air Has Shaped Life*.

Chapter 5: Climate, Cognition, and Human Evolution

1. BBC, "Did Climate Change Make Us Intelligent?" https://www.bbc.co.uk/teach/did-climate-change-make-us-intelligent/zj2qrj6. This site hosts the clip from Brian Cox's BBC 2 series, *Human Universe*.

2. See, for instance, National Research Council, *Understanding Climate's Influence on Human Evolution*.

3. See, for example, Morrone, "Cladistic Biogeography." On Matthew, see Watson, "William Diller Matthew"; Colbert, *William Diller Matthew, Paleontologist*, 108–9. Some of what follows draws on Livingstone, "Changing Climate, Human Evolution and the Revival of Environmental Determinism."

4. Fleming, "T.C. Chamberlin, Climate Change and Cosmogony."

5. Matthew, "Climate and Evolution," 178–79. Later, in 1939, this was published as a monograph: *Climate and Evolution*.

6. Letter, Theodore Roosevelt to William Diller Matthew, reprinted in Colbert, *William Diller Matthew, Paleontologist*, 108–9. Roosevelt's own thinking on the application of evolution to human affairs appears in Roosevelt, *Biological Analogies in History*.

7. Rainger, "Just before Simpson." See also Bowler, *Theories of Human Evolution*, 174–76.

8. Matthew, "Climate and Evolution," 180.

9. Bowler, *The Fontana History of the Environmental Sciences*, 345–46.

10. Matthew, "Climate and Evolution," 180, 210. This aspect of Matthew's analysis was doubtless of interest to Roosevelt. On Roosevelt's racial thought, see Dyer, *Theodore Roosevelt and the Idea of Race*.

11. Matthew, "Climate and Evolution," 211.

12. Quoted in Strange and Bashford, *Griffith Taylor*, 89.

13. Taylor, "Climatic Cycles and Evolution," 289, 304, 305, 307, 308, 313, 300.

14. Taylor, "Climatic Cycles and Evolution," 300–1, 290.

15. Christie, "Environment and Race," 431, 452. See also Winlow, "Mapping the Contours of Race."

16. Taylor, "The Ecological Basis of Anthropology," 224.

17. Taylor, *Environment and Race*, 39, 40, 41, 106, 220.

18. Taylor, *Environment and Race*, 12.

19. Taylor, "Preface," vi.

20. Taylor, *Environment and Race*, 4.

21. Huntington, *The Character of Races*, 76, 23.

22. Ellsworth Huntington, "The Role of Deserts in Evolution" (typescript), Huntington Papers, Yale University Library, box 64, folder 607.

23. Huntington, *Character of Races*, 33, 39, 75, 40, 36, 51.

24. Huntington, *World-Power and Evolution*, 107, 108, 116, 179, 123, 240.

25. Here Huntington drew on the work of Australian psychologist Stanley D. Porteus, who connected different brain weights to the length of the period of immaturity in different organisms, including human races. See Porteus, *Temperament and Race*. Porteus was favorably disposed toward Huntington's climatic theory. In more recent times, Porteus' work has come under severe criticism and his name has been removed from the Social Sciences building at the University of Hawaii. He was also the author of *Primitive Intelligence and Environment*.

26. Huntington, *The Pulse of Progress*, 173.

27. Huntington, *Character of Races*, 1.

28. Clark, *God—or Gorilla*, 21.

29. On the Scopes trial, see Larsen, *Summer for the Gods*.

30. Rainger, *An Agenda for Antiquity*.

31. See Bowler, *The Eclipse of Darwinism*.

32. Osborn, *Man Rises to Parnassus*, 156.

33. Osborn, "Why Central Asia?"

34. Rainger, "What's the Use," 125.

35. Osborn, *Men of the Old Stone Age*, 489.

36. See Leidy, "Prefatory Remarks," xviii.

37. Osborn, *Man Rises to Parnassus*, 160–61, 167.

38. Osborn, *Man Rises to Parnassus*, 169–70.

39. See Osborn, "The Approach to the Immigration Problem through Science"; Osborn, "Race Progress in Relation to Social Progress"; Regal, *Henry Fairfield Osborn*.

40. Osborn, "Preface."

41. Clark, *God—or Gorilla*, 117.

42. Tambe, "Climate, Race Science and the Age of Consent."

43. See, for example, Rigg, "Climatic Determinism."

44. Schneider and Londer, *The Coevolution of Climate and Life*, 102, 117, 119.

45. See Coon, "Some Problems of Human Variability."

46. Beals, "Head Form and Climatic Stress," 85, 90.

47. In the meantime, a similar climate–cranium argument was put forward by Weninger, "As to the Influence of Climate on Head Form."

48. On this subject, Beals called attention to the work of Schreider, "Ecological Rules, Body-Heat Regulation, and Human Evolution."

49. Beals et al., "Climate and the Evolution of Brachycephalization," 436.

50. Beals et al., "Brain Size, Cranial Morphology, Climate," 307, 320.

51. Comments by Kathleen R. Gibson, after Beals et al., "Brain Size, Cranial Morphology, Climate," 320–21.

52. Blumenberg, "The Evolution of the Advanced Hominid Brain."

53. Lewthwaite, "Environmentalism and Determinism."

54. See Fleming, *Historical Perspectives on Climate Change*.

55. For an overview of earlier techniques, see Imbrie and Imbrie, *Ice Ages: Solving the Mystery*.

56. Gribbin and Gribbin, *Children of the Ice*, 63, 87, 93, 111.

57. Potts, "Variability Selection in Hominid Evolution," 82.

58. Burckle, "Current Issues in Pliocene Paleoclimatology," 6.

59. Vrba, "On the Connections between Paleoclimate and Evolution," 24–45.

60. Wood, "Evolution of the Early Hominin Masticatory System."

61. Wesselman, "Of Mice and Almost-Men."

62. Avery, "Southern Savannas and Pleistocene Hominid Adaptation," 474.

63. Kimbel, "Hominid Speciation and Pliocene Climatic Change," 425.

64. Vrba, "Ecological and Adaptive Changes," 63, 67, 70, 71. Italics in original.

65. See, for example, Vrba, "Role of Environmental Stimuli in Hominid Origins."

66. Vrba et al., "Climatic Influences on Early Hominid Behavior"; Vrba, "Climate, Heterochrony, and Human Evolution."

67. Vrba et al., "Climatic Influences," 128, 146.

68. Stanley, *Children of the Ice Age*, 154, 126. Here he built on his earlier analysis: Stanley, "An Ecological Theory for the Origin of *Homo*."

69. Stanley, *Children of the Ice*, 11, 12.

70. Stanley, *Children of the Ice*, 13, 14.

71. Stanley, *Children of the Ice*, 153–54, 15, 4, 14.

72. Stanley, *Children of the Ice*, 168, 175.

73. Calvin, "The Great Climate Flip-Flop."

74. See, for instance, Dennett, *Intuition Pumps*.

75. Calvin, *A Brain for All Seasons*, 5.

76. Calvin, *A Brain for All Seasons*, 47, 24.

77. Calvin, *A Brain for All Seasons*, 18, 91. Heatherington and Reid interpret the links between human brain size, behavioral flexibility, and rapidly changing climate in the context of a nongradualist understanding of emergent evolution with little or no role accorded to natural selection. See Hetherington and Reid, *The Climate Connection*, 75–77.

78. See Marks, *Tales of the Ex-Apes*.

79. Calvin, *The Ascent of Mind*, 3, 27, 13, 3.

80. Calvin, *The Ascent of Mind*, 69–70.

81. Calvin, *The Ascent of Mind*, 128, 132, 148.

82. Different mechanisms have been proposed. Richerson and Boyd, for example, suggested that social learning and the capacity for imitation were vital elements in the development of human cognitive complexity, particularly during the "exceedingly noisy" Pleistocene climate deterioration. Richerson and Boyd, "Climate, Culture, and the Evolution of Cognition," 343.

83. Holmes, "Cool Times, Smart Times."

84. Swartzman and Middendorf, "Biospheric Cooling and the Emergence of Intelligence," 425, 428.

85. Swartzman et al., "Was Climate the Prime Releaser for Encephalization?" 439, 443.

86. Kleidon, "Climatic Constraints on Maximum Levels," 405, 413, 407, 425.

87. Kleidon, "Climatic Constraints on Maximum Levels," 406.

88. Pondering on the future economic implications of climate change for human metabolic activity, Heal and Park observe, "One of the principal mechanisms through which temperature affects performance appears to be the ability of the brain to dispose of waste heat: on average the brain generates 20% of all the heat generated by the human body, and its performance is temperature-sensitive, so that it needs to dispose of waste heat. This becomes harder as the ambient temperature rises." See Heal and Park, "Feeling the Heat."

89. Science Daily, "Global Warming Could Be Reversing a Trend That Led to Bigger Human Brains."

90. It is worth noting that other students of "the big brain" find little room for climatic explanations in their accounts of its origin. See, for example, Lynch and Granger, *Big Brain*.

91. Sample, "Climate Swings Shaped Human Evolution." This item reported research on "African Paleoclimate and Human Evolution" in a special issue of the *Journal of Human Evolution* 53, no. 5 (2007): 446–634, edited by Beth A. Christensen and Mark M. Maslin.

92. Sherriff, "Researchers Link Human Skull Size and Climate."

93. Behrensmeyer, "Climate Change and Human Evolution."

94. Oppenheimer, *Out of Eden*, 3, 7, 8, 345, 277.

95. Finlayson, *Neanderthals and Modern Humans*, 72, 193; Finlayson, *The Humans Who Went Extinct*.

96. Finlayson, *Neanderthals and Modern Humans*, 85.

97. Stringer's work is introduced in Trinkaus and Shipman, *The Neanderthals*, 359–61.

98. Stringer, *Homo Britannicus*, 132, 133, 162–63.

99. Fagan, *Cro-Magnon*, 156, 159.

100. Shapiro, "The Role of the American Museum of Natural History," 6.

101. Heads, "Darwin's Changing Views on Evolution," 1020; Grehan and Schwartz, "Evolution of the Second Orangutan."

102. Briggs, "Centrifugal Speciation and Centres of Origin." See also the discussion in Cecca, "La Dimension Biogéographique de l'Évolution de la Vie."

103. E. O. Wilson, "Island Biogeography in the 1960s," 1.

104. Here Pearson called on John Chappell's efforts to resurrect Huntington's vision. See Chappell, "Climatic Pulsations in Inner Asia."

105. Pearson, *Climate and Evolution*, 23, 24, 213.

106. Pearson, *Climate and Evolution*, 1, 23.

107. Vrba, "Ecological and Adaptive Changes," 70.

108. Stringer, *Homo Britannicus*, 133.

109. Vrba, "On the Connections," 27, 28, 41.

110. Taylor, *Environment, Race, and Migration*, 459. Here Taylor also resorted to a traffic light analogy.

111. Oppenheimer, *Out of Eden*, 206, 197, 195, 197, 200, 347. On scientific racism, see Haller, *Outcasts from Evolution*; Stepan, *The Idea of Race in Science*; Livingstone, *Adam's Ancestors*.

112. The article from which the quotations are drawn appeared as Graves, "New Models and Metaphors for the Neanderthal Debate," 513, 514–15.

113. I have discussed this in Livingstone, "Cultural Politics and the Racial Cartographics of Human Origins." See also Lieberman and Jackson, "Race and Three Models of Human Origin"; Goodrum, "The History of Human Origins Research."

114. Stringer, *Homo Britannicus*, viii, xii, 165–66, 180, 181.

115. Calvin, *Brain for All Seasons*, 277, 278.

116. Finlayson, *Humans Who Went Extinct*, 219, 220, 220.

117. Clive Finlayson, "Climate and Humans."

118. Coombes and Barber, "Environmental Determinism in Holocene Research," 304. See also de Menocal, "Cultural Responses to Climate Change during the Late Holocene."

119. Chambers and Brain, "Paradigm Shifts in Late-Holocene Climatology?"

120. Oldfield, Frank, "Forward to the Past," 19.

121. Stoczkowski, *Explaining Human Origins*, 180.

Chapter 6: Mind, Mood, and Meteorology

1. Clayton et al., *Mental Health and Our Changing Climate,* 24, 31.

2. Eshelman, "The Mental Health Toll of Climate Change."

3. Rodman, "Weather, Climate Changes Multiply Health Concerns."

4. Borchard, "Weather and Mood."

5. Taylor, "Can Rainy Days Really Get You Down?"

6. Haslam, "Here Comes the Sun."

7. Ferreira and Smith, "Hotter Weather Brings More Stress."

8. Hippocrates, *On Airs, Waters and Places,* parts 19, 23.

9. This essay was published in French for the first time in 1892 and translated into English by Melvin Richter. See Richter, "An Introduction to Montesquieu's 'An Essay on the Causes That May Affect Men's Minds and Characters.'" Another English translation is available as "An Essay on the Causes Affecting Minds and Characters," in Carrithers, *The Spirit of Laws by Montesquieu.*

10. Montesquieu, "An Essay on the Causes That May Affect Men's Minds and Characters," 153.

11. Baron de Montesquieu, *The Spirit of the Laws* (1989), book XIV, §5, §6.

12. Schaub, "Montesquieu on Slavery," 71.

13. Richter, "An Introduction," 132.

14. Montesquieu, "An Essay on the Causes," 142, 140, 143.

15. Montesquieu, *The Spirit of the Laws* (1989), book XIV, §2.

16. Montesquieu, *The Spirit of the Laws* (1989), book XV, §1.

17. Montesquieu, "An Essay on the Causes," 143, 144, 155.

18. Borsay, "Falconer, William."

19. Falconer, *Remarks on the Influence of Climate,* iv, v, 4, 6.

20. Falconer, *Remarks on the Influence of Climate,* 4, 19, 26, 30, 45, 47.

21. Falconer, *Remarks on the Influence of Climate,* 51, 52, 71, 72.

22. Falconer, *Remarks on the Influence of Climate,* 130–34 passim.

23. Falconer, *Remarks on the Influence of Climate,* 135, 141, 143, 140.

24. Lind, *An Essay on Diseases Incidental to Europeans in Hot Climates,* 170–71.

25. Johnson and Martin, *The Influence of Tropical Climates on European Constitutions,* 7, 8, 627.

26. Van Deusen, "Observations on a Form of Nervous Prostration, (Neurasthenia)"; Beard, "Neurasthenia, or Nervous Exhaustion."

27. Woodruff, "The Soldier in the Tropics."

28. Woodruff, *The Effects of Tropical Light,* v, 4, 311, 321, 314–15, 299.

29. Woodruff, *The Effects of Tropical Light,* 191, 192, 195, 210.

30. Woodruff, *The Effects of Tropical Light,* 225, 312, 271–72.

31. Kennedy, "The Perils of the Midday Sun," 124.

32. Charles, "Neurasthenia, and Its Bearing," 2, 6, 9.

33. Kennedy, "The Perils of the Midday Sun," 123.

34. See Stewart, "Edwin Grant Dexter."

35. Abbe, "Introduction," xii, xiii, xxii, xx, xxix, xxx, xxxi.

36. Dexter, *Weather Influences,* 59, 60, 92.

37. Dexter, *Weather Influences,* 60, 143, 74.

38. Dexter, *Weather Influences*, 55, 269, 275, 273.

39. See Stewart, "Edwin Grant Dexter," 753.

40. Ward, *Climate*, 310.

41. Huntington, *The Pulse of Asia*, 363, 364, 365.

42. Huntington, *Civilization and Climate*, 134, 136.

43. Huntington, *Mainsprings of Civilization*, 343.

44. Huntington, *The Pulse of Progress*, 146.

45. Further details of the Düll's' research can be found in Halberg et al., "27-Day Cycles in Human Mortality."

46. Balfour, "Problems of Acclimatization," part 1, 84.

47. Balfour, "Sojourners in the Tropics," 1329.

48. Balfour, "Problems of Acclimatisation," part 2, 243.

49. Kames, *Sketches of the History of Man*, book 1, 26. Kames made this comment in the context of those attributing racial difference to the power of the climate. Sorokin quoted it in his *Contemporary Sociological Theories*, 99.

50. Sorokin, *Contemporary Sociological Theories*, 101, 142–43, 161, 150, 193. Sorokin's detailed assessment of Huntington's work also incorporates critical commentary on his climatic theories of health, race, civilization, economics, birth rate, and other matters. It is therefore also relevant to other aspects of Huntington's thinking that are treated elsewhere in this book.

51. Miller, *Climatology*, 1, 2.

52. See Stehr, "The Ubiquity of Nature"; Klautke, "Defining the *Volk*."

53. Mills, *Climate Makes the Man*, 104, 106.

54. Mills, *Medical Climatology*, 176, 183, 184.

55. Landsberg, *Weather and Health*, 92, 94.

56. See Henderson, "Helmut Landsberg and the Evolution of 20th Century American Climatology." Henderson comments that few climate modelers "appeared to exhibit the kind of restraint that he valued."

57. Thomas, *Under the Weather*, 40, 72, 112.

58. Quetelet, *A Treatise on Man*; Lebeau, "Crime and Climate." Quetelet's continuing influence is clear from the Pinkerton Crime Index, "The Seasonality of Crime."

59. Rosenthal, *Winter Blues*.

60. See Geddes, *Chasing the Sun*.

61. Overy and Tansey, eds., *The Recent History of Seasonal Affective Disorder*, 3, 22.

62. Among the works they examined were Cunningham, "Weather, Mood and Helping Behavior"; Griffitt, "Environmental Effects on Interpersonal Affective Behavior"; Goldstein "Weather, Mood and Internal-External Control"; Sanders and Brizzolara, "Relationships between Mood and Weather."

63. Howarth and Hoffman, "A Multidimensional Approach to the Relationship between Mood and Weather," 21, 22.

64. Keller et al., "A Warm Heart and a Clear Head."

65. Chand and Murthy, "Climate Change and Mental Health," 44, 46.

66. Klimstra et al., "Come Rain or Come Shine," 1495.

67. Connolly, "Some Like It Mild and Not Too Wet," 459.

68. Smith, *Tides in the Affairs of Men*, 23, 28, 29, 33.

69. Parker, *Physioeconomics*, 193, 244.

70. The thinking of both Smith and Parker is further examined in chapter 9.

71. Grohol, "Weather Can Change Your Mood"; Grohol, "Can Weather Affect Your Mood?"

72. *Report of the National Advisory Commission on Civil Disorders*, 71, 327.

73. Baron, "Aggression as a Function of Ambient Temperature," 183.

74. Baron, "Aggression as a Function of Ambient Temperature," 187.

75. Baron and Bell, "Aggression and Heat," 255. Later still, with two other fellow investigators, he published on the influence of atmospheric ions on human behavior. See Baron et al., "Negative Ions and Behavior."

76. Anderson's later work on the effects of video games on aggression was controversial. Critics found his conclusions there to be overstated and blind to the limitations of his data. Some felt that the causal chain he identified linking video games with subsequent aggression was not empirically well grounded.

77. Baron and Ransberger, "Ambient Temperature and the Occurrence of Collective Violence."

78. Carlsmith and Anderson, "Ambient Temperature and the Occurrence of Collective Violence," 337.

79. Anderson, "Temperature and Aggression: Effects of Quarterly, Yearly and City Rates," 1161, 1162, 1172.

80. Anderson, "Temperature and Aggression: Ubiquitous Effects of Heat," 93, 94.

81. Anderson et al., "Hot Years and Serious and Deadly Assault," 1222.

82. Anderson, "Heat and Violence," 37.

83. Plante et al., "Effects of Rapid Climate Change on Violence and Conflict." See Anderson, "Climate Change and Violence," 128–32; Plante and Anderson, "Global Warming and Violent Behavior."

84. Kenrick, "Ambient Temperature and Horn Honking."

85. See Reifman et al., "Temper and Temperature on the Diamond"; Larrick et al., "Temper, Temperature, and Temptation."

86. Auliciems and DiBartolo, "Domestic Violence in a Subtropical Environment."

87. Vrij et al., "Aggression of Police Officers as a Function of Temperature," 369.

88. See Cohn and Rotton, "Assault as a Function of Time and Temperature"; Rotton and Cohn, "Violence Is a Curvilinear Function of Temperature."

89. Cohn, "The Prediction of Police Calls for Service." Earlier work on this theme included Michael and Zumpe, "Annual Rhythms in Human Violence and Sexual Aggression"; Michael and Zumpe, "Sexual Violence in the United States and the Role of Season"; Michael and Zumpe, "An Annual Rhythm in the Battering of Women."

90. Lebeau, "Crime and Climate," 307.

91. Cohn and Rotton, "Assault as a Function of Time and Temperature," 1332.

92. Ranson, "Crime, Weather, and Climate Change," 287.

93. Harries and Stadler, "Determinism Revisited," 254.

94. Ranson, "Crime, Weather, and Climate Change," 287.

95. Agnew, "Dire Forecast," 21.

96. Mares, "Climate Change and Levels of Violence," 778.

97. Wei et al., "Regional Ambient Temperature," 890.

98. They drew on Digman, "Higher-Order Factors of the Big Five"; DeYoung, "Higher-Order Factors of the Big Five in a Multi-Informant Sample."

99. Wei et al., "Regional Ambient Temperature," 890, 891, 892. The various theories the team supplemented are discussed in Talhelm et al., "Large-Scale Psychological Differences"; Rentfrow et al., "A Theory of the Emergence, Persistence, and Expression of Geographical Variation"; Uskul et al., "Ecocultural Basis of Cognition."

100. Choi, "Could Climate Change Affect People's Personalities?"

101. de Vliert, "Climato-Economic Habitats," 465.

102. Van de Vliert had earlier developed aspects of his climato-economic theory of culture in de Vliert, "Climatoeconomic Roots of Survival"; de Vliert, *Climate, Affluence, and Culture*; de Vliert, "Climato-Economic Origins of Variation."

103. Van de Vliert, "Climato-Economic Habitats," 466.

104. Van de Vliert, "Climato-Economic Habitats," 467, 473, 476, 478.

105. Allik and Realo, "How Is Freedom Distributed across the Earth?" 482.

106. Güss, "What about Politics and Culture?"

107. Baumeister et al., "Individual Identity and Freedom of Choice," 485.

108. For a brief overview, see Cianconi et al., "The Impact of Climate Change on Mental Health."

109. Clayton, "Psychology and Climate Change."

110. See the website for the Dawn Wellness Centre and Rehab Thailand at https://thedawnrehab.com/blog/warm-climate-help-recovery/

111. Health Scotland, "Beating the Winter Blues."

112. See, for example, Finnegan, "'An Aid to Mental Health.'" The appropriate physical environments in which asylums were to be located are the focus of attention in Philo, "'Fit Localities for an Asylum.'"

113. de Vliert and Murray, "Climate and Creativity," 17, 19, 26.

114. Dorn, *The Geography of Science.*

115. So, for instance, Pennebaker et al., "Stereotypes of Emotional Expressiveness of Northerners and Southerners."

Chapter 7: Weather, Wealth, and Zonal Economics

1. Mishra, "Can the Climate of a Country Determine Its Wealth?"; Pittis, "The Economic Advantages of Life in a Cold Country"; Kestenbaum, "Hot Climates May Create Sluggish Economies."

2. Landes, *The Wealth and Poverty of Nations*, 7.

3. Diamond, *Guns, Germs and Steel*, chap. 5.

4. Pittis, "The Economic Advantages of Life in a Cold Country."

5. Mishra, "Can the Climate of a Country Determine Its Wealth?"

6. Quoted in Kestenbaum, "Hot Climate May Create Sluggish Economies."

7. Jones et al., "Does Climate Change Affect Economic Growth?"

8. Nordhaus, "Geography and Macroeconomics," 3510, 3511, 3514.

9. Gates, "The Spread of Ibn Khaldūn's Ideas on Climate and Culture," 415.

10. Nordhaus, "Geography and Macroeconomics," 3511, 3514, 3515.

11. See, from different perspectives, Schmidt, *Ibn Khaldûn, Historian, Sociologist and Philosopher*; Fromherz, *Ibn Khaldun, Life and Times*; Irwin, *Ibn Khaldun: An Intellectual Biography*. The influence of Ibn Khaldūn on nineteenth- and twentieth-century American and European liberals and on the intercultural origins of international law is charted in Gamarra, "Ibn Khaldun (1332–1406)."

12. Gates, "The Spread of Ibn Khaldûn's Ideas on Climate and Culture," 419.

13. Ibn Khaldūn, *The Muqaddimah*, vol. 1, 104.

14. Fromherz, *Ibn Khaldun, Life and Times*, 138.

15. Ibn Khaldūn, *The Muqaddimah*, vol. 1, 167–68.

16. Boulakia, "Ibn Khaldûn: A Fourteenth Century Economist," 1109.

17. Ibn Khaldūn, *The Muqaddimah*, vol. 1, 172.

18. Ibn Khaldūn was inconsistent in his depiction of the zones. Elsewhere, reflecting on architecture—"the first and oldest craft of sedentary civilization"—he noted that dwellings constructed by "the inhabitants of the second, third, fourth, fifth, and sixth zones" were characterized by "moderation." Ibn Khaldūn, *The Muqaddimah*, vol. 2, 357.

19. Ibn Khaldūn, *The Muqaddimah*, vol. 1, 168, 169.

20. Ibn Khaldūn, *The Muqaddimah*, vol. 1, 174. On the subject of medical spirits more generally, see Bond, "Medical Spirits and the Medieval Language of Life." See also Rosenthal, *Humor in Early Islam*.

21. Ibn Khaldūn, *The Muqaddimah*, vol. 1, 105.

22. Ibn Khaldūn, *The Muqaddimah*, vol. 2, 136, 137.

23. Fromherz, *Ibn Khaldun, Life and Times*, 129, 127–28.

24. Fromherz emphasizes this point warning against the dangers of decontextualizing "Ibn Khaldun's logical and innovative explanations of history" because these are the parts of the *Muqaddimah* that "seem most attuned to the modern ear." Fromherz, *Ibn Khaldun, Life and Times*, 5.

25. Ibn Khaldūn, *The Muqaddimah*, vol. 3, 262.

26. Asatrian, "Ibn Khaldūn on Magic and the Occult," 110.

27. White, "Ibn Khaldūn in World Philosophy of History," 113.

28. Fromherz, *Ibn Khaldun, Life and Times*, 114.

29. Ibn Khaldūn, *The Muqaddimah*, vol. 1, 169, 170.

30. White, "Ibn Khaldūn in World Philosophy of History," 112, 118.

31. Gellner, Review of Ibn Khaldūn, *The Muqaddimah*. On the significance of Rosenthal's translation, see also Rabīʿ, *The Political Theory of Ibn Khaldūn*.

32. Gellner, "From Ibn Khaldun to Karl Marx," 386.

33. Gellner, *Muslim Society*.

34. Ibn Khaldūn, *Prolégomènes d'Ebn-Khaldoun*.

35. Toynbee, *A Study of History*, vol. 3, 322, 321.

36. Toynbee, *A Study of History*, vol. 3, 324.

37. See McNeill, "Historians, Superhistory, and Climate Change"; Irwin, "Toynbee and Ibn Khaldun."

38. Chardin, *Sir John Chardin's Travels*, 249. This famous passage is regularly quoted. Du Bos cited it in *Réflexions Critiques sur la Poésie et sur la Peinture*, vol. 2, 285. More recently, it is quoted in French by Gates, "The Spread of Ibn Khaldûn's Ideas on Climate and Culture," 418, and in English, by Glacken, *Traces on the Rhodian Shore*, 553, and Gaukroger, *The Natural and the Human*, 252.

39. Gates, "The Spread of Ibn Khaldûn's Ideas on Climate and Culture," 421.

40. Eurich, "Chardin, Sir John [Jean]."

41. Chardin, *Sir John Chardin's Travels*, 193, 227.

42. Voltaire, "Climate." This same phrasing, without attribution, and referring to the "bad atmosphere of a great portion of Portugal" appears in [Ashe], *History of the Azores or Western Islands*, 219.

43. The original French reads, "on pourroit croire que la Zone torride & les deux Glaciales, ne font pas fort propres pour les Sciences." Fontenelle, *Poésies Pastorales*, 144.

44. Du Bos, *Critical Reflections on Poetry*, vol. 2, 213.

45. Du Bos, *Critical Reflections on Poetry*, vol. 2, 102, 112.

46. Du Bos, *Critical Reflections on Poetry*, vol. 1, 145; vol. 2, 111, 112.

47. See Tooley, "Bodin and the Medieval Theory of Climate." See also the discussion of Bodin in Boia, *The Weather in the Imagination*, 33–38; Glacken, *Traces on the Rhodian Shore*, 434–47.

48. Bodin, *Six Books of the Commonwealth*, 149.

49. Bodin, *Method for the Easy Comprehension of History*, 85.

50. Bodin, *Six Books of the Commonwealth*, 150.

51. Tooley, "Bodin and the Medieval Theory of Climate," 79.

52. Bodin, *Six Books of the Commonwealth*, 155.

53. Bodin, *Method for the Easy Comprehension of History*, 138.

54. de Dainville, *La Géographie des Humanistes*, 25–27.

55. Bodin, *Six Books of the Commonwealth*, 158 159, 157.

56. Gates, "The Spread of Ibn Khaldûn's Ideas on Climate and Culture," 416.

57. Glacken, *Traces on the Rhodian Shore*, 434.

58. Voltaire, "Climate," 203.

59. See the discussion of Arbuthnot's *Essay Concerning the Effects of Air* in chapter 2.

60. See Shackleton, "The Evolution of Montesquieu's Theory of Climate."

61. Montesquieu, *The Spirit of the Laws* (1989), book XIV, §2.

62. Montesquieu, *The Spirit of the Laws* (1989), book XIV, §2.

63. Montesquieu, *The Spirit of the Laws* (1989), book XVI, §4.

64. Montesquieu, *The Spirit of the Laws* (1989), book XIV, §4, §5.

65. Spector, "Europe."

66. More generally, on this theme, see Livingstone, "Race, Space and Moral Climatology."

67. Montesquieu, *The Spirit of the Laws* (1989), book XIV, §2.

68. Montesquieu, *The Spirit of the Laws* (1989), book XIV, §10.

69. See Devletoglou, "Montesquieu and the Wealth of Nations."

70. Montesquieu, *The Spirit of the Laws* (1989), book XIV, §5.

71. See the discussion in Macfarlane, *Montesquieu and the Making of the Modern World.*

72. Montesquieu, *The Spirit of the Laws* (1989), book XV, §1, §7; book XVI, §2; book XVI, §7.

73. Montesquieu, *The Spirit of the Laws* (1989), book XXI, §3.

74. The role of geography and race in Enlightenment thinking is discussed in Withers, *Placing the Enlightenment*, notably chap. 7, "Geographies of Human Difference."

75. On Kant's geographical interests, see Elden and Mendieta, eds., *Reading Kant's Geography*. On the anthropological thinking of Kant and Herder, see Zammito, *Kant, Herder, and the Birth of Anthropology*.

76. Kant, "On the Different Races of Man," 23. On Kant's racial thinking, see Bernasconi, "Who Invented the Concept of Race?" See also Livingstone, "Race, Space and Moral Climatology"; Kleingeld, "Kant's Second Thoughts on Race"; Mikkelsen, ed., *Kant and the Concept of Race.*

77. Kant, "On National Characteristics," 55.

78. Kant, "On National Characteristics," 50.

79. Kant, "On Countries That Are Known and Unknown to Europeans," 63, 64.

80. David Hume, "Of National Characters," 207, 209, 213, 215.

81. Gaukroger, *The Natural and the Human*, 254. See also Ilany, *In Search of the Hebrew People.*

82. Smith, *An Inquiry into the Nature and Causes of the Wealth of Nations*, 472, 473.

83. Cousin, *Introduction to the History of Philosophy*, 251, 252. See also Nacci, "'Tel Climat Donné, Tel Peuple Suit'"; Sörlin, "Environmental Times."

84. Herder, *Outlines of a Philosophy of the History of Man*, book 7, chap. 5, 184.

85. Berlin, *Vico and Herder*, 176.

86. Herder, *Outlines of a Philosophy of the History of Man*, book 7, chap. 5, 184.

87. So, for example, Bahr et al., "On Diversity, Empathy, and Community."

88. On what might be called Herder's ambivalent pluralism, see Linker, "The Reluctant Pluralism of J.G. Herder."

89. Herder, *Outlines*, book 7, chap. 1, 163, 166. See Zammito, *Kant, Herder*, 345.

90. See the discussion of Herder's historicism in Beiser, *The German Historicist Tradition*, 98–166.

91. Herder, *Outlines*, 176.

92. Herder, *Outlines*, 169, 173, 12, 22.

93. On Humboldt and the Humboldt phenomenon more generally, see Rupke, *Alexander von Humboldt*.

94. Humboldt and Bonpland, *Essai sur la Géographie des Plantes*, 141, 139.

95. Humboldt and Bonpland, *Personal Narrative of Travels*, vol. 3, 14–15.

96. On the influence of Ratzel on Semple, see Keighren, *Bringing Geography to Book*. Alfred Kroeber insisted that Ratzel "was far from being the crass environmentalist which Semple's misrepresentatively selected adaptation makes him out to be." See Kroeber, *Cultural and Natural Areas of Native North America*, 7.

97. Wanklyn, *Friedrich Ratzel*; Lowie, *The History of Ethnological Theory*, 119–27.

98. I have discussed something of this project in Livingstone, *The Geographical Tradition*, 196–202. For a recent attempt to find contemporary resonances in Ratzel's thinking and to appropriate his biogeographical politics for latter-day purposes, see Barua, "Ratzel's Biogeography."

99. Ratzel, *The History of Mankind*, vol. 1, 27.

100. Ratzel, *The History of Mankind*, vol. 1, 28, 29.

101. Althusser, *Politics and History*, 29.

102. White, "Ibn Khaldūn in World Philosophy of History," 115. Shortly after publishing this assessment, White resisted scientific history by urging that historians should be alert to the similarities between art and science and seek to transcend them. See White, "The Burden of History."

103. White, "Ibn Khaldūn in World Philosophy of History," 124, 125.

Chapter 8: Slavery, Sustenance, and Tropical Supervision

1. Biographical details are available in Dana, "Memoir of Arnold Guyot"; Libbey, "The Life and Scientific Work of Arnold Guyot"; Ferrell, "Arnold Henry Guyot." See also Koelsch, "Seedbed of Reform." I discuss Guyot's contribution to matters of science and religion in Livingstone, *Dealing with Darwin*.

2. Guyot, *The Earth and Man*, viii.

3. Guyot, *The Earth and Man*, 246, 242.

4. Guyot, *The Earth and Man*, 306.

5. Guyot, *The Earth and Man*, 308.

6. On the whole issue of scientific theodicy, see Young, "The Historiographic and Ideological Contexts of the Nineteenth-Century Debate"; Moore, "Theodicy and Society."

7. Guyot, *The Earth and Man*, 209.

8. Guyot, *The Earth and Man*, 232.

9. Guyot, *The Earth and Man*, 194, 275. On Guyot's abolitionist sentiments, see Wilson, "Influences of 'Kosmos' in 'Earth and Man.'"

10. Jefferson, *Notes on the State of Virginia*, 319, 270.

11. See Magnis, "Thomas Jefferson and Slavery."

12. Van Evrie, *Negroes and Negro Slavery*.

13. Van Evrie, *White Supremacy and Negro Subordination*, 148, 212.

14. George M. Fredrickson, *The Black Image in the White Mind*, 92. Van Evrie is discussed on 91–96.

15. Van Evrie, *White Supremacy and Negro Subordination*, 30–31.

16. Van Evrie, *White Supremacy and Negro Subordination*, 315, 264.

17. See Agassiz, "The Diversity of Origin"; Agassiz, "Geographical Distribution of Animals." See also Lurie, "Louis Agassiz and the Races of Man."

18. Van Evrie, *White Supremacy*, 245, 256.

19. See Livingstone, *Nathaniel Southgate Shaler*.

20. Shaler, *Nature and Man in America*, 211, 244.

21. Shaler, *Kentucky*, 232–33.

22. Shaler, *The Story of Our Continent*, 186.

23. Shaler, *The Citizen*, 224.

24. Shaler, "The Negro Problem," 707.

25. Shaler, *Nature and Man in America*, 202.

26. Shaler, "The Negro Problem," 697, 698.

27. Shaler, *The Citizen*, 223.

28. Shaler, *The Neighbor*, 147, 187.

29. Shaler, *The Neighbor*, 188.

30. Shaler, "Science and the African Problem," 42.

31. Shaler, *The Neighbor*, 191, 133.

32. See Fisher, "Antebellum Black Climate Science," 474. See also Kass, "Dr. Thomas Hodgkin, Dr. Martin Delany, and the 'Return to Africa'"; Levine, *Martin Delany, Frederick Douglass and the Politics of Representative Identity*; Asaka, *Tropical Freedom*.

33. Emerson, "Geographic Influences in American Slavery." A brief biographical sketch was published by Brigham, "Memoir of Frederick Valentine Emerson."

34. The synthesizing of racial thinking and the science of meteorology in what has been called "racial climatology" in nineteenth-century America is discussed in Baker, "Meteorological Frontiers."

35. Abbe, "Introduction," xxix.

36. See Semmel, "H.T. Buckle."

37. Buckle, *History of Civilization*, vol. 1, 8.

38. Buckle, *History of Civilization*, vol. 1, 9, 19, 33, 34.

39. See Semmel, "H.T. Buckle"; Bowler, *The Invention of Progress*, 27–32.

40. Buckle, *History of Civilization*, vol. 1, 41.

41. Buckle, *History of Civilization*, vol. 1, 55, 56.

42. Buckle, *History of Civilization*, vol. 1, 63, 64.

43. Buckle, *History of Civilization*, vol. 1, 32.

44. A history of the wage fund theory is available in Stirati, *The Theory of Wages in Classical Economics*.

45. Buckle, *History of Civilization*, vol. 1, 43, 44.

46. Buckle, *History of Civilization*, vol. 1, 81, 69, 71.

47. Buckle, *History of Civilization*, vol. 1, 39, 131.

48. Lyell, *Principles of Geology*, vol. 2, 232, 233.

49. Buckle, *History of Civilization*, vol. 1, 123, 124.

50. See Wells, "The Critics of Buckle"; Heyck, "Buckle, Henry Thomas." Wells' account is, in part, a defence of Buckle against his critics, particularly Dilthey, Collingwood, and Croce.

51. Quoted in Semmel, "H.T. Buckle," 370.

52. Lavery, "Situating Eugenics"; Cannato, "Immigration and the Brahmins."

53. See Solomon, *Ancestors and Immigrants*, 99–101.

54. Ward, "Fallacies of the Melting Pot Idea," 230–31. I have discussed this in Livingstone, "'Never Shall Ye Make the Crab Walk Straight.'"

55. See Putnam, *Memories of a Publisher*.

56. Ward, *Climate*, iv.

57. That view was shared by Friedrich Ratzel and later by the Cornell geologist and mathematician William H. Norton. See Norton, "The Influence of the Desert on Early Islam."

58. Ward, *Climate*, 258, 280.

59. See Clayton, "The Influence of Rainfall on Commerce"; Moore, *Economic Cycles*; Pettersson, *Climatic Variation in Historic and Prehistoric Time*. The work of Clayton, Moore, and Pettersson on climate and economy is more fully discussed in chapter 9.

60. Ward, *Climate*, 227.

61. Ward, *Climate*, 227–28. On John R. Commons, see Cherry, "Racial Thought and the Early Economics Profession"; Leonard, "More Merciful and Not Less Effective"; Ramstad and Starkey, "The Racial Theories of John R. Commons"; Gonce, "John R. Commons's 'Five Big Years.'"

62. Ward, *Climate*, 228. Here he quoted from Guyot, *The Earth and Man*, 246.

63. Ward, *Climate*, 229, 230.

64. Ward, *Climate*, 272, 273, 274.

65. Kidd, *The Control of the Tropics*, 52.

66. The major biographical treatment is Crook, *Benjamin Kidd*. See also Sturt, "Kidd, Benjamin."

67. Hofstadter, *Social Darwinism*, 100–1.

68. See Bannister, *Social Darwinism*, 150–58.

69. Quoted in Jones, *Social Darwinism in English Thought*, 1, from an interview Kidd gave to the *Daily Chronicle* on June 20, 1894.

70. For further details, see Crook, *Benjamin Kidd*, 81–141.

71. Kidd, *The Control of the Tropics*, 15.

72. Kidd, *The Control of the Tropics*, 24.

73. The debates over human acclimatization are discussed in chapter 3.

74. Kidd, *The Control of the Tropics*, 30, 48.

75. Kidd, *The Control of the Tropics*, 78, 53.

76. Kidd, *The Control of the Tropics*, 83, 85–86.

77. Quoted in Crook, *Benjamin Kidd*, 121. I have relied on Crook's survey of responses to Kidd's volume on 121–24.

78. Ireland, "Is Tropical Colonization Justifiable?" 336–37.

79. Ireland, *Tropical Colonization*, 33.

80. Ireland, *Tropical Colonization*, 156–57.

81. Ireland, *Tropical Colonization*, 164.

82. Ireland, "Is Tropical Colonization Justifiable?" 337–38.

83. Johnston, *A History of the Colonization of Africa*, 101–2.

84. Johnston, *A History of the Colonization of Africa*, facing 275.

85. See Livingstone, *Livingstone's Lives*, 156–57; McMullen, "Kirk, Sir John"; Hazell, *The Last Slave Market*.

86. Johnston, "Sir John Kirk."

87. Kirk, "The Extent to Which Tropical Africa Is Suited for Development," 525, 526.

88. Kirk, "The Extent to Which Tropical Africa Is Suited for Development," 534, 535.

89. Biographical treatments include Mellor, *Lugard in Hong Kong*; Perham, *Lugard: The Years of Adventure 1858–1898*; Perham, *Lugard: The Years of Authority 1898–1945*.

90. See Bain, *Between Anarchy and Society*.

91. Kirk-Greene, "Lugard, Frederick John Dealtry."

92. Ofcansky, "A Bio-Bibliography of F.D. Lugard."

93. Kirk, "Extent," 535.

94. Lugard, *The Dual Mandate*, 20, 357, 358, 44, 507.

95. Lugard, *The Dual Mandate*, 42, 282, 43.

Chapter 9: Climate, Capital, Civilization

1. Huntington, *Civilization and Climate*, chap. 4. This work was favorably regarded by some medical practitioners. See, for instance, James Alexander Miller's presidential address to the American Climatological and Clinical Association in 1916, "Some Physiological Effects of Various Atmospheric Conditions." Miller was also president of the New York Academy of Medicine. See Goodridge and Van Ingen, "James Alexander Miller—In Memoriam."

2. Huntington, *Civilization and Climate*, 57–58.

3. Huntington, *Civilization and Climate*, 81–82.

4. Huntington, *Civilization and Climate*, chap. 6, "Work and Weather," passim.

5. Huntington, *Civilization and Climate*, 135, 136.

6. Huntington and Williams, *Business Geography*, 92, 70.

7. Huntington, *Civilization and Climate*, 49, 149.

8. Huntington, *Civilization and Climate*, 161.

9. Letter, J. Russell Smith to Ellsworth Huntington, July 3, 1914, Huntington Papers, box 6, folder 34, Yale University Archives.

10. Letter, Franz Boas to Ellsworth Huntington, November 5, 1913, Huntington Papers, box 6, folder 34, Yale University Archives. Parts of this letter appeared anonymously in *Civilization and Climate*, 158.

11. Huntington, *Civilization and Climate*, 162, 163.

12. A. L. Kroeber to Ellsworth Huntington, December 6, 1913, Huntington Papers, box 6, folder 34, Yale University Archives.

13. Huntington, *Civilization and Climate*, 157; Letter, George G. Chisholm to Ellsworth Huntington, November 13, 1913, Huntington Papers, box 6, folder 34, Yale University Archives.

14. Huntington, *Civilization and Climate*, 156, 157.

15. Huntington, *Civilization and Climate*, 205.

16. Huntington and Williams, *Business Geography*, 77, 78.

17. Huntington, *Civilization and Climate*, 218–19.

18. Huntington, *Civilization and Climate*, 270.

19. See Powell, *Griffith Taylor and 'Australia Unlimited'*; David Oldroyd, "Griffith Taylor and His Views on Race."

20. See Christie, "'Pioneering for a Civilized World,'" 129.

21. Biographical details are available in Strange and Bashford, *Griffith Taylor*.

22. Taylor, *Australia in Its Physiographic and Economic Aspects*, 248, 253.

23. Strange and Bashford, *Griffith Taylor*, 82.

24. Apparently, Taylor was not the first to use this term. Carolyn Strange observes that it was first used in a 1910 article by Ball, "Climatological Diagrams." See Strange, "Transgressive Transnationalism," 35.

25. In turn, Huntington found Taylor's diagrams "most illuminating" (as he put it in a 1917 letter to Taylor) and used a modified version of the climograph in his own 1919 *World Power and Evolution* to depict conditions in a wide variety of different locations. See Strange and Bashford, *Griffith Taylor*, 86.

26. Taylor, "The Control of Settlement by Humidity and Temperature," 24, 32.

27. Griffith Taylor, "Geographical Factors," 55. This article was reprinted as "The Settlement of Tropical Australia" in 1919.

28. Taylor, "The Distribution of Future White Settlement," 375, 377.

29. Taylor, "The Distribution of Future White Settlement," 392, 393.

30. Taylor, "The Distribution of Future White Settlement," 402.

31. Taylor, *Australia: A Study of Warm Environments*, 75.

32. Taylor, *Australia: A Study of Warm Environments*, 444, 445.

33. Taylor, *Canada: A Study of Cool Continental Environments*, 517.

34. See Christie, "'Pioneering for a Civilized World,'" 117.

35. Benedict, "Isothermic Anthropology."

36. Taylor's work informed Bowman's "The Pioneer Fringe" and his much more extensive study, *The Pioneer Fringe*. In this latter publication, Bowman's technological possibilism plainly surfaces. For while drawing on Taylor's work on settlement, he resisted climatic reductionism by insisting that "man" "measurably conquers his environment with the tools or the techniques at his command and every now and then improves both of them." Such actions, he went on, effectively "checkmate the climate" (76). On Bowman more generally, see Smith, "Bowman's New World"; Wright and Carter, "Isaiah Bowman 1878–1950"; Smith, *American Empire: Roosevelt's Geographer*. Strange and Bashford cite a letter Bowman sent to Taylor in 1928, in which he wrote, "It is with some amazement that I see you are accepting in what seems to be an uncritical way the loose generalities of my friend Ellsworth Huntington. . . . I regard a large part of Huntington's writing a waste of time to read." Strange and Bashford, *Griffith Taylor*, 99.

37. Strange, "Transgressive Transnationalism," 29, 39.

38. Huntington, "Climatic Variations and Economic Cycles."

39. Clayton was the author of *World Weather*; "World Weather and Solar Activity"; *The Sunspot Period*; "The Sunspot-Period and Spring Rainfall in the United States."

40. Rawson was the author of *Report on the Rainfall of Barbados*. See Morton-Gittens, "Sir Rawson William Rawson." Hall was the author of *The Rainfall of Jamaica* and *The Meteorology of Jamaica*. Robert De Courcy Ward provided an obituary for Hall in the *Daily Gleaner*, September 8, 1920.

41. Clayton, "The Influence of Rainfall on Commerce," 163, 164.

42. Ward, "A Year of Weather and Trade," 448.

43. Moore, *Economic Cycles*, 135.

44. More generally, Mirowski comments that Moore "was much more knowledgeable in the scientific literature of his time than any of his peers." Mirowski, "Problems in the Paternity of Econometrics," 600.

45. Stigler, "Henry L. Moore and Statistical Economics."

46. Mirowski, "Problems in the Paternity of Econometrics," 599.

47. Moore, *Economic Cycles*, 124.

48. Mirowski, "Problems in the Paternity of Econometrics," 598.

49. Moore, *Economic Cycles*, 110.

50. Jevons, "Commercial Crises and Sunspots." For further discussion, see Gallegati, "Jevons, Sunspot Theory."

51. Stigler, "Henry L. Moore and Statistical Economics," 11.

52. Barnhart, "Rainfall and the Populist Party."

53. Elton, "Periodic Fluctuations in the Numbers of Animals." See also Erickson, "Knowing Nature through Markets."

54. Erickson and Mitman, *When Rabbits Became Humans*.

55. Huntington, "The Matamek Conference on Biological Cycles," 231. The substance of this report, together with resolutions that were passed at the conference and a list of participants, can be accessed at https://cyclesresearchinstitute.wordpress.com/2015/06/29/the-matamak-conference-on-cycles-in-1931/

56. Jacoby, Review of *The Causes of Economic Fluctuations*.

57. King, *The Causes of Economic Fluctuations*, 223.

58. Smith, *Tides in the Affairs of Men*, 11, 70.

59. In chapter 6, we explored something of the influence climate was thought to exert over the human mind.

60. Smith was also influenced by the sunspot theories of Harlan True Stetson (1885–1964), an astronomer who directed the MIT Cosmic Terrestrial Research Laboratory. Smith came to believe that variations in radiation also had psychological effects, which fed through into buying and selling waves, thereby determining fluctuations in commodity prices. See Stetson, *Sunspots and Their Effects*.

61. Blair, *Climatology*, 102, 108.

62. Miller, *Climatology*, 1–4.

63. Wheeler, "Climatic Phases and Business Cycles," 146, 186.

64. I discuss Mills' biometeorological work in chapter 4. He was also the author of "Weather Stimulation, Human Energy and Business Cycles," chap. 11 of Mills, *Climate Makes the Man*.

65. Wheeler, "The Effect of Climate on Animals and Human Beings," 50.

66. Wheeler, "Climatic Belts of the World," 11.

67. It is important to point out that some who repudiated climatic determinism could still adopt a zonal account of civilization. In his *Les Pays Tropicaux*, French geographer Pierre Gourou affirmed that great civilizations had only existed in the temperate world and that in tropical zones, any germ of civilization that took shape had come from outside the tropics. See Gourou, *Les Pays Tropicaux*. For Gourou's thoughts on environmental determinism, see Gourou, "Le Déterminisme Physique dans 'l'Esprit des Lois.'" A thorough analysis of Gourou's tropicality is available in Bowd and Clayton, *Impure and Worldly Geography*.

68. Wheeler, "Climatic Influence and the Industrial Centres of the World," 15, 25.

69. Wheeler, "The Changeableness of Climate," 55. See also Wheeler's "Evidence Dealing with the Climate of Historic Time."

70. Wheeler, "The Principal Climatic Cycles in History," 113.

71. Wheeler, "Climatic Conditioning in History," 127.

72. On Pettersson and his circle, see Thompson, "Dr Otto Pettersson"; Bergwik, "Father, Son, and the Entrepreneurial Spirit."

73. Pettersson, *Climatic Variations in Historic and Prehistoric Time*.

74. Wheeler, "Presidential Address," 35. The citationary architecture of Wheeler's project is dramatically on view in this piece as he reprises the work of such climatic determinists as Bodin, Arbuthnot, Montesquieu, Guyot, Ward, Huntington, Mills, Peterson, Moore, and Edgar Lawrence Smith.

75. Smith, *Industrial and Commercial Geography*, 10.

76. Trevor Barnes has recently argued that the stern environmental determinist beginnings of Smith's text were not characteristic of the entire work. See Barnes, "'In the Beginning Was Economic Geography.'"

77. Wheeler, "Presidential Address," 45.

78. Sörlin, "The Global Warming That Did Not Happen," 93.

79. See Douglass, "Weather Cycles in the Growth of Big Trees." Huntington had examined the sunspot theory in "The Solar Hypothesis of Climatic Changes."

80. Huntington, *Civilization and Climate*, 270.

81. Huntington, *The Pulse of Asia*, 385, 382. Here Huntington drew on Eduard Brückner's claims about a thirty-five-year climatic cycle, to bolster his views on climatic fluctuations. On Brückner's analyses, see Lehmann, "Whither Climatology?"

82. See Butlin, "George Adam Smith"; Aiken, *Scriptural Geography*.

83. Huntington, *Palestine and Its Transformation*, 4.

84. In coming to these judgments, Huntington had called on the testimony of Irish missionaries such as Josiah Leslie Porter, whose maps of the ancient cities of Damascus and Palymra he used, and William Wright, author of *The Empire of the Hittites* and *An Account of Palmyra and Zenobia, with Travels and Adventures in Bashan and the Desert*. Wright's Hittite project is discussed in Mathieson, "Irish Missions, Science, and Scripture." For the use of missionaries in understanding other dimensions of climate knowledge, see Endfield and Nash, "Drought, Desiccation and Discourse"; Endfield and Nash, "Missionaries and Morals."

85. Huntington, "Changes of Climate and History." Huntington had in his cross-hairs Olmstead's "Climate and History."

86. See Ellsworth Huntington, "The Greenest of Deserts"; Huntington, "Physical Environment as a Factor in the Present Condition of Turkey"; MacDougal, "North American Deserts"; Huntington, "The Arabian Desert and Human Character"; Huntington, "The Fluctuating Climate of North America."

87. Huntington, *The Pulse of Progress*, 1.

88. Huntington, *Mainsprings of Civilization*.

89. See the discussion in Martin, *Ellsworth Huntington*, 102–11; Kullmer, "The Latitude Shift of the Storm Track"; Kullmer, "A Remarkable Reversal."

90. Huntington, *Mainsprings of Civilization*, 518, 519, 520.

91. Fleming, "Civilization, Climate, and Ozone," 225.

92. By this I do not mean to imply that concern with environmental influence entirely disappeared from the academy. It remained in the *Annales* school, in the writings of certain climate historians, and among some archaeologists. Coombes and Barber remark, "Such studies were to remain infrequent until the 1990s, when concern over anthropogenic impacts on global warming and a growing appreciation of the magnitude and rapidity of Holocene climatic change combined to produce an intellectual climate suddenly sympathetic to the idea of environmentally triggered catastrophes." Coombes and Barber, "Environmental Determinism in Holocene Research," 304. On this revival, see also Chambers and Brain, "Paradigm Shifts in Late-Holocene Climatology?" Hubert Horace Lamb, himself a major figure in the development of the study of climate change's influence on human history, remarked in 1981 that historians and archaeologists had only recently begun to seriously take account of climate change. Lamb, "An Approach to the Study of the Development of Climate," 291–309.

93. Landes, *The Wealth and Poverty of Nations*, 3, 4.

94. Landes, *The Wealth and Poverty of Nations*, 4, 5.

95. Landes, *The Wealth and Poverty of Nations*, 14

96. Landes, *The Wealth and Poverty of Nations*, 17, 18.

97. Boia, *The Weather in the Imagination*, 114.

98. Diamond, *Guns, Germs and Steel*, 94.

99. Diamond, *Guns, Germs and Steel*, 184.

100. Diamond, *Guns, Germs and Steel*, 15, 22, 307, 313, 298.

101. Blaut, "Environmentalism and Eurocentrism," 396.

102. Merrett, "Debating Destiny," 803, 804. Another critical essay is Judkins et al., "Determinism within Human-Environment Research."

103. Davis, "The World Until Yesterday by Jared Diamond."

104. Diamond, *Guns, Germs and Steel*, 178.

105. Fagan, *Floods, Famines, and Emperors*, 108.

106. Fagan, *Floods, Famines, and Emperors*, 114.

107. Fagan, *The Little Ice Age*, xiii.

108. Fagan, *The Long Summer*, 70–71.

109. Fagan, *Floods, Famines, and Emperors*, 160, 165.

110. Fagan, *The Great Warming*, 11.

111. In a popular piece for the *Scientific American*, Daniel Grossman identified Richardson Gill as an example of "the pendulum swinging back" to earlier geophysical explanations for the collapse of civilizations of the sort (though now much more sophisticated) promoted by Ellsworth Huntington. Gill is the author of *The Great Maya Droughts*. See Grossman, "Parched Turf Battle."

112. Some attribute Mayan conflicts less to drought than to increased summer temperatures. For instance, Carleton et al., "Increasing Temperature Exacerbated Classic Maya Conflict."

113. Haug et al., "Climate and the Collapse of Maya Civilization."

114. Hodell et al., "Possible Role of Climate in the Collapse of Classic Maya Civilization."

115. Hodell, "Maya Megadrought?"

116. Aimers, "Drought and the Maya."

117. Butzer, "Collapse, Environment, and Society," with supplemental information at www.pnas.org/lookup/suppl/doi:10.1073/pnas.1114845109/-/DCSupplemental

118. See, for instance, Endfield, *Climate and Society in Colonial Mexico*; Endfield, "The Resilience and Adaptive Capacity"; Butzer and Endfield, "Critical Perspectives on Historical Collapse."

119. Erickson, "Neo-environmental Determinism and Agrarian 'Collapse.'" The "nondeterministic nature of human responses to past climate change" are also emphasized in Armit et al., "Rapid Climate Change Did Not Cause Population Collapse," 17045. See also Plunkett and Swindles, "Bucking the Trend." Similar reservations have been expressed over linking cooling conditions and societal change during the Late Antique Little Ice Age of the sixth and seventh centuries. See Büntgen et al., "Cooling and Societal Change."

120. Joseph A. Tainter issues some warnings about the popularization of collapse narratives, especially when called upon to predict future trends. See Tainter, "Collapse, Sustainability, and the Environment."

121. These titles may be found, respectively, at https://co2cards.com/ancient-civilizations-collapsed; https://www.nytimes.com/2014/05/28/opinion/climate-change-doomed-the-ancients.html; https://nypost.com/2019/06/03/climate-change-could-end-human-civilization-by-2050-report/

122. For example, Klimenko, "Thousand-Year History."

123. Nikonov, *Civilization's Temperature*, 50, 133, 136.

124. Behringer, *A Cultural History of Climate*, 98.

125. Lamb, *Climate. Present, Past and Future*, vol. 1; Lamb, *Climate. Present, Past and Future*, vol. 2; Grove, *The Little Ice Age*; Pfister, "The Little Ice Age"; Pfister, *Klimageschichte der Schweiz 1525–1860*. Pfister and Brázdil have provided a useful overview in Pfister and Brázdil, "Climatic Variability in Sixteenth-Century Europe."

126. Behringer, "Weather, Hunger and Fear," 8. See also Behringer, "Climatic Change and Witch-hunting."

127. Quoted in Behringer, "Weather, Hunger and Fear," 9.

128. Behringer, *A Cultural History of Climate*, 128.

129. Oster, "Witchcraft, Weather and Economic Growth," 221.

130. Kwiatkowska, "The Light Was Retreating Before Darkness," 34.

131. Apps, "Witchcraft: A Problem for All Times?" 492.

132. Behringer, *Cultural History of Climate*, 123.

133. Behringer, *Cultural History of Climate*, 133, 146.

134. So, for example, Fagan, *Little Ice Age*, 91; Parker, *Global Crisis*, 8ff; Brooke, *Climate Change and the Course of Global History*, 451.

135. Behringer, "Weather, Hunger and Fear," 1, 18.

136. Hulme, Review of *A Cultural History of Climate* by Wolfgang Behringer. See also the critical assessment by Haude, "'Keep Calm'? A Critique of Wolfgang Behringer's 'A Cultural History of Climate.'"

137. Though Buckle's name is conspicuous by its absence, there are, to my mind, strong resonances between Parker's proposals and Buckle's philosophy of food.

138. Parker, *Physioeconomics*, viii.

139. Parker, *Physioeconomics*, 28, 29.

140. Parker, *Physioeconomics*, ix vii.

141. Parker, *Physioeconomics*, 177, 200.

142. Heal and Park, "Feeling the Heat," 2.

143. See, for example, Wyon, "The Effects of Moderate Heat Stress"; Zivin et al., "Temperature and Human Capital"; Hancock et al., "A Meta-Analysis of Performance Response under Thermal Stressors."

144. These would include Palutikof, "The Impact of Weather and Climate on Industrial Production"; Cao and Wei, "Stock Market Returns"; Worthington, "Whether the Weather"; Dell et al., "Temperature and Income"; Burke et al., "Global Non-Linear Effect of Temperature on Economic Production"; Griffin et al., "Extreme High Surface Temperature Events."

145. Gallup et al., "Geography and Economic Development," 182. This team has repeatedly argued the same point. For a more popular audience, without the complex statistics, see Sachs et al., "The Geography of Poverty and Wealth."

146. Mellinger et al., "Climate, Water Navigability, and Economic Development."

147. Gallup and Sachs, "Agriculture, Climate, and Technology," 734.

148. Sachs, "Institutions Don't Rule: Direct Effects of Geography on Per Capita Income," 2. A contrary view can be found in Rodrik et al., "Institutions Rule."

149. Sachs, "Tropical Underdevelopment," 10, 28.

150. Thus, Sachs, *The End of Poverty*.

151. Rosen, "It's the Politics, Stupid"; Correia, "Climate Change Catastrophism."

152. Sachs, "Helping the World's Poorest."

153. Chappell considers that "Huntington . . . made the most decisive step since Hippocrates towards something new and conclusive in environmental-causation thinking." See Chappell, "Environmental Causation," 166.

154. Harris and Mann, "Global Temperature Trends."

155. Michael, *Fugitive Pieces*; Flannery, *The Future Eaters*, 363.

Chapter 10: Climate Wars

1. These snippets are from brief comments printed in the book's front matter.

2. Dyer, *Climate Wars*.

3. Welzer, *Climate Wars*. This book had earlier appeared in German under the title *Klimakriege: Wofür im 21. Jahrhundert getötet wird* (Frankfurt am Main: S. Fischer Verlag GmbH, 2008).

4. Welzer, *Climate Wars*, 12, 22, 64, 65.

5. Welzer, *Climate Wars*, 75, 81.

6. It is something similar with other writers too. James Lee, for example, comments that he has no intention of giving "new life" to "environmental determinism" but immediately goes on to insist that "climate change is a strong and potent factor in determining the destiny of societies." And while he maintains that his forecasts are "not specific predictions," the odor of historical inevitability certainly lingers in the air. A sample: "there will be an intensification and expansion of an existing area of climate change and conflict, the Equatorial Tension Belt"; "a new Tension Belt, roughly located around the Polar Circles, will arise"; "Africans will increasingly become caught up in livelihood wars, which will pose a threat to the territorial integrity of countries"; "the Sahel will be a zone for intermittent chaos." Lee, *Climate Change and Armed Conflict*, 5, 117, 7, 140.

7. Welzer, *Climate Wars*, 95, 180.

8. Hippocrates, *On Airs, Waters and Places*, parts 16, 23.

9. Herodotus, *The Histories*, book 9.122.3. This brief passage, it has been suggested, may have been inserted by Herodotus as a warning that a Persian threat still existed. See Glacken, *Traces on the Rhodian Shore*, 90–91.

10. Forsdyke, "Athenian Democratic Ideology," 346.

11. Thomas, *Herodotus in Context*, 68.

12. Strabo, *Geography*, book II, chap. 5, §26, 191, 192. See also Glacken, *Traces on the Rhodian Shore*, 104.

13. Vegetius, *Epitome of Military Science*, 3.

14. Floyd-Wilson, *English Ethnicity and Race*, 37f; Blair, *The Theater of Nature: Jean Bodin*.

15. Bodin, *Six Books of the Commonwealth*, 151, 152.

16. Montesquieu, *The Spirit of Laws* (1750), book XIV, §2.

17. Falconer, *Remarks on the Influence of Climate*, 23, 4, 5, 9, 14, 15.

18. Dunbar, *Essays on the History of Mankind*, 207, 221, 222, 232, 218, 219.

19. Draper, *History of the American Civil War*, vol. 1, iii, 37.

20. Fleming, *John William Draper and the Religion of Science*, 120, 116.

21. Draper, *History of the Conflict between Religion and Science*. The "conflict model" is discussed in many places. See, for example, Brooke, *Science and Religion*; Russell, "The Conflict of Science and Religion," 12–16; Cantor, "What Shall We Do with the 'Conflict Thesis?'"; Ungureanu, *Science, Religion, and the Protestant Tradition*.

22. Draper, *The Influence of Physical Agents on Life*, 6. See Draper, *Introductory Lecture, to the Course of Chemistry*.

23. Draper, *History of the American Civil War*, vol. 1, 20, 21–22.

24. Draper, *History of the American Civil War*, vol. 1, 101–2.

25. Draper, *History of the American Civil War*, vol. 1, 110, 100.

26. Draper, *History of the American Civil War*, vol. 1, 242, 342.

27. Draper, *History of the American Civil War*, vol. 1, 361.

28. Draper, *History of the American Civil War*, vol. 1, 469.

29. Fleming, *John William Draper*, 119. Draper's thinking on science and religion is relevant here. See Reuben, *The Making of the Modern University*.

30. Keighren, *Bringing Geography to Book*.

31. Semple, *Influences of Geographic Environment*, 608, 610, 611, 616, 618.

32. Semple, *Influences of Geographic Environment*, 619, 620, 622.

33. Semple, *American History and Its Geographic Conditions*, 239, 346, 280.

34. Shaler, *Kentucky*, 232. See Livingstone, "Science and Society: Nathaniel S. Shaler and Racial Ideology."

35. Semple, *American History and Its Geographic Conditions*, 285.

36. Huntington, *Civilization and Climate*, 22, 23.

37. Huntington, *Mainsprings of Civilization*, 289, 290.

38. Huntington, *Civilization and Climate*, 251.

39. Huntington, *World-Power and Evolution*, 23, 24, 227, 228.

40. See chapter 7. Later, with Stephen Visher, Huntington produced a full-length treatment of the science of climatic oscillations over historic and prehistoric time scales. See Huntington and Visher, *Climatic Changes: Their Nature and Causes*.

41. Huntington, *The Pulse of Asia*, 16, 6.

42. Fleming, *Historical Perspectives on Climate Change*, 97.

43. Huntington, *The Pulse of Asia*, 5, 6, 382.

44. Huntington, "Changes of Climate and History," 213, 215.

45. For a general historical examination of the desiccation thesis, see Grove, *Green Imperialism*. Mike Davis discusses Huntington's thinking on climatic pulsations and reviews a number of severe critics at the time. See Davis, "The Coming Desert." See also Davis, *The Arid Lands*.

46. Huntington, *The Pulse of Asia*, 13, 14, 374, 379.

47. Huntington, *The Pulse of Progress*, 134, 130.

48. Huntington, *Mainsprings of Civilization*, 562, 567, 568; Huntington, "Changes of Climate and History," 226. The fortunes of Genghis Khan have continued to be traced to climatic influence and will be examined in more detail in chapter 11.

49. Huntington, *Mainsprings of Civilization*, 224.

50. See Fleming, *Historical Perspectives on Climate Change*, 98–99.

51. David Arnold, noting the skepticism with which Huntington's analyses came to be regarded, nonetheless tellingly observed that in his own time, Huntington's extremism "was not atypical." See Arnold, *The Problem of Nature*, 34.

52. See Toynbee's foreword to Martin, *Ellsworth Huntington*, ix.

53. Toynbee, *A Study of History*, vol. 1, 292.

54. Toynbee's reliance on Huntington was severely criticized by Spate, "Reflections on Toynbee's *A Study of History*," 287–304. This analysis originally appeared in *Historical Studies: Australia and New Zealand* 5, no. 20 (1953): 324–37.

55. Toynbee, *Study of History*, vol. 2, 413; vol. 1, 278.

56. Toynbee, *Study of History*, vol. 3, 435, 440.

57. Wright believed that civilized societies could "substitute 'rational' for 'necessary' solutions."

58. Wright, *A Study of War*, 63, 354 (table 9).

59. Wright, *A Study of War*, 468, 469.

60. Markham, *Climate and the Energy of Nations*, x, 10, 21.

61. Cox, *Losing Our Cool*, 87.

62. Markham, *Climate and the Energy of Nations*, 20.

63. Cowie, *Climate Change: Biological and Human Aspects*, 344.

64. Markham, *Climate and the Energy of Nations*, 180, 163, 88.

65. Recall that the subtitle of *The Pulse of Asia* was *A Journey in Central Asia Illustrating the Geographic Basis of History*.

66. Zhang et al., "Climate Change and War Frequency," 403.

67. Huang and Su, "Climate Change and Zhou Relocations," 310.

68. Chen, "Climate Induced Migration and Conflict."

69. Mazo, *Climate Conflict*, 37.

70. Hsiang et al., "Quantifying the Influence of Climate on Human Conflict"; Huntington, "Climatic Change and Agricultural Exhaustion."

71. Wheeler, "Effect of Climate on Human Behavior in History," 37.

72. Wheeler, "Effect of Climate on Human Behavior in History," 48.

73. Wheeler, "Climatic Conditioning in History," 129.

74. Zahorchak, ed., *Climate*, 245.

75. In 2015, it was reproduced in a piece, "Raymond H. Wheeler's 100 Year Cycle of Climate Change, Regime Change and War," for *Time-Price Research*, http://time-price-research-astrofin.blogspot.com/2015/06/raymond-h-wheelers-100-year-cycle-of.html; Harris and Mann, "Global Temperature Trends."

76. Booker, "Global Crisis, by Geoffrey Parker."

77. Degroot, "Review—Global Crisis"; Simms, "When Winter Came for Kings"; Cushman, Review of *Global Crisis* by Geoffrey Parker, 1429.

78. Parker, *Global Crisis*, 1. Parker had previously outlined the central thrust of his argument, including the Voltaire quotation, in Parker, "Crisis and Catastrophe."

79. Cushman, Review of *Global Crisis* by Geoffrey Parker, 1430.

80. Parker, *Global Crisis*, xix.

81. Parker, "Crisis and Catastrophe," 1073.

82. Parker, *Global Crisis*, 14.

83. Degroot, "Review—Global Crisis"; Booker, "Global Crisis by Geoffrey Parker"; Cushman, Review of *Global Crisis* by Geoffrey Parker, 1430.

84. De Vries, "The Crisis of the Seventeenth Century," 375.

85. Pagden, Review of *Global Crisis*, 516.

86. Pagden, Review of *Global Crisis*, 516; De Vries, "The Crisis of the Seventeenth Century," 376.

87. More recently, it has been argued that the whole idea of a Little Ice Age is misconstrued and that its proposed characteristics are actually an artifact of the statistical procedures employed in historical reconstruction. See Kelly and Ó Gráda, "The Waning of the Little Ice Age"; Ó Gráda, "Is Europe's Little Ice Age a Myth?"

88. Dustcover blurb on Parker, *Global Crisis*.

89. Fagan, *The Little Ice Age*, 21.

90. Fagan, *The Little Ice Age*, xviii, 59, 84.

91. Degroot, *The Frigid Golden Age*, 12.

92. Degroot, *The Frigid Golden Age*, 16, 3, 5, 155, 195.

93. Degroot, *The Frigid Golden Age*, 195, 167, 168.

94. Jardine, "Droughts, Deluges and Raging Debates"; Degroot, "Review—Global Crisis."

95. Parker, *Global Crisis*, 689, 695.

96. Fagan, *The Little Ice Age*, 213.

97. Degroot, *The Frigid Golden Age*, 308.

Chapter 11: Securitizing Climate Change

1. Harvey, "Climate Change Will Increase Threat of War."

2. Huhne, "The Geopolitics of Climate Change."

3. Chew, "Did Climate Change Spark the War in Syria?"

4. Iowa State University, "Researchers Present Study on How Global Climate Change Affects Violence."

5. Black, "Climate 'Is a Major Cause' of Conflict."

6. Schiermeier, "Climate Cycles Drive Civil War."

7. Parry, "Climate Fluctuations May Increase Civil Violence."

8. Skirble, "Climate Is Major Violence Trigger."

9. Wallace-Wells, "The Uninhabitable Earth."

10. Hsiang et al., "Civil Conflicts Are Associated with the Global Climate," 438.

11. Parry, "Climate Fluctuations May Increase Civil Violence."

12. Hsiang et al., "Civil Conflicts," 440.

13. Skirble, "Climate Is Major Violence Trigger." The wide-ranging powers that Hsiang ascribes to climate in shaping modern society are manifest in Carleton and Hsiang, "Social and Economic Impacts of Climate."

14. Hsiang et al., "Quantifying the Influence of Climate on Human Conflict."

15. Morin, "Violence Will Rise as Climate Changes."

16. Plumer, "Will Global Warming Lead to More War?" This piece includes links to a range of newspaper reports of the article.

17. Zhang et al., "Global Climate Change, War, and Population Decline," 19214.

18. Zhang et al., "Global Climate Change, War, and Population Decline," 19215.

19. Zhang et al., "Global Climate Change, War, and Population Decline," 19218, 19214, 19218, 19219, 19216.

20. Zhang et al., "Climate Change and War," 405.

21. Tol and Wagner, "Climate Change and Violent Conflict," 77.

22. Burke et al., "Warming Increases the Risk of Civil War in Africa," 20673.

23. Homer-Dixon, *Environment, Scarcity, and Violence*.

24. Hendrix and Glaser, "Trends and Triggers"; Hendrix and Salehyan, "Climate Change, Rainfall, and Social Conflict."

25. Hendrix and Glaser, "Trends and Triggers," 703.

26. Smith, "Genghis Khan's Secret Weapon Was Rain."

27. Zielinksi, "Warm, Wet Times Spurred Medieval Mongol Rise."

28. Rice, "Genghis Khan Rode Climate Change."

29. Connor, "How Climate Change Helped Genghis Khan."

30. Massey, "Rise of Genghis Khan Linked to Unusual Rains."

31. Reynolds, "'Climate Change' Drove Rise of Genghis Khan."

32. BBC, "Genghis Khan: Good Weather 'Helped Him to Conquer.'"

33. Pederson et al., "Pluvials, Droughts, the Mongol Empire and Modern Mongolia."

34. The works he referenced were Huntington, *Mainsprings of Civilization*; Lattimore, *Inner Asian Frontiers of China*; Lattimore, "Caravan Routes of Inner Asia."

35. Jenkins, "A Note on Climatic Cycles and the Rise of Chinggis Khan," 221.

36. Pederson et al., "Pluvials, Droughts, the Mongol Empire and Modern Mongolia."

37. Sheffer, "Ancient Tree Rings Suggest Good Weather Helped Genghis Khan."

38. Stinson, "Genghis Khan Aided by Climate Change."

39. Walsh, "How Climate Change Drove the Rise of Genghis Khan."

40. "Mild Weather May Have Propelled Genghis Khan."

41. Zielinski, "Warm, Wet Conditions Spurred."

42. Walsh, "How Climate Change Drove."

43. Pederson and Hessl, "Wet Climate Helped Genghis Khan Conquer Asia."

44. National Science Foundation, "Climate of Genghis Khan's Ancient Time Extends Long."

45. Walsh, "How Climate Change Drove."

46. Massey, "Rise of Genghis Khan Linked to Unusual Rains."

47. Zielinksi, "Warm, Wet Times Spurred Medieval Mongol Rise."

48. Davis, *The Arid Lands*, chap. 4; Huntington, *Mainsprings of Civilization*, 568.

49. Huntington, *The Pulse of Asia*, 379.

50. Netburn, "Did Rainy Climate Aid Genghis Khan?" Weatherford's account is also recorded on the blog History's Shadow. See Lang, "Climate Change Insufficient to Explain Genghis Khan's Greatness."

51. So far as proposed links among climate change, conflict, and migration are concerned, the underlying scholarship is claimed to clearly point in a different direction regardless of claims by the news media. Brzoska and Frölich, for example, observe that "academic research into the links between climate change, migration and conflict has predominantly been critical about such predictions. There is, so far, limited evidence both for the proposition that climate change will lead to major population movements as well as that modern migration movements generally trigger violent conflict. The theoretical foundation and empirical support for such propositions are thin." Brzoska and Frölich, "Climate Change, Migration and Violent Conflict," 192.

52. Dyer, *Climate Wars*, 10.

53. Dyer, "Excerpt: 'Climate Wars' by Gwynne Dyer."

54. Hampson, "The Climate for War," 9.

55. The key publication he had in mind was Homer-Dixon, "On the Threshold."

56. Kaplan, "The Coming Anarchy" 7, 8. Available at http://www.theatlantic.com/magazine/archive/1994/02/the-coming-anarchy/304670/

57. Kaplan, *The Revenge of Geography*, 37, 56, 69. The work was critically reviewed by geographers such as Johnston, Review of *The Revenge of Geography*.

58. Vali Nasr, blurb on the book's dustjacket.

59. Kaplan, "The Revenge of Geography," 100, 105.

60. See the survey in Brauch and Scheffran, "Introduction: Climate Change, Human Security, and Violent Conflict."

61. Federal Ministry for the Environment (BMU), *Climate Change and Conflict*. It is similar with the Global Humanitarian Forum's Human Impact Report, *Climate Change*, which emphasizes issues of global social inequality and what it calls "climate justice."

62. Trombetta, "Climate Change and the Environmental Conflict Discourse," 157.

63. Schwartz and Randall, *An Abrupt Climate Change Scenario*, 1, 7, 5, 17, 22.

64. See Nordås and Gleditsch, "Climate Change and Conflict."

65. Testimony of Rt. Hon. Margaret Beckett before a Joint Hearing of the House Permanent Select Committee on Intelligence, Subcommittee on Intelligence Community Management, and House Select Committee on Energy Independence and Global Warming, U.S. House of Representatives, June 24, 2008. The record can be downloaded at https://www.fas.org/irp/congress/2008_hr/climate.pdf

66. CNA, *National Security and the Threat of Climate Change*, 6, 20, 26, 37, 40, 9.

67. Campbell et al., *The Age of Consequences*, 59, 77, 85, 109.

68. Intergovernmental Panel on Climate Change, *Climate Change 2007*.

69. Intergovernmental Panel on Climate Change, *Climate Change 2014*, 771.

70. See https://climateandsecurity.org/about/

71. Kirkup, "Chris Huhne: Climate Change Threatens UK Security."

72. Schäfer et al., "Securitization of Media Reporting on Climate Change?"

73. The prevalence of geographical determinism among national security policymakers is noted in Dalby, "Environment: From Determinism to the Anthropocene." See also Dalby, "Rethinking Geopolitics."

74. Hulme, "Climate Security: The New Determinism."

75. Sutton et al., "Does Warming Increase the Risk of Civil War in Africa?"

76. Burke et al., "Reply to Sutton et al."

77. Buhaug, "Climate Not to Blame for African Civil Wars," 16480, 16477, 16481. Buhaug discussed other dimensions of war in Buhaug and Gates, "The Geography of Civil War."

78. I have examined the idea of the geography of knowledge in Livingstone, *Putting Science in its Place*.

79. Gleditsch, "Whither the Weather?"

80. Slettebak, "Don't Blame the Weather!" 163.

81. Theisen, "Climate Clashes?" 81, 88, 93, 81.

82. Theisen et al., "Climate Wars?" 79, 80, 81, 99.

83. Theisen, "Blood and Soil?" 811, 814; Salehyan, "From Climate Change to Conflict?"

84. See also Adano et al., "Climate Change, Violent Conflict and Local Institutions"; Koubi et al., "Climate Variability, Economic Growth, and Civil Conflict."

85. Peluso and Watts, "Violent Environments," 5, 6, 14. This strongly resonates with the conclusion to a study of climate in nineteenth-century Zululand that argues that the potential for conflict and violence in the wake of extreme weather events is markedly influenced by the role of political leaders and different ideological systems. See Klein et al., "Climate, Conflict and Society."

86. Theisen, "Blood and Soil?" 814.

87. Owain and Maslin, "Assessing the Relative Contribution." For a more popular audience, Maslin contributed a piece to *The Conversation* (April 24, 2018) entitled "Climate Change Is Not a Key Cause of Conflict."

88. Theisen et al., "Is Climate Change a Driver of Armed Conflict?" 622.

89. Alves, "The New Wars of Climate Change."

90. Editorial, "Don't Jump to Conclusions about Climate Change and Civil Conflict." The contrast between Jordan and Syria has been made many times. One other example, whose author is sympathetic to climate change as an aggravating factor in situations of conflict, is the leader article "How Climate Change Can Fuel Wars" for *The Economist*, May 25, 2019.

91. See, for example, O'Loughlin et al., "Climate Variability and Conflict Risk in East Africa."

92. Raleigh et al., "Extreme Temperatures and Violence." See also Brown and Crawford, "Climate Change: A New Threat to Stability in West Africa?"

93. Scheffran et al., "Climate Change and Violent Conflict," 870.

94. Barnett, "Security and Climate Change," 10, 14, 13; see also Barnett and Adger, "Climate Change, Human Security and Violent Conflict." Emily Gilbert notes that the U.S. military "is the world's single largest energy consumer." Gilbert, "The Militarization of Climate Change," 7.

95. Barnett, "Destabilising the Environment-Conflict Thesis," 274. See also Barnett, *The Meaning of Environmental Security*.

96. Eckersley, "Environmental Security, Climate Change, and Globalizing Terrorism"; Hartmann, "Rethinking Climate Refugees and Climate Conflict."

97. Gemenne et al., "Climate and Security," 1, 2.

98. Editorial, "Don't Jump to Conclusions about Climate Change and Civil Conflict."

99. Adams et al., "Sampling Bias in Climate-Conflict Research," 203. See also Verhoeven, "Gardens of Eden or Hearts of Darkness?"

100. Selby, "Positivist Climate Conflict Research," 829, 842, 845.

101. Mach et al., "Climate as a Risk Factor for Armed Conflict."

102. Parenti, *Tropic of Chaos*, 8, 11, 13. For some critical thoughts on Parenti's project, see Clayton, "Tropicality and the Choc en Retour."

103. "How to Think about Global Warming and War."

104. Hobbes, *Leviathan*, 64. On Hobbes' thinking about the genesis of war, see Abizadeh, "Hobbes on the Causes of War."

Chapter 12: Immortal Bird

1. Spate, "Quantity and Quality in Geography," 382. See also his analyses in Spate, "Toynbee and Huntington: A Study in Determinism"; Spate, "Environmentalism." Something of the ongoing legacy of geographical determinism may be gleaned from the titles of a range of recent books such as Kaplan, *The Revenge of Geography*; Marshall, *Prisoners of Geography*; and Morris, *Geography Is Destiny*. Critical dimensions of this resurgence are discussed in Immerwahr, "Are We Really Prisoners of Geography?"

2. Chakrabarty, "The Climate of History," 207. Comparably, see Robin and Steffen, "History for the Anthropocene."

3. Crosby, "The Past and Present of Environmental History," 1185.

4. King, "Climate Change Science: Adapt, Mitigate, or Ignore?"

5. Reflecting for the *Independent* newspaper on Earth Day 2021, Joe Sommerland remarked on how, even before global warming had gripped the public imagination, novelists and movie makers had evoked the nightmare of "Mother Earth turning against humanity and seeking retribution for the evils of pollution and deforestation." Sommerland, "Earth Day 2021: The 'Revenge of Nature.'" On the use of confessional vocabulary at COP26, see Mathiesen, "Leaders Confess Climate Sins." See also Johnston et al., eds., *Nature's Revenge*; Pearce, *The Last Generation: How Nature Will Take Her Revenge*.

6. Hulme, "Climate Security: The New Determinism."

7. Donner, "The Ugly History of Climate Determinism Is Still Evident Today."

8. Sweet, "The Degeneration Thesis," 127, 128.

9. Harper, "How Climate Change and Plague." See also Harper, "The Environmental Fall of the Roman Empire." On Gibbon, see Wolloch, "Edward Gibbon's Cosmology."

10. Bremner, "French Drown Their Sorrows in Spring."

11. See the illuminating discussions in Harley, "Silences and Secrecy," and Harley, "Deconstructing the Map."

12. The crudely ideological interpretation of Richard Peet is a case in point. See Peet, "The Social Origins of Environmental Determinism."

13. Intergovernmental Panel on Climate Change, *Climate Change 2022*. Commenting on this issue, Ilan Kelman, professor of disasters and health at University College London, noted that the report stated "with high confidence" that disasters and violent conflicts were "not significantly influenced by human-caused climate change" but that this news was not prominently advertised in press releases. Kelmann, "IPCC Report."

14. Hulme, *Weathered*, 83. Among the counternarratives he has in mind are Sen, *Poverty and Famines*; Davis, *Late Victorian Holocausts*.

15. Degroot, *The Frigid Golden Age*, 308.

16. For instance, Webster, "Climate Change Report Sets Out an Apocalyptic Vision of Britain."

17. Žižek, *Living in the End Times*, x, 328.

18. Smith and Howe, *Climate Change as Social Drama*, 58, 63; Stern, *The Economics of Climate Change*.

19. Oreskes and Conway, *The Collapse of Western Civilization*; Scranton, *Learning to Die in the Anthropocene*.

20. Garrard, "Environmentalism and the Apocalyptic Tradition," 50.

21. Ghosh, *The Great Derangement*, 153, 154; Ghosh, "Where Is the Fiction about Climate Change?" The term "catastrophozoic" was coined in 1996 by the conservation biologist Michael Soulé.

22. Baysal, ed., *Apocalyptic Visions in the Anthropocene*.

23. Clarke, "Reading Climate in J.G. Ballard."

24. See Dillon, "Imagining Apocalypse"; Trexler, *Anthropocene Fictions*, 108–18.

25. Trojanow, *The Lamentations of Zeno*. Colum McCann's comments come from his dustjacket blurb. The other observations are from Goodbody, "Cli-Fi—Genre of the Twenty-First Century?" 137.

26. Skrimshire, "Climate Change and Apocalyptic Faith," 234. See also Skrimshire, ed., *Future Ethics*; Skrimshire, "Eschatology."

27. Garrard, "Environmentalism and the Apocalyptic Tradition," 31, 64.

28. Trexler, *Anthropocene Fictions*, 117.

29. Benner et al., *Violent Climate Imaginaries*, 24.

30. This distinction was earlier proposed by O'Leary, *Arguing the Apocalypse*.

31. Foust and Murphy, "Revealing and Reframing Apocalyptic Tragedy in Global Warming Discourse," 152, 156.

32. Skrimshire, "Climate Change and Apocalyptic Faith," 235, 238.

33. Žižek and Gunjević, *God in Pain: Inversions of Apocalypse*, 71.

34. Swyngedouw, "Apocalyse Now!" 11. See also Swyngedouw, "Apocalypse Forever?"; Sturm and Lustig, "Variegated Environmental Apocalypses."

35. Barnett, "The Theology of Climate Change," 218, 227, 219. See also Barnett, *After the Flood*.

36. On the relevance of historical research for climate-change adaptation, see Adamson et al., "Rethinking the Present."

37. Locher and Fressoz argue that simplistic views of the past have fueled the mistaken conception that "for about two generations we have been experiencing a complete transformation of our relationship with the environment." They contend that "a historical understanding of past environmental discourses is essential for contemporary social and green theory because the dominant narratives used to reflect upon the contemporary environmental crisis are too simple." Locher and Fressoz, "Modernity's Frail Climate," 580.

BIBLIOGRAPHY

A F.R.G.S. [Burton, Richard F.] *Wanderings in West Africa. From Liverpool to Fernando Po.* London: Bradbury and Evans, 1863.

Abbe, Cleveland. "Introduction." In Edwin Grant Dexter, *Weather Influences: An Empirical Study of the Mental and Physiological Effects of Definite Meteorological Conditions*, xi–xxxi. New York: Macmillan, 1904.

Abercromby, Ralph. *Seas and Skies in Many Latitudes or Wanderings in Search of Weather.* London, 1888.

Abizadeh, Arash. "Hobbes on the Causes of War: A Disagreement Theory." *American Political Science Review* 105 (2011): 298–315.

Ackerknecht, Erwin H. "Changing Attitudes toward History and Geography of Disease." In *History and Geography of the Most Important Diseases*, chap. 1, 1–6. New York: Hafner, 1972.

Adams, Courtland, Tobias Ide, Jon Barnett, and Adrien Detges. "Sampling Bias in Climate–Conflict Research." *Nature Climate Change* 8 (March 2018): 200–3.

Adamson, George C. D. "'The Languor of the Hot Weather': Everyday Perspectives on Weather and Climate in Colonial Bombay, 1819–1828." *Journal of Historical Geography* 38 (2012): 143–54.

Adamson, George C. D., Matthew J. Hannaford, and Eleonora J. Rohland. "Re-thinking the Present: The Role of a Historical Focus in Climate Change Adaptation Research." *Global Environmental Change* 38 (2018): 195–205.

Adano, Wario R., Ton Dietz, Karen Witsenburg, and Fred Zaal. "Climate Change, Violent Conflict and Local Institutions in Kenya's Drylands." *Journal of Peace Research* 49 (2012): 65–80.

Agassiz, Louis. "The Diversity of Origin of the Human Races." *Christian Examiner and Religious Miscellany* 49 (1850): 110–45.

Agassiz, Louis. "Geographical Distribution of Animals." *Christian Examiner and Religious Miscellany* 48 (1850): 181–204.

Agnew, Robert. "Dire Forecast: A Theoretical Model of the Impact of Climate Change on Crime." *Theoretical Criminology* 16 (2011): 21–42.

Aiken, Edwin James. *Scriptural Geography: Portraying the Holy Land.* London: I. B. Taurus, 2010.

Aimers, James. "Drought and the Maya: The Story of the Artefacts." *Nature* 479 (November 2011): 44.

Allen, John Adams. *Medical Examinations for Life Insurance.* Chicago: Horton and Leonard, 1867.

Allik, Jüri, and Anu Realo. "How Is Freedom Distributed across the Earth?" *Journal of Cross-Cultural Psychology* 42 (2011): 482–83.

Althusser, Louis. *Politics and History: Montesquieu, Rousseau, Marx*, translated from the French by Ben Brewster. London: Verso, 2007. First published in Paris in 1959.

Alves, Luis. "The New Wars of Climate Change." *Ovi Magazine*, January 17, 2008. http://www.ovimagazine.com/art/2530

Anderson, Craig A. "Climate Change and Violence." In *The Encyclopedia of Peace Psychology*, edited by Daniel J. Christie, 128–32. Hoboken, NJ: Wiley-Blackwell, 2012.

Anderson, Craig A. "Heat and Violence." *Current Directions in Psychological Science* 10 (2001): 33–38.

Anderson, Craig A. "Temperature and Aggression: Effects of Quarterly, Yearly and City Rates of Violent and Nonviolent Crime." *Journal of Personality and Social Psychology* 52 (1987): 1161–73.

Anderson, Craig A. "Temperature and Aggression: Ubiquitous Effects of Heat on Occurrence of Human Violence." *Psychological Bulletin* 106 (1989): 74–96.

Anderson, Craig A., Brad Bushman, and Ralph W. Groom. "Hot Years and Serious and Deadly Assault: Empirical Tests of the Heat Hypothesis." *Journal of Personality and Social Psychology* 73 (1997): 1213–23.

Anderson, Warwick. "Climates of Opinion: Acclimatization in 19th-Century France and England." *Victorian Studies* 35 (1992): 135–57.

Anderson, Warwick. *Colonial Pathologies: American Tropical Medicine, Race and Hygiene in the Philippines*. Durham, NC: Duke University Press, 2006.

Anderson, Warwick. "Immunities of Empire: Race, Disease, and the New Tropical Medicine, 1900–1920." *Bulletin of the History of Medicine* 70 (1996): 94–118.

Apps, Lara. "Witchcraft: A Problem for All Times?" *Histoire Sociale/Social History* 39 (2006): 487–95.

Arbuthnot, John. *An Essay Concerning the Effects of Air on Human Bodies*. London, 1751.

Aristotle. *Problems*, translated by W. S. Hett. London: Heinemann, 1936–37.

Armit, Ian, Graeme T. Swindles, Katharina Becker, Gill Plunkett, and Maarten Blaauw. "Rapid Climate Change Did Not Cause Population Collapse at the End of the European Bronze Age." *Proceedings of the National Academy of Sciences* 111 (2014): 17045–49.

Arnold, David. "India's Place in the Tropical World, 1770–1930." *Journal of Imperial and Commonwealth History* 26 (1998): 1–21.

Arnold, David. "Introduction: Tropical Medicine before Manson." In *Warm Climates and Western Medicine: The Emergence of Tropical Medicine, 1500–1900*, Wellcome Institute Series in the History of Medicine, edited by David Arnold, 1–19. Atlanta, GA: Rodopi, 1996.

Arnold, David. *The Problem of Nature: Environment, Culture and European Expansion*. Oxford: Blackwell, 1996.

Arnold, David, ed. *Warm Climates and Western Medicine: The Emergence of Tropical Medicine, 1500–1900*, Wellcome Institute Series in the History of Medicine. Atlanta, GA: Rodopi, 1996.

Asatrian, Mushegh. "Ibn Khaldūn on Magic and the Occult." *Iran and the Caucasus* 7 (2003): 73–123.

[Ashe, Thomas.] *History of the Azores or Western Islands; Containing an Account of the Government, Laws, and Religion, the Manners, Ceremonies, and Character of the Inhabitants: and Demonstrating the Importance of these Valuable Islands to the British Empire*. London: Sherwood, Neely and Jones, 1813.

Audy, J. Ralph, and Frederick L. Dunn. "Health and Disease." In *Human Ecology*, edited by Frederick Sargent II, 325–43. Amsterdam: North-Holland Publishing Company, 1974.

Auliciems, Andris. "Human Adaptation within a Paradigm of Climatic Determinism and Change." In *Biometeorology for Adaptation to Climatic Variability and Change*, edited by Kristie L. Ebi, Ian Burton, and Glenn McGregor, 235–67. Dordrecht: Springer, 2009.

Auliciems, Andris, and L. DiBartolo. "Domestic Violence in a Subtropical Environment: Police Calls and Weather in Brisbane." *International Journal of Biometeorology* 39 (1995): 34–39.

Avery, D. Margaret. "Southern Savannas and Pleistocene Hominid Adaptation: The Micromammalian Perspective." In *Paleoclimate and Evolution with Emphasis on Human Origins*, edited by Elisabeth S. Vrba, George H. Denton, Timothy C. Partridge, and Lloyd H. Burckle, 459–78. New Haven, CT: Yale University Press, 1995.

Baer, Ferdinand. "Helmut E. Landsberg, 1906–1985." *Bulletin of the American Meteorological Society* 67, no. 12 (1986): 1522–23.

Baer, Ferdinand, Norman L. Canfield, and J. Murray Mitchell, eds. *Climate in Human Perspective: A Tribute to Helmut E. Landsberg*. Dordrecht: Kluwer Academic Publishers, 1991.

Bahr, Howard M., Marie B. Durrant, Matthew T. Evans, and Suzanne L. Maughan. "On Diversity, Empathy, and Community: The Relevance of Johann Gottfried Herder." *Rural Sociology* 73 (2008): 503–27.

Baigent, Elizabeth. "Jackson, Robert Edmund Scoresby- (*bap.* 1833, *d.* 1867)." In *Oxford Dictionary of National Biography*. Oxford: Oxford University Press, 2004. https://doi.org/10.1093/ref:odnb/14548

Baikie, Robert. *Observations on the Neilgherries, Including an Account of Their Topography, Climate, Soil and Productions and of the Effects of the Climate on the European Constitution*. Calcutta: Baptist Mission Press, 1834.

Bain, William. *Between Anarchy and Society: Trusteeship and the Obligations of Power*. Oxford: Oxford University Press, 2003.

Baker, Zeke. "Meteorological Frontiers: Climate Knowledge, the West, and US Statecraft, 1800–1850." *Social Science History* 42 (2018): 731–61.

Balfour, Andrew. "Problems of Acclimatisation" [Part 1]. *The Lancet* 205, no. 5211 (1923): 84–88.

Balfour, Andrew. "Problems of Acclimatisation" [Part 2]. *The Lancet* 205, no. 5214 (1923): 243–47.

Balfour, Andrew. "Sojourners in the Tropics." *The Lancet* 204, no. 5209 (1923): 1329–34.

Ball, John. "Climatological Diagrams." *Cairo Scientific Journal* 4 (1910): 280–81.

Bannister, Robert C. *Social Darwinism: Science and Myth in Anglo-American Social Thought*. Philadelphia: Temple University Press, 1979.

Barkhuus, Arne. "Medical Geographies." *Ciba Symposia* 6 (1944–45): 1997–2016.

Barnes, Trevor J. "'In the Beginning Was Economic Geography'—A Science Studies Approach to Disciplinary History." *Progress in Human Geography* 25 (2001): 521–44.

Barnett, Jon. "Destabilising the Environment-Conflict Thesis." *Review of International Studies* 26 (2000): 271–88.

Barnett, Jon. *The Meaning of Environmental Security: Ecological Politics and Policy in the New Security Era*. London: Zed Books, 2001.

Barnett, Jon. "Security and Climate Change." *Global Environmental Change* 13 (2003): 7–17.

Barnett, Jon, and W. Neil Adger. "Climate Change, Human Security and Violent Conflict." *Political Geography* 26 (2007): 639–55.

Barnett, Lydia. *After the Flood: Imagining the Global Environment in Early Modern Europe.* Baltimore: Johns Hopkins University Press, 2019.

Barnett, Lydia. "The Theology of Climate Change: Sin as Agency in the Enlightenment's Anthropocene." *Environmental History* 20 (2015): 217–37.

Barnhart, John D. "Rainfall and the Populist Party in Nebraska." *American Political Science Review* 19 (1925): 527–40.

Baron, Robert A. "Aggression as a Function of Ambient Temperature and Prior Anger Arousal." *Journal of Personality and Social Psychology* 21 (1972): 183–89.

Baron, Robert A., and Paul A. Bell. "Aggression and Heat: The Influence of Ambient Temperature, Negative Affect, and a Cooling Drink on Physical Aggression." *Journal of Personality and Social Psychology* 33 (1976): 245–55.

Baron, Robert A., and V. M. Ransberger. "Ambient Temperature and the Occurrence of Collective Violence: The 'Long, Hot Summer' Revisited." *Journal of Personality and Social Psychology* 36 (1978): 351–60.

Baron, Robert A., Gordon W. Russell, and Robert L. Arms. "Negative Ions and Behavior: Impact on Mood, Memory, and Aggression among Type A and Type B Persons." *Journal of Personality and Social Psychology* 48 (1985): 746–54.

Barrett, Frank A. "Alfred Haviland's Nineteenth-Century Map Analysis of the Geographical Distribution of Diseases in England and Wales." *Social Science and Medicine* 46 (1998): 767–81.

Barrett, Frank A. "August Hirsch: As Critic of, and Contributor to Geographical Medicine and Medical Geography." In *Medical Geography in Historical Perspective*, edited by Nicolaas A. Rupke, 98–117. London: Wellcome Trust Centre for the History of Medicine, 2000.

Barrett, Frank A. "Daniel Drake's Medical Geography." *Social Science and Medicine* 42 (1996): 791–800.

Barrett, Frank A. *Disease and Geography: The History of an Idea*, York University–Atkinson College Geographical Monographs no. 23. Toronto: Becker Associates, 2000.

Bartholomew, Michael. "Lind, James (1716–1794)." In *Oxford Dictionary of National Biography*. Oxford: Oxford University Press, 2004. https://doi.org/10.1093/ref:odnb/16669

Barua, Maan. "Ratzel's Biogeography: A More-Than-Human Encounter." *Journal of Historical Geography* 61 (2018): 102–8.

Bashford, Alison. "Anticolonial Climates: Physiology, Ecology, and Global Population, 1920s–1950s." *Bulletin of the History of Medicine* 86 (2012): 596–626.

Bashford, Alison. "'Is White Australia Possible?' Race, Colonialism and Tropical Medicine." *Ethnic and Racial Studies* 23 (2000): 248–71.

Bashford, Alison, and Sarah W. Tracy. "Introduction: Modern Airs, Waters, and Places." *Bulletin of the History of Medicine* 86 (2012): 495–514.

Bashford, Alison, and Sarah W. Tracy, eds. "Modern Airs, Waters and Places." *Bulletin of the History of Medicine* 86, no. 4 (2012).

Baum, Marsha L. *When Nature Strikes: Weather Disasters and the Law.* Westport, CT: Praeger, 2007.

Baumeister, Roy F., Jina Park, and Sarah E. Ainsworth. "Individual Identity and Freedom of Choice in the Context of Environmental and Economic Conditions." *Journal of Cross-Cultural Psychology* 42 (2011): 484–521.

Baysal, Kübra, ed. *Apocalyptic Visions in the Anthropocene and the Rise of Climate Fiction*. Newcastle upon Tyne: Cambridge Scholars Publishing, 2021.

BBC. "Did Climate Change Make Us Intelligent?" BBC Teach website. https://www.bbc.co.uk/teach/did-climate-change-make-us-intelligent/zj2qrj6

BBC. "Genghis Khan: Good Weather 'Helped Him to Conquer.'" BBC News, March 11, 2014. http://www.bbc.co.uk/news/world-asia-26523524

Beals, Kenneth L. "Head Form and Climatic Stress." *American Journal of Physical Anthropology* 37 (1972): 85–92.

Beals, Kenneth L., Courtland L. Smith, and Stephen M. Dodd. "Brain Size, Cranial Morphology, Climate, and Time Machines." *Current Anthropology* 25 (1984): 301–30.

Beals, Kenneth L., Courtland L. Smith, and Stephen M. Dodd. "Climate and the Evolution of Brachycephalization." *American Journal of Physical Anthropology* 62 (1983): 425–37.

Beard, George. "Neurasthenia, or Nervous Exhaustion." *Boston Medical and Surgical Journal* 80, no. 13 (1869): 217–21.

Beattie, James, Edward Melillo, and Emily O'Gorman, eds. *Eco-Cultural Networks and the British Empire: New Views on Environmental History*. London: Bloomsbury, 2014.

Behrensmeyer, Anna K. "Climate Change and Human Evolution." *Science* 311 (January 27, 2006): 476–78.

Behringer, Wolfgang. "Climatic Change and Witch-hunting: the Impact of the Little Ice Age on Mentalities." *Climatic Change* 43 (1999): 335–51.

Behringer, Wolfgang. *A Cultural History of Climate*, translated by Patrick Camiller. Cambridge: Polity Press, 2010.

Behringer, Wolfgang. "Weather, Hunger and Fear: Origins of the European Witch-Hunts in Climate, Society and Mentality." *German History* 13 (1995): 1–27.

Beiser, Frederick C. *The German Historicist Tradition*. Oxford: Oxford University Press, 2015.

Belgrave, Michael. "Thomson, Arthur Saunders, 1816–1860." In *Dictionary of New Zealand Biography. Te Ara—the Encyclopedia of New Zealand*. First published in 1990, updated July 2014. https://teara.govt.nz/en/biographies/1t95/thomson-arthur-saunders

Benedict, Ruth. "Isothermic Anthropology." *New York Herald-Tribune*, January 22, 1928.

Benner, Ann-Kathrin, Delf Rothe, Sara Ullström, and Johannes Stripple. *Violent Climate Imaginaries: Science-Fiction-Politics*. Hamburg: Institute for Peace Research, University of Hamburg, 2019.

Bergwik, Staffan. "Father, Son, and the Entrepreneurial Spirit: Otto Pettersson, Hans Pettersson, and the Early Twentieth-Century Inheritance of Oceanography." In *Domesticity in the Making of Modern Science*, edited by Donald L. Opitz, Staffan Bergwik, and Brigitte Van Tiggelen, 192–214. New York: Palgrave Macmillan, 2016.

Berlin, Isaiah. *Vico and Herder*. New York: The Viking Press, 1976.

Bernasconi, Robert. "Who Invented the Concept of Race? Kant's Role in the Enlightenment Construction of Race." In *Race*, edited by Robert Bernasconi, 11–36. Oxford: Blackwell, 2001.

Biddiss, Michael D. "The Politics of Anatomy: Dr Robert Knox and Victorian Racism." *Proceedings of the Royal Society of Medicine* 69 (1976): 245–50.

"Biographical Notice of the Late Dr. William Currie, A Physician of Philadelphia." *Register of Pennsylvania* 6, no. 13 (September 1830): 204–6.

Bischof, Marco. "Introduction to Integrative Biophysics." In *Integrative Biophysics: Biophotonics*, edited by Fritz-Albert Popp and Lev Belonssov, 1–112. Dordrecht: Kluwer, 2003.

Bisset, Charles. *Medical Essays and Observations*. Newcastle-upon-Tyne: A. Millar & D. Wilson, 1766.

Black, Richard. "Climate 'Is a Major Cause' of Conflict in Africa." BBC News Channel, November 24, 2009. http://news.bbc.co.uk/1/hi/sci/tech/8375949.stm

Blair, Ann. *The Theater of Nature: Jean Bodin and Renaissance Science*. Princeton, NJ: Princeton University Press, 1997.

Blair, Thomas A. *Climatology. General and Regional*. New York: Prentice-Hall, 1951. First published 1942.

Blaut, James M. "Environmentalism and Eurocentrism." *Geographical Review* 89 (1999): 391–408.

Blumenberg, Bennett. "The Evolution of the Advanced Hominid Brain." *Current Anthropology* 24 (1983): 589–623.

Bodin, Jean. *Method for the Easy Comprehension of History*, translated by Beatrice Reynolds. New York: Norton, 1969.

Bodin, Jean. *Six Books of the Commonwealth*, abridged and translated by M. J. Tooley. Oxford: Blackwell, 1955.

Boia, Lucian. *The Weather in the Imagination*. London: Reaktion Books, 2005.

Bond, James J. "Medical Spirits and the Medieval Language of Life." *Traditio* 40 (1984): 91–130.

Booker, Christopher. "Global Crisis, by Geoffrey Parker—Review." *Spectator*, June 1, 2013.

Booth, Christopher C. "Hillary, William (1697–1763)." *Oxford Dictionary of National Biography*. Oxford: Oxford University Press, 2004. https://doi.org/10.1093/ref:odnb/13318

Borchard, Therese. "Weather and Mood: Rainy with a Chance of Depression." *Everyday Health*, May 25, 2016. http://web.archive.org/web/20160529175429/https://www.everydayhealth.com/columns/therese-borchard-sanity-break/does-weather-affect-your-mood/

Bordier, Arthur. *La Géographie Médicale*. Paris: Beinwald, 1884.

Borsay, Anne. "Falconer, William (1744–1824)." *Oxford Dictionary of National Biography*. Oxford: Oxford University Press, 2004. https://doi.org/10.1093/ref:odnb/9118

Boudin, Jean Christian Marc François. *Essai de Géographie Médicale, Lois et Distribution Géographique des Maladies*. Paris: Germer-Bailliére, 1843.

Boudin, Jean Christian Marc François. *Traité de Géographie et de Statistique Médicales et des Maladies Endémiques*. Paris: J-B Bailliére, 1857.

Boulakia, Jean David C. "Ibn Khaldûn: A Fourteenth Century Economist." *Journal of Political Economy* 79 (1971): 1105–18.

Bouma, J. J. S. H. J. W. "A Short History of Human Biometeorology." *Experientia* 43 (1987): 1–6.

Bowd, Gavin, and Daniel Clayton. *Impure and Worldly Geography: Pierre Gourou and Tropicality*. Abingdon: Routledge, 2019.

Bowler, Peter J. *The Eclipse of Darwinism: Anti-Darwinian Evolution Theories in the Decades around 1900*. Baltimore: Johns Hopkins University Press, 1983.

Bowler, Peter J. *The Fontana History of the Environmental Sciences*. London: Fontana, 1992.

Bowler, Peter J. *The Invention of Progress: The Victorians and the Past*. Oxford: Blackwell, 1989.

Bowler, Peter J. *Theories of Human Evolution: A Century of Debate, 1844–1944*. Oxford: Blackwell, 1986.

Bowman, Isaiah. "The Pioneer Fringe." *Foreign Affairs* 6 (October 1927): 49–66.

Bowman, Isaiah. *The Pioneer Fringe, American Geographical Society Special Publication 13*. New York: American Geographical Society, 1931.

Brandt, Allan. *No Magic Bullet: A Social History of Venereal Disease in the United States Since 1880*. New York: Oxford University Press, 1985.

Brauch, Hans Günther, and Jürgen Scheffran. "Introduction: Climate Change, Human Security, and Violent Conflict in the Anthropocene." In *Climate Change, Human Security and Violent Conflict: Challenges for Societal Stability. Hexagon Series on Human and Environmental Security and Peace Vol. 8*, edited by Jürgen Scheffran, Michael Brzoska, Hans Günther Brauch, Peter Michael Link, and Janpeter Schilling, 3–40. Berlin: Springer, 2012.

Bremner, C. "French Drown Their Sorrows in Spring." *The Times*, May 5, 2001, 17.

Briggs, John C. "Centrifugal Speciation and Centres of Origin." *Journal of Biogeography* 27 (2000): 1183–88.

Brigham, Albert Perry. "Memoir of Frederick Valentine Emerson." *Annals of the Association of American Geographers* 10 (1920): 149–52.

Brinton, William. *On the Medical Selection of Lives for Assurance*. New York: John Hopper, 1863 (orig. 1856).

Brooke, John Hedley. *Science and Religion: Some Historical Perspectives*. Cambridge: Cambridge University Press, 1991.

Brooke, John L. *Climate Change and the Course of Global History*. New York: Cambridge University Press, 2014.

Brooks, David. "The Philosophy of Data." *New York Times*, February 5, 2013, Section A, 23.

Brown, Oli, and Alec Crawford. "Climate Change: A New Threat to Stability in West Africa? Evidence from Ghana and Burkina Faso." *African Security Review* 17, no. 3 (2008): 39–57.

Brzoska, Michael, and Christiane Frölich. "Climate Change, Migration and Violent Conflict: Vulnerabilities, Pathways and Adaptation Strategies." *Migration and Development* 5 (2016): 190–210.

Buckle, Henry Thomas. *History of Civilization in England*. New ed., 3 vols. London: Longmans, Green, and Co., 1873.

Buhaug, Halvard. "Climate Not to Blame for African Civil Wars." *Proceedings of the National Academy of Sciences* 107, no. 38 (2010): 16477–82.

Buhaug, Halvard, and Scott Gates. "The Geography of Civil War." *Journal of Peace Research* 39 (2002): 417–33.

Büntgen, Ulf, Vladimir S. Myglan, Fredrik Charpentier Ljungqvist, Michael McCormick, Nicola Di Cosmo, Michael Sigl, Johann Jungclaus, Sebastian Wagner, Paul J. Krusic, Jan Esper, Jed O. Kaplan, Michiel A. C. de Vaan, Jürg Luterbacher, Lukas Wacker, Willy Tegel, and Alexander V. Kirdyanov, "Cooling and Societal Change during the Late Antique Little Ice Age from 536 to around 660 A.D." *Nature Geoscience* 9 (February 2016): 231–36.

Burckle, Lloyd H. "Current Issues in Pliocene Paleoclimatology." In *Paleoclimate and Evolution with Emphasis on Human Origins*, edited by Elisabeth S. Vrba, George H. Denton, Timothy C. Partridge, and Lloyd H. Burckle, 3–7. New Haven, CT: Yale University Press, 1995.

Burke, Marshall, Solomon M. Hsiang, and Edward Miguel. "Global Non-Linear Effect of Temperature on Economic Production." *Nature* 527 (2015): 235–39.

Burke, Marshall B., Edward Miguel, Shanker Satyanath, John A. Dykema, and David B. Lobell. "Reply to Sutton et al.: Relationship between Temperature and Conflict Is Robust." *Proceedings of the National Academy of Sciences* 107, no. 25 (June 2010): E103.

Burke, Marshall B., Edward Miguel, Shanker Satyanath, John A. Dykema, and David B. Lobell. "Warming Increases the Risk of Civil War in Africa." *Proceedings of the National Academy of Sciences* 106, no. 49 (2009): 20670–74.

Burton, Richard F. *The Lake Regions of Central Africa. A Picture of Exploration*. Vol. 1. London: Longman, Green, Longman, and Roberts, 1860.

Butlin, Robin. "George Adam Smith and the Historical Geography of the Holy Land: Contents, Contexts and Connections." *Journal of Historical Geography* 14 (1988): 381–404.

Butzer, Karl W. "Collapse, Environment, and Society." *Proceedings of the National Academy of Science* 109, no. 10 (2012): 3632–39.

Butzer, Karl W., and Georgina H. Endfield. "Critical Perspectives on Historical Collapse." *Proceedings of the National Academy of Science* 109, no. 10 (2012): 3628–31.

Callaway, Helen. "Dressing for Dinner in the Bush: Rituals of Self Definition and British Imperial Authority." In *Dress and Gender: Making and Meaning in Cultural Contexts*, edited by Ruth Barnes and Joanne B. Eicher, 232–47. New York: Berg, 1993.

Calvin, William H. *The Ascent of Mind: Ice Age Climates and the Evolution of Intelligence*. New York: Bantam Books, 1991.

Calvin, William H. *A Brain for All Seasons: Human Evolution and Abrupt Climate Change*. Chicago: University of Chicago Press, 2002.

Calvin, William H. "The Great Climate Flip-Flop." *The Atlantic Monthly* 281 (1998): 47–64. https://www.theatlantic.com/magazine/archive/1998/01/the-great-climate-flip-flop/308313/

Campbell, John A., and David N. Livingstone. "Neo-Lamarckism and the Development of Geography in the United States and Great Britain." *Transactions of the Institute of British Geographers* 8 (1983): 267–94.

Campbell, Kurt M., Jay Gulledge, J. R. McNeill, John Podesta, Peter Ogden, Leon Fuerth, R. James Woolsey, Alexander T. J. Lennon, Julianne Smith, Richard Weitz, and Derek Mix. *The Age of Consequences: The Foreign Policy and National Security Implications of Global Climate Change*. Washington, D.C.: Center for Strategic and International Studies, and Center for a New American Security, 2007.

Cannato, Vincent J. "Immigration and the Brahmins: An Influx of Undesirables at the End of the Nineteenth Century Hit Boston's Elite Rather Hard." *Humanities* 30 (May/June, 2009): 12–17. https://www.neh.gov/humanities/2009/mayjune/feature/immigration-and-the-brahmins

Cannon, Walter B. "Biographical Memoir of Lawrence Joseph Henderson, 1878–1942." *National Academy of Sciences* 23 (1943): 30–58.

Cantor, Geoffrey. "What Shall We Do with the 'Conflict Thesis?'" In *Science and Religion: New Historical Perspectives*, edited by Thomas Dixon, Geoffrey Cantor, and Stephen Pumphrey, 283–98. Cambridge: Cambridge University Press, 2010.

Cao, Melanie, and Jason Wei. "Stock Market Returns: A Note on Temperature Anomaly." *Journal of Banking and Finance* 29 (2005): 1559–73.

Carey, Mark. "Climate and History: A Critical Review of Historical Climatology and Climate Change Historiography." *WIREs Climate Change* 3 (2012): 233–49.

Carey, Mark. "Climate, Medicine, and Peruvian Health Resorts." *Science, Technology and Human Values* 39 (2014): 795–818.

Carey, Mark. "Inventing Caribbean Climates: How Science, Medicine, and Tourism Changed Tropical Weather from Deadly to Healthy." *Osiris* 26 (2011): 129–41.

Carleton, Tamma A., and Solomon M. Hsiang. "Social and Economic Impacts of Climate." *Science* 353, no. 6304 (September 2016): aad9837.

Carleton, W. Christopher, David Campbell, and Mark Collard. "Increasing Temperature Exacerbated Classic Maya Conflict over the Long Term." *Quaternary Science Reviews* 163 (2017): 209–18.

Carlsmith, J. M., and C. A. Anderson. "Ambient Temperature and the Occurrence of Collective Violence: A New Analysis." *Journal of Personality and Social Psychology* 37 (1979): 337–34.

Carrithers, David Wallace, ed. *The Spirit of Laws by Montesquieu. A Compendium of the First English Edition*. Berkeley: University of California Press, 1977.

Castree, Noel. "A Climate of Ill Health." *American Scientist* 99 (2011): 418–20.

Cecca, Fabrizio. "La Dimension Biogéographique de l'Évolution de la Vie." *Comtes Rendus Palevol* 8 (2009): 119–32.

Centers for Disease Control and Prevention. "Health Effects." November 2010. https://web.archive.org/web/20140430173924/https://www.cdc.gov/climateandhealth/effects/

Chakrabarty, Dipesh. "The Climate of History: Four Theses." *Critical Inquiry* 35 (2009): 197–222.

Chalmers, Lionel. *An Account of the Weather and Diseases of South-Carolina*. 2 vols. London: Edwards and Charles Dilly, 1776.

Chambers, F. M., and S. A. Brain. "Paradigm Shifts in Late-Holocene Climatology?" *The Holocene* 12 (2002): 239–49.

Chand, Prabat Kumar, and Pratima Murthy. "Climate Change and Mental Health." *Regional Health Forum* 12 (2008): 43–48.

Chang, Jiat-Hwee. *A Genealogy of Tropical Architecture: Colonial Networks, Nature and Technoscience*. London: Routledge, 2016.

Chang, Jiat-Hwee, and Anthony D. King. "Towards a Genealogy of Tropical Architecture: Historical Fragments of Power-Knowledge, Built Environment and Climate in the British Colonial Territories." *Singapore Journal of Tropical Geography* 32 (2011): 283–300.

Chapman, Carleton B. "Edward Smith (?1818–1874): Physiologist, Human Ecologist, Reformer." *Journal of the History of Medicine and Allied Sciences* 22 (1967): 1–26.

Chappell, John E., Jr. "Climatic Pulsations in Inner Asia and Correlations between Sunspots and Weather." *Palaeogeography, Palaeoclimatology, Palaeoecology* 10 (1971): 177–97.

Chappell, John E., Jr. "Environmental Causation." In *Themes in Geographic Thought*, edited by Milton E. Harvey and Brian P. Holly, 163–86. London: Croom Helm, 1981.

Chardin, Sir John. *Sir John Chardin's Travels in Persia*, with an Introduction by Sir Percy Sykes. London: The Argonaut Press, 1927. First published 1724.

Charles, Richard Havelock. "Neurasthenia, and Its Bearing on the Decay of Northern Peoples in India." *Transactions of the Royal Society of Tropical Medicine and Hygiene* 7 (1913): 2–31.

Chen, Ed. "Climate Induced Migration and Conflict: Historical Evidence, and Like Future Outlook." *Science 2.0*, September 21, 2009. https://www.science20.com/alchemist/blog/climate_induced_migration_and_conflict_historical_evidence_and_likely_future_outlook

Cherry, Robert D. "Racial Thought and the Early Economics Profession." *Review of Social Economy* 34 (1976): 147–62.

Chew, Kristina. "Did Climate Change Spark the War in Syria?" *Care 2 News Network*, March 13, 2013. https://web.archive.org/web/20130315061604/http://www.care2.com/causes/did-climate-change-spark-war-in-syria.html

Choi, Charles Q. "Could Climate Change Affect People's Personalities?" *LiveScience*, November 27, 2017. https://www.livescience.com/61033-climate-temperature-personality.html

Chowdhury, Rituparna Ray. "Climatic Determinism and the Conceptualization of the Tropics in British India." In *Oxford Research Encyclopedia of Climate Science*, 2021. https://oxfordre.com/climatescience/view/10.1093/acrefore/9780190228620.001.0001/acrefore-9780190228620-e-836

Christie, Nancy J. "Environment and Race: Geography's Search for a Darwinian Synthesis." In *Darwin's Laboratory: Evolutionary Theory and Natural History in the Pacific*, edited by Roy MacLeod and Philip F. Rehbock, 426–73. Honolulu: University of Hawai'i Press, 1994.

Christie, Nancy J. "'Pioneering for a Civilized World': Griffith Taylor and the Ecology of Geography." *Scientia Canadensis: Canadian Journal of the History of Science, Technology and Medicine* 17 (1993): 103–54.

Cianconi, Paolo, Sophia Betro, and Luigi Janiri. "The Impact of Climate Change on Mental Health: A Systematic Descriptive Review." *Frontiers in Psychiatry*, March 6, 2020. https://www.frontiersin.org/articles/10.3389/fpsyt.2020.00074/full

Clark, Constance Areson. *God—or Gorilla: Images of Evolution in the Jazz Age*. Baltimore: Johns Hopkins University Press, 2008.

Clarke, Jim. "Reading Climate in J.G. Ballard." *Critical Survey* 25, no. 2 (2013): 7–21.

Clayton, Daniel. "Tropicality and the Choc en Retour of Covid-19 and Climate Change." *eTropic: Electronic Journal of Studies in the Tropics* 20, no. 1 (2021): 54–93. https://journals.jcu.edu.au/etropic/article/view/3787/3649

Clayton, H. Helm. "The Influence of Rainfall on Commerce and Politics." *Popular Science Monthly* 60 (1901): 158–65.

Clayton, Henry H. *The Sunspot Period*. Washington, D.C.: Smithsonian Institution, 1939.

Clayton, Henry H. "The Sunspot-Period and Spring Rainfall in the United States." *Transactions American Geophysical Union* 23 (1942): 284–86.

Clayton, Henry H. *World Weather*. New York: Macmillan, 1923.

Clayton, Henry H. "World Weather and Solar Activity." *Smithsonian Miscellaneous Collections* 89, no. 15 (1934): 1–52.

Clayton, Susan. "Psychology and Climate Change." *Psychologist Papers* 40 (2019): 167–73.

Clayton, Susan, Christie Manning, Kirra Krygsman, and Meighen Speiser. *Mental Health and Our Changing Climate: Impacts, Implications, and Guidance*. Washington, D.C.: American Psychological Association, and ecoAmerica, 2017.

CNA. *National Security and the Threat of Climate Change*. Alexandria, VA: CAN Corporation, 2007.

Cohn, Ellen G. "The Prediction of Police Calls for Service: The Influence of Weather and Temporal Variables on Rape and Domestic Violence." *Journal of Environmental Psychology* 13 (1993): 71–83.

Cohn, Ellen G., and James Rotton. "Assault as a Function of Time and Temperature: A Moderator-Variable Time-Series Analysis." *Journal of Personality and Social Psychology* 72 (1997): 1322–34.

Colbert, Edwin H. *William Diller Matthew, Paleontologist: The Splendid Drama Observed*. New York: Columbia University Press, 1992.

Collingham, Elizabeth M. *Imperial Bodies: The Physical Experience of the Raj 1800–1947*. Oxford: Polity Press, 2001.

Connolly, Marie. "Some Like It Mild and Not Too Wet: The Influence of Weather on Subjective Well-being." *Journal of Happiness Studies* 14 (2013): 457–73.

Connor, Steve. "How Climate Change Helped Genghis Khan." *The Independent*, March 10, 2014. http://www.independent.co.uk/news/science/how-climate-change-helped-genghis-khan-scientists-believe-a-sudden-period-of-warmer-weather-allowed-the-mongols-to-invade-with-such-success-9182580.html

Cook, Harold J. "Sydenham, Thomas (*bap.* 1624, *d.* 1689)." In *Oxford Dictionary of National Biography*. Oxford: Oxford University Press, 2004. https://doi.org/10.1093/ref:odnb/26864

Coombes, Paul, and Keith Barber. "Environmental Determinism in Holocene Research: Causality or Coincidence?" *Area* 37 (2005) 303–11.

Coon, Carleton S. "Some Problems of Human Variability and Natural Selection in Climate and Culture." *American Naturalist* 89, no. 848 (1955): 257–79.

Correia, David. "Climate Change Catastrophism: The New Environmental Determinism." *La Jicarita: An Online Magazine of Environmental Politics in New Mexico*, January 26, 2013. https://lajicarita.wordpress.com/2013/01/16/climate-change-catastrophism-the-new-environmental-determinism/

Corvalán, C. F., H. N. B. Gopalan, and P. Llansó. "Conclusions and Recommendations for Action." In *Climate Change and Human Health*, edited by A. J. McMichael, D. H. Campbell-Lendrum, C. F. Corvalán, K. L. Ebi, A. K. Githeko, J. D. Scheraga, and A. Woodward, 267–81. Geneva: World Health Organization, 2003.

Cousin, Claude Henri Victor. *Introduction to the History of Philosophy*, translated by H. G. Linberg. Boston: Hilliard, Gray, Little, and Wilkins, 1832.

Cowie, Jonathan. *Climate Change: Biological and Human Aspects*. Cambridge: Cambridge University Press, 2007.

Cox, Stan. *Losing Our Cool: Uncomfortable Truths about Our Air-Conditioned World*. New York: The New Press, 2012.

Crawfurd, John. *History of the Indian Archipelago Containing an Account of the Manners, Arts, Languages, Religions, Institutions, and Commerce of its Inhabitants*. 3 vols. Edinburgh: Constable, 1820.

Crawfurd, John. "On the Connexion between Ethnology and Physical Geography." *Transactions of the Ethnological Society of London* 2 (1863): 4–23.

Crook, D. P. *Benjamin Kidd: Portrait of a Social Darwinist*. New York: Cambridge University Press, 1984.

Crosby, Alfred W. "The Past and Present of Environmental History." *American Historical Review* 100 (1995): 1177–89.

Crozier, Anna. "What Was Tropical about Tropical Neurasthenia? The Utility of the Diagnosis in the Management of British East Africa." *Journal of the History of Medicine and Allied Sciences* 64 (2009): 518–48.

Cullimore, Daniel Henry. *The Book of Climates. Acclimatization; Climatic Diseases; Health Resorts and Mineral Springs; Sea Sickness; Sea Voyages; and Sea Bathing. Second Edition, with a Chapter on the Climate of Africa as it Affects Europeans by Surgeon Parke*. London: Baillière, Tindall, and Cox, 1891.

Cullimore, Daniel Henry. "On Tropical and Sub-Tropical Climates and the Acclimatisation of the Fair Races in Hot Countries." *The Medical Press*, October 31, 1888, 436–39, 461–64.

Cunningham, Michael R. "Weather, Mood and Helping Behavior: Quasi Experiments with the Sunshine Samaritan." *Journal of Personality and Social Psychology* 37 (1979): 1947–56.

Currie, William. *An Historical Account of the Climates and Diseases of the United States of America; and of the Remedies and Methods of Treatment, which have been found Most Useful and Efficacious, Particularly in those Diseases which Depend upon Climate and Situation Collected Principally from Personal Observation, and the Communications of Physicians of Talents and Experience, Residing in the Several States*. Philadelphia: T. Dobson, 1792.

Currie, William. *Memoirs of the Yellow Fever which Prevailed in Philadelphia, and Other Parts of the United States of America, in the Summer and Autumn of the Present Year, 1798 . . . To Which is Added A Collection of Facts Respecting the Origin of the Fever*. Philadelphia: Thomas Dobson, 1798.

Curtin, Philip D. *Death by Migration: Europe's Encounter with the Tropical World in the Nineteenth Century*. Cambridge: Cambridge University Press, 1989.

Cushman, Gregory T. Review of *Global Crisis* by Geoffrey Parker. *American Historical Review* 120 (2015): 1429–31.

Dalby, Simon. "Environment: From Determinism to the Anthropocene." In *The Wiley Blackwell Companion to Political Geography*, edited by John Agnew, Virginie Mamadouh, Anna J. Secor, and Joanne Sharp, 451–61. London: John Wiley, 2015.

Dalby, Simon. "Rethinking Geopolitics: Climate Security in the Anthropocene." *Global Policy* 5, no. 1 (2014): 1–9.

Dalén, Per. *Season of Birth: A Study of Schizophrenia and Other Mental Disorders*. New York: Elsevier, 1975.

Dana, James D. "Memoir of Arnold Guyot, 1807–1884." *Biographical Memoirs of the National Academy of Sciences* 2 (1886): 309–47.

Darwin, Charles. *The Descent of Man and Selection in Relation to Sex*. London: John Murray, 1871.

Davidson, Andrew. *Geographical Pathology: An Inquiry into the Geographical Distribution of Infective and Climatic Diseases*. Edinburgh: Young J. Pentland, 1892.

Davis, Diana K. *The Arid Lands: History, Power, Knowledge*. Cambridge, MA: MIT Press, 2016.

Davis, Mike. "The Coming Desert: Kropotkin, Mars and the Pulse of Asia." *New Left Review* 97 (2016): 23–43.

Davis, Mike. *Late Victorian Holocausts: El Niño Famines and the Making of the Third World*. London: Verso, 2001.

Davis, Wade. "The World Until Yesterday by Jared Diamond—Review." *Guardian*, January 9, 2013. https://www.theguardian.com/books/2013/jan/09/history-society

de Dainville, François. *La Géographie des Humanistes*. Paris: Beauchesne et Ses Fils, 1940.

de Menocal, Peter B. "Cultural Responses to Climate Change during the Late Holocene." *Science* 292 (2001): 667–73.

de Vliert, Evert Van. *Climate, Affluence, and Culture*. Cambridge: Cambridge University Press, 2009.

de Vliert, Evert Van. "Climato-Economic Habitats Support Patterns of Human Needs, Stresses, and Freedoms." *Behavioral and Brain Sciences* 36 (2013): 465–521.

de Vliert, Evert Van. "Climato-Economic Origins of Variation in Ingroup Favoritism." *Journal of Cross-Cultural Psychology* 42 (2011): 494–515.

de Vliert, Evert Van. "Climatoeconomic Roots of Survival versus Self-Expression Cultures." *Journal of Cross-Cultural Psychology* 38 (2007): 156–72.

de Vliert, Evert Van, and Damian R. Murray. "Climate and Creativity: Cold and Heat Trigger Invention and Innovation in Richer Populations." *Creativity Research Journal* 30 (2018): 17–28.

de Vries, Jan. "The Crisis of the Seventeenth Century: The Little Ice Age and the Mystery of the 'Great Divergence.'" *Journal of Interdisciplinary History* 44 (2014): 369–77.

de Waal, Alex. "Is Climate Change the Culprit for Darfur?" *African Arguments*, June 25, 2007. https://africanarguments.org/2007/06/is-climate-change-the-culprit-for-darfur/

DeBlieu, Jan. *Wind: How the Flow of Air Has Shaped Life, Myth, and the Land.* Boston: Houghton Mifflin, 1998.

Degroot, Dagomar. *The Frigid Golden Age: Climate Change, the Little Ice Age, and the Dutch Republic, 1560–1720.* Cambridge: Cambridge University Press, 2018.

Degroot, Dagomar. "Review—Global Crisis: War, Climate Change & Catastrophe in the 17th Century." *E-International Relations*, December 12, 2014. https://www.e-ir.info/2014/12/12/review-global-crisis-war-climate-change-catastrophe-in-the-17th-century/

DeLacy, Margaret. *The Germ of an Idea: Contagionism, Religion, and Society in Britain, 1660–1730.* London: Palgrave, 2016.

Dell, Melissa, Benjamin F. Jones, and Benjamin A. Olken. "Temperature and Income: Reconciling New Cross-Sectional and Panel Estimates." *American Economic Review: Papers and Proceedings* 99 (2009): 198–204.

Deming, David. *Science and Technology in World History. Volume 2: Early Christianity, the Rise of Islam and the Middle Ages.* Jefferson, NC: MacFarland, 2010.

Dennett, Daniel C. *Intuition Pumps and Other Tools for Thinking.* London: Penguin, 2013.

Devletoglou, Nicos E. "Montesquieu and the Wealth of Nations." *Canadian Journal of Economics and Political Science* 29 (1963): 1–25.

Dewhurst, Kenneth. *Dr Thomas Sydenham (1624–1689): His Life and Original Writings.* Berkeley: University of California Press, 1966.

Dewhurst, Kenneth. "Thomas Sydenham (1624–1689): Reformer of Clinical Medicine." *Medical History* 6 (1962): 101–18.

Dexter, Edwin Grant. *Weather Influences. An Empirical Study of the Mental and Physiological Effects of Definite Meteorological Conditions.* New York: Macmillan, 1904.

DeYoung, C. G. "Higher-Order Factors of the Big Five in a Multi-Informant Sample." *Journal of Personality and Social Psychology* 91 (2006): 1138–51.

Diamond, Jared. *Guns, Germs and Steel: A Short History of Everybody for the Last 12,000 Years.* London: Chatto and Windus, 1997.

Diaz, Henry F., R. Sari Kovats, Anthony J. McMichael, and Neville Nicholls. "Climate and Human Health Linkages on Multiple Timescales." In *History and Climate: Memories of the Future?* edited by P. D. Jones, A. E. J. Ogilvie, T. D. Davies, and K. R. Briffa, 267–89. Dordrecht: Kluwer, 2001.

Digman, J. M. "Higher-Order Factors of the Big Five." *Journal of Personality and Social Psychology*, 73 (1997): 1246–56.

Dillon, Sarah. "Imagining Apocalypse: Maggie Gee's *The Flood*." *Contemporary Literature* 48 (2007): 374–97.

Donner, Simon. "The Ugly History of Climate Determinism Is Still Evident Today." *Scientific American*, June 24, 2020. https://www.scientificamerican.com/article/the-ugly-history-of-climate-determinism-is-still-evident-today/

Dorn, Harold. *The Geography of Science*. Baltimore: Johns Hopkins University Press, 1991.

Dorn, Michael Leverett. "Climate, Alcohol, and the American Body Politic: The Medical and Moral Geographies of Daniel Drake, 1785–1852." PhD dissertation, University of Kentucky, 2002.

Douglass, A. E. "Weather Cycles in the Growth of Big Trees." *Weather Review* 37 (1010): 225–37.

Drake, Daniel. *A Discourse on Intemperance*. Cincinnati, OH: Looker and Reynolds, 1828.

Drake, Daniel. *Notices Concerning Cincinnati*. Cincinnati, OH: Privately Printed, 1810. Reprinted in *Quarterly Publication of the Historical and Philosophical Society of Ohio* 3 (1908), parts 1 and 2.

Drake, Daniel. "Some Account of the Epidemic Diseases Which Prevail at Mays-Lick, in Kentucky." *The Philadelphia Medical and Physical Journal* 3 (1808): 85–90.

Drake, Daniel. *A Systematic Treatise, Historical, Etiological, and Practical, on the Principal Diseases of the Interior Valley of North America, as They Appear in the Caucasian, African, Indian, and Esquimaux Varieties of its Population*. Cincinnati, OH: Winthrop B. Smith, 1850.

Draper, John William. *History of the American Civil War*. 3 vols. New York: Harper, 1867–70.

Draper, John William. *History of the Conflict between Religion and Science*. New York: Appleton, 1875.

Draper, John William. *The Influence of Physical Agents on Life: Being an Introductory Lecture to the Course on Chemistry and Physiology in the University of New York*. New York: John A. Gray, 1850.

Draper, John William. *Introductory Lecture, to the Course Of Chemistry: On the Relations Of Atmospheric Air to Animals and Plants*. New York: Jennings, 1844–45.

Droessler, E. G. "Helmut Landsberg, 1906–1985." *Eos* 67, no. 19 (1986): 457.

Du Bos, Jean-Baptiste, l'Abbé. *Critical Reflections on Poetry, Painting and Music*, translated by Thomas Nugent. 2 vols. London: Nourse, 1748. First published 1719.

Du Bos, Jean-Baptiste, l'Abbé. *Réflexions Critiques sur la Poésie et sur la Peinture*. Paris, 1760.

Dunbar, James. *Essays on the History of Mankind in Rude and Cultivated Ages*. London, 1780.

Duncan, James S. *In the Shadows of the Tropics: Climate, Race and Biopower in Nineteenth Century Ceylon*. London: Ashgate, 2007.

Duncan, James S. "The Struggle to be Temperate: Climate and 'Moral Masculinity' in Mid-Nineteenth Century Ceylon." *Singapore Journal of Tropical Geography* 21 (2000): 34–47.

Dyer, Gwynne. *Climate Wars: The Fight for Survival as the World Overheats*. Oxford: Oneworld Publications, 2010.

Dyer, Gwynne. "Excerpt: 'Climate Wars' by Gwynne Dyer." *CTV News*, November 21, 2008. https://www.ctvnews.ca/excerpt-climate-wars-by-gwynne-dyer-1.339415

Dyer, Thomas G. *Theodore Roosevelt and the Idea of Race*. Baton Rouge: Louisiana State University Press, 1980.

Eckersley, Robyn. "Environmental Security, Climate Change, and Globalizing Terrorism." In *Rethinking Insecurity, War and Violence: Beyond Savage Globalization*, edited by Damian Greenfel and Paul James, 85–97. London: Routledge, 2009.

Editorial. "Climate Change—The New Bioterrorism." *The Lancet* 358, no. 9294 (November 17, 2001): 1657.

Editorial. "Don't Jump to Conclusions about Climate Change and Civil Conflict: Many Studies That Link Global Warming to Civil Unrest Are Biased and Exacerbate Stigma about the Developing World." *Nature* 554 (February 13, 2018): 275–76. https://www.nature.com/articles/d41586-018-01875-9

Elden, Stuart, and Eduardo Mendieta, eds. *Reading Kant's Geography*. Albany: State University of New York Press, 2011.

Elton, Charles S. "Periodic Fluctuations in the Numbers of Animals: Their Causes and Effects." *Journal of Experimental Biology* 2 (1924): 119–63.

Emerson, F. V. "Geographic Influences in American Slavery." *Bulletin of the American Geographical Society* 43 (1911): 13–26.

Endfield, Georgina H. "The British Women's Emigration Association and the Climate(s) of South Africa." In *Weather, Climate, and the Geographical Imagination: Placing Atmospheric Knowledges*, edited by Martin Mahony and Samuel Randalls, 132–51. Pittsburgh: University of Pittsburgh Press, 2020.

Endfield, Georgina H. *Climate and Society in Colonial Mexico: A Study in Vulnerability*. Oxford: Blackwell, 2008.

Endfield, Georgina H. "Reculturing and Particularizing Climate Discourses: Weather, Identity, and the Work of Gordon Manley." *Osiris* 26, no. 1 (2011): 142–62.

Endfield, Georgina H. "The Resilience and Adaptive Capacity of Social-Environmental Systems in Colonial Mexico." *Proceedings of the National Academy of Science* 109, no. 10 (2012): 3676–681.

Endfield, Georgina H., and David J. Nash. "Drought, Desiccation and Discourse: Missionary Correspondence and Nineteenth-Century Climate Change in Central Southern Africa." *Geographical Journal* 168 (2002): 33–47.

Endfield, Georgina H., and David J. Nash. "'A Good Site for Health': Missionaries and the Pathological Geography of Central Southern Africa." *Singapore Journal of Tropical Geography* 28 (2007): 142–57.

Endfield, Georgina H., and David J. Nash. "'Happy Is the Bride the Rain Falls On': Climate, Health and 'the Woman Question' in Nineteenth-Century Missionary Documentation." *Transactions of the Institute of British Geographers* 30 (2005): 368–86.

Endfield, Georgina H., and David J. Nash. "Missionaries and Morals: Climatic Discourse in Nineteenth-Century Central Southern Africa." *Annals of the Association of American Geographers* 92 (2002): 727–42.

Endfield, Georgina H., and Samuel Randalls. "Climate and Empire." In *Eco-Cultural Networks and the British Empire: New Views on Environmental History*, edited by James Beattie, Edward Melillo and Emily O'Gorman, 21–43. London: Bloomsbury, 2014.

Epstein, Paul R. "Climate and Health." *Science* 285, no. 5426 (1999): 347–48.

Epstein, Paul R., and Dan Ferber. *Changing Planet, Changing Health: How the Climate Crisis Threatens Our Health and What We Can Do about It*. Berkeley: University of California Press, 2011.

Erickson, Clark L. "Neo-environmental Determinism and Agrarian 'Collapse' in Andean Prehistory." *Antiquity* 73 (1999): 634–42.

Erickson, Paul. "Knowing Nature through Markets: Traces, Populations, and the History of Ecology." *Science as Culture* 19 (2010): 529–51.

Erickson, Paul, and Gregg Mitman. *When Rabbits Became Humans (and Humans, Rabbits): Stability, Order, and History in the Study of Populations.* Working Papers on the Nature of Evidence: How Well do "Facts" Travel? 19/07. London: Department of Economic History, London School of Economics and Political Science, 2007. http://eprints.lse.ac.uk/22517/

Eshelman, Robert S. "The Mental Health Toll of Climate Change Could be Dire." *Seeker*, March 30, 2017. https://www.seeker.com/health/the-mental-health-toll-of-climate-change-could-be-dire

Eurich, Amanda. "Chardin, Sir John [Jean] (1643–1712)." In *Oxford Dictionary of National Biography.* Oxford: Oxford University Press, 2004. https://doi.org/10.1093/ref:odnb/5138

Eylers, Eva. "Planning the Nation: The Sanitorium Movement in Germany." *The Journal of Architecture* 19 (2014): 667–92.

Eze, Emmanuel Chukwudi, ed. *Race and the Enlightenment: A Reader.* Oxford: Wiley-Blackwell, 1997.

Fagan, Brian. *Cro-Magnon: How the Ice Age Gave Birth to the First Modern Humans.* New York: Bloomsbury, 2010.

Fagan, Brian. *Floods, Famines, and Emperors: El Niño and the Fate of Civilizations.* London: Basic Books, 1999.

Fagan, Brian. *The Great Warming: Climate Change and the Rise and Fall of Civilizations.* New York: Bloomsbury Press, 2008.

Fagan, Brian. *The Little Ice Age: How Climate Made History, 1300–1850.* London: Basic Books, 2000.

Fagan, Brian. *The Long Summer: How Climate Changed Civilization.* London: Granta Books, 2005.

Falconer, William. *Remarks on the Influence of Climate, Situation, Nature of Country, Population, Nature of Food, and Way of Life on the Disposition and Temper, Manners and Behaviour, Intellects, Laws and Customs, Form of Government, and Religion, of Mankind.* London: Dilly, 1781.

Farwell, Byron. *Burton: A Biography of Sir Richard Francis Burton.* London: Longmans, Green and Co., 1963.

Federal Ministry for the Environment (BMU). *Climate Change and Conflict.* Berlin: Federal Ministry for the Environment, Nature Conservation and Nuclear Safety, 2002. http://www.afes-press.de/pdf/Brauch_ClimateChange_BMU.pdf

Felkin, Robert. "Can Europeans Become Acclimatised in Tropical Africa?" *Scottish Geographical Magazine* 2 (1886): 647–57.

Felkin, Robert. "The Distribution of Disease in Africa." *Report of the British Association for the Advancement of Science* (1893), Section E, 839.

Felkin, Robert. "On Acclimatisation." *Scottish Geographical Magazine* 7 (1891): 647–56.

Felkin, Robert. "Tropical Highlands: Their Suitability for European Settlement." *Transactions of the Seventh International Congress on Hygiene and Demography* 10 (1892): 155–64.

Ferng, Jennifer, Jiat-Hwee Chang, Erik L'Heureux, and Daniel J. Ryan. "Climatic Design and Its Others: 'Southern' Perspectives in the Age of the Anthropocene." *Journal of Architectural Education* 74 (2020): 250–62.

Ferreira, Susana, and Travis Smith. "Hotter Weather Brings More Stress, Depression and Other Mental Health Problems." *The Conversation*, March 25, 2020. https://theconversation.com/hotter-weather-brings-more-stress-depression-and-other-mental-health-problems-134325

Ferrell, Edith. "Arnold Henry Guyot, 1807–1884." *Geographers: Biobibliographical Studies* 5 (1981): 63–71.

Finke, Leonhard Ludwig. *Versuch einer Allgemeinen Medicinisch-Praktischen Geographie*. 3 vols. Leipzig: Weidmann, 1792–95.

Finlayson, Clive. "Climate and Humans: The Long View. Viewpoint," *BBC News*, December 24, 2009. http://news.bbc.co.uk/1/hi/sci/tech/8430276.stm

Finlayson, Clive. *The Humans Who Went Extinct: Why Neanderthals Died Out and We Survived*. New York: Oxford University Press, 2009.

Finlayson, Clive. *Neanderthals and Modern Humans: An Ecological and Evolutionary Perspective*. Cambridge: Cambridge University Press, 2004.

Finnegan, Diarmid. "'An Aid to Mental Health': Natural History, Alienists and Therapeutics in Victorian Scotland." *Studies in History and Philosophy of the Biological and Biomedical Sciences* 39 (2008): 326–37.

Fisher, Colin. "Antebellum Black Climate Science: The Medical Geography and Emancipatory Politics of James McCune Smith and Martin Delany." *Environmental History* 26 (2021): 461–83.

Flannery, Tim. *The Future Eaters: An Ecological History of the Australasian Lands and People*. Sydney: New Holland, 1994.

Fleming, Donald. *John William Draper and the Religion of Science*. Philadelphia: University of Pennsylvania Press, 1950.

Fleming, James Rodger. "Civilization, Climate, and Ozone: Ellsworth's Huntington's 'Big' Views on Biophysics, Biocosmics, and Biocracy." In *Weather, Climate, and the Geographical Imagination: Placing Atmospheric Knowledges*, edited by Martin Mahony and Samuel Randalls, 215–31. Pittsburgh: University of Pittsburgh Press, 2020.

Fleming, James Rodger. *Historical Perspectives on Climate Change*. New York: Oxford University Press, 1998.

Fleming, James Rodger. "T.C. Chamberlin, Climate Change and Cosmogony." *Studies in the History and Philosophy of Modern Physics* 31B, no. 3 (2000): 293–308.

Floyd-Wilson, Mary. *English Ethnicity and Race in Early Modern Drama*. Cambridge: Cambridge University Press, 2006.

Folk, G. Edgar. "The International Society of Biometeorology: A Fifty Year History." https://uwm.edu/biometeorology/wp-content/uploads/sites/439/2017/06/ISB_50YearHistoryofBiometeorology.pdf

Fontenelle, Bernard le Bovier de. *Poésies Pastorales; avec Un Traité sur la Nature de l'Eglogue; Une Digression sur les Anciens, et les Modernes; et Un Recueil de Poésies Diverses*. Nouvelle edition, A La Haye, 1728.

Forsdyke, Sara. "Athenian Democratic Ideology and Herodotus's *Histories*." *American Journal of Philology* 122 (2001): 329–25.

Foust, Christina R., and William O'Shannon Murphy. "Revealing and Reframing Apocalyptic Tragedy in Global Warming Discourse." *Environmental Communication: A Journal of Nature and Culture* 3, no. 2 (2009): 151–67.

Fredrickson, George M. *The Black Image in the White Mind: The Debate on Afro-American Character and Destiny, 1817–1914*. New York: Harper and Row, 1972.

Fromherz, Allen James. *Ibn Khaldun, Life and Times*. Edinburgh: Edinburgh University Press, 2010.

Frumkin, Howard, Jeremy Hess, George Luber, Josephine Malilay, and Michael McGeehin. "Climate Change: The Public Health Response." *American Journal of Public Health* 98 (2008): 435–45.

Gallegati, Mauro. "Jevons, Sunspot Theory and Economic Fluctuations." *History of Economic Ideas* 11 (1994): 23–40.

Gallup, John Luke, and Jeffrey D. Sachs. "Agriculture, Climate, and Technology: Why Are the Tropics Falling Behind?" *American Journal of Agricultural Economics* 82 (2000): 731–37.

Gallup, John Luke, Jeffrey D. Sachs, and Andrew D. Mellinger. "Geography and Economic Development." *International Regional Science* Review 22 (1999): 179–232.

Gamarra, Yolanda. "Ibn Khaldun (1332–1406): A Precursor of Intercivilizational Discourse." *Leiden Journal of International Law* 28 (2015): 4414–56.

Garrard, Greg. "Environmentalism and the Apocalyptic Tradition." *Green Letters: Studies in Ecocriticism* 3, no. 1 (2001): 27–68.

Gates, Warren. "The Spread of Ibn Khaldūn's Ideas on Climate and Culture." *Journal of the History of Ideas* 28 (1967): 415–22.

Gaukroger, Stephen. *The Natural and the Human: Science and the Shaping of Modernity, 1739–1841.* Oxford: Oxford University Press, 2016.

Geddes, Linda. *Chasing the Sun; the New Science of Sunlight and How It Shapes Our Bodies and Minds.* London: Wellcome Collection, 2019.

Gelder, John. "Design with Climate in Ancient Rome: Vitruvius meets Olgyay." In *Engaging Architectural Science,* 52nd International Conference of the Architectural Science Association, edited by P. Rajagopalan and M. M. Andamon, 745–52. Melbourne: RMIT Press, 2018.

Gellner, Ernest. "From Ibn Khaldun to Karl Marx." *Political Quarterly* 32 (1961): 385–92.

Gellner, Ernest. *Muslim Society*. Cambridge: Cambridge University Press, 1981.

Gellner, Ernest. Review of *The Muqaddimah,* by Ibn Khaldūn, translated from the Arabic (and with an Introduction) by Franz Rosenthal. *Philosophy* 36 (1961): 255–56.

Gemenne, François, Jon Barnett, W. Neil Adger, and Geoffrey D. Dabelko. "Climate and Security: Evidence, Emerging Risks, and a New Agenda." *Climatic Change* 124 (2014): 1–9.

Gentilcore, David. "Louis Sambon and the Clash of Pellagra Etiologies in Italy and the United States, 1905–14." *Journal of the History of Medicine and Allied Sciences* 71 (2015): 19–42.

Gerbi, Antonello. *The Dispute of the New World: The History of a Polemic, 1750–1900*. Pittsburgh: University of Pittsburgh Press, 1973.

Ghosh, Amitav. *The Great Derangement: Climate Change and the Unthinkable*. Chicago: University of Chicago Press, 2016.

Ghosh, Amitav. "Where Is the Fiction about Climate Change?" *Guardian,* October 28, 2016. https://www.theguardian.com/books/2016/oct/28/amitav-ghosh-where-is-the-fiction-about-climate-change-

Gilbe, Emily. "The Militarization of Climate Change." *ACME: An International E-Journal for Critical Geographies* 11, no. 1 (2012): 1–14.

Gill, Richardson. *The Great Maya Droughts: Water, Life, and Death*. Albuquerque: University of New Mexico Press, 2000.

Githeko, A. K., and A. Woodward. "International Consensus on the Science of Climate and Health: The IPCC Third Assessment Report." In *Climate Change and Human Health,* edited by A. J. McMichael, D. H. Campbell-Lendrum, C. F. Corvalán, K. L. Ebi, A. K. Githeko, J. D. Scheraga, and A. Woodward, 43–60. Geneva: World Health Organization, 2003.

Glacken, Clarence J. *Traces on the Rhodian Shore: Nature and Culture in Western Thought from Ancient Times to the End of the Eighteenth Century*. Berkeley: University of California Press, 1967.

Gleditsch, Nils Petter. "Whither the Weather? Climate Change and Conflict." *Journal of Peace Research* 49 (2012): 3–9.

Global Humanitarian Forum. *Climate Change: The Anatomy of a Silent Crisis*. Geneva: Global Humanitarian Forum, 2009.

Goldstein, K. M. "Weather, Mood and Internal-External Control." *Perceptual and Motor Skills* 35 (1972): 786.

Golinski, Jan. *British Weather and the Climate of Enlightenment*. Chicago: University of Chicago Press, 2007.

Gonce, Richard A. "John R. Commons's 'Five Big Years': 1899–1904." *American Journal of Economics and Sociology* 61 (2002): 755–77.

Goodbody, Alex. "Cli-Fi—Genre of the Twenty-First Century? Narrative Strategies in Contemporary Climate Fiction and Film." In *Green Matters: Ecocultural Functions of Literature*, edited by Maria Löschnigg and Melanie Braunecker, 131–53. Leiden: Brill Rodopi, 2020.

Goodridge, Malcolm, and Philip Van Ingen. "James Alexander Miller—In Memoriam." *Bulletin of the New York Academy of Medicine* 24 (1948): 743–46.

Goodrum, Matthew R. "The History of Human Origins Research and Its Place in the History of Science: Research Problems and Historiography." *History of Science* 47 (2009): 337–57.

Gourou, Pierre. "Le Déterminisme Physique dans 'l'Esprit des Lois.'" *L'Homme* 3, no. 3 (1963): 5–11.

Gourou, Pierre. *Les Pays Tropicaux: Principles d'une Géographie Humaine et Économique*. Paris: Presses Universitaires de France, 1947.

Grant, Ben. *Postcolonialism, Psychoanalysis and Burton: Power Play of Empire*. London: Routledge, 2009.

Grant, Madison. *The Passing of the Great Race, or, the Racial Basis of European History*. New York: Scribner, 1916.

Graves, Paul. "New Models and Metaphors for the Neanderthal Debate." *Current Anthropology* 32 (1991): 513–41.

Greenfel, Damian, and Paul James, eds. *Rethinking Insecurity, War and Violence: Beyond Savage Globalization*. London: Routledge, 2009.

Grehan, John R., and Jeffrey H. Schwartz. "Evolution of the Second Orangutan: Phylogeny and Biogeography of Hominid Origins." *Journal of Biogeography* 36 (2009): 1823–44.

Gribbin, John, and Mary Gribbin. *Children of the Ice: Climate and Human Origins*. Oxford: Blackwell, 1990.

Griffin, Paul, David Lont, and Martien Lubberink. "Extreme High Surface Temperature Events and Equity-Related Physical Climate Risk." *Weather and Climate Extremes* 26 (2019): article 100220.

Griffitt, William. "Environmental Effects on Interpersonal Affective Behavior: Ambient Effective Temperature and Attraction." *Journal of Personality and Social Psychology* 15 (1970): 240–44.

Grohol, John M. "Can Weather Affect Your Mood?" PsychCentral, August 29, 2014. https://psychcentral.com/blog/can-weather-affect-your-mood

Grohol, John M. "Weather Can Change Your Mood." PsychCentral, November 9, 2008. https://psychcentral.com/blog/weather-can-change-your-mood

Grossman, Daniel. "Parched Turf Battle." *Scientific American* 287 (December 2002): 32–33.

Grove, Jean M. *The Little Ice Age*. London: Methuen, 1988.

Grove, Richard. *Green Imperialism: Colonial Expansion, Tropical Island Edens and the Origins of Environmentalism, 1600–1860*. Cambridge: Cambridge University Press, 1995.

Guldi, Jo, and David Armitage. *The History Manifesto*. Cambridge: Cambridge University Press, 2014.

Güss, C. Dominik. "What about Politics and Culture?" *Journal of Cross-Cultural Psychology* 42 (2011): 490–91.

Guyot, Arnold. *The Earth and Man: Lectures on Comparative Physical Geography, in Its Relation to the History of Mankind*, translated from the French by C. C. Felton. Boston: Gould, Kendall, and Lincoln, 1849.

Haines, A., R. S. Kovats, D. Campbell-Lendrum, and C. Corvalan. "Climate Change and Human Health: Impacts, Vulnerability and Public Health." *Public Health* 120 (2006): 585–96.

Halberg, F., N. Düll-Pfaff, L. Gumarova, T. A. Zenchenko, O. Schwartzkopff, E. M. Freytag, J. Freytag, and G. Cornelissen. "27-Day Cycles in Human Mortality: Traute and Bernhard Düll." *History of Geo- and Space Sciences* 4 (2013): 47–59.

Hall, Maxwell. *The Meteorology of Jamaica*. Kingston: Institute of Jamaica, 1904.

Hall, Maxwell. *The Rainfall of Jamaica*. Kingston: Institute of Jamaica, 1892.

Haller, John, Jr. *Outcasts from Evolution: Scientific Attitudes of Racial Inferiority, 1859–1900*. Urbana: University of Illinois Press, 1971.

Hampson, Fen Osler. "The Climate for War: The Economic and Political Consequences of Climate Change Will Be a New Source of Conflict among Nations." *Canadian Institute for International Peace and Security* 3, no. 3 (1988): 8–9.

Han, Byung-Chul. *Psychopolitics: Neoliberalism and the New Technologies of Power*. London: Verso, 2017.

Hancock, P. A., Jennifer M. Ross, and James L. Szalma. "A Meta-Analysis of Performance Response under Thermal Stressors." *Human Factors* 49 (2007): 851–77.

Hannaway, Caroline. "Environment and Miasmata." In *Companion Encyclopedia of the History of Medicine*, edited by W. F. Bynum and Roy Porter, 292–308. Abingdon: Routledge, 1993.

Harley, J. Brian. "Deconstructing the Map." *Cartographica* 26, no. 2 (1989): 1–20.

Harley, J. Brian. "Silences and Secrecy: The Hidden Agenda of Cartography in Early Modern Europe." *Imago Mundi* 40 (1988): 57–76.

Harper, Kyle. "The Environmental Fall of the Roman Empire." *Daedalus* 145 (2016): 101–11.

Harper, Kyle. "How Climate Change and Plague Helped Bring Down the Roman Empire." *Smithsonian Magazine*, December 19, 2017. https://www.smithsonianmag.com/science-nature/how-climate-change-and-disease-helped-fall-rome-180967591/

Harries, Keith D., and Stephen J. Stadler. "Determinism Revisited: Assault and Heat Stress in Dallas, 1980." *Environment and Behavior* 15 (1983): 235–56.

Harris, Chris, and Randy Mann. "Global Temperature Trends from 2500 B.C. to 2040 A.D." LongRangeWeather Harris-Mann Climatology. http://www.longrangeweather.com/global_temperatures.htm

Harrison, Mark. *Climates and Constitutions: Health, Race, Environment and British Imperialism in India 1600–1850*. Oxford: Oxford University Press, 1999.

Harrison, Mark. "'The Tender Frame of Man': Disease, Climate and Racial Difference in India and the West Indies, 1760–1860." *Bulletin of the History of Medicine* 70 (1996): 68–93.

Harrison, Mark. "Tropical Medicine in 19th-Century India." *British Journal for the History of Science* 25 (1992): 299–318.

Hartmann, Betsy. "Rethinking Climate Refugees and Climate Conflict: Rhetoric, Reality and the Politics of Policy Discourse." *Journal of International Development* 22 (2010): 233–46.

Harvey, Fiona. "Climate Change Will Increase Threat of War, Chris Huhne to Warn: UK Climate Secretary to Tell Defence Experts That Conflict Caused by Climate Change Risks Reversing the Progress of Civilisation." *The Guardian*, July 6, 2011. http://www.guardian.co.uk/environment/2011/jul/06/climate-change-war-chris-huhne

Haslam, Nick. "Here Comes the Sun: How Weather Affects our Mood." *The Conversation*, October 22, 2013. https://theconversation.com/here-comes-the-sun-how-the-weather-affects-our-mood-19183

Haude, Rüdiger. "'Keep Calm'? A Critique of Wolfgang Behringer's 'A Cultural History of Climate.'" *Journal of Environmental Studies and Sciences* 9 (2019): 397–408.

Haug, Gerald H., Detlef Günther, Larry C. Peterson, Daniel M. Sigman, Konrad A. Hugen, and Beat Aeschlimann. "Climate and the Collapse of Maya Civilization." *Science* 299 (2003): 1731–73.

Haviland, Alfred. *Climate, Weather and Disease: Being a Sketch of the Opinions of the Most Celebrated Antient and Modern Writers With Regard to the Influence of Climate and Weather in Producing Disease*. London: John Churchill, 1855.

Haviland, Alfred. *The Geographical Distribution of Disease in Great Britain*. 2nd ed. London: Swan Sonnenschein, 1892.

Haviland, Alfred. *The Geographical Distribution of Heart Disease and Dropsy in England and Wales*. London, 1871.

Hazell, Alastair. *The Last Slave Market: Dr John Kirk and the Struggle to End the East African Slave Trade*. London: Constable, 2012.

Heads, Michael. "Darwin's Changing Views on Evolution: From Centres of Origin and Teleology to Vicariance and Incomplete Lineage Sorting." *Journal of Biogeography* 26 (2009): 1018–26.

Heal, Geoffrey, and Jisung Park. "Feeling the Heat: Temperature, Physiology and the Wealth of Nations." Working Paper 19725, National Bureau of Economic Research, 2013. http://www.nber.org/papers/w19725

Health Scotland. "Beating the Winter Blues." NHS Health Scotland. https://www.nhsinform.scot/healthy-living/mental-wellbeing/low-mood-and-depression/beating-the-winter-blues

Henderson, Gabriel D. "The Dilemma of Reticence: Helmut Landsberg, Stephen Schneider, and Public Communication of Climate Risk, 1971–1976." *History of Meteorology* 6 (2014): 53–78.

Henderson, Gabriel D. "Helmut Landsberg and the Evolution of 20th Century American Climatology: Envisioning a Climatological Renaissance." *WIREs Climate Change* 8 (2017) 8:e442. doi: 10.1002/wcc.442.

Hendrix, Cullen S., and Sarah M. Glaser. "Trends and Triggers: Climate, Climate Change and Civil Conflict in Sub-Saharan Africa." *Political Geography* 26 (2007): 695–715.

Hendrix, Cullen S., and Idean Salehyan. "Climate Change, Rainfall, and Social Conflict in Africa." *Journal of Peace Research* 49 (2012): 35–50.

Herder, Johann Gottfried. *Outlines of a Philosophy of the History of Man,* translated from the German *Ideen zur Philosophie der Geschichte der Menschheit* by T. Churchill. New York: Bergman Publishers, 1800. First published 1784.

Herodotus. *The Histories,* edited by A. D. Godley. Cambridge, MA: Harvard University Press, 1920–25.

Hetherington, Renée, and Robert G. B. Reid. *The Climate Connection: Climate Change and Modern Human Evolution.* Cambridge: Cambridge University Press, 2010.

Heyck, Thomas William. "Buckle, Henry Thomas." In *Oxford Dictionary of National Biography.* Oxford: Oxford University Press, 2004. https://doi.org/10.1093/ref:odnb/3861

Higham, John. *Strangers in the Land: Patterns of American Nativism 1860–1925.* New York: Atheneum, 1973.

Hillary, William. *Observations on the Changes of the Air, and the Concomitant Epidemical Diseases in the Island of Barbadoes. To which is added, A Treatise on the Putrid Bilious Fever, Commonly Called the Yellow Fever; and Such Other Diseases as are Indigenous Or Endemial, in the West India Islands, or in the Torrid Zone.* Philadelphia, 1811. First published 1759.

Hillary, William. *A Treatise on Such Diseases as are the Most Frequent in, or are Peculiar to the West-India Islands, or the Torrid Zone, both Acute and Chronical.* Philadelphia: Aitken, 1811.

Hippocrates. *On Airs, Waters, and Places,* translated by Francis Adams. Internet Classics Archive. http://classics.mit.edu/Hippocrates/airwatpl.1.1.html

Hirsch, August. *Handbook of Historical and Geographical Pathology,* translated from the Second German Edition by Charles Creighton. 3 vols. London: The Sydenham Society, 1883, 1885, 1886.

Hobbes, Thomas. *Leviathan.* London: J. M. Dent, 1914. First published 1651.

Hodell, David A. "Maya Megadrought?" *Nature* 479 (November 2011): 45.

Hodell, David A., Jason H. Curtis, and Mark Brenner. "Possible Role of Climate in the Collapse of Classic Maya Civilization." *Nature* 375 (June 1995): 391–94.

Hofstadter, Richard. *Social Darwinism in American Thought.* Philadelphia: University of Pennsylvania Press, 1944.

Holmes, Bob. "Cool Times, Smart Times: Did an Ice Age Allow the Boom in Brain Size That Made Humans What They Are Today?" *New Scientist* 2719 (August 2009): 6–7.

Homer-Dixon, Thomas F. *Environment, Scarcity, and Violence.* Princeton, NJ: Princeton University Press, 2001.

Homer-Dixon, Thomas F. "On the Threshold: Environmental Changes as Causes of Acute Conflict." *International Security* 19, no. 2 (1991): 76–116.

Horine, Emmet F. *Daniel Drake, 1785–1852: Pioneer Physician of the Midwest.* Philadelphia: University of Pennsylvania Press, 1961.

Horton, James Africanus B. *Diseases of Tropical Climates and their Treatment, with Hints for the Preservation of Health in the Tropics.* London: Churchill, 1874.

Horton, James Africanus B. *Physical and Medical Climate and Meteorology of the West Coast of Africa with Valuable Hints to Europeans for the Preservation of Health in the Tropics.* London: John Churchill, 1867.

Horton, James Africanus B. *West African Countries and Peoples, British and Native. With the Requirements Necessary for Establishing that Self Government Recommended by the Committee of the House of Commons, 1865; and a Vindication of the African Race.* London: W. J. Johnson, 1868.

"How Climate Change Can Fuel Wars." *The Economist* 431, no. 9114 (May 2019): 60–62.

"How to Think about Global Warming and War." *The Economist* 431, no. 9114 (May 2019): 17–18.

Howarth, E., and M. S. Hoffman. "A Multidimensional Approach to the Relationship between Mood and Weather." *British Journal of Psychology* 75 (1984): 15–23.

Howell, Jessica. *Exploring Victorian Travel Literature: Disease, Race and Climate*. Edinburgh: Edinburgh University Press, 2014.

Howsam, Leslie. "An Experiment with Science for the Nineteenth-Century Book Trade: The International Scientific Series." *British Journal for the History of Science* 33 (2000): 187–207.

Hsiang, Solomon M., Marshall Burke, and Edward Miguel. "Quantifying the Influence of Climate on Human Conflict." *Science* 341 (September 2013): 1235367-1–1235367-14. http://www.sciencemag.org/content/341/6151/1235367.full#xref-ref-8-1

Hsiang, Solomon M., Jyle C. Meng, and Mark A. Cane. "Civil Conflicts Are Associated with the Global Climate." *Nature* 476 (August 2011): 438–41.

Huang, Chun Chang, and Hongxia Su. "Climate Change and Zhou Relocations in Early Chinese History." *Journal of Historical Geography* 35 (2009): 297–310.

Huhne MP, The Rt Hon Chris. "The Geopolitics of Climate Change," July 7, 2011. https://www.gov.uk/government/speeches/the-rt-hon-chris-huhne-mp-the-geopolitics-of-climate-change-speech-to-future-maritime-operations-conference-at-the-royal-united-services-institute

Hull, Edmund C. P. *The European in India or Anglo-Indian's Vade-Mecum: A Handbook of Practical Information for Those Proceeding to, or Residing in the East Indies, Relating to Outfits, Routes, Time for Departure, Indian Climate and Seasons Housekeeping, Servants, etc. Also an Account of Anglo-Indian Social Customs and Native Character*. London: Henry S. King, 1871.

Hulme, Mike. "Climate Security: The New Determinism." *Open Democracy*, December 20, 2007. https://www.opendemocracy.net/article/climate_security_the_new_determinism

Hulme, Mike. "Reducing the Future to Climate: A Story of Climate Determinism and Reductionism." *Osiris Klima* 26 (2011): 245–66.

Hulme, Mike. Review of *A Cultural History of Climate* by Wolfgang Behringer, Review 925, *Reviews in History*, June 2010. http://www.history.ac.uk/reviews/review/925

Hulme, Mike. *Weathered: Cultures of Climate*. London: Sage, 2017.

Hulme, Mike. *Why We Disagree about Climate Change: Understanding Controversy, Inaction and Opportunity*. Cambridge: Cambridge University Press, 2009.

Humboldt, A. de, and A. Bonpland. *Essai sur la Géographie des Plantes Accompagné d'un Tableau Physique des Régions Équinoxiales*. Paris: Chez Levrault, Schoell et Compagnie, 1805.

Humboldt, Alexander von, and Aimé Bonpland. *Personal Narrative of Travels to the Equinoctial Regions of the New Continent, During the Years 1799–1804*, translated into English by Helen Maria Williams. 7 vols. London: Longman, 1818.

Hume, David. "Of National Characters." In *Essays: Moral, Political and Literary*, edited and with a foreword by Eugene F. Miller, 197–215. Indianapolis: Liberty Fund, 1987.

Hunt, James. "On Ethno-Climatology; or the Acclimatization of Man." *Transactions of the Ethnological Society of London* 2 (1863): 50–83.

Huntington, Ellsworth. "The Adaptability of the White Man to Tropical America." *Journal of Race Development* 5 (1914): 185–211.

Huntington, Ellsworth. "The Arabian Desert and Human Character." *Journal of Geography* 10 (1912): 169–75.

Huntington, Ellsworth. *Civilization and Climate*. New Haven, CT: Yale University Press, 1915.

Huntington, Ellsworth. "Changes of Climate and History." *American Historical Review* 18 (1913): 213–32.

Huntington, Ellsworth. *The Character of Races as Influenced by Physical Environment, Natural Selection and Historical Development*. New York: Scribner, 1924.

Huntington, Ellsworth. "Climatic Change and Agricultural Exhaustion as Elements in the Fall of Rome." *Quarterly Journal of Economics* 31 (1917): 173–208.

Huntington, Ellsworth. "Climatic Variations and Economic Cycles." *Geographical Review* 1 (1916): 192–202.

Huntington, Ellsworth. "The Control of Pneumonia and Influenza by the Weather." *Ecology* 1 (1920): 6–23.

Huntington, Ellsworth. "The Effect of Overpopulation on Chinese Character." *Birth Control Review* 10, no. 7 (July 1926): 221–22.

Huntington, Ellsworth. "The Fluctuating Climate of North America." *Geographical Journal* 40 (1912): 264–80, 392–411.

Huntington, Ellsworth. "The Greenest of Deserts." *Harper's Magazine* 123, no. 733 (1911): 50–58.

Huntington, Ellsworth. "Influenza and the Weather in the United States in 1918." *The Scientific Monthly* 17, no. 5 (1923): 462–71.

Huntington, Ellsworth. *Mainsprings of Civilization*. New York: John Wiley, 1945.

Huntington, Ellsworth. "The Matamek Conference on Biological Cycles, 1932." *Science* 74, no. 1914 (September 4, 1931): 229–35.

Huntington, Ellsworth. *Palestine and Its Transformation*. Boston: Houghton Mifflin, 1911.

Huntington, Ellsworth. "Physical Environment as a Factor in the Present Condition of Turkey." *Journal of Race Development* 1 (1911): 460–81.

Huntington, Ellsworth. *The Pulse of Asia: A Journey in Central Asia Illustrating the Geographic Basis of History*. Boston and New York: Houghton Mifflin, 1907.

Huntington, Ellsworth. *The Pulse of Progress, Including a Sketch of Jewish History*. New York: Charles Scribner's, 1926.

Huntington, Ellsworth. Review of *Health and Social Progress* by R. M. Binder. *Geographical Review* 12 (1922): 156–57.

Huntington, Ellsworth. Review of *Medical Climatology* by Clarence Mills. *Science* 90 (1939): 540–42.

Huntington, Ellsworth. *Season of Birth: Its Relation to Human Abilities*. New York: John Wiley & Sons, 1938.

Huntington, Ellsworth. "The Solar Hypothesis of Climatic Changes." *Bulletin of the Geological Society of America* 25 (1914): 477–590.

Huntington, Ellsworth. *Tomorrow's Children: the Goal of Eugenics*. New York: John Wiley, 1935.

Huntington, Ellsworth. *Weather and Health: A Study of Daily Mortality in New York City*. Washington, D.C.: Bulletin of the National Research Council, no. 75, 1930, 1–161.

Huntington, Ellsworth. *World-Power and Evolution*. New Haven, CT: Yale University Press, 1919.

Huntington, Ellsworth, and Stephen Sargent Visher. *Climatic Changes: Their Nature and Causes.* New Haven, CT: Yale University Press, 1922.

Huntington, Ellsworth, and Leon F. Whitney. *Builders of America.* New York: Morrow, 1927.

Huntington, Ellsworth, and Frank E. Williams. *Business Geography.* New York: Wiley, 1922.

Huntington, Ellsworth, Frank E. Williams, and Samuel van Valkenburg. *Economic and Social Geography.* New York: Wiley, 1933.

Hutchinson, Thomas J. *Impressions of Western Africa with Remarks on the Diseases of the Climate.* London: Longman, Brown, Green, Longmans and Roberts, 1858.

Huxham, John. *An Essay on Fevers, and their Various Kinds, as Depending on Different Constitutions of the Blood.* Austen: London, 1750.

Huxham, John. "Observations on the Air and Epidemic Diseases. Part I. From the Year 1728 to 1737 Inclusive, Translated from the Latin Original." In *The Works of John Huxham in Two Volumes.* London, 1788.

Ibn Khaldūn. *The Muqaddimah: An Introduction to History,* translated by Franz Rosenthal. 3 vols. Princeton, NJ: Princeton University Press, 1967.

Ibn Khaldūn. *Prolégomènes d'Ebn-Khaldoun. Notices et Extraits des Manuscripts de la Bibliothèque Impériale.* Paris: Imprimerie Impériale, 1863–68.

Ikuko, Asaka. *Tropical Freedom: Climate, Settler Colonialism, and Black Exclusion in the Age of Emancipation.* Durham, NC: Duke University Press, 2017.

Ilany, Ofri. *In Search of the Hebrew People: Bible and Nation in the German Enlightenment.* Bloomington: Indiana University Press, 2018.

Imbrie, John, and Katherine Palmer Imbrie. *Ice Ages: Solving the Mystery.* Cambridge, MA: Harvard University Press, 1986.

Immerwahr, Daniel. "Are We Really Prisoners of Geography?" *Guardian,* November 10, 2022. https://www.theguardian.com/world/2022/nov/10/are-we-really-prisoners-of-geography-maps-geopolitics

Intergovernmental Panel on Climate Change. *Climate Change 2007: Impacts, Adaptation and Vulnerability—Working Group II Contribution to the Fourth Assessment Report of the Intergovernmental Panel on Climate Change.* New York: Cambridge University Press, 2007.

Intergovernmental Panel on Climate Change. *Climate Change 2014: Synthesis Report—Contribution of Working Groups I, II and III to the Fifth Assessment Report of the Intergovernmental Panel on Climate Change.* New York: Cambridge University Press, 2014.

Intergovernmental Panel on Climate Change. *Climate Change 2022: Impacts, Adaptation and Vulnerability, Summary for Policy Makers. Working Group II contribution to the Sixth Assessment Report of the Intergovernmental Panel on Climate Change.* Cambridge: Cambridge University Press, 2023. https://www.ipcc.ch/report/ar6/wg2/

Iowa State University. "Researchers Present Study on How Global Climate Change Affects Violence." *Newswise.* March 19, 2010. http://www.newswise.com/articles/iowa-state-researchers-present-study-on-how-global-climate-change-affects-violence

Ireland, Alleyne. "Is Tropical Colonization Justifiable?" *Annals of the American Academy of Political and Social Science* 19 (1902): 331–39.

Ireland, Alleyne. *Tropical Colonization: An Introduction to the Study of the Subject.* New York: Macmillan, 1899.

Irwin, Robert. *Ibn Khaldun: An Intellectual Biography*. Princeton: Princeton University Press, 2018.

Irwin, Robert. "Toynbee and Ibn Khaldun." *Middle Eastern Studies* 33 (1997): 461–79.

Ishizuka, Hisao. "'Fibre Body': The Concept of Fibre in Eighteenth-Century Medicine, c. 1700–40." *Medical History* 56 (2012): 562–84.

Jacoby, Neil H. Review of *The Causes of Economic Fluctuations, Possibilities of Anticipation and Control* by Willford I. King. *Journal of Business of the University of Chicago*, 12 (1939): 216–18.

Janerich, Dwight T., Ian H. Porter, and Vito Logrillo. "Season of Birth and Neonatal Mortality." *American Journal of Public Health*, 61 (1971): 1119–25.

Janković, Vladimir. *Confronting the Climate: British Airs and the Making of Environmental Medicine*. New York: Palgrave Macmillan, 2010.

Janković, Vladimir. "The Last Resort: A British Perspective on the Medical South, 1815–1870." *Journal of Intercultural Studies* 27 (2006): 271–98.

Jardine, Lisa. "Droughts, Deluges and Raging Debates." *Financial Times*, April 27, 2013, 9.

Jefferson, Thomas. *Notes on the State of Virginia*. Philadelphia: R. T. Rawle, 1801.

Jeffreys, Julius. *The British Army in India: Its Preservation by an Appropriate Clothing, Housing, Locating, Recreative, Employment and Hopeful Encouragement of the Troops*. London: Longman, Brown, Green, Longmans & Roberts, 1858.

Jenkins, Gareth. "A Note on Climatic Cycles and the Rise of Chinggis Khan." *Central Asiatic Journal* 18 (1974): 217–26.

Jennings, Eric T. *Curing the Colonizers: Hydrotherapy, Climatology, and French Colonial Spas*. Durham, NC: Duke University Press, 2006.

Jevons, William Stanley. "Commercial Crises and Sunspots." *Nature* 19 (1878): 33–37.

Johnson, James. *The Influence of Tropical Climates on European Constitutions: Being a Treatise on the Principal Diseases Incidental to Europeans in the East and West Indies, Mediterranean, and Coast of Africa*. London, 1821.

Johnson, James, and James Ranald Martin. *The Influence of Tropical Climates on European Constitutions*. 6th ed. London: Highley, 1841.

Johnson, Ryan. "Commodity Culture: Tropical Health and Hygiene in the British Empire." *Endeavour* 32, no. 2 (2008): 70–74.

Johnson, Ryan. "European Cloth and 'Tropical' Skin: Clothing Material and British Ideas of Health and Hygiene in Tropical Climates." *Bulletin of the History of Medicine* 83 (2009): 530–60.

Johnston, Harry H. *A History of the Colonization of Africa by Alien Races*. Cambridge: Cambridge University Press, 1899.

Johnston, Harry H. "Sir John Kirk." *Geographical Journal* 59 (1922): 225–28.

Johnston, Josée, Michael Gismondi, and James Goodman, eds. *Nature's Revenge: Reclaiming Sustainability in an Age of Corporate Globalization*. New York: Broadview Press, 2006.

Johnston, Ron. Review of *The Revenge of Geography* by Robert Kaplan. *The AAG Review of Books* 1 (2013): 1–3.

Jones, Benjamin, Benjamin Olken, and Melissa Dell. "Does Climate Change Affect Economic Growth?" *VoxEU The Centre for Economic Policy Research Policy Portal*, June 6, 2009. http://www.voxeu.org/article/does-climate-change-affect-economic-growth

Jones, Greta. *Social Darwinism in English Thought: The Interaction between Biological and Social Theory*. Sussex: Harvester Press, 1980.

Jones, P. D., A. E. J. Ogilvie, T. D. Davies, and K. R. Briffa, eds. *History and Climate: Memories of the Future?* Dordrecht: Kluwer, 2001.

Judkins, Gabriel, Marissa Smith, and Eric Keys. "Determinism within Human-Environmental Research and the Rediscovery of Environmental Causation." *Geographical Journal* 174 (2008): 17–29.

Kames, Henry Home, Lord. *Sketches of the History of Man*. Edinburgh: W. Creech, 1774.

Kant, Immanuel. "On Countries That Are Known and Unknown to Europeans." In *Race and the Enlightenment: A Reader*, edited by Emmanuel Chukwudi Eze, 58–64. Oxford: Wiley-Blackwell, 1997.

Kant, Immanuel. "On National Characteristics." In *Race and the Enlightenment: A Reader*, edited by Emmanuel Chukwudi Eze, 49–57. Oxford: Wiley-Blackwell, 1997.

Kant, Immanuel. "On the Different Races of Man." In *This Is Race: An Anthology Selected From the International Literature on the Races of Man*, edited by Earl W. Count, 16–24. New York: Schuman, 1950.

Kaplan, Robert D. "The Coming Anarchy: How Scarcity, Crime, Overpopulation, Tribalism, and Disease Are Rapidly Destroying the Social Fabric of Our Planet." *Atlantic*, February 1994. http://www.theatlantic.com/magazine/archive/1994/02/the-coming-anarchy/304670/

Kaplan, Robert D. "The Revenge of Geography." *Foreign Policy* 172 (May/June, 2009): 96–105.

Kaplan, Robert. *The Revenge of Geography. What the Map Tells Us about Coming Conflicts and the Battle Against Fate*. New York: Random House, 2012.

Kass, Amalie M. "Dr. Thomas Hodgkin, Dr. Martin Delany, and the 'Return to Africa.'" *Medical History* 27 (1983): 373–93.

Keele, Kenneth D. "The Sydenham-Boyle Theory of Morbific Particles." *Medical History* 18 (1974): 240–48.

Keighren, Innes. *Bringing Geography to Book: Ellen Semple and the Reception of Geographical Knowledge*. London: I. B. Taurus, 2010.

Keller, Matthew C., Barbara L. Fredrickson, Oscar Ybarra, Stéphane Côté, Karen Johnson, Joe Mikels, Anne Conway, and Tor Wager. "A Warm Heart and a Clear Head: The Contingent Effects of Weather on Mood and Cognition." *Psychological Science* 16 (2005): 724–31.

Kelly, Morgan, and Cormac Ó Gráda. "The Waning of the Little Ice Age: Climate Change in Early Modern Europe." *Journal of Interdisciplinary History* 44 (2014): 301–25.

Kelman, Ilan. "IPCC Report: How Politics—Not Climate Change—Is Responsible for Disasters and Conflict." *The Conversation*, February 20, 2022. https://theconversation.com/ipcc-report-how-politics-not-climate-change-is-responsible-for-disasters-and-conflict-178071

Kennedy, Dane. "Diagnosing the Colonial Dilemma: Tropical Neurasthenia and the Alienated Briton." In *Decentering Empire: Britain, India, and the Transcolonial World*, edited by Dane Kennedy and Durba Ghosh, 157–81. Hyderabad: Orient Longman, 2006.

Kennedy, Dane. *The Highly Civilized Man: Richard Burton and the Victorian World*. Cambridge, MA: Harvard University Press, 2005.

Kennedy, Dane. *The Magic Mountains: Hill Stations and the British Raj*. Berkeley: University of California Press, 1996.

Kennedy, Dane. "The Perils of the Midday Sun: Climatic Anxieties in the Colonial Tropics." In *Imperialism and the Natural World*, edited by John M. Mackenzie, 118–40. Manchester: Manchester University Press, 1990.

Kenny, Judith T. "Claiming the High Ground: Theories of Imperial Authority and the British Hill Stations in India." *Political Geography* 16 (1997): 655–73.

Kenny, Judith T. "Climate, Race and Imperial Authority: the Symbolic Landscape of the British Hill Station in India." *Annals of the Association of American Geographers* 85 (1995): 694–71.

Kenrick, Douglas T. "Ambient Temperature and Horn Honking: A Field Study of the Heat/Aggression Relationship." *Environment and Behavior* 18 (1986): 179–91.

Kestenbaum, David. "Hot Climates May Create Sluggish Economies." *National Public Radio*, July 17, 2009. https://www.npr.org/2009/07/17/106697286/hot-climates-may-create-sluggish-economies?t=1539342032076

Kidd, Benjamin. *The Control of the Tropics*. New York: Macmillan, 1898.

Kimbel, William H. "Hominid Speciation and Pliocene Climatic Change." In *Paleoclimate and Evolution with Emphasis on Human Origins*, edited by Elisabeth S. Vrba, George H. Denton, Timothy C. Partridge, and Lloyd H. Burckle, 425–37. New Haven, CT: Yale University Press, 1995.

King, David A. "Climate Change Science: Adapt, Mitigate, or Ignore?" *Science* 303, no. 5655 (2004): 176–77.

King, Willford Isbell. *The Causes of Economic Fluctuations: Possibilities of Anticipation and Control*. New York: Ronald Press, 1938.

Kiple, Kenneth F., and Kriemhild Coneè Ornelas. "Race, War and Tropical Medicine in the Eighteenth-Century Caribbean." In *Warm Climates and Western Medicine: The Emergence of Tropical Medicine, 1500–1900*, Wellcome Institute Series in the History of Medicine, edited by David Arnold, 65–79. Atlanta, GA: Rodopi, 1996.

Kirk, John. "The Extent to Which Tropical Africa Is Suited for Development by the White Race, or under Their Superintendence." In *Report of the Sixth International Geographical Congress Held in London, 1895*, 523–35. London: John Murray, 1896.

Kirk-Greene, A. H. M. "Lugard, Frederick John Dealtry, Baron Lugard (1858–1945)." In *Oxford Dictionary of National Biography*. Oxford: Oxford University Press, 2004. https://doi.org/10.1093/ref:odnb/34628

Kirkup, James. "Chris Huhne: Climate Change Threatens UK Security." *Telegraph*, July 7, 2011. http://www.telegraph.co.uk/news/8622945/Chris-Huhne-Climate-change-threatens-UK-security.html

Klautke, Egbert. "Defining the *Volk*: Willy Helpach's *Völkerpsychologie* between National Socialism and Liberal Democracy, 1934–1954." *History of European Ideas* 39 (2013): 693–708.

Kleidon, Axel. "Climatic Constraints on Maximum Levels of Human Metabolic Activity and Their Relation to Human Evolution and Global Change." *Climatic Change* 95 (2009): 405–31.

Klein, Jørgen, David Nash, Kathleen Pribyl, Georgina H. Endfield, and Matthew Hannaford. "Climate, Conflict and Society: Changing Responses to Weather Extremes in Nineteenth Century Zululand." *Environment and History* 24 (2018): 377–401.

Kleingeld, Pauline. "Kant's Second Thoughts on Race." *Philosophical Quarterly* 57 (2007): 573–92.

Klimenko, Vladimir. "Thousand-Year History of Northeastern Europe Exploration in the Context of Climatic Change: Medieval to Early Modern Times." *Holocene* 26 (2016): 365–79.

Klimstra, Theo A., Jaap J. A. Denissen, Hans M. Koot, Pol A. C. van Lier, Tom Frijns, Loes Keijsers, Quinten A. W. Raaijmakers, Marcel A. G. van Aken, and Wim H. J. Meeus. "Come Rain or Come Shine: Individual Differences in How Weather Affects Mood." *Emotion* 11 (2011) 1495–99.

Kneale, James, and Samuel Randalls. "Imagined Geographies of Climate and Race in Anglophone Life Assurance, c. 1840–1930." In *Weather, Climate, and the Geographical Imagination: Placing Atmospheric Knowledges*, edited by Martin Mahony and Samuel Randalls, 115–31. Pittsburgh: University of Pittsburgh Press, 2020.

Kneale, James, and Samuel Randalls. "Invisible Atmospheric Knowledges in British Insurance Companies, 1830–1914." *History of Meteorology* 6 (2014): 35–52.

Knox, Robert. *The Races of Men: A Philosophical Enquiry into the Influence of Race over the Destinies of Nations*. 2nd ed. London: Henry Renshaw, 1862.

Koelsch, William A. "Seedbed of Reform: Arnold Guyot and School Geography in Massachusetts, 1849–1855." *Journal of Geography* 107 (2008): 35–42.

Koubi, Vally, Thomas Bernauer, Anna Kalbhenn, and Gabriele Spilker. "Climate Variability, Economic Growth, and Civil Conflict." *Journal of Peace Research* 49 (2012): 113–27.

Kroeber, Alfred L. *Cultural and Natural Areas of Native North America*. Berkeley: University of California Press, 1939.

Kullmer, Charles Julius. "The Latitude Shift of the Storm Track in the 11-Year Solar Period: Storm Frequency Maps of the United States, 1883–1930." *Smithsonian Miscellaneous Collections* 89, no. 2 (1933): 1–34.

Kullmer, Charles Julius. "A Remarkable Reversal in the Distribution of Storm Frequency in The United States in Double Hale Solar Cycles of Interest in Long-Range Forecasting." *Smithsonian Miscellaneous Collections* 103, no. 10 (1943): 1–20.

Kwiatkowska, Teresa. "The Light Was Retreating before Darkness: Tales of the Witch Hunt and Climate Change." *Medievalia* 42 (2010): 30–37.

LaFauci, Lauren. "Climate Theories." In *Climate and American Literature*, edited by Michael Boyden, 41–57. Cambridge: Cambridge University Press, 2021.

Lamb, Hubert H. "An Approach to the Study of the Development of Climate and Its Impact in Human Affairs." In *Climate and History: Studies in Past Climates and Their Impact on Man*, edited by T. M. L. Wigley, M. J. Ingram, and G. Farmer, 291–309. Cambridge: Cambridge University Press, 1981.

Lamb, Hubert H. *Climate. Present, Past and Future. Vol. 1. Fundamentals and Climate Now*. London: Methuen, 1972.

Lamb, Hubert H. *Climate. Present, Past and Future. Vol. 2. Climatic History and the Future*. London: Methuen, 1977.

Landes, David S. *The Wealth and Poverty of Nations: Why Some Are So Rich and Some So Poor*. London: Abacus, 1998.

Landsberg, Helmut E. "Atmospheric Variability and Climatic Determinism." *Yearbook of Pacific Coast Geographers* 30 (1968): 13–23.

Landsberg, Helmut E. *Weather and Health: An Introduction to Biometeorology*. New York: Doubleday, 1969.

Lane, Joan. "Wintringham, Clifton (*bap.* 1689, *d.* 1748)." In *Oxford Dictionary of National Biography*. Oxford: Oxford University Press, 2004. https://doi.org/10.1093/ref:odnb/29781

Lang, Stefan. "Climate Change Insufficient to Explain Genghis Khan's Greatness," March 11, 2014, *History's Shadow*. http://historysshadow.wordpress.com/2014/03/11/climate-change-insufficient-to-explain-genghis-khans-greatness/

Larrick, Richard P., Thomas A. Timmerman, Andrew M. Carton, and Jason Abrevaya. "Temper, Temperature, and Temptation. Heat-Related Retaliation in Baseball." *Psychological Science* 22 (2011): 423–28.

Larsen, Edward J. *Summer for the Gods: The Scopes Trial and America's Continuing Debate over Science and Religion*. New York: Basic Books, 1997.

Latham, Robert Gordon. *The Varieties of the Human Species: Being a Manual of Ethnography, Introductory to the Study of History*. London: Houlston and Stoneman, 1856.

Lattimore, Owen. "Caravan Routes of Inner Asia." In *Studies in Frontier History: Collected Papers, 1928–1958*. London: Oxford University Press, 1962. Originally published in the *Geographical Journal* 71 (1928): 497–528.

Lattimore, Owen. *Inner Asian Frontiers of China*. Boston: Beacon Press, 1962.

Lavery, Colm. "Geography and Eugenics in Britain and the United States, 1900–1950." PhD dissertation, Queen's University Belfast, 2015.

Lavery, Colm. "The Power of Racial Mapping: Ellsworth Huntington, Immigration, and Eugenics in the Progressive Era." *Journal of the Gilded Age and Progressive Era* 21 (2022): 262–78.

Lavery, Colm. "Situating Eugenics: Robert DeCourcy Ward and the Immigration Restriction League of Boston." *Journal of Historical Geography* 53 (2016): 54–62.

Lavery, Colm. "Stephen S. Visher: The Geography and Eugenics of a (Forgotten) Adopted Indianan." *Indiana Magazine of History* 115 (2019): 1–19.

Lavie, P. "Two 19th-Century Chronobiologists: Thomas Laycock and Edward Smith." *Chronobiology International* 9 (1992): 83–96.

Le Roy Ladurie, Emmanuel, *Times of Feast, Times of Famine: A History of Climate Since the Year 1000*, translated by Barbara Bray. London: George Allen & Unwin, 1971.

Lebeau, James. "Crime and Climate." In *Encyclopedia of World Climatology*, edited by John E. Oliver, 307. Dordrecht: Springer, 2005.

Lee, James R. *Climate Change and Armed Conflict: Hot and Cold Wars*. London: Routledge, 2009.

Lehmann, Philipp N. "Whither Climatology? Brückner's *Climate Oscillations*, Data Debates, and Dynamic Climatology." *History of Meteorology* 7 (2015): 337–40.

Leidy, Joseph. "Prefatory Remarks." In Josiah C. Nott and George R. Gliddon, *Indigenous Races of the Earth*. Philadelphia: Trübner, J. B. Lippincott, 1857.

Leonard, Thomas C. "More Merciful and Not Less Effective: Eugenics and Progressive-Era American Economics." *History of Political Economy* 35 (2003): 709–34.

Levine, Robert S. *Martin Delany, Frederick Douglass and the Politics of Representative Identity*. Chapel Hill: University of North Carolina Press, 1997.

Lewthwaite, Gordon R. "Environmentalism and Determinism: A Search for Clarification." *Annals of the Association of American Geographers* 56 (1966): 1–23.

Li, Shang-Jen. "Natural History of Parasitic Disease: Patrick Manson's Philosophical Method." *Isis* 93 (2002): 206–28.

Libbey, William. "The Life and Scientific Work of Arnold Guyot." *Journal of the American Geographical Society of New York* 16 (1884): 194–221.

Lieberman, L., and F. L. C. Jackson. "Race and Three Models of Human Origin." *American Anthropologist* 97 (1995): 231–42.

Liebowitz, Ruth Prelowski. "Landsberg, Helmut Erich." *Encyclopedia.com*, 2008. http://www.encyclopedia.com/doc/1G2-2830905841.html

Liggins, Felicity. "Climate Change and Health." Met Office. http://web.archive.org/web/20120403182405/http://www.metoffice.gov.uk/media/pdf/2/j/transcript-climate-change-health.pdf

Lind, James. *An Essay on Diseases Incidental to Europeans in Hot Climates. With the Method of Preventing Their Fatal Consequences.* London: Printed for T. Becket and P.A. De Hondt, 1768.

Linker, Damon. "The Reluctant Pluralism of J.G. Herder." *The Review of Politics* 62 (2000): 267–93.

"Links Fascism with Weather. Increasingly Warmer Temperatures May Produce Such Trend." *Mason City Globe-Gazette*, March 27, 1941.

Lister, Thomas D. *Medical Examination for Life Insurance.* London: Edward Arnold, 1921.

Livingstone, David N. *Adam's Ancestors: Race, Religion and the Politics of Human Origins.* Baltimore: Johns Hopkins University Press, 2008.

Livingstone, David N. "Changing Climate, Human Evolution and the Revival of Environmental Determinism." *Bulletin of the History of Medicine* 86 (2012): 564–95.

Livingstone, David N. "Cultural Politics and the Racial Cartographics of Human Origins." *Transactions of the Institute of British Geographers* 35 (2010): 204–21.

Livingstone, David N. *Dealing with Darwin: Place, Politics and Rhetoric in Religious Engagements with Evolution.* Baltimore: Johns Hopkins University Press, 2014.

Livingstone, David N. *The Geographical Tradition: Episodes in the History of a Contested Enterprise.* Oxford: Blackwell, 1992.

Livingstone, David N. "Human Acclimatization: Perspectives on a Contested Field of Inquiry in Science, Medicine and Geography." *History of Science* 25 (1987): 359–94.

Livingstone, David N. "The Moral Discourse of Climate: Historical Considerations on Race, Place and Virtue." *Journal of Historical Geography* 17 (1991): 413–34.

Livingstone, David N. *Nathaniel Southgate Shaler and the Culture of American Science.* Tuscaloosa: University of Alabama Press, 1987.

Livingstone, David N. "'Never Shall Ye Make the Crab Walk Straight': An Inquiry into the Scientific Sources of American Racial Geography." In *Nature and Science: Essays in the History of Geographical Knowledge*, edited by Felix Driver and Gillian Rose, 37–48. London: Institute of British Geographers, Historical Geography Research Series, No 28, 1992.

Livingstone, David N. *Putting Science in Its Place: Geographies of Scientific Knowledge*, Chicago: University of Chicago Press, 2003.

Livingstone, David N. "Race, Space and Moral Climatology: Notes toward a Genealogy." *Journal of Historical Geography* 28 (2002): 159–80.

Livingstone, David N. "Science and Society: Nathaniel S. Shaler and Racial Ideology." *Transactions of the Institute of British Geographers* 9 (1984): 181–210.

Livingstone, David N. "Tropical Climate and Moral Hygiene: The Anatomy of a Victorian Debate." *British Journal for the History of Science* 32 (1999): 93–110.

Livingstone, Justin D. *Livingstone's Lives: A Metabiography of a Victorian Icon.* Manchester: Manchester University Press, 2014.

Locher, Fabien, and Jean-Baptiste Fressoz. "Modernity's Frail Climate: A Climate History of Environmental Reflexivity." *Critical Inquiry* 38 (2012): 579–98.

Lombard, Henri-Clermond. *Les Climats De Montagnes Considerérés au Point de Vue Médical.* Genève: Joël Cherbuliez, 1858.

Lombard, Henri-Clermond. *Traité de Climatologie Médicale Comprenant la Météorologie Médicale et l'Étude des Influences Physiologiques, Pathologiques, Prophylactiques et Thérapeutiques du Climat sur la Santé.* Paris: Bailliere, 1877–80.

Lombroso, Cesare. *The Man of Genius.* London: Walter Scott, 1891.

Löschnigg, Maria, and Melanie Braunecker. *Green Matters: Ecocultural Functions of Literature.* Leiden: Brill Rodopi, 2020.

Lossio, Jorge. "British Medicine in the Peruvian Andes: The Travels of Archibald Smith, M.D. (1820–1870)." *História Ciências Saúde-Manguinhos* 13 (2006): 833–50.

Lovell, Mary S. *A Rage to Live: A Biography of Richard and Isabel Burton.* New York: W. W. Norton, 1998.

Lowie, Robert H. *The History of Ethnological Theory.* New York: Holt, Rinehart and Winston, 1937.

Luczak, Ewa Barbara. *Breeding and Eugenics in the American Literary Imagination: Heredity Rules in the Twentieth Century.* London: Palgrave, 2015.

Lugard, F. D. *The Dual Mandate in Tropical Africa.* Edinburgh: Blackwood, 1922.

Lurie, Edward. "Louis Agassiz and the Races of Man." *Isis* 45 (1954): 227–42.

Lyell, Charles. *Principles of Geology: Being an Inquiry How Far the Former Changes of the Earth's Surface Are Referable to Causes Now in Operation.* 4 vols, 3rd ed. London: John Murray, 1834.

Lynas, Mark. *Six Degrees: Our Future on a Hotter Planet.* London: Harper Perennial, 2008.

Lynch, Gary, and Richard Granger. *Big Brain: The Origins and Future of Human Intelligence.* New York: Palgrave Macmillan, 2008.

MacDougal, D. "North American Deserts." *Geographical Journal* 39 (1912): 105–20.

Macfarlane, Alan. *Montesquieu and the Making of the Modern World.* Cambridge: Cam Rivers Publishing, 2018.

Mach, Katharine J., Caroline M. Kraan, W. Neil Adger, Halvard Buhaug, Marshall Burke, James D. Fearon, Christopher B. Field, Cullen S. Hendrix, Jean-Francois Maystadt, John O'Loughlin, Philip Roessler, Jürgen Scheffran, Kenneth A. Schultz, and Nina von Uexkull. "Climate as a Risk Factor for Armed Conflict." *Nature* 571 (July 2019): 193–97.

Magnis, Nicholas E. "Thomas Jefferson and Slavery: An Analysis of His Racist Thinking as Revealed by His Writings and Political Behavior." *Journal of Black Studies* 29 (1999): 491–509.

Mahony, Martin, and Georgina Endfield. "Climate and Colonialism." *WIREs Climate Change* 9, 2 (2018) e510. https://doi.org/10.1002/wcc.510

Mahony, Martin, and Mike Hulme. "Epistemic Geographies of Climate Change: Science, Space and Politics." *Progress in Human Geography* 42 (2018): 395–424.

Mahony, Martin, and Samuel Randalls, eds. *Weather, Climate, and the Geographical Imagination: Placing Atmospheric Knowledges.* Pittsburgh: University of Pittsburgh Press, 2020.

Mair, R. S. *Medical Guide for Anglo-Indians, Being a Compendium of Advice to Europeans in India Relating to the Preservation and Regulation of their Health.* London: Henry S. King, 1874.

Manson-Bahr, Philip H. *Patrick Manson: The Father of Tropical Medicine.* London: Nelson, 1962.

Mares, Dennis. "Climate Change and Levels of Violence in Socially Disadvantaged Neighborhood Groups." *Journal of Urban Health* 90, no. 4 (2013): 768–83.

Markham, Sydney Frank. *Climate and the Energy of Nations*. New York: Oxford University Press, 1944. First published 1942.

Marks, Elias. *The Aphorisms of Hippocrates*, translated from the Latin Version of Verhoofd. New York: Collins, 1817.

Marks, Jonathan. *Tales of the Ex-Apes: How We Think about Human Evolution*. Oakland: University of California Press, 2015.

Marrum, James A. "Hippocrates and the Hippocratic Tradition: Impact on Development of Medical Knowledge and Practice?" In *Handbook of the Philosophy of Medicine*, edited by Thomas Schramme and Steven Edwards, 821–37. Dordrecht: Springer, 2017.

Marshall, Tim. *Prisoners of Geography: Ten Maps That Explain Everything about the World*. New York: Scribner, 2016.

Martens, Pim. "How Will Climate Change Affect Human Health?" *American Scientist* 87, no. 6 (1999): 534.

Martin, Craig. "Experience of the New World and Aristotelian Revisions of the Earth's Climates during the Renaissance." *History of Meteorology* 3 (2016): 1–16.

Martin, Geoffrey J. *Ellsworth Huntington: His Life and Thought*. Hamden, CT: Archon Books, 1973.

Martin, James Ranald. *The Influence of Tropical Climates on European Constitutions, Including Practical Observations on the Nature and Treatment of the Diseases of Europeans on Their Return from Topical Climates. A New Edition*. London: John Churchill, 1856.

Martin, James Ranald. *Notes on the Medical Topography of Calcutta*. Calcutta: G. H. Huttmann, 1837.

Martin, James Ranald. *Official Report on the Medical Topography and Climate of Calcutta*. Calcutta: G. H. Huttmann, 1839.

Martin, R. Montgomery. *History of the British Colonies Vol IV. Possessions in Africa and Austral-Asia*. London: James Cochrane, 1835.

Maslin, Mark Andrew. "Climate Change Is Not a Key Cause of Conflict." *The Conversation*, April 24, 2018. https://theconversation.com/climate-change-is-not-a-key-cause-of-conflict-finds-new-study-94331

Massey, Nathanael. "Rise of Genghis Khan Linked to Unusual Rains in Mongolia. Changing Precipitation Patterns Helped Genghis Khan Rise to Power—and Rise to Victory across Eurasia." *Scientific American*, March 12, 2014. http://www.scientificamerican.com/article/rise-of-genghis-khan-linked-to-unusual-rains-in-mongolia/

Mathiesen, Karl. "Leaders Confess Climate Sins at COP26." *Politico*, November 2, 2021. https://www.politico.eu/article/climate-change-cop26-carbon-emissions-commitments/

Mathieson, Stuart. "Irish Missions, Science, and Scripture in the Holy Land, 1841-75." Forthcoming.

Matthew, William Diller. "Climate and Evolution." *Annals of the New York Academy of Sciences* 24 (February 1915): 171–318.

Matthew, William Diller. *Climate and Evolution*. New York: Special Publications of the New York Academy of Sciences, 1939.

Mazo, Jeffrey. *Climate Conflict: How Global Warming Threatens Security and What to Do about It*. London: Routledge, 2010.

McMichael, Anthony J. "Global Climate Change and Health: An Old Story Writ Large." In *Climate Change and Human Health*, edited by A. J. McMichael, D. H. Campbell-Lendrum, C. F. Corvalán, K. L. Ebi, A. K. Githeko, J. D. Scheraga, and A. Woodward, 1–17. Geneva: World Health Organization, 2003.

McMichael, Anthony J., D. H. Campbell-Lendrum, C. F. Corvalán, K. L. Ebi, A. K. Githeko, J. D. Scheraga, and A. Woodward, eds. *Climate Change and Human Health*. Geneva: World Health Organization, 2003.

McMichael, Anthony J., and Rosalie E. Woodruff. "Climate Change and Human Health." In *Encyclopedia of World Climatology*, edited by John E. Oliver, 209–13. Dordrecht: Springer, 2005.

McMullen, Michael D. "Kirk, Sir John (1832–1922)." In *Oxford Dictionary of National Biography*. Oxford: Oxford University Press, 2004. https://doi.org/10.1093/ref:odnb/34336

McNeill, J. R. "Historians, Superhistory, and Climate Change." In *Methods in World History: A Critical Approach*, edited by Arne Jarrick, Janken Myrdal, and Maria Wallenberg Bondesson, 19–43. Lund: Nordic Academic Press, 2016.

Mellinger, Andrew D., Jeffrey Sachs, and John L. Gallup. "Climate, Water Navigability, and Economic Development." CID Working Paper 24, Center for International Development at Harvard University, 1999. https://www.researchgate.net/publication/5061613_Climate_Water_Navigability_and_Economic_Development

Mellor, Bernard. *Lugard in Hong Kong: Empires, Education and a Governor at Work 1907–1912*. Hong Kong: Hong Kong University Press, 1992.

Merrett, Christopher D. "Debating Destiny: Nihilism or Hope in *Guns, Germs, and Steel*?" *Antipode* 35 (2003): 801–6.

Met Office. "Climate Change and Healthcare," November 4, 2010. https://web.archive.org/web/20101209141355/http://www.metoffice.gov.uk/health/professionals/climatechange

Metcalf, Thomas R. "Architecture and the Representation of Empire: India, 1860–1910." *Representations* 6 (1984): 37–65.

Meulendijks, Max. "The Descent of Malady: On the Darwinisation of British Medicine, 1890–1920." PhD thesis, Queen's University Belfast, 2019.

Meyer, William B. *Americans and Their Weather*. Oxford: Oxford University Press, 2000.

Meyer, William B., and Dylan M. T. Guss. *Neo-Environmental Determinism: Geographical Critiques*. Basingstoke: Palgrave Macmillan, 2017.

Michael, Anne. *Fugitive Pieces*. London: Bloomsbury, 1997.

Michael, Richard P., and Doris Zumpe. "An Annual Rhythm in the Battering of Women." *American Journal of Psychiatry* 143 (1980): 637–40.

Michael, Richard P., and Doris Zumpe. "Annual Rhythms in Human Violence and Sexual Aggression in the United States and the Role of Temperature." *Social Biology* 30 (1983): 263–78.

Michael, Richard P., and Doris Zumpe. "Sexual Violence in the United States and the Role of Season." *American Journal of Psychiatry* 140 (1983): 883–86.

Midgley, Mary. *The Myths We Live By*. London: Routledge, 2004.

Mikkelsen, Jon M., ed. *Kant and the Concept of Race*. Albany: State University of New York Press, 2013.

"Mild Weather May Have Propelled Genghis Khan to Power, Study Says." *The Japan Times*, March 12, 2014. https://web.archive.org/web/20150415040235/http://www.japantimes.co

.jp/news/2014/03/12/world/mild-weather-may-have-propelled-genghis-khan-to-power-study-says/#.VS3i6GDP1R4

Miller, Austin A. *Climatology*. 9th ed. London: Methuen, 1969.

Miller, Genevieve. "'Airs, Waters, Places' in History." *Journal of the History of Medicine* 17 (1962): 129–40.

Miller, James Alexander. "Some Physiological Effects of Various Atmospheric Conditions." *Transactions of the American Climatological and Clinical Association* 32 (1916): 1–14.

Mills, Clarence A. *Air Pollution and Community Health*. Boston: Christopher Publishing House, 1954.

Mills, Clarence A. *Climate Makes the Man*. London: Victor Gollancz, 1946.

Mills, Clarence A. *Living with the Weather*. Cincinnati, OH: The Caxton Press, 1934.

Mills, Clarence A. *Medical Climatology: Climatic and Weather Influences in Health and Disease*. London: Bailliére, Tindall & Cox, 1939.

Mills, Clarence A. "Temperature Dominance over Human Life." *Science* 110 (September 16, 1949): 267–71.

Mills, Clarence A. *This Air We Breathe*. Boston: Christopher Publishing House, 1962.

Mills, Clarence A. *World Power and Shifting Climates*. Boston: Christopher Publishing House, 1963.

Mirowski, Philip. "Problems in the Paternity of Econometrics: Henry Ludwell Moore." *History of Political Economy* 22 (1990): 587–609.

Mishra, Shreshtha. "Can the Climate of a Country Determine Its Wealth?" *Qrius*, January 17, 2020. https://qrius.com/climate-determine-wealthy-countries/

Miura, T. "The Influence of Seasonal Atmospheric Factors on Human Reproduction." *Experientia* 43 (1987): 48–54.

Montesquieu, Charles-Louis De Secondat, Baron de. "An Essay on the Causes that May Affect Men's Minds and Characters." *Political Theory* 4 (1976): 139–62.

Montesquieu, Charles-Louis De Secondat, Baron de. *The Spirit of the Laws*, Cambridge Texts in the History of Political Thought, translated and edited by Anne M. Cohler, Basia Carolyn Miller, and Harold Samuel Stone. Cambridge: Cambridge University Press, 1989.

Montesquieu, Charles-Louis De Secondat, Baron de. *The Spirit of Laws*, translated from the French by Thomas Nugent. London: Printed for J. Nourse and P. Vaillant, 1750.

Moore, Henry Ludwell. *Economic Cycles: Their Law and Cause*. New York: Macmillan Company, 1914.

Moore, James R. "Theodicy and Society: The Crisis of the Intelligentsia." In *Victorian Faith in Crisis*, edited by Richard J. Helmstadter and Bernard Lightman, 153–86. London: Palgrave Macmillan, 1990.

Moore, Norman (rev. Richard Hankins). "Huxham, John (*c.*1692–1768)." In *Oxford Dictionary of National Biography*. Oxford: Oxford University Press, 2004. https://doi.org/10.1093/ref:odnb/14319

Moore, Sir William. "Is the Colonisation of Tropical Africa by Europeans Possible?" *Journal of the Epidemiological Society* 10 (1891): 27–45.

Morello, Lauren. "Impacts of Climate Change Extend to Human Health." *Scientific American*, April 22, 2010. http://www.scientificamerican.com/article/climate-change-human-health/

Morin, Monte. "Violence Will Rise as Climate Changes, Scientists Predict." *Los Angeles Times*, August 1, 2013. https://web.archive.org/web/20130810043748/http://articles.latimes.com/2013/aug/01/science/la-sci-climate-change-conflict-20130802

Morris, Ian. *Geography Is Destiny: Britain and the World, a 10,000 Year History*. London: Profile Books, 2022.

Morrone, Juan J. "Cladistic Biogeography: Identity and Place." *Journal of Biogeography* 32 (2005): 1281–86.

Morton-Gittens, Dane. "Sir Rawson William Rawson: Governor of Barbados, 1869–1875." In *Ideology, Regionalism, and Society in Caribbean History*, edited by Shane J. Pantin, and Jerome Teelucksingh, 179–205. New York: Palgrave Macmillan, 2017.

Moss, Nan. *Weather Shamanism: Harmonizing Our Connection with the Elements*. Rochester, VT: Bear & Company, 2008.

Mühry, Adolf. *Die Geographischen Verhältnisse der Krankheiten oder Grundzüge der Noso-Geographie*. Leipzig: C. F. Winter, 1856.

Nacci, Michela. "'Tel Climat Donné, Tel Peuple Suit': La Natura, i Popoli e Victor Cousin." *Secondo Natura* 4 (2021): 211–22.

Naraindas, Harish. "Poisons, Putresence and the Weather: A Genealogy of the Advent of Tropical Medicine." *Contributions to Indian Sociology* 30 (1996): 1–35.

National Geographic. "Six Degrees Could Change the World." National Geographic Channel, August 2008. http://www.natgeotv.com/ca/six_degrees

National Research Council. *Understanding Climate's Influence on Human Evolution*. Washington, D.C.: National Academies Press, 2010.

National Science Foundation. "Climate of Genghis Khan's Ancient Time Extends Long Shadow over Asia of Today." *National Science Foundation*, March 11, 2014, Press Release 14-032. https://www.nsf.gov/news/news_summ.jsp?cntn_id=130669

Natural Resources Defense Council. "Extreme Weather: Record-Breaking Events in 2011." http://www.nrdc.org/health/climate/extreme-weather.asp

Netburn, Deborah. "Did Rainy Climate Aid Genghis Khan?" *Los Angeles Times*, March 15, 2014. https://www.latimes.com/science/la-sci-genghis-khan-climate-20140315-story.html#axzz2wPH5dMjV

Nikonov, Alexander. *Civilization's Temperature: Effect of Climate on Humankind's History*, translated from Russian by Paul R. Friedman. n.p.: Xlibris Corporation, 2010.

Njoh, Ambe J. "Colonial Philosophies, Urban Space, and Racial Segregation in British and French Colonial Africa." *Journal of Black Studies* 38 (2008): 579–99.

Nonaka, K. "Effect of Delivery Season on Subsequent Birth Interval in Early 20th Century in Japan." *International Journal of Biometeorology* 33 (1989): 238–45.

Nordås, Ragnhild, and Nils Petter Gleditsch. "Climate Change and Conflict." *Political Geography* 26 (2007): 627–38.

Nordhaus, William D. "Geography and Macroeconomics: New Data and New Findings." *Proceedings of the National Academy of Sciences* 103, no. 10 (2006): 3510–17.

Norris, Albert S., and J. R. Chowning. "Season of Birth and Mental Illness: A Critical Examination." *Archives of General Psychiatry* 7 (1962): 206–12.

Norton, William Harmon. "The Influence of the Desert on Early Islam." *Journal of Religion* 4 (1924): 383–96.

Nova, Alessandro. "The Role of the Winds in Architectural Theory from Vitruvius to Scamozzi." In *Aeolian Winds and the Spirit in Renaissance Architecture*, edited by Barbara Kenda, 70–86. London: Routledge, 2006.

Numbers, Ronald L. "Medical Science before Scientific Medicine: Reflections on the History of Medical Geography." In *Medical Geography in Historical Perspective*, edited by Nicolaas A. Rupke, 217–20. London: Wellcome Trust Centre for the History of Medicine, 2000.

"Obituary: William F. Petersen, M.D., 1887–1950." *A.M.A. Archives of Pathology* 51 (1951): 130–32.

Ó Gráda, Cormac. "Is Europe's Little Ice Age a Myth?" *World Economic Forum*, March 2015. https://www.weforum.org/agenda/2015/03/is-europes-little-ice-age-a-myth/

Ofcansky, Thomas P. "A Bio-Bibliography of F.D. Lugard." *History in Africa* 9 (1982): 209–19.

Okihiro, Gary Y. *Pineapple Culture: A History of the Tropical and Temperate Zones*. Berkeley: University of California Press, 2009.

Oldfield, Frank. "Forward to the Past: Changing Approaches to Quaternary Palaeoecology." In *Climate Change and Human Impact on the Landscape*, edited by F. M. Chambers, 13–21. London: Chapman and Hall, 1993.

Oldroyd, David. "Griffith Taylor and His Views on Race, Environment and Settlement and the Peopling of Australia." In *Useful and Curious Geological Enquiries beyond the World: Pacific-Asia Historical Themes. The Nineteenth International INHIGEO Symposium*, edited by David F. Branagan and G.H. McNally 251–74. Springwood, NSW: International Commission on the History of Geological Sciences, 1994.

O'Leary, Stephen D. *Arguing the Apocalypse: A Theory of Millennial Rhetoric*. Oxford: Oxford University Press, 1994.

Olmstead, A. T. "Climate and History." *Journal of Geography* 10 (1912): 163–68.

O'Loughlin, John, Frank D. W. Witmer, Andrew M. Linke, Ariene Laing, Andrew Gettleman, and Jimy Dudhia. "Climate Variability and Conflict Risk in East Africa, 1990–2009." *Proceedings of the National Academy of Sciences* 109, no. 45 (November 2012): 18344–49.

Oppenheimer, Stephen. *Out of Eden: The Peopling of the World*. London: Robinson, 2004.

Oreskes, Naomi, and Erik Conway. *The Collapse of Western Civilization: A View from the Future*. New York: Columbia University Press, 2014.

Osborn, Henry Fairfield. "The Approach to the Immigration Problem through Science." *Proceedings of the National Immigration Conference, Special Paper* 26 (1924): 44–53.

Osborn, Henry Fairfield. *Man Rises to Parnassus: Critical Epochs in the Prehistory of Man*. Princeton, NJ: Princeton University Press, 1927.

Osborn, Henry Fairfield. *Men of the Old Stone Age: Their Environment, Life and Art*. New York: Scribner, 1915.

Osborn, Henry Fairfield. "Preface." In Madison Grant, *The Passing of the Great Race, or, the Racial Basis of European History*. New York: Scribner, 1916.

Osborn, Henry Fairfield. "Race Progress in Relation to Social Progress." *Journal of the National Institute of Social Sciences* 9 (1924): 8–18.

Osborn, Henry Fairfield. "Why Central Asia?" *Natural History* 26 (June 25, 1926): 266–67.

Osborne, Michael A. "The Geographical Imperative in Nineteenth-Century French Medicine." In *Medical Geography in Historical Perspective*, edited by Nicolaas A. Rupke, 31–50. London: Wellcome Trust Centre for the History of Medicine, 2000.

Osborne, Michael A. *The Emergence of Tropical Medicine in France*. Chicago: University of Chicago Press, 2014.

Osborne, Michael A. "Resurrecting Hippocrates: Hygienic Science and the French Scientific Expeditions to Egypt, Morea and Algeria." In *Warm Climates and Western Medicine: The*

Emergence of Tropical Medicine, 1500–1900, Wellcome Institute Series in the History of Medicine, edited by David Arnold, 80–98. Atlanta, GA: Rodopi, 1996.

Osborne, Michael A., and Richard S. Fogarty. "Medical Climatology in France: The Persistence of Neo-Hippocratic Ideas in the First Half of the Twentieth Century." *Bulletin of the Historical of Medicine* 86 (2012): 543–63.

Oster, Emile. "Witchcraft, Weather and Economic Growth in Renaissance Europe." *Journal of Economic Perspectives* 18 (2004): 215–28.

Ostrow, Joanne. "'Six Degrees' Charts Climate Apocalypse in HD Television." *Denver Post*, August 2, 2008. http://www.denverpost.com/ostrow/ci_8190284

Overy, C., and E. M. Tansey, eds. *The Recent History of Seasonal Affective Disorder (SAD)*. Wellcome Witnesses to Contemporary Medicine, vol. 51. London: Queen Mary, University of London, 2014.

Owain, Erin Llwyd, and Mark Andrew Maslin. "Assessing the Relative Contribution of Economic, Political and Environmental Factors on Past Conflict and the Displacement of People in East Africa." *Palgrave Communications* 4 (2018), article number 47. https://www.nature.com/articles/s41599-018-0096-6

Pagden, Anthony. Review of *Global Crisis: War, Climatic Change, and Catastrophe in the Seventeenth Century* by Geoffrey Parker. *Common Knowledge* 21 (2015): 515–16.

Palmer, Alan. "Seacole [*née* Grant], Mary Jane (1805–1881)." In *Oxford Dictionary of National Biography*. Oxford: Oxford University Press, 2004. https://doi.org/10.1093/ref:odnb/41194

Palutikof, Jean. "The Impact of Weather and Climate on Industrial Production in Great Britain." *International Journal of Climatology* 3 (1983): 65–79.

Paravisini-Gebert, Lizabeth. "Mrs. Seacole's *Wonderful Adventures in Many Lands* and the Consciousness of Transit." In *Black Victorians/Black Victoriana*, edited by Gretchen Holbrook Gerzina, 71–87. New Brunswick, NJ: Rutgers University Press, 2003.

Parenti, Christian. *Tropic of Chaos: Climate Change and the New Geography of Violence*. New York: Nation Books, 2011.

Parker, Geoffrey. "Crisis and Catastrophe: The Global Crisis of the Seventeenth Century Reconsidered." *American Historical Review* 113 (2008): 1053–79.

Parker, Geoffrey. *Global Crisis: War, Climate Change and Catastrophe in the Seventeenth Century*. New Haven, CT: Yale University Press, 2013.

Parker, Philip M. *Physioeconomics: The Basis for Long-Run Economic Growth*. Cambridge, MA: MIT Press, 2000.

Parrish, Susan Scott. "Climate and Race." In *Climate and American Literature*, edited by Michael Boyden, 75–90. Cambridge: Cambridge University Press, 2021.

Parry, Wynne. "Climate Fluctuations May Increase Civil Violence." *Live Science*, August 24, 2011. http://www.livescience.com/15739-climate-el-nino-violent-conflict.html

Patz, Jonathan A., Diarmid Campbell-Lendrum, Tracey Holloway, and Jonathan A. Foley. "Impact of Regional Climate Change on Human Health." *Nature* 438 (November 17, 2005): 310–17.

Patz, Jonathan A., Michael A. McGeehin, Susan M. Bernard, Kristie L. Ebi, Paul R. Epstein, Anne Grambsch, Duane J. Gubler, Paul Reiter, Isabelle Romieu, Joan B. Rose, Jonathan M. Samet, and Juli Trtanj. "The Potential Health Impacts of Climate Variability and Change for the United States: Executive Summary of the Report of the Health Sector of the U.S. National Assessment." *Environmental Health Perspectives* 108 (2000): 367–76.

Paul, Bimal Kanti. "Approaches to Medical Geography: An Historical Perspective." *Social Science and Medicine* 20 (1985): 399–409.

Pearce, Fred. *The Last Generation: How Nature Will Take Her Revenge for Climate Change.* London: Eden Project Books, 2006.

Pearson, Ronald. *Climate and Evolution.* London: Academic Press, 1978.

Pederson, Neil, and Amy E. Hessl. "Wet Climate Helped Genghis Khan Conquer Asia." *The Conversation,* March 10, 2012. http://theconversation.com/wet-climate-helped-genghis-khan-conquer-asia-24185

Pederson, Neil, Amy E. Hessl, Nachin Baatarbileg, Kevin J. Anchukaitis, and Nicola Di Cosmo. "Pluvials, Droughts, the Mongol Empire, and Modern Mongolia." *Proceedings of the National Academy of Science* 111, no. 12 (2014): 4375–79.

Peet, Richard. "The Social Origins of Environmental Determinism." *Annals of the Association of American Geographers* 75 (1985): 309–33.

Peluso, Nancy Lee, and Michael Watts. "Violent Environments." In *Violent Environments,* edited by Nancy Lee Peluso and Michael Watts, 3–38. Ithaca, NY: Cornell University Press, 2001.

Pennebaker, James W., Bernard Rimé, and Virginia E. Blankenship. "Stereotypes of Emotional Expressiveness of Northerners and Southerners: A Cross-Cultural Test of Montesquieu's Hypotheses." *Journal of Personality and Social Psychology* 70 (1996): 372–80.

Perham, Margery. *Lugard: The Years of Adventure 1858–1898.* London: Collins, 1956.

Perham, Margery. *Lugard: The Years of Authority 1898–1945.* London: Collins, 1960.

Petersen, William F. *Hippocratic Wisdom, For Him Who Wishes to Pursue Properly the Science of Medicine: A Modern Appreciation of Ancient Scientific Achievement.* Springfield, IL: Charles C Thomas, 1946.

Petersen, William F. *Lincoln–Douglas: The Weather as Destiny.* Springfield. IL: Charles C Thomas, 1943.

Petersen, William F. *Man–Weather–Sun.* Springfield, IL: Charles C Thomas, 1947.

Petersen, William F. *The Patient and the Weather. Vol. 1, Part 2. Autonomic Integration.* Ann Arbor, MI: Edward Bros, 1936.

Petersen, William F. *The Patient and the Weather. Vol. 2. Autonomic Dysintegration.* Ann Arbor, MI: Edward Bros, 1934.

Petersen, William F. *The Patient and the Weather. Vol. 3. Mental and Nervous Diseases.* Ann Arbor, MI: Edward Bros, 1937.

Petersen, William F. *Protein Therapy and Non-Specific Reactions.* New York: Macmillan, 1922.

Petersen, William F., and Samuel A. Levinson. *The Skin Reactions, Blood Chemistry and Physical Status of 'Normal' Men and Clinical Patients.* Chicago: American Medical Association, 1930.

Petrie, W. M. Flinders. *The Revolutions of Civilisation.* London: Harper, 1922.

Pettersson, Otto. *Climatic Variations in Historic and Prehistoric Time.* Göteborg: W. Zachrissons, 1914.

Pfister, Christian. *Klimageschichte der Schweiz 1525–1860. Das Klima der Schweiz von 1525–1860 und seine Bedeutung in der Geschichte von Bevölkerung und Landwirtschaft.* 2 vols. Bern: Verlag Paul Haupt, 1988.

Pfister, Christian. "The Little Ice Age: Thermal and Wetness Indices for Central Europe." *Journal of Interdisciplinary History* 10 (1980): 665–96.

Pfister, Christian, and Rudolf Brázdil. "Climatic Variability in Sixteenth-Century Europe and its Social Dimension: A Synthesis." *Climatic Change* 43 (1999): 5–53.

Philo, Chris. "'Fit Localities for an Asylum': The Historical Geography of the Nineteenth-Century 'Mad-Business' in England as Viewed through the Pages of the *Asylum Journal*." *Journal of Historical Geography* 13 (1987): 398–415.

Pinkerton Crime Index. "The Seasonality of Crime. Perspectives in Crime: Why Does Crime Spike in Summer?" https://pinkerton.com/our-insights/blog/the-seasonality-of-crime

Pittis, Don. "The Economic Advantages of Life in a Cold Country." CBC News, January 7, 2010. https://www.cbc.ca/news/business/the-economic-advantages-of-life-in-a-cold-country-1.871761

Plante, Courtney, Johnie J. Allen, and Craig A. Anderson. "Effects of Rapid Climate Change on Violence and Conflict." In *The Oxford Research Encyclopedia of Climate Science*, April 2017. https://oxfordre.com/climatescience/view/10.1093/acrefore/9780190228620.001.0001/acrefore-9780190228620-e-344

Plante, Courtney, and Craig A. Anderson. "Global Warming and Violent Behavior." *Observer* 30, no. 2 (2017): 29–32.

Plumer, Brad. "Will Global Warming Lead to More War? It's Not That Simple." *Washington Post*, August 5, 2013. http://www.washingtonpost.com/blogs/wonkblog/wp/2013/08/05/will-global-warming-lead-to-more-war-its-not-that-simple/

Plunkett, Gill, and Graeme T. Swindles. "Bucking the Trend: Population Resilience in a Marginal Environment." *PLoS One* 17, no. 4 (2022): e0266680. https://doi.org/10.1371/journal.pone.0266680

Poincaré, Léon. *Prophylaxie et Géographie Médicale des Principales Maladies Tributaires de l'Hygiène*. Paris: G. Masson, 1884.

Porter, Roy. *The Greatest Benefit to Mankind: A Medical History of Humanity from Antiquity to the Present*. London: HarperCollins, 1997.

Porteus, S. D. *Primitive Intelligence and Environment*. New York: Macmillan, 1937.

Porteus, S. D. *Temperament and Race*. Boston: Richard G. Badger, 1926.

Portier, Christopher J., Kimberly Thigpen Tart, Sarah R. Carter, Caroline H. Dilworth, Anne E. Grambsch, Julia Gohlke, Jeremy Hess, Sandra N. Howard, George Luber, Jeffrey T. Lutz, Tanya Maslak, Natasha Prudent, Meghan Radtke, Joshua P. Rosenthal, Teri Rowles, Paul A. Sandifer, Joel Scheraga, Paul J. Schramm, Daniel Strickman, Juli M. Trtanj, and Pai-Yei Whung. *A Human Health Perspective On Climate Change: A Report Outlining the Research Needs on the Human Health Effects of Climate Change*. Research Triangle Park, NC: Environmental Health Perspectives/National Institute of Environmental Health Sciences, 2010. www.niehs.nih.gov/climatereport

Potts, Richard. "Variability Selection in Hominid Evolution." *Evolutionary Anthropology* 27 (1998): 81–96.

Powell, Joseph M. *Griffith Taylor and 'Australia Unlimited.'* Brisbane: University of Queensland Press, 1993.

Pratt, Mary Louise. *Imperial Eyes: Travel Writing and Transculturation*. London: Routledge, 1992.

Putnam, George Haven. *Memories of a Publisher 1865–1915*. Honolulu, HI: University Press of the Pacific, 2001. First published 1916.

Quetelet, Adolphe. *A Treatise on Man and the Development of His Faculties*. New York: Franklin, 1842.

Rabī', Maḥmūd Muḥammad. *The Political Theory of Ibn Khaldūn*. Leiden: E. J. Brill, 1967.

Rainger, Ronald. *An Agenda for Antiquity: Henry Fairfield Osborn and Vertebrate Paleontology at the American Museum of Natural History, 1890–1935*. Tuscaloosa: University of Alabama Press, 1991.

Rainger, Ronald. "Just before Simpson: William Diller Matthew's Understanding of Evolution." *Proceedings of the American Philosophical Society* 130 (1986): 453–74.

Rainger, Ronald. "What's the Use: William King Gregory and the Functional Morphology of Fossil Vertebrates." *Journal of the History of Biology* 22 (1989): 103–39.

Raleigh, Clionadh, Andrew Linke, and John O'Loughlin. "Extreme Temperatures and Violence." *Nature Climate Change* 4 (2014): 76–77.

Ramstad, Yngve, and James L. Starkey. "The Racial Theories of John R. Commons." *Research in the History of Economic Thought and Methodology* 13 (1995): 1–74.

Ranson, Matthew. "Crime, Weather, and Climate Change." *Journal of Environmental Economics and Management* 67 (2014): 274–302.

Ratzel, Friedrich. *The History of Mankind*, translated from the second German edition by A. J. Butler. 3 vols. London: Macmillan, 1896–98.

Ravenstein, E. G. "Lands of the Globe Still Available for European Settlement." *Proceedings of the Royal Geographical Society* 13 (1891): 27–35.

Rawson, R. W. *Report on the Rainfall of Barbados and Upon Its Influence on the Sugar Crops 1847–71, with Two Supplements, 1873–1874*. Barbados: House of Assembly, 1874.

Regal, Brian. *Henry Fairfield Osborn: Race and the Search for the Origins of Man*. Burlington, VT: Ashgate, 2002.

Reifman, Alan S., Richard P. Larrick, and Steven Fein. "Temper and Temperature on the Diamond: The Heat-Aggression Relationship in Major League Baseball." *Personality and Social Psychology Bulletin* 17 (1991): 580–85.

Rentfrow, Peter J., Samuel D. Gosling, and Jeff Potter. "A Theory of the Emergence, Persistence, and Expression of Geographical Variation in Psychological Characteristics." *Perspectives on Psychological Science* 3 (2008): 339–69.

Report of the National Advisory Commission on Civil Disorders. Washington, D.C.: U.S. Government Printing Office, 1968.

Reuben, Julie A. *The Making of the Modern University: Intellectual Transformation and the Marginalization of Morality*. Chicago: University of Chicago Press, 1996.

Review of *Medical Climatology: Climatic and Weather Influences in Health and Disease* by Clarence A. Mills. *Journal of the American Medical Association* 113 (1939): 2263.

Reynolds, Emma. "'Climate Change' Drove Rise of Genghis Khan." *The Australian*, March 11, 2014. https://www.theaustralian.com.au/search-results?q=rise+genghis+khan

Rice, Doyle. "Genghis Khan Rode Climate Change to Take Over Asia." *USA Today*, March 10, 2014. http://www.usatoday.com/story/weather/2014/03/10/genghis-khan-climate-change-drought-mongolia/6259279/

Richards, Evelleen. "The 'Moral Anatomy' of Robert Knox: The Interplay between Biological and Social Thought in Victorian Scientific Naturalism." *Journal of the History of Biology* 22 (1989): 373–436.

Richerson, Peter J., and Robert Boyd. "Climate, Culture, and the Evolution of Cognition." In *The Evolution of Cognition*, edited by Cecelia Heyes and Ludwig Huber, 329–46. Cambridge, MA: MIT Press, 2000.

Richter, Melvin. "An Introduction to Montesquieu's 'An Essay on the Causes That May Affect Men's Minds and Characters.'" *Political Theory* 4 (1976): 132–38.

Rigg, J. B. "Climatic Determinism." *Weather* 16 (1961): 255–60, 298–303, 327–33.

Riley, James C. *The Eighteenth-Century Campaign to Avoid Disease*. Basingstoke: Macmillan, 1987.

Robin, Libby, and Will Steffen. "History for the Anthropocene." *History Compass* 5, no. 5 (2007): 1694–719.

Robinson, Amy. "Authority and the Public Display of Identity: *Wonderful Adventures of Mrs. Seacole in Many Lands.*" *Feminist Studies* 20 (1994): 535–57.

Robinson, Jane. *Mary Seacole: The Charismatic Black Nurse Who Became a Heroine of the Crimea*. London: Constable, 2005.

Roddis, H. *James Lind: Founder of Nautical Medicine*. New York: H. Schuman, 1950.

Rodman, Kristen. "Weather, Climate Changes Multiply Health Concerns." *AccuWeather*, April 9, 2014. https://web.archive.org/web/20140730152806/http://www.accuweather.com:80/en/weather-news/climate-weather%20mentalhealth/16843803

Rodrik, Dani, Arvind Subramanian, and Francesco Trebbi. "Institutions Rule: The Primacy of Institutions over Geography and Integration in Economic Development." *Journal of Economic Growth* 9 (2004): 131–65.

Rohli, Robert V., and Gregory D. Bierly. "The Lost Legacy of Robert DeCourcy Ward in American Geographical Climatology." *Progress in Physical Geography* 35 (2011): 547–64.

Roosevelt, Theodore. *Biological Analogies in History*. New York: Oxford University Press, 1910.

Rosen, Armin. "It's the Politics, Stupid: What Jeffrey Sachs' Development Work Is Missing." *The Atlantic*, January 10, 2013. https://www.theatlantic.com/international/archive/2013/01/its-the-politics-stupid-what-jeffrey-sachs-development-work-is-missing/267054/

Rosen, Stephen. "Guestwords: Spring Fever." *East Hampton Star*, April 17, 2013. https://www.easthamptonstar.com/archive/guestwords-spring-fever

Rosen, Stephen. *Weathering: How the Atmosphere Conditions Your Body, Your Mind, Your Moods—and Your Health*. New York: M. Evans, 1979.

Rosen, Stephen. *Youth, Middle-Age, and You-Look-Great*. New York: Prospect Press, 2013.

Rosenberg, Charles E. "Epilogue: *Airs, Waters, Places*. A Status Report." *Bulletin of the History of Medicine* 86 (2012): 661–70.

Rosenthal, Franz. *Humor in Early Islam*. Leiden: E. J. Brill, 1956.

Rosenthal, Norman. *Winter Blues*. New York: Guilford Press, 1993.

Ross, Angus. "Arbuthnot, John (*bap.* 1667, *d.* 1735)." In *Oxford Dictionary of National Biography*. Oxford: Oxford University Press, 2004. https://doi.org/10.1093/ref:odnb/610

Rotton, James, and Ellen G. Cohn. "Violence Is a Curvilinear Function of Temperature in Dallas: A Replication." *Journal of Personality and Social Psychology* 78 (2000): 1074–81.

Rupke, Nicolaas A. "Adolf Mühry (1810–1888): "Göttingen's Humboldtian Medical Geographer." In *Medical Geography in Historical Perspective*, edited by Nicolaas A. Rupke, 86–97. London: Wellcome Trust Centre for the History of Medicine, 2000.

Rupke, Nicolaas A. *Alexander von Humboldt: A Metabiography*. Chicago: Chicago University Press, 2008.

Rupke, Nicolaas A. "Humboldtian Medicine." *Medical History* 40 (1996): 293–310.

Rupke, Nicolaas A., ed. *Medical Geography in Historical Perspective*. London: Wellcome Trust Centre for the History of Medicine, 2000.

Russell, Colin. "The Conflict of Science and Religion." In *Science and Religion: A Historical Introduction*, edited by Gary B. Ferngren, 3–12. Baltimore: Johns Hopkins University Press, 2002.

Rutty, John. *A Chronological History of the Weather and Seasons, and of the Prevailing Diseases in Dublin, with their Various Periods, Successions, and Revolutions, during the Space of Forty Years*. London, 1770.

Sachs, Jeffrey D. *The End of Poverty: How We Can Make It Happen in Our Lifetime*. London: Penguin, 2005.

Sachs, Jeffrey D. "Helping the World's Poorest." *The Economist*, August 14, 1999, 18.

Sachs, Jeffrey D. "Institutions Don't Rule: Direct Effects of Geography on Per Capita Income." Working Paper 9490, National Bureau of Economic Research, 2003. https://www.researchgate.net/publication/5197462_Institutions_Don%27%27t_Rule_Direct_Effect_of_Geography_on_Per_Capita_Income

Sachs, Jeffrey D. "Tropical Underdevelopment." Working Paper 8119, National Bureau of Economic Research, 2001. https://www.nber.org/papers/w8119

Sachs, Jeffrey D., Andrew D. Mellinger, and John L. Gallup. "The Geography of Poverty and Wealth." *Scientific American*, March 2001, 70–75.

Salehyan, Idean. "From Climate Change to Conflict? No Consensus Yet." *Journal of Peace Research* 45 (2008): 315–26.

Sambon, L. Westenra. "Acclimatization of Europeans in Tropical Lands." *Geographical Journal* 12 (1898): 589–99.

Sambon, L. Westenra. "Climate and Colonisation." *The Quarterly Review* 190, no. 379 (1899): 268–88.

Sambon, L. Westenra. "L'acclimatation des Européens dans les Régions Tropicales." *Bulletin de la Société de Géographie, Lyon* 17 (1901): 44–66.

Sambon, L. Westenra. "The Possibility of the Acclimatization of Europeans in Tropical Regions." *British Medical Journal* 1 (1897): 61–66.

Sample, Ian. "Climate Swings Shaped Human Evolution, Researchers Claim." *Guardian*, November 19, 2007.

Sanders, J. L., and M. S. Brizzolara. "Relationships between Mood and Weather." *Journal of General Psychology* 107 (1982): 157–58.

Sargent, Frederick II, ed. *Human Ecology*. Amsterdam: North-Holland, 1974.

Sargent, Frederick, II. *Hippocratic Heritage: A History of Ideas about Weather and Human Health*. New York: Pergamon Press, 1982.

Sargent, Frederick, II. "Nature and Scope of Human Ecology." In *Human Ecology*, edited by Frederick Sargent II, 1–25. Amsterdam: North-Holland, 1974.

Sargent, Frederick, II. "The William F. Petersen Foundation." *International Journal of Biometeorology* 7 (1963): 105.

Satchell, Max. "Rutty, John (1698–1775)." In *Oxford Dictionary of National Biography*. Oxford: Oxford University Press, 2004. https://doi.org/10.1093/ref:odnb/24380

Savage, Victor R. *Western Impressions of Nature and Landscape in Southeast Asia*. Singapore: Singapore University Press, 1984.

Schäfer, Mike S., Jürgen Scheffran, and Logan Penniket. "Securitization of Media Reporting on Climate Change? A Cross-National Analysis in Nine Countries." *Security Dialogue* 47 (2016): 76–96.

Schaub, Diana J. "Montesquieu on Slavery." *Perspectives on Political Science* 34 (2005): 70–78.

Scheffran, Jürgen, Michael Brzoska, Hans Günther Brauch, Peter Michael Link, and Janpeter Schilling, eds. *Climate Change, Human Security and Violent Conflict: Challenges for Societal Stability. Hexagon Series on Human and Environmental Security and Peace Vol. 8*. Berlin: Springer, 2012.

Scheffran, Jürgen, Michael Brzoska, Jasmin Kominek, P. Michael Link, and Janpeter Schilling. "Climate Change and Violent Conflict." *Science* 326 (May 18, 2010): 869–71.

Schiermeier, Quirin. "Climate Cycles Drive Civil War: Tropical Conflicts Double during El Niño Years." *Nature*, published online August 24, 2011. http://www.nature.com/news/2011/110824/full/news.2011.501.html

Schmidt, Nathaniel. *Ibn Khaldûn, Historian, Sociologist and Philosopher*. New York: Columbia University Press, 1930.

Schneider, Stephen H., and Randi Londer. *The Coevolution of Climate and Life*. San Francisco: Sierra Book Club, 1984.

Schreider, Eugéne. "Ecological Rules, Body-Heat Regulation, and Human Evolution." *Evolution* 18 (1964): 1–9.

Schwartz, Peter, and Doug Randall. *An Abrupt Climate Change Scenario and Its Implications for United States National Security*, October 2003. https://www.iatp.org/documents/an-abrupt-climate-change-scenario-and-its-implications-for-unitedstates-national-securit-0

Science Daily. "Global Warming Could Be Reversing a Trend That Led to Bigger Human Brains." *ScienceDaily*, March 23, 2007. http://www.sciencedaily.com/releases/2007/03/070322142633.htm

Scoresby-Jackson, Robert Edmund. *Medical Climatology: A Topographical and Meteorological Description of Localities Resorted to in Winter and Summer by Invalids of Various Classes, Both at Home and Abroad*. London: John Churchill, 1862.

Scranton, Roy. *Learning to Die in the Anthropocene: Reflections on the End of a Civilization*. San Francisco: City Lights Publishers, 2015.

Seacole, Mary. *Wonderful Adventures of Mrs Seacole in Many Lands: With an Introductory Preface by W.H. Russell*. Cambridge: Cambridge University Press, 2014. First published 1858.

Seaton, Helen J. "Another Florence Nightingale? The Rediscovery of Mary Seacole." *The Victorian Web: Literature, History, and Culture in the Age of Victoria*. http://www.victorianweb.org/history/crimea/seacole.html

Selby, Jan. "Positivist Climate Conflict Research: A Critique." *Geopolitics* 19 (2014): 829–56.

Semmel, Bernard. "H.T. Buckle: The Liberal Faith and the Science of History." *British Journal of Sociology* 27 (1976): 371–86.

Semple, Ellen Churchill. *American History and Its Geographic Conditions*. Boston: Houghton, Mifflin, 1903.

Semple, Ellen Churchill. *Influences of Geographic Environment on the Basis of Ratzel's System of Anthropo-geography*. New York: Henry Holt, 1911.

Sen, Amartya. *Poverty and Famines: An Essay on Entitlement and Deprivation*. Oxford: Oxford University Press, 1981.

Shackleton, Robert. "The Evolution of Montesquieu's Theory of Climate." *Revue Internationale de Philosophie* 9 (1955): 317–29.

Shaler, Nathaniel Southgate. *The Citizen: A Study of the Individual and the Government*. New York: Barnes, 1907.

Shaler, Nathaniel Southgate. *Kentucky: A Pioneer Commonwealth*. Boston: Houghton, Mifflin, 1884.

Shaler, Nathaniel Southgate. *Nature and Man in America*. New York: Scribner's, 1891.

Shaler, Nathaniel Southgate. "The Negro Problem." *Atlantic Monthly* 54 (1884): 696–709.

Shaler, Nathaniel Southgate. *The Neighbor: The Natural History of Human Contacts*. Boston: Houghton, Mifflin, 1904.

Shaler, Nathaniel Southgate. "Science and the African Problem." *Atlantic Monthly* 66 (1890): 36–45.

Shaler, Nathaniel Southgate. *The Story of Our Continent: A Reader in the Geography and Geology of North America*. Boston: Ginn, 1894.

Shapiro, Harry L. "The Role of the American Museum of Natural History in 20th Century Paleoanthropology." In *Ancestors: The Hard Evidence*, edited by Eric Delson, 6–8. New York: Alan R. Liss, 1985.

Sheffer, Sarah. "Ancient Tree Rings Suggest Good Weather Helped Genghis Khan Build His Empire." *PBS Newshour*, March 12, 2104. http://www.pbs.org/newshour/rundown/ancient-tree-rings-suggest-good-weather-helped-genghis-khan-build-empire/

Sheridan, Richard B. *Doctors and Slaves: A Medical and Demographic History of Slavery in the British West Indies, 1680–1834*. Cambridge: Cambridge University Press, 1985.

Sherriff, Lucy. "Researchers Link Human Skull Size and Climate." *The Register*, March 26, 2007. https://www.theregister.com/2007/03/26/research_skulls/

Shuttleton, David E. "'A Modest Examination': John Arbuthnot and the Scottish Newtonians." *Journal for Eighteenth Century Studies* 18 (1995): 46–62.

Simms, Brendan. "When Winter Came for Kings." *Wall Street Journal*, June 1, 2013, C.7.

Skirble, Rosanne. "Climate Is Major Violence Trigger." *Voice of America*, August 25, 2011. http://www.voanews.com/content/study-climate-cycles-drive-civil-conflict-128462148/169685.html

Skrimshire, Stefan. "Climate Change and Apocalyptic Faith." *WIREs Climate Change* 5 (2014): 233–46.

Skrimshire, Stefan. "Eschatology." In *Systematic Theology and Climate Change*, edited by P. Scott and M. Northcott, 154–74. London: Routledge, 2014.

Skrimshire, Stefan, ed. *Future Ethics: Climate Change and Apocalyptic Imagination*. London: Continuum, 2010.

Slettebak, Rune T. "Don't Blame the Weather! Climate-Related Natural Disasters and Civil Conflict." *Journal of Peace Research* 49 (2012): 163–76.

Sluyter, Andrew. "Neo-Environmental Determinism, Intellectual Damage Control, and Nature/Society Science." *Antipode* 35 (2003): 813–17.

Smith, Adam. *An Inquiry into the Nature and Causes of the Wealth of Nations*. Edinburgh: Black and Tate, new edition, 1838.

Smith, Edgar Lawrence. *Tides in the Affairs of Men: An Approach to the Appraisal of Economic Change*. New York: Macmillan Press, 1939.

Smith, Edward. *The Aortic System Anatomically and Physiologically Considered with a View to Exemplify or Set Forth . . . the Wisdom, Power, and Goodness of God as Revealed in Hold Writ*. London: Simkin, Marshall and Co., 1840.

Smith, Edward. *Foods*. New York: Appleton, 1873.

Smith, Edward. *Health and Disease as Influenced by the Daily, Seasonal, and Other Cyclical Changes in the Human System*. London: Walton and Maberly, 1861.

Smith, J. Russell. *Industrial and Commercial Geography*. New York: Henry Holt, 1913.

Smith, Neil. *American Empire: Roosevelt's Geographer and the Prelude to Globalisation*. Berkeley: University of California Press, 2003.

Smith, Neil. "Bowman's New World and the Council on Foreign Relations." *Geographical Review* 76 (1986): 438–60.

Smith, Philip, and Nicolas Howe. *Climate Change as Social Drama: Global Warming in the Public Square*. New York: Cambridge University Press, 2015.

Smith, Roff. "Genghis Khan's Secret Weapon Was Rain." *National Geographic Daily News*, March 10, 2014. https://web.archive.org/web/20140311210105/http://news.nationalgeographic.com/news/2014/03/140310-genghis-khan-mongols-mongolia-climate-change/

Smith, T. Roger. "Architectural Art in India." *Journal of the Society of Arts*, March 1873, 278–86.

Smith, T. Roger. "On Buildings for European Occupation in Tropical Climates, Especially India." *Papers Read at the Royal Institute of British Architects* (1867–68): 197–208.

Solomon, Barbara Maria. *Ancestors and Immigrants: A Changing New England Tradition*. Cambridge, MA: Harvard University Press, 1956.

Sommerland, Joe. "Earth Day 2021: The 'Revenge of Nature' in Fiction, from Day of the Triffids to Godzilla and Swamp Thing." *Independent*, April 21, 2021. https://www.independent.co.uk/climate-change/news/climate-change-fiction-nature-revenge-b1831306.html

Sörlin, Sverker. "Environmental Times: Synchronizing Human-Earth Temporalities from Annales to Anthropocene, 1920s–2020s." In *Times of History, Times of Nature: Temporalization and the Limits of Modern Knowledge*, edited by Anders Ekström and Staffan Bergwik, 64–101. New York: Berghahn, 2022.

Sörlin, Sverker. "The Global Warming That Did Not Happen: Historicizing Glaciology and Climate Change." In *Nature's End: History and the Environment*, edited by Sverker Sörlin and Paul Warde, 93–114. New York: Palgrave Macmillan, 2009.

Sorokin, Pitirim. *Contemporary Sociological Theories*. New York: Harper & Brothers, 1928.

Spate, Oskar H. K. "Environmentalism." In *International Encyclopedia of the Social Sciences*, vol. 5, edited by David Sills, 93–96. New York: The Free Press, 1968.

Spate, Oskar H. K. "Quantity and Quality in Geography." *Annals of the Association of American Geographers* 50 (1960): 377–94.

Spate, Oskar H. K. "Reflections on Toynbee's *A Study of History*: A Geographer's View." In *Toynbee and History: Critical Essays and Reviews*, edited by M. F. Ashley Montagu, 287–304. Boston: Porter Sargent, 1956.

Spate, Oskar H. K. "Toynbee and Huntington: A Study in Determinism." *Geographical Journal* 118 (1952): 406–28.

Spector, Céline. "Europe," translated by Philip Stewart. In *A Montesquieu Dictionary* [online], directed by Catherine Volpilhac-Auger, ENS Lyon, September 2013. http://dictionnaire-montesquieu.ens-lyon.fr/en/article/1380009044/en

Spengler, Joseph J. Review of *Medical Climatology: Climatic and Weather Influences in Health and Disease* by Clarence A. Mills. *Social Forces* 18 (1940): 447–48.

Stanley, Steven M. *Children of the Ice Age: How a Global Catastrophe Allowed Humans to Evolve*. New York: W. H. Freeman, 1998.

Stanley, Steven M. "An Ecological Theory for the Origin of *Homo*." *Paleobiology* 18 (1992): 237–57.

Stehr, Nico. "The Ubiquity of Nature: Climate and Culture." *Journal of the History of the Behavioral Sciences* 32 (1996): 151–59.

Stepan, Nancy. *The Idea of Race in Science: Great Britain 1800–1960*. London: Macmillan, 1982.

Stern, Nicholas. *The Economics of Climate Change: The Stern Review*. Cambridge: Cambridge University Press, 2007. First published 2006.

Stetson, Harlan True. *Sunspots and Their Effects*. New York: McGraw-Hill, 1934.

Stewart, Alan E. "Edwin Grant Dexter: An Early Researcher in Human Behavioral Biometeorology." *International Journal of Biometeorology* 59 (2015): 745–58.

Stigler, George J. "Henry L. Moore and Statistical Economics." *Econometrica* 30 (1962): 1–21.

Stinson, Matthew. "Genghis Khan Aided by Climate Change to Conquer Asia." *Liberty Voice*, March 12, 2014. http://guardianlv.com/2014/03/genghis-khan-aided-by-climate-change-to-conquer-asia/

Stirati, Antonella. *The Theory of Wages in Classical Economics: A Study of Adam Smith, David Ricardo, and Their Contemporaries*. London: Elgar, 1994.

Stocking, George W., Jr. "The Persistence of Polygenist Thought in Post-Darwinian Anthropology." In *Race, Culture, and Evolution: Essays in the History of Anthropology*, 42–68. New York: Free Press, 1968.

Stocking, George W., Jr. *Victorian Anthropology*. New York: Free Press, 1987.

Stocking, George W., Jr. "What's in a Name? The Origins of the Royal Anthropological Institute (1837–71)." *Man* 6 (1971): 369–90.

Stoczkowski, Wiktor. *Explaining Human Origins: Myth, Imagination and Conjecture*, translated from the French by Mary Turton. Cambridge: Cambridge University Press, 1994.

Strabo. *Geography*, translated by H. C. Hamilton and W. Falconer. London: Bohn, 1856.

Strange, Carolyn. "Transgressive Transnationalism: Griffith Taylor and Global Thinking." *Australian Historical Studies* 41 (2010): 25–40.

Strange, Carolyn, and Alison Bashford. *Griffith Taylor: Visionary, Environmentalist, Explorer*. Toronto: University of Toronto Press, 2008.

Stringer, Chris. *Homo Britannicus: The Incredible Story of Human Life in Britain*. London: Penguin, 2006.

Sturm, Tristan, and Nicholas Ferris Lustig. "Variegated Environmental Apocalypses: Post-Politics, the Contestatory, and an Eco-Precariat Manifesto for a Radical Apocalyptics." In *Imagining Apocalyptic Politics in the Anthropocene*, edited by Earl Harper and Doug Specht, 213–34. London: Routledge, 2021.

Sturt, Henry. "Kidd, Benjamin (1858–1916)." In *Oxford Dictionary of National Biography*. Oxford: Oxford University Press, 1927.

Sullivan, John. *The Endemic Diseases of Tropical Climates with Their Treatment*. London: J.&A. Churchill, 1877.

Sutton, Alexandra E., Justin Dohn, Kara Loyd, Andrew Tredennick, Gabriela Bucini, Alexandro Solórzano, Lara Prihodko, and Niall P. Hanan. "Does Warming Increase the Risk of Civil War in Africa?" *Proceedings of the National Academy of Sciences* 107, no. 25 (June 2010): E102.

Swartzman, David, and George Middendorf. "Biospheric Cooling and the Emergence of Intelligence." In *A New Era in Bioastronomy*, edited by G. Lemarchland and K. Meech, 425–29. ASP Conference Series, Vol. 213, 2000.

Swartzman, David, George Middendorf, and Miranda Armour-Chelu. "Was Climate the Prime Releaser for Encephalization? An Editorial Comment." *Climatic Change* 95 (2009): 439–47.

Sweet, Timothy. "The Degeneration Thesis." In *Climate and American Literature*, edited by Michael Boyden, 126–41. Cambridge: Cambridge University Press, 2021.

Swyngedouw, Erik. "Apocalypse Forever? Post-Political Populism and the Spectre of Climate Change." *Theory Culture Society* 27 (2010): 213–32.

Swyngedouw, Erik. "Apocalypse Now! Fear and Doomsday Pleasures." *Capitalism Nature Socialism* 24 (2013): 9–18.

Sydenham, Thomas. *The Works of Thomas Sydenham, M.D. Translated from the Latin Edition of Dr. Greenhill with a Life of the Author by R. G. Latham. In Two Volumes*. London: Sydenham Society, 1848.

Tainter, Joseph A. "Collapse, Sustainability, and the Environment: How Authors Choose to Fail or Succeed." *Reviews in Anthropology* 37 (2008): 342–71.

Talhelm, T., X. Zhang, S. Oishi, C. Shimin, D. Duan, X. Lan, and S. Kitayama. "Large-Scale Psychological Differences within China Explained by Rice versus Wheat Agriculture." *Science* 344 (2014): 603–8.

Tambe, Ashwini. "Climate, Race Science and the Age of Consent in the League of Nations." *Theory, Culture and Society* 28 (2011): 109–30.

Tavárez, Fidel. "'The Moral Miasma of the Tropics': American Imperialism and the Failed Annexation of the Dominican Republic, 1869–1871." *Nuevo Mundo Mundos Nuevos* (2011). https://journals.openedition.org/nuevomundo/61771

Taylor, Griffith. *Australia: A Study of Warm Environments and Their Effect on British Settlement*. London: Methuen, 1959. First published 1940.

Taylor, Griffith. *Australia in Its Physiographic and Economic Aspects*. Oxford: Clarendon Press, 1911.

Taylor, Griffith. *Canada: A Study of Cool Continental Environments and Their Effects on British and French Settlement*. London: Methuen, 1947.

Taylor, Griffith. "Climatic Cycles and Evolution." *Geographical Review* 8 (1919): 289–328.

Taylor, Griffith. "The Control of Settlement by Humidity and Temperature (with Special Reference to Australia and the Empire). An Introduction to Comparative Climatology, Illustrated by 70 Climographs." *Bulletin 14, Commonwealth Bureau of Meteorology, Australia* (1916): 6–32.

Taylor, Griffith. "The Distribution of Future White Settlement: A World Survey Based on Physiographic Data." *Geographical Review* 12 (1922): 375–402.

Taylor, Griffith. "The Ecological Basis of Anthropology." *Ecology* 15 (1934): 223–42.

Taylor, Griffith. *Environment and Race: A Study of the Evolution, Migration, Settlement and Status of the Races of Man*. Oxford: Oxford University Press, 1927.

Taylor, Griffith. *Environment, Race, and Migration. Fundamentals of Human Distribution: With Special Sections on Racial Classification, and Settlement in Canada and Australia*. Chicago: University of Chicago Press, 1937.

Taylor, Griffith. "Geographical Factors Controlling the Settlement of Tropical Australia." *Queensland Geographical Journal* 32–33 (1918): 1–67.

Taylor, Griffith. "Preface." In *Geography in the Twentieth Century: A Study of Growth, Fields, Techniques, Aims and Trends*, 3rd ed., edited by Griffith Taylor. New York: Philosophical Library, 1957.

Taylor, Griffith. "The Settlement of Tropical Australia." *Geographical Review* 8 (1919): 84–115.

Taylor, Julie. "Can Rainy Days Really Get You Down?" WebMD. https://www.webmd.com/balance/features/can-rainy-days-really-get-you-down

Tenney, A. A. Review of *Civilization and Climate* by Ellsworth Huntington. *Political Science Quarterly* 31 (1916): 633–35.

Theisen, Ole Magnus. "Blood and Soil? Resource Scarcity and Internal Armed Conflict Revisited." *Journal of Peace Research* 45 (2008): 801–18.

Theisen, Ole Magnus. "Climate Clashes? Weather Variability, Land Pressure, and Organized Violence in Kenya, 1989–2004." *Journal of Peace Research* 49 (2012): 81–96.

Theisen, Ole Magnus, Nils Peter Gleditsch, and Halvard Buhaug. "Is Climate Change a Driver of Armed Conflict?" *Climatic Change* 117 (2013): 613–25.

Theisen, Ole Magnus, Helge Holtermann, and Halvard Buhaug. "Climate Wars? Assessing the Claim That Drought Breeds Conflict." *International Security* 36 (Winter 2011/2012): 79–106.

Thomas, Pat. *Under the Weather: How the Weather and Climate Affect our Health.* London: Fusion Press, 2004.

Thomas, Rosalind. *Herodotus in Context: Ethnography, Science and the Art of Persuasion.* Cambridge: Cambridge University Press, 2002.

Thompson, D'Arcy Wentworth. "Dr Otto Pettersson." *Nature* 3736 (June 7, 1941): 701–2.

Thomson, Arthur S. "Could the Natives of a Temperate Climate Colonize and Increase in a Tropical Country and Vice Versa?" *Transactions of the Medical and Physiological Society of Bombay* 6 (1843): 112–38.

Thomson, Arthur S. "On the Doctrine of Acclimatization." *Madras Quarterly Medical Review* 2 (1840): 69–76.

Thomson, Joseph. "Niger and Central Sudan Sketches." *Scottish Geographical Magazine* 2 (1886): 577–96.

Tisdale, Sallie. "Confessions of a Cosmic Resonator." *Outside Magazine,* December 1995. http://www.outsideonline.com/1837011/confessions-cosmic-resonator

Tol, Richard S. J., and Sebastian Wagner. "Climate Change and Violent Conflict in Europe Over the Last Millennium." *Climate Change* 99 (2010): 65–79.

Tooley, Marian J. "Bodin and the Medieval Theory of Climate." *Speculum* 28 (1953): 64–83.

Tout, David G. "Biometeorology." *Progress in Physical Geography* 11 (1987): 473–86.

Tower, Walter S. Review of *Civilization and Climate* by Ellsworth Huntington. *Journal of Political Economy* 25 (1917): 748–50.

Toynbee, Arnold J. *A Study of History*. Oxford: Oxford University Press, 1934.

Trapham, Thomas. *A Discourse of the State of Health in the Island of Jamaica with a Provision Therefore Calculated from the Air, the Place, and the Water.* London, 1679.

Trexler, Adam. *Anthropocene Fictions: The Novel in a Time of Climate Change.* Charlottesville: University of Virginia Press, 2015.

Trinkaus, Erik, and Pat Shipman, *The Neanderthals: Changing the Image of Mankind.* London: Jonathan Cape, 1993.

Trojanow, Ilija. *The Lamentations of Zeno. A Novel,* translated by Philip Boehm. London: Verso, 2016.

Trombetta, Maria J. "Climate Change and the Environmental Conflict Discourse." In *Climate Change, Human Security and Violent Conflict: Challenges for Societal Stability. Hexagon Series on Human and Environmental Security and Peace Vol. 8*, edited by Jürgen Scheffran, Michael Brzoska, Hans Günther Brauch, Peter Michael Link, and Janpeter Schilling, 151–64. Berlin: Springer, 2012.

Tromp, S. W. *Biometeorology: The Impact of the Weather and Climate on Humans and Their Environment*. London: Heyden, 1980.

Ungureanu, James C. *Science, Religion, and the Protestant Tradition: Retracing the Origins of Conflict*. Pittsburgh: University of Pittsburgh Press, 2019.

U.S. Environmental Protection Agency. "Human Health Impacts and Adaptation." June 2012. https://web.archive.org/web/20120610041007/http://www.epa.gov/climatechange/impacts-adaptation/health.html

Uskul, A. K., S. Kitayama, and R. E. Nisbett. "Ecocultural Basis of Cognition: Farmers and Fishermen Are More Holistic Than Herders." *Proceedings of the National Academy of Sciences* 105 (2008): 8552–56.

Vaj, Daniela. "Medical Geography and Phthisic Immunity in the High Altitudes: The Origins of a Therapeutic Hypothesis." *Journal of Alpine Research* 93 (2005): 34–42.

Valenčius, Conevery Bolton. "Histories of Medical Geography." In *Medical Geography in Historical Perspective*, edited by Nicolaas A. Rupke, 3–28. London: Wellcome Trust Centre for the History of Medicine, 2000.

Van Deusen, E. H. "Observations on a Form of Nervous Prostration, (Neurasthenia), Culminating in Insanity." *American Journal of Insanity* 25, no. 4 (1869): 445–61.

Van Evrie, J. H. *Negroes and Negro Slavery: The First, an Inferior Race—the Latter, Its Normal Condition*. Baltimore: John D. Toy, ca. 1853.

Van Evrie, J. H. *White Supremacy and Negro Subordination or, Negroes a Subordinate Race, and (So-called) Slavery Its Normal Condition*. New York: Van Evrie, Horton & Co., 1868.

Vegetius. *Epitome of Military Science*, translated by N. P. Milner. Liverpool: Liverpool University Press, 1993.

Verhoeven, Harry. "Gardens of Eden or Hearts of Darkness? The Genealogy of Discourses on Environmental Insecurity and Climate Wars in Africa." *Geopolitics* 19 (2014): 784–805.

Viseltear, Arthur J. "C.-E.A. Winslow and the Early Years of Public Health at Yale, 1915–1925." *The Yale Journal of Biology and Medicine* 55 (1982): 137–51.

Visher, Stephen S. "Ecology of American Notables." *Human Biology* 1 (1929): 544–54.

Visher, Stephen S. "Geography of American Notables: A Statistical Study of Birthplaces, Training, Distribution: An Effort to Evaluate Various Environmental Factors." *Indiana University Studies* 20 (1928): 1–128.

Vitruvius. *The Architecture of Marcus Vitruvius Pollio*, translated by Joseph Gwilt. London: Lockwood, 1874.

Voltaire. "Climate." *A Philosophical Dictionary*. In *The Works of Voltaire: A Contemporary Version*, vol. 3, 203–4. Paris: E. R. DuMont, 1901.

von Tschudi, J. J. *Travels in Peru, during the Years 1838–1842*, translated by Thomasina Ross. London: David Bogue, 1847.

Vrba, Elisabeth S. "Climate, Heterochrony, and Human Evolution." *Journal of Anthropological Research* 52 (1996): 1–28.

Vrba, Elisabeth S. "Ecological and Adaptive Changes Associated with Early Hominid Evolution." In *Ancestors: The Hard Evidence*, edited by Eric Delson, 63–71. New York: Alan R. Liss, 1985.

Vrba, Elisabeth S. "On the Connections between Paleoclimate and Evolution." In *Paleoclimate and Evolution with Emphasis on Human Origins*, edited by Elisabeth S. Vrba, George H. Denton, Timothy C. Partridge, and Lloyd H. Burckle, 24–45. New Haven, CT: Yale University Press, 1995.

Vrba, Elisabeth S. "Role of Environmental Stimuli in Hominid Origins." In *Handbook of Palaeoanthropology, Volume 3: Phylogeny of Hominins*, edited by W. Henke, H. Rothe, and I. Tattersall, 1–41. New York: Springer-Verlag, 2006.

Vrba, Elisabeth S., G. H. Denton, and M. L. Denton. "Climatic Influences on Early Hominid Behavior." *Ossa* 14 (1989): 127–56.

Vrba, Elisabeth S., George H. Denton, Timothy C. Partridge, and Lloyd H. Burckle, eds. *Paleoclimate and Evolution with Emphasis on Human Origins*. New Haven, CT: Yale University Press, 1995.

Vrij, Aldert, Jaap Van Der Steen, and Leendert Koppelaar. "Aggression of Police Officers as a Function of Temperature: An Experiment with the Fire Arms Training System." *Journal of Community and Applied Social Psychology* 4 (1994): 365–70.

Walker, Mike, and John Lowe. "Quaternary Science 2007: A 50-Year Retrospective." *Journal of the Geological Society, London* 164 (2007): 1073–92.

Wallace-Wells, David. "The Uninhabitable Earth." *New York Magazine Intelligencer*, July 10, 2017. https://nymag.com/intelligencer/2017/07/climate-change-earth-too-hot-for-humans.html

Walsh, Bryan. "How Climate Change Drove the Rise of Genghis Khan." *Time*, March 10, 2014. http://time.com/18147/climate-change-genghis-khan-mongolia/

Walsh, Lynda. *Scientists as Prophets: A Rhetorical Genealogy*. Oxford: Oxford University Press, 2013.

Wanklyn, Harriet. *Friedrich Ratzel: A Biographical Memoir and Bibliography*. Cambridge: Cambridge University Press, 1961.

Ward, Robert DeCourcy. "The Acclimatization of the White Race in the Tropics." *New England Journal of Medicine* 201 (1929): 617–27.

Ward, Robert DeCourcy. "The Classification of Climates: I." *Bulletin of the American Geographical Society* 38 (1906): 401–41.

Ward, Robert De Courcy. *Climate Considered Especially in Relation to Man*. New York: G. P. Putnam's Sons, 1908.

Ward, Robert DeCourcy. "The Crisis in Our Immigration Policy." *The Institution Quarterly* 4 (1913): 37–50.

Ward, Robert DeCourcy. "Fallacies of the Melting Pot Idea and America's Traditional Immigration Policy." In *The Alien in our Midst or 'Selling our Birthright for a Mess of Industrial Pottage'*, edited by Madison Grant and Charles Steward Davison, 230–36. New York: Galton Publishing Co., 1930.

Ward, Robert DeCourcy. "Fog in the United States." *Geographical Review* 13 (1923): 576–82.

Ward, Robert DeCourcy. "Immigration and Eugenics—Second Report of the Subcommittee on Selective Immigration of the Eugenics Committee of the United States." *Journal of Heredity* 16 (1925): 287–93.

Ward, Robert DeCourcy. "Immigration and the Three Percent Restrictive Law." *Journal of Heredity* 12 (1921): 319–25.

Ward, Robert DeCourcy. "The Immigration of Orientals." *Journal of Heredity* 10 (1919): 110.

Ward, Robert DeCourcy. "Land and Sea Breezes." *Monthly Weather Review* 42 (1914): 274–77.

Ward, Robert DeCourcy. "Meteorology and War-Flying. United States." *Monthly Weather Review* 45 (1918): 591–600.

Ward, Robert DeCourcy. "Physiological Effects of Diminished Air Pressure." *Science* 14 (1901): 814.

Ward, Robert DeCourcy. "The Second Year of the New Immigration Law." *Journal of Heredity* 18 (1927): 3–18.

Ward, Robert DeCourcy. "Weather Controls Over the Fighting during the Autumn of 1918." *The Scientific Monthly* 8 (1919): 5–15.

Ward, Robert DeCourcy. "Weather Controls Over the Fighting in Mesopotamia, in Palestine, and Near the Suez Canal." *The Scientific Monthly* 6 (1918): 289–304.

Ward, Robert De Courcy. "A Year of Weather and Trade in the United States." *Popular Science Monthly* 61 (1902): 439–48.

Waterhouse, Paul (rev. John Elliott). "Smith, Thomas Roger (1830–1903)." In *Oxford Dictionary of National Biography*. Oxford: Oxford University Press, 2004. https://doi.org/10.1093/ref:odnb/36163

Watson, D. M. S. "William Diller Matthew–1871–1930." *Obituary Notices of Fellows of the Royal Society* 1 (1932): 71–74.

Wear, Andrew. "Place, Health, and Disease: The *Airs, Waters, Places* Tradition in Early Modern England and North America." *Journal of Medieval and Early Modern Studies* 38 (2008): 443–65.

Wear, Andrew. "The Prospective Colonist and Strange Environments: Advice on Health and Prosperity." In *Cultivating the Colonies: Colonial States and Their Environmental Legacies*, edited by Christina Folke Ax, Niels Brimnes, Niklas Thode Jensen, and Karen Oslund, 19–46. Athens: Ohio University Press, 2011.

Webb, W. W. (rev. Roy Bridges). "Parke, Thomas Heazle (1857–1893)." In *Oxford Dictionary of National Biography*. Oxford: Oxford University Press, 2004. https://doi.org/10.1093/ref:odnb/21289

Webster, Ben. "Climate Change Report Sets Out and Apocalyptic Vision of Britain." *Times*, February 26, 2010.

Webster, Ben. "Curb Climate Global Warming or Face a Health Catastrophe, Expert Warns." *Times*, June 23, 2015.

Wei, Wenqi, Jackson G. Lu, Adam D. Galinsky, Han Wu, Samuel D. Gosling, Peter J. Rentfrow, Wenjie Yuan, Qi Zhang, Yongyu Guo, Ming Zhang, Wenjing Gui, Xiao-Yi Guo, Jeff Potter, Jian Wang, Bingtan Li, Xiaojie Li, Yang-Mei Han, Meizhen Lv, Xiang-Qing Guo, Yera Choe, Weipeng Lin, Kun Yu, Qiyu Bai, Zhe Shang, Ying Han, and Lei Wang. "Regional Ambient Temperature Is Associated with Human Personality." *Nature Human Behaviour* 1 (2017): 890–95. https://pubmed.ncbi.nlm.nih.gov/31024181/

Weihe, Wolf H. "Review on the History of the International Society of Biometeorology." *International Journal of Biometeorology* 40 (1997): 9–15.

Wells, G. A. "The Critics of Buckle." *Past and Present* 9 (1956): 75–89.

Welzer, Harald. *Climate Wars: Why People Will Be Killed in the Twenty-First Century*, translated by Patrick Camiller, originally published in German in 2008. Cambridge: Polity, 2012.

Weninger, Margarete. "As to the Influence of Climate on Head Form." *Anthropologischer Anzeiger* 37 (1979): 18–26.

Wesselman, Henry B. "Of Mice and Almost-Men: Regional Paleoecology and Human Evolution in the Turkana Basin." In *Paleoclimate and Evolution with Emphasis on Human Origins*, edited by Elisabeth S. Vrba, George H. Denton, Timothy C. Partridge, and Lloyd H. Burckle, 354–67. New Haven, CT: Yale University Press, 1995.

Wheeler, Raymond H. "The Changeableness of Climate." In *Climate: The Key to Understanding Business Cycles, with a Forecast of Trends into the 21st Century. The Raymond H. Wheeler Papers*, compiled by Michael Zahorchak, 51–59. Linden, NJ: Tide Press, 1983.

Wheeler, Raymond H. "Climatic Belts of the World." In *Climate: The Key to Understanding Business Cycles, with a Forecast of Trends into the 21st Century. The Raymond H. Wheeler Papers*, compiled by Michael Zahorchak, 1–11. Linden, NJ: Tide Press, 1983.

Wheeler, Raymond H. "Climatic Conditioning in History." In *Climate: The Key to Understanding Business Cycles, with a Forecast of Trends into the 21st Century. The Raymond H. Wheeler Papers*, compiled by Michael Zahorchak, 119–31. Linden, NJ: Tide Press, 1983.

Wheeler, Raymond H. "Climatic Influence and the Industrial Centres of the World." In *Climate: The Key to Understanding Business Cycles, with a Forecast of Trends into the 21st Century. The Raymond H. Wheeler Papers*, compiled by Michael Zahorchak, 13–25. Linden, NJ: Tide Press, 1983.

Wheeler, Raymond H. "Climatic Phases and Business Cycles." In *Climate: The Key to Understanding Business Cycles, with a Forecast of Trends into the 21st Century. The Raymond H. Wheeler Papers*, compiled by Michael Zahorchak, 141–97. Linden, NJ: Tide Press, 1983.

Wheeler, Raymond H. "The Effect of Climate on Animals and Human Beings." In *Climate: The Key to Understanding Business Cycles, with a Forecast of Trends into the 21st Century. The Raymond H. Wheeler Papers*, compiled by Michael Zahorchak, 35–50. Linden, NJ: Tide Press, 1983.

Wheeler, Raymond H. "Evidence Dealing with the Climate of Historic Time." In *Climate: The Key to Understanding Business Cycles, with a Forecast of Trends into the 21st Century. The Raymond H. Wheeler Papers*, compiled by Michael Zahorchak, 61–83. Linden, NJ: Tide Press, 1983.

Wheeler, Raymond H. "The Principal Climatic Cycles in History." In *Climate: The Key to Understanding Business Cycles, with a Forecast of Trends into the 21st Century. The Raymond H. Wheeler Papers*, compiled by Michael Zahorchak, 109–17. Linden, NJ: Tide Press, 1983.

Wheeler, Raymond H. "The Effect of Climate on Human Behavior in History." *Kansas Academy of Science* 46 (1943): 33–51.

Whitbeck, R. H. Review of *Civilization and Climate* by Ellsworth Huntington. *American Historical Review* 21 (1916): 781–82.

White, Hayden V. "The Burden of History." *History and Theory* 5 (1966): 111–34.

White, Hayden V. "Ibn Khaldūn in World Philosophy of History." *Comparative Studies in Society and History* 2 (1959): 110–25.

White, Sam, Christian Pfister, and Franz Mauelshagen, eds. *The Palgrave Handbook of Climate History*. London: Palgrave Macmillan, 2018.

White, Silva. "On the Comparative Value of African Lands." *Scottish Geographical Magazine* 7 (1891): 191–95.

Wigley, T. M. L., M. J. Ingram, and G. Farmer, eds. *Climate and History: Studies in Past Climates and Their Impact on Man*. Cambridge: Cambridge University Press, 1981.

Willinsky, John. *Learning to Divide the World: Education at Empire's End*. Minneapolis: University of Minnesota Press, 1998.

Wilson, Edward O. "Island Biogeography in the 1960s." In *The Theory of Island Biogeography Revisited*, edited by Jonathan B. Losos and Robert E. Ricklefs, 1–12. Princeton, NJ: Princeton University Press, 2009.

Wilson, Philip K. "Influences of 'Kosmos' in 'Earth and Man.'" In *Alexander von Humboldt: From the Americas to the Cosmos*, coordinated by Raymond Erickson, Mauricio A. Font, and Brian Schwartz, 371–84. New York: Bildner Center for Western Hemisphere Studies, CUNY, 2017.

Winlow, Heather. "Mapping the Contours of Race: Griffith Taylor's Zones and Strata Theory." *Geographical Research* 47 (2009): 390–407.

Winslow, C.-E. A., and L. P. Herrington, *Temperature and Human Life*. Princeton, NJ: Princeton University Press, 1949.

Winther, Paul C. *Anglo-European Science and the Rhetoric of Empire: Malaria, Opium and British Rule in India, 1756–1895*. Lanham, MD: Lexington Books, 2003.

Withers, Charles W. J. *Placing the Enlightenment: Thinking Geographically in the Age of Reason*. Chicago: University of Chicago Press, 2007.

Wolloch, Nathaniel. "Edward Gibbon's Cosmology." *International Journal of the Classical Tradition* 17 (2010): 164–77.

Wood, Bernard A. "Evolution of the Early Hominin Masticatory System: Mechanisms, Events, and Triggers." In *Paleoclimate and Evolution with Emphasis on Human Origins*, edited by Elisabeth S. Vrba, George H. Denton, Timothy C. Partridge, and Lloyd H. Burckle, 438–48. New Haven, CT: Yale University Press, 1995.

Woodruff, Charles E. *The Effects of Tropical Light on White Men*. New York: Rebman, 1905.

Woodruff, Charles E. "The Soldier in the Tropics: His Food, Alcohol, and Acclimatization." *Philadelphia Medical Journal* 5 (1900): 768–82.

Worboys, Michael. "Germs, Malaria and the Invention of Mansonian Tropical Medicine: from 'Diseases in the Tropics' to 'Tropical Diseases.'" In *Warm Climates and Western Medicine: The Emergence of Tropical Medicine, 1500–1900*, Wellcome Institute Series in the History of Medicine, edited by David Arnold, 181–207. Atlanta, GA: Rodopi, 1996.

Worboys, Michael. "Manson, Ross and Colonial Medical Policy: Tropical Medicine in London and Liverpool, 1899–1914." In *Disease, Medicine, and Empire: Perspectives on Western Medicine and the Experience of European Expansion*, edited by Roy Macleod and Milton Lewis, 21–37. London: Routledge, 1988.

World Health Organization. "Climate and Health." Fact Sheet, July 2005. https://web.archive.org/web/20051214075919/http://www.who.int/globalchange/news/fsclimandhealth/en/

Worthington, A. C. "Whether the Weather: A Comprehensive Assessment of Climate Effects in the Australian Stock Market." Faculty of Business, Accounting and Finance Working Papers, University of Wollongong, 2006. https://ro.uow.edu.au/accfinwp/33/

Wright, John K., and George F. Carter. "Isaiah Bowman 1878–1950." *Biographical Memoirs of the National Academy of Sciences* 33 (1959): 37–64.

Wright, Quincy. *A Study of War. Second Edition, with a Commentary on War since 1942*. Chicago: University of Chicago Press, 1965.

Wyon, D. P. "The Effects of Moderate Heat Stress on Typewriting Performance." *Ergonomics* 17 (1974): 309–17.

Young, Robert M. "The Historiographic and Ideological Contexts of the Nineteenth-Century Debate on Man's Place in Nature." In *Changing Perspectives in the History of Science: Essays in Honour of Joseph Needham*, edited by Mikuláš Teich and Robert Young, 344–438. London: Heinemann, 1973.

Zahorchak, Michael, ed. *Climate: The Key to Understanding Business Cycles, with a Forecast of Trends into the 21st Century. The Raymond H. Wheeler Papers*. Linden, NJ: Tide Press, 1983.

Zammito, John H. *Kant, Herder, and the Birth of Anthropology*. Chicago: Chicago University Press, 2002.

Zhang, David D., Peter Brecke, Harry F. Lee, Yuan-Qing He, and Jane Zhang. "Global Climate Change, War, and Population Decline in Recent Human History." *Proceedings of the National Academy of Sciences* 104, no. 49 (2007): 19214–19.

Zhang, David D., Jane Zhang, Harry F. Lee, and Yuan-Qing He, "Climate Change and War Frequency in Eastern China over the Last Millennium." *Human Ecology* 35 (2007): 403–14.

Zielinksi, Sarah. "Warm, Wet Times Spurred Medieval Mongol Rise. Genghis Khan—and his Army of Men on Horseback—Benefitted from Boom in Grasslands." *Smithsonian Magazine*, March 10, 2014. http://www.smithsonianmag.com/science-nature/warm-wet-times-spurred-medieval-mongol-rise-180950030/?no-ist

Zivin, Joshua Graff, Solomon M. Hsiang, and Matthew Neidell. "Temperature and Human Capital in the Short and Long Run." *Journal of the Association of Environmental and Resources Economists* 5 (2018): 77–105.

Žižek, Slavoj. *Living in the End Times*. London: Verso, 2011.

Žižek, Slavoj, and Boris Gunjević, *God in Pain: Inversions of Apocalypse*. London: Verso, 2012.

Zuck, David. "Jeffreys, Julius (1800–1877)." *Oxford Dictionary of National Biography*. Oxford: Oxford University Press, 2004. https://doi.org/10.1093/ref:odnb/14706

Zuck, David. "Julius Jeffreys, Pioneer of Humidification." *Proceedings of the History of Anaesthesia Society* 8b (1990): 70–80.

INDEX

A NOTE ON THE TYPE

This book has been composed in Arno, an Old-style serif typeface in the classic Venetian tradition, designed by Robert Slimbach at Adobe.